D0308483

Design of
logic systems

Difficulty at the beginning

Times of growth are beset with
difficulties. They resemble a first
birth. But these difficulties arise
from the very profusion of all that
is struggling to attain form.

I. CHING

Design of logic systems

DOUGLAS LEWIN
Professor of Computer Science and Information Engineering,
University of Sheffield

Van Nostrand Reinhold (UK) Co. Ltd

To Yasi, tattycat

First published in 1985 by
Van Nostrand Reinhold (UK) Co. Ltd
Molly Millars Lane, Wokingham,
Berkshire, England

Reprinted 1985, 1986, 1987

Typeset in 10 on 11 pt Times by
Graphicraft, Hong Kong

Printed in Great Britain by
T. J. Press (Padstow) Ltd, Padstow, Cornwall

Library of Congress Cataloging in Publication Data

Lewin, Douglas.
    Design of logic systems.

    Includes bibliographies and index.
    1. Switching circuits.   2. Integrated circuits.
3. Logic design.   I. Title.
TK7868.S9L397   1985   621.3819′5835   84-21981
ISBN 0-442-30606-7 (pbk.)

Preface

Then you better start swimmin'
Or you'll sink like a stone
For the times they are a-changin'
BOB DYLAN

Logic design is the process of interconnecting logic elements or modules to perform a specific function or task; as such it is fundamental to all digital engineering applications ranging from microprocessors through to information and communication systems. Switching theory, which is the formal basis of logic design, is concerned with the theory of producing viable, working structures in terms of logic elements and their interconnections to meet a particular functional input–output specification.

Since the last edition of this book in 1974, advances in microelectronics have brought about a considerable change in digital design practice. The major causal factors are the commercial availability of 8- and 16-bit programmable microprocessors, which can replace hardwired logic circuits by software, and the use of semi-custom and custom LSI and VLSI circuits. It may be thought that the versatility of the microprocessor has made switching theory obsolete, replacing the need for logic design with programming techniques, but this is far from the case. A digital system does not consist solely of a microprocessor chip, but involves many other logic circuits in interfacing it to the external world. Moreover, the design of the microprocessor chip itself demands considerable logic design expertise.

Thus, the need still remains for engineers and computer scientists to have a firm foundation in switching theory and logic design—as shown by their inclusion in most engineering core curricula.

The objective of the first edition of this book was 'to describe, from an engineering viewpoint, those methods of designing logical circuits which have evolved as useful and practical techniques, from the vast amount of published work on switching theory'. This objective is still appropriate for the third edition, particularly so in the case of the fundamental aspects of switching theory which continue to remain highly relevant.

However, with the requirement to design systems using MSI/LSI circuit modules, and more recently the ability to custom design LSI/VLSI circuits —systems on a chip—the application of the theory, particularly the criteria adopted for minimization etc., has radically changed. For example, when designing using subsystem modules, the criterion changes from minimizing the number of gates and inputs to reducing the number of module types and their interconnections. Similarly, at the silicon chip level, the prime requirements are to utilize regular structures and minimize surface area and interconnections.

There is an obvious and urgent need to develop new logical configurations and methods of representation and design which are more appropriate to realizing complex logic circuits in silicon (and any other) technologies. However, though many of these problems are difficult and their solutions still being sought, there are numerous ways in which existing theory can be suitably extended or modified to assist in the design process.

Thus, the major change in this new edition is the underlying emphasis on how switching theory can be applied to the design of systems using LSI and VLSI circuits. In particular, new chapters have been included on Design Methods and Tools, Logic Circuit Testing and Reliable Design including the state-machine approach to design, logic simulation and testing, failure-tolerant and self-testing circuits, linear circuits, array and cellular structures including uncommitted and programmable logic arrays (ULAs and PLAs). In addition the opportunity has been taken to update the basic material, including the bibliography and reference sources, and to extend the tutorial problems section.

The book is primarily intended as a text for undergraduate courses on logic and digital systems design in universities and polytechnics and as an introduction to the topic for postgraduate students. However, since the book requires no previous knowledge of the subject and is written in a tutorial style with numerous worked examples, it would be suitable for anyone who wished to acquire a working knowledge of logic design.

Finally I would like to express my thanks to all my students and colleagues who both knowingly and unknowingly have helped towards the development of this text. In particular I would like to thank Mrs Anne Steven without whose care and expert typing skills the final manuscript would not have been produced.

D. L.

Contents

1 Introduction to digital systems

1.1 Introduction

Over the last decade the use of digital techniques has penetrated most branches of engineering and science. With the availability of cheap and reliable digital integrated circuits for the processing and storage of information and data the range of electronic systems and equipment that exploits digital methods has increased enormously and is still expanding. The most spectacular changes have occurred in the communication and control fields where the marrying together of these two disciplines by computer science (in the image of a microprocessor) has given birth to information technology.

It might be thought that, with the emergence of programmable logic systems such as the microcomputer being produced on a single chip of silicon, logical design has now become obsolete. This is far from the case; certainly particular design philosophies and algorithms now need to be revised in the light of large scale and very large scale integrated circuit technology (LSI/VLSI) but the fundamental concepts are just as relevant.

Moreover, as computer systems grow more complex the problem of providing efficient communication with the computer becomes greater. The equipment required to input and output data ranges over a wide spectrum, from simple keyboards and printers to sophisticated audio and visual input/output devices. Again there is a need to incorporate mass storage systems and specialized data collection and control devices for real-time operation. The interfacing of these devices is not trivial and requires a considerable knowledge of both hardware and software techniques.

The engineering of all these systems, including the computer itself, depends primarily on the design of suitable logical networks and their interconnection paths. The process of connecting together switching or gating circuits[1,2] in a logical and systematic manner to fulfil a given functional requirement is called *logical design*. It is in the area of interface design that most engineers make their first acquaintance with logical design using standard logic modules available as small or medium scale integrated circuits (SSI/MSI).

It is of course possible to design a digital system using intuitive methods based on experience and 'cut and try' techniques. In some cases this can be successful but with complex systems involving large numbers of variables the approach can spell disaster! For example, a particular combination of inputs or operating sequences can easily be overlooked in the initial design phases. Due to the difficulty of checking out large logic systems (exhaustive testing is prohibitively expensive) these design errors could remain un-detected until the system goes operational. This state of affairs could obviously have serious and deadly consequences if the system was working in a real-time environment such as an air defence complex or nuclear power station. The use of formal logic design methods based on *switching theory* coupled with a top-down *structured design* approach can obviate many of these problems.

The digital system design process can be considered as consisting of four basic stages—the *behavioural* or *algorithmic* level where the specification for the required system is defined in a generalized abstract form as a set of logical procedures. This is followed by the *functional* or *architectural* level where the required logical processes are expressed as specific logical func-tions or subsystems for such tasks as counting, shifting, searching etc. At the *structural* or *logic* level the subsystem functions are realized in terms of logic which can be either hardware or software. Finally, in the case of inte-grated circuit realizations, comes the *physical* or *circuit* level where the structural representation is translated into physical circuits. In this book we shall be concerned primarily with the translation from the functional to the structural levels—the *logic design stage*.

The logic design process consists of two main activities:

(a) Enumerating the problem and, in so doing, forming a set of formal input/output specifications for the digital system.
(b) Implementing these specifications in terms of suitable logic circuits or devices (assuming a hardware realization), taking into account such constraints as performance, cost and reliability. This would involve consideration of the type of switching circuit, number of gates and/or modules, testing procedures, number of interconnec-tions, circuit delays, silicon area (in the case of integrated circuits) etc.

The outcome of the design procedure should be an optimum, that is cost effective, interconnection diagram for the logic modules. Alternatively, if the design is to be realized in silicon, a geometrical logical structure (such as a *stick diagram*) which can be mapped directly into an integrated circuit layout. The use of switching theory enables the problem to be completely understood and formally specified in mathematical terms. Moreover, by applying certain rules (algorithms) the specification may be reduced to a simpler or more convenient form which can then be implemented directly in terms of hardware. Thus switching theory enables a feasible design to be developed, in the sense that the input/output specifications are completely fulfilled. However, the theory cannot produce, for various reasons, an optimum design for all cases, and it is when this happens that the designer has to rely once again on his own initiative.

Unfortunately, then, with the present state of the art, switching theory is not the panacea it might seem, for two main reasons. First, the theory of large complex switching systems, or automata, is not yet fully understood.

Second, for systems involving a large number of variables, the amount of computation required becomes intractable even when large and powerful computers are used. As we shall see later, this is a major problem in *computer aided* design and is a result of the algorithms being such that the computation time required increases exponentially with the number of variables. However, a system may be decomposed into smaller subsystem blocks which then become amenable to theoretical design with the necessary computation being performed using computer programs—*software tools*.

Switching theory is a powerful tool (and the only possible one at the present time) for the definition and design of large digital systems. Logic designers must be able to apply the theory, to appreciate its limitations, and in so doing arrive at a cost effective and reliable system design. In the following chapters we shall explain the theory and techniques of logic design, including the use of software tools, but we must start at the very beginning with the coding of the basic information.

1.2 Number systems

In order to manipulate, display or transmit numbers using electrical or mechanical devices it is necessary to represent each symbol (0–9 in the case of decimal numbers) by a distinct state or condition of the device. For example, consider a decimal number counter consisting of mechanical gearwheels, of the type found in many car mileometers. The ten symbols are represented by ten cogs on the gearwheels, and each decade has its own individual gearwheel. Thus, to represent a five-digit decimal number, we require five gearwheels each with ten cogs. Each complete revolution of a gearwheel (count of 10, 100, 1000 etc.) causes the next gearwheel, representing the next highest power of ten, to enmesh so producing the effect of a carry. To perform the same task electronically we would need a ten-state device: in the simplest sense, ten on/off switches each connected to a display device such as an LED (light emitting diode) to represent one decade. As naturally occurring ten-state devices are very rare, and when specially made tend to be expensive in components, it would appear obvious to use a number system with fewer symbols. Furthermore, there are numerous examples of readily available two-state devices, such as switches (on/off), relay contacts (made/unmade), transistor circuits (conducting/cut-off), magnetic materials etc. Thus, if we could use a method of counting which required only two symbols (the *binary system*) we could utilize these two-state devices to devise economical hardware representations.

This idea is perfectly feasible, since there is nothing unique about the decimal system with its ten symbols and place value method of representation, i.e. units, tens, hundreds etc. In fact, the only reason for this choice of *base* (or *radix*) seems to be the anatomical fact that we have ten fingers.

A number system based on a positive integer radix may be defined mathematically in terms of the polynomial

$$N = a_n q^n + a_{n-1} q^{n-1} + \cdots a_2 q^2 + a_1 q^1 + a_0 q^0 + a_{-1} q^{-1} + \cdots a_{-m} q^{-m}$$

where N is a positive real number, q the radix, and a represents the symbols. That this is a place value system is apparent from the polynomial, the *radix point* (decimal point for radix 10) occurring between terms with positive and negative radix indices. Movement of the radix point left or right produces division and multiplication respectively by the radix.

As an example of the application of this polynomial let us express the decimal number 175.5 using various number systems:

Decimal: $q = 10$, symbols 0, 1, 2, 3, 4, 5, 6, 7, 8, 9

$$(175.5)_{10} = 1 \times 10^2 + 7 \times 10^1 + 5 \times 10^0 + 5 \times 10^{-1}$$

Hexadecimal: $q = 16$, symbols 0, 1, 2, 3, 4, 5, 6, 7, 8, 9, A, B, C, D, E, F

$$(175.5)_{10} = (AF.8)_{16} = A \times 16^1 + F \times 16^0 + 8 \times 16^{-1}$$

Note that new symbols need to be invented when the radix is greater than 10, i.e. $A = 10$ decimal, $B = 11$ decimal etc.
Octal: $q = 8$, symbols 0, 1, 2, 3, 4, 5, 6, 7

$$(175.5)_{10} = (257.4)_8 = 2 \times 8^2 + 5 \times 8^1 + 7 \times 8^0 + 4 \times 8^{-1}$$

Binary: $q = 2$, symbols 0, 1

$$(175.5)_{10} = (10101111.1)_2$$
$$= 1 \times 2^7 + 0 \times 2^6 + 1 \times 2^5 + 0 \times 2^4 + 1 \times 2^3$$
$$1 \times 2^2 + 1 \times 2^1 + 1 \times 2^0 + 1 \times 2^{-1}$$

Ternary: $q = 3$, symbols 0, 1, 2

$$(175.5)_{10} = (20111.111 \cdots)_3$$
$$= 2 \times 3^4 + 0 \times 3^3 + 1 \times 3^2 + 1 \times 3^1 + 1 \times 3^0$$
$$+ 1 \times 3^{-1} + 1 \times 3^{-2} + 1 \times 3^{-3} \cdots$$

Table 1.1 shows a selection of number systems expressed in the more familiar concept of allocating separate columns to the various powers of the radix. Each time we count up to the radix times the column power we add 1 to the next left-hand column, i.e. a carryover. Some interesting points arise from considering these examples:

(a) The number length depends on the magnitude of the chosen radix. For example, binary numbers require many more symbols than their decimal or hexadecimal equivalent.

(b) The factors of the radix determine the ease and accuracy of representing common fractions. In a decimal system, the fraction $\frac{1}{3}$ is $0.333\cdots$ (recurring) since 3 is not a factor of the base 10. However in the ternary system $\frac{1}{3}$ becomes 0.1 and for duodecimal (radix of 12) 0.4, since 3 is a factor of both 12 and 3.

(c) If a radix is chosen with more than ten symbols it is necessary to invent new ones to represent 10, 11, 12 etc. At the risk of confusion it has become standard practice to use the letters of the alphabet, A, B, C etc. for this purpose.

(d) The complexity of the multiplication (and addition) tables increases

Table 1.1 Number systems

Column groups: *Decimal, radix of ten* (cols 1–6) · *Binary, radix of two* (cols 7–16) · *Hexadecimal, radix of sixteen* (cols 17–22) · *Octal, radix of eight* (cols 23–28).

10^2 100	10^1 10	10^0 1	.	10^{-1} $\frac{1}{10}$	10^{-2} $\frac{1}{100}$	2^6 64	2^5 32	2^4 16	2^3 8	2^2 4	2^1 2	2^0 1	.	2^{-1} $\frac12$	2^{-2} $\frac14$	16^2 256	16^1 16	16^0 1	.	16^{-1} $\frac{1}{16}$	16^{-2} $\frac{1}{256}$	8^2 64	8^1 8	8^0 1	.	8^{-1} $\frac18$	8^{-2} $\frac{1}{64}$
		0	.									0	.					0	.					0	.		
		0	.	5	0							0	.	1	0			0	.	8	0			0	.	4	0
		1	.									1	.					1	.					1	.		
		2	.								1	0	.					2	.					2	.		
		3	.								1	1	.					3	.					3	.		
		4	.							1	0	0	.					4	.					4	.		
		5	.							1	0	1	.					5	.					5	.		
		6	.							1	1	0	.					6	.					6	.		
		7	.							1	1	1	.					7	.					7	.		
		8	.						1	0	0	0	.					8	.				1	0	.		
		9	.						1	0	0	1	.					9	.				1	1	.		
	1	0	.						1	0	1	0	.					A	.				1	2	.		
	1	1	.						1	0	1	1	.					B	.				1	3	.		
	1	2	.						1	1	0	0	.					C	.				1	4	.		
	1	3	.						1	1	0	1	.					D	.				1	5	.		
	1	4	.						1	1	1	0	.					E	.				1	6	.		
	1	5	.						1	1	1	1	.					F	.				1	7	.		
	1	6	.					1	0	0	0	0	.				1	0	.				2	0	.		
1	0	0	.			1	1	0	0	1	0	0	.				6	4	.			1	4	4	.		

with the size of the radix. For example, compare the binary multiplication table with that for octal (Table 1.2).

Although the decimal number system is the only representation used in everyday life all the others, particularly the binary system, are used extensively in the design and application of digital and computer systems.

1.3 The binary system

Let us now discuss the binary system in more detail. It is apparent that with only two symbols it is easier to represent numbers (or any information) more economically in terms of hardware by using two-state devices such as

Table 1.2 Arithmetic tables

×	0	1
0	0	0
1	0	1

Binary multiplication

+	0	1
0	0	1
1	1	0 (carry 1)

Binary addition

×	0	1	2	3	4	5	6	7
0	0	0	0	0	0	0	0	0
1	0	1	2	3	4	5	6	7
2	0	2	4	6	10	12	14	16
3	0	3	6	11	14	17	22	25
4	0	4	10	14	20	24	30	34
5	0	5	12	17	24	31	36	43
6	0	6	14	22	30	36	44	52
7	0	7	16	25	34	43	52	61

Octal multiplication

+	0	1	2	3	4	5	6	7
0	0	1	2	3	4	5	6	7
1	1	2	3	4	5	6	7	10
2	2	3	4	5	6	7	10	11
3	3	4	5	6	7	10	11	12
4	4	5	6	7	10	11	12	13
5	5	6	7	10	11	12	13	14
6	6	7	10	11	12	13	14	15
7	7	10	11	12	13	14	15	16

Octal addition

+	0	1	2	3	4	5	6	7	8	9	A	B	C	D	E	F
0	0	1	2	3	4	5	6	7	8	9	A	B	C	D	E	F
1	1	2	3	4	5	6	7	8	9	A	B	C	D	E	F	10
2	2	3	4	5	6	7	8	9	A	B	C	D	E	F	10	11
3	3	4	5	6	7	8	9	A	B	C	D	E	F	10	11	12
4	4	5	6	7	8	9	A	B	C	D	E	F	10	11	12	13
5	5	6	7	8	9	A	B	C	D	E	F	10	11	12	13	14
6	6	7	8	9	A	B	C	D	E	F	10	11	12	13	14	15
7	7	8	9	A	B	C	D	E	F	10	11	12	13	14	15	16
8	8	9	A	B	C	D	E	F	10	11	12	13	14	15	16	17
9	9	A	B	C	D	E	F	10	11	12	13	14	15	16	17	18
A	A	B	C	D	E	F	10	11	12	13	14	15	16	17	18	19
B	B	C	D	E	F	10	11	12	13	14	15	16	17	18	19	1A
C	C	D	E	F	10	11	12	13	14	15	16	17	18	19	1A	1B
D	D	E	F	10	11	12	13	14	15	16	17	18	19	1A	1B	1C
E	E	F	10	11	12	13	14	15	16	17	18	19	1A	1B	1C	1D
F	F	10	11	12	13	14	15	16	17	18	19	1A	1B	1C	1D	1E

Hexadecimal addition

switches. However, the numbers so represented are much longer than their decimal counterparts. Thus, if we wish to represent the decimal number 10^n as a binary number, where n can be any integer, it follows that

$$2^b \geq 10^n$$

and

$$b = \left[\frac{n}{\log_{10} 2}\right] = \frac{n}{0.301}$$

where b is the number of binary digits (*bits*) required in the representation; b must either be an integer or the next largest integer. To represent decimal numbers in binary thus requires approximately three times as many digits, e.g. to obtain an accuracy of 1 part in 10^3 requires a 10-bit binary number. This makes the manual transcription and processing of large binary numbers a time consuming and error prone process.

One convenient way of overcoming this problem is to use the octal or hexadecimal systems. Since 2 is a factor of both 8 and 16, conversion from binary to octal (or hexadecimal) and vice versa is a very simple process. For example, take the binary number

110110111010011110101101101

To find its octal equivalent, we split the binary number into groups of three ($2^3 = 8$) starting from the right-hand, least significant, digit, and then write down the octal equivalent of each three-bit group:

1	101	101	110	100	111	101	101	101
1	5	5	6	4	7	5	5	5

Again, to perform a hexadecimal conversion we split the number into groups of four ($2^4 = 16$), this time writing down the hexadecimal equivalent of the four-bit group:

1	1011	0111	0100	1111	0110	1101
1	B	7	4	F	6	D

Thus $(110110111010011110101101101)_2$

$$= (155647555)_8 = (1B74F6D)_{16}$$

Conversion from octal (or hexadecimal) to binary is the direct inverse of this operation.

Octal and hexadecimal numbers are used in digital computers for allocating codes to machine-code orders and memory addresses on account of this easy conversion to binary, which is the usual internal representation adopted for the hardware.

Integral binary numbers may be converted into decimal numbers either by directly adding the relevant power of two, or by successive division by binary ten and converting each remainder into a decimal number: in the last method, the equivalent decimal number appears with the least significant number first. Consider the binary number 101101101; this is equivalent to

$$1 \times 2^8 + 0 \times 2^7 + 1 \times 2^6 + 1 \times 2^5 + 0 \times 2^4$$
$$+ 1 \times 2^3 + 1 \times 2^2 + 0 \times 2^1 + 1 \times 2^0$$

which is equal to $256 + 64 + 32 + 8 + 4 + 1 = 365$. Using the alternative method we divide the binary number by binary ten:

(a)
$$\frac{.100100}{1010 \overline{)101101101}}$$
$$\underline{1010}$$
$$1011$$
$$\underline{1010}$$

Remainder 101 = 5

(b)
$$\frac{11}{1010 \overline{)100100}}$$
$$\underline{1010}$$
$$10000$$
$$\underline{1010}$$

Remainder 110 = 6

(c)
$$1010 \overline{)11}$$

Remainder 11 = 3

Thus, decimal equivalent = 365.

The reverse procedure (that is, decimal to binary) is accomplished by successive division by 2, noting the remainder at each stage; again the least significant digit appears first. Using the same example, we have:

$$2 \overline{)365}$$
$$2 \overline{)182} \quad \text{Remainder} \quad 1 \quad \text{least significant digit}$$

	Remainder
91	0
45	1
22	1
11	0
5	1
2	1
1	0
	1

The equivalent binary number is, then, 101101101.

Fractional numbers may be converted in a similar way. For example, to convert decimal fractions to binary, the fraction is multiplied successively by 2 and the integral (whole number) part of each product (either 0 or 1) is retained as the binary fraction. For instance, consider the decimal fraction 0.45678. Multiplying by 2 we have

Binary point	
	0.45678
0	0.91356
1	1.82712
1	1.65424
1	1.30848
0	0.61696, etc.

Thus the binary equivalent of 0.45678 is $0.01110\cdots$. To convert binary fractions to decimal, a similar procedure is followed; the binary fraction is repeatedly multiplied by binary ten, and after each operation the integral part is converted to its decimal equivalent.

Similar procedures may be followed for converting between decimal and any other radix, for example to convert decimal 365 to hexadecimal we must successively divide by 16, thus:

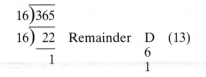

Thus the equivalent hexadecimal number is 16D. This may be converted back into decimal by simply adding the appropriate powers, i.e.

$$1 \times 16^2 + 6 \times 16^1 + D \times 16^0 = 256 + 96 + 13 = 365$$

Arithmetic using binary numbers is a far simpler procedure than the corresponding decimal process, due to the very elementary rules of addition and multiplication (see Table 1.2). However, the long numbers can still be a handicap, especially in the case of protracted carries. Consider the addition sums below:

(a) Augend	101101	45	(b)	↓↓ ↓ 10101111	175
Addend	10110	22		100101	37
Sum	1000011	67		111101	61
				110010	50
				101000011	323

Note that in example (b), columns 3, 5 and 6, the sum obtained is effectively 4, i.e. binary 100, and the carryover is to the second column up. This would happen, of course, in decimal addition if a column of numbers summed to a value greater than 99, an unusual occurrence due to the short numbers involved.

Subtraction is carried out by following the normal method except that the borrow now becomes the next power of 2:

(a) Minuend	101110	46	(b)	111100	60
Subtrahend	10001	17		110111	55
Difference	011101	29		000101	5

An alternative method of subtraction is to add a negative number to the minuend, i.e. $46 - 17 = 46 + (-17)$. To use this method we must have some means of representing negative numbers in the binary system. There are two ways of doing this:

(a) to express the number as a magnitude with a minus sign attached;
(b) to use a complement notation.

Method 1 is familiar and easy to use except that, as well as doing the actual arithmetic, we have also to deduce the correct sign. This can lead to problems in machine implementation.

The most convenient method is that in which we use the 2s *complement* of the number; this is defined as $2^n - N$, where N is the binary number and 2^n is the *next* highest power of 2. In order to distinguish negative numbers from positive numbers, the range of N is restricted and the most significant digit (m.s.d.) is used to represent the sign (positive 0, negative 1) of the number. For example, let N be an eight-bit number; the maximum number that can be represented is $N = 2^8 - 1 = 255$. If we now use the m.s.d. as a *sign digit*, the number range is restricted to $-2^7 \leq N < 2^7$ with a maximum positive number of $2^7 - 1$. To find the 2s complement of N, say 17, we must subtract this from the next highest power of 2, i.e. 2^8.

$$
\begin{array}{r|l}
1 & 00000000 \\
  & 00010001 \quad 17 \\
\hline
1 & 11101111 \quad -17 \\
\hline
\end{array}
$$

Disregard overflow

$\llcorner$ sign digit

Now we may perform the subtraction $46 - 17$ as

$$
\begin{array}{r|l}
  & 00101110 \quad 46 \\
  & 11101111 \quad -17 \\
\hline
1 & 00011101 \quad 29 \\
\hline
\end{array}
$$

The 2s complement may be found easily by taking the 1s complement (that is, the inverse—replace 0s by 1s and vice versa) and adding $+1$ to the least significant digit. Note also that using this method the correct sign digit appears automatically as a result of including it in the arithmetic operations. The idea of complements applies to any number system, and for this reason the binary 2s complement is also referred to as the radix complement.

Multiplication is performed in the usual way by multiplying and then shifting one place to the left, finally adding the partial products. Care must be exercised in the addition due to the frequent formation of protracted carries as in the following example. Note that since the multiplier can only be 0 or 1, the partial product is either zero or equal to the multiplicand.

Multiplicand	1101101	109
Multiplier	1011	11

$$
\begin{array}{r}
1101101 \\
1101101 \\
1101101 \\
\end{array}
$$

Product   10010101111   1199

Again, the process of division is very similar to standard decimal arithmetic, but simplified because it is only possible to divide once or not at all.

$$
\begin{array}{ll}
\qquad\qquad\qquad\quad \overline{1101100} \quad \text{Quotient} & 108 \\
\text{Divisor} \quad 11011\,\overline{)101101101101} \quad \text{Dividend} & 27\,\overline{)2925} \\
\qquad\qquad\qquad \underline{11011} \qquad\qquad\ \text{Remainder} & 9
\end{array}
$$

$$
\begin{array}{l}
100101 \\
\underline{11011} \\
101001 \\
\underline{11011} \\
\quad 11101 \\
\quad \underline{11011}
\end{array}
$$

Remainder   1001

The most convenient and usual way of representing numbers in a digital system is to use a *fixed point* notation in which all numbers are treated as binary fractions within the range $-1 \le x < 1$. The binary point lies to the immediate right of the most significant digit, which is also the sign digit when using the 2s complement notation for negative numbers. Integers and real numbers may be represented in this system by using a suitable predetermined scaling factor. Thus an 8-bit word may be considered as

2^0	2^{-1}	2^{-2}	2^{-3}	2^{-4}	2^{-5}	2^{-6}	2^{-7}	
0	1	0	1	0	0	0	0	$0.625 \times 2^0 = 80 \times 2^{-7}$
0	0	0	0	1	1	1	0	14×2^{-7}
0	0	1	0	0	1	0	0	4.5×2^{-4}
1	1	1	1	1	1	1	1	-1×2^{-7}
1	0	1	1	0	0	0	0	-0.625×2^0

Considerable problems can arise, however, when performing arithmetic operations. Care must be taken, for example, to ensure that addition and subtraction are carried out using numbers with the same scaling factor and that their sum or difference is within the number range of the machine. The same problem arises in multiplication and division. For example, if we perform the multiplication $(2 \times 2^{-7}) \times (15 \times 2^{-7})$ we obtain the *double length* product 30×2^{-14} which effectively shifts the product right out of a single length register. Thus to perform single length multiplication it is first necessary to scale the multiplicand and multipler such that the product can be represented in a single length register i.e. $(15 \times 2^{-5}) \times (2 \times 2^{-2}) = 30 \times 2^{-7}$.

To overcome the scaling difficulties encountered with fixed point numbers a *floating point*[3] system can be used. Floating point numbers are expressed in the form:

$$x = a \times 2^b$$

where a is called the *mantissa* (the fractional part) and b the *exponent* (the power to which the mantissa is to be raised). This method of representation is analogous to the usual scientific notation of expressing very large or very small decimal numbers in the form

$$89764.54 = 0.8976456 \times 10^5$$
$$0.0000027 = 0.27 \times 10^{-5}$$

Note that the field of the mantissa (the number of bits used to represent it in the binary notation) defines the accuracy and the exponent field the size of the number that can be represented.

1.4　Binary coded decimals

In the preceding section we have seen how numbers represented in the pure binary system may be manipulated arithmetically in much the same way as the more familiar decimal numbers. In fact there is no great difference in technique except for the change of radix. Though this is ideal for computing machines, the human operator still likes to think and communicate in the decimal system. Furthermore, there is a need to transmit and process data for basic communication purposes as well as for numerical computation. Thus there is a need to code numbers and alphabetical characters (often abbreviated to alphanumerics) in terms of binary symbols or bits. In this way we can communicate in our normal language, the data being encoded into some form of binary code for the convenience of the hardware, or logic circuits, comprising the digital system.

In order to represent the ten decimal numbers 0 to 9 we need four bits, giving 2^4 or 16 possible combinations, of which only ten are used. Each code, then, has four bits and these bits can be arranged in any way to represent the decimal digits. They are, however, generally assigned values —*weights*—which when summed give the decimal number represented by the four-bit combination; the most frequently used set of weights is the 8–4–2–1 of pure binary. Such codes are called *arithmetic codes* (see Table 1.3). For example, to represent the decimal number 9873 in 8421 binary-coded decimal, we would code up as follows:

　　　1001　　　1000　　　0111　　　0011

Many different weighted codes are possible, but the weights must be chosen in such a way that their sums are not greater than 15 and not less than 9; moreover, one of the weights must be 1 and another must be either 1 or 2. Weights can also be either positive or negative. Examples of possible combinations are

　　　3321　　　5321　　　7421　　　8421　　　5421　　　5211

Some binary-coded decimal codes have additional useful properties. The 7421 code, for example, has a minimum number of 1s in its representation and, if used in such a way that an electrical device must be in a power consuming state to indicate a 1, results in minimum power consumption. Arithmetic codes in which the sum of the weights is exactly 9 have the property that the 9s complement of the number (analogous to the 1s complement mentioned above, i.e. $9 - N$, where N is the number) can be obtained by simply inverting the binary equivalent. For example, in the

Table 1.3 Binary codes

Decimal number	Pure binary $2^3\ 2^2\ 2^1\ 2^0$	Binary coded decimal 7421	5421	5211	Excess three code	Reflected or Gray code $g_4\ g_3\ g_2\ g_1$
0	0 0 0 0	0000	0000	0000	0011	0 0 0 0
1	0 0 0 1	0001	0001	0001	0100	0 0 0 1
2	0 0 1 0	0010	0010	0100	0101	0 0 1 1
3	0 0 1 1	0011	0011	0110	0110	0 0 1 0 ↑
4	0 1 0 0	0100	0100	0111 ↑	0111 ↑	0 1 1 0 ↓
5	0 1 0 1	0101	1000	1000 ↓	1000 ↓	0 1 1 1
6	0 1 1 0	0110	1001	1001	1001	0 1 0 1
7	0 1 1 1	1000	1010	1011	1010	0 1 0 0 ↑
8	1 0 0 0	1001	1011	1110	1011	1 1 0 0 ↓
9	1 0 0 1	1010	1100	1111	1100	1 1 0 1

5211 code shown in Table 1.3, decimal $4 \equiv 0111$, $9 - 4 = 5 \equiv 1000$. A further requirement is that the arithmetic code must be symmetrically organized about its centre; this is apparent by inspection of the code and the example above. Self-complementing codes such as these (and also the excess three code, which is obtained by adding 3 to each group of the pure binary code) are very useful when performing decimal or binary-coded decimal arithmetic.

The reflected binary or Gray code (also called a cyclic progressive code) is used chiefly in digital shaft position encoders and has the merit of incurring only one digit change when passing from any one combination to the next. A code is said to be reflecting when its mirror-image is reproduced (excluding the most significant digit) about the midpoint of a complete ascending tabulation of the code. For example, in the Gray code (Table 1.3) the first eight combinations form a three-bit Gray code of length eight (ignoring digit g_4). That this reflects can be seen by noting that digits $g_1 g_2$ are symmetrical about the midpoint, with the most significant digit g_3 inverted.

One of the most common BCD codes used for transmitting data to and from a digital system, such as a microcomputer system with VDU, keyboard and printer, is the 8-bit ASCII code (American Standard Code for Information Interchange). A typical selection of coded characters is shown in Table 1.4; note that the code is essentially 7-bit with the 8th bit normally being reserved for parity checking (see next section) or simply set to zero. For convenience in decoding, the ASCII code is divided into four groups of characters: control characters such as ESC, CR etc., special and numeric symbols and upper and lower case alphabets; the bits b_6 and b_7 are used to define these groups. The actual coding for special symbols and control characters, as with the decision on parity, will vary with particular systems.

Table 1.4 ASCII character codes

Character	Binary $b_8\ b_7\ b_6\ b_5\ b_4\ b_3\ b_2\ b_1$	Hex	Character	Binary $b_8\ b_7\ b_6\ b_5\ b_4\ b_3\ b_2\ b_1$	Hex
A	1 0 0 0 0 0 1	41	0	0 1 1 0 0 0 0	30
B	1 0 0 0 0 1 0	42	1	0 1 1 0 0 0 1	31
C	1 0 0 0 0 1 1	43	2	0 1 1 0 0 1 0	32
D	1 0 0 0 1 0 0	44	3	0 1 1 0 0 1 1	33
E	1 0 0 0 1 0 1	45	4	0 1 1 0 1 0 0	34
F	1 0 0 0 1 1 0	46	5	0 1 1 0 1 0 1	35
G	1 0 0 0 1 1 1	47	6	0 1 1 0 1 1 0	36
H	1 0 0 1 0 0 0	48	7	0 1 1 0 1 1 1	37
I	1 0 0 1 0 0 1	49	8	0 1 1 1 0 0 0	38
J	1 0 0 1 0 1 0	4A	9	0 1 1 1 0 0 1	39
K	1 0 0 1 0 1 1	4B	:	0 1 1 1 0 1 0	3A
L	1 0 0 1 1 0 0	4C	;	0 1 1 1 0 1 1	3B
M	1 0 0 1 1 0 1	4D	<	0 1 1 1 1 0 0	3C
N	1 0 0 1 1 1 0	4E	=	0 1 1 1 1 0 1	3D
O	1 0 0 1 1 1 1	4F	>	0 1 1 1 1 1 0	3E
P	1 0 1 0 0 0 0	50	?	0 1 1 1 1 1 1	3F
Q	1 0 1 0 0 0 1	51	CR	0 0 0 1 1 0 1	OD
R	1 0 1 0 0 1 0	52	ESC	0 0 1 1 0 1 1	1B
S	1 0 1 0 0 1 1	53	ACK	0 0 0 0 1 1 0	06
T	1 0 1 0 1 0 0	54	LF	0 0 0 1 0 1 0	0A
U	1 0 1 0 1 0 1	55	a	1 1 0 0 0 0 1	61
V	1 0 1 0 1 1 0	56	b	1 1 0 0 0 1 0	62
W	1 0 1 0 1 1 1	57	c	1 1 0 0 0 1 1	63
X	1 0 1 1 0 0 0	58	d	1 1 0 0 1 0 0	64
Y	1 0 1 1 0 0 1	59	e	1 1 0 0 1 0 1	65
Z	1 0 1 1 0 1 0	5A	f	1 1 0 0 1 1 0	66
[	1 0 1 1 0 1 1	5B	g	1 1 0 0 1 1 1	67
\	1 0 1 1 1 0 0	5C	h	1 1 0 1 0 0 0	68
]	1 0 1 1 1 0 1	5D	i	1 1 0 1 0 0 1	69
∧	1 0 1 1 1 1 0	5E	j	1 1 0 1 0 1 0	6A
–	1 0 1 1 1 1 1	5F	k	1 1 0 1 0 1 1	6B

1.5 Error detecting and correcting codes

If codes are used which utilize all possible combinations, such as representing the decimal numbers 0–15 by pure binary equivalents, any error which may occur (such as picking up or dropping digits) will go undetected, since the incorrect combination will still represent a valid number. To overcome this problem, *redundancy* must be introduced by adding extra bits to the code. All of the 4-bit codes described above, if used solely to represent the decimal digits 0–9, contain some redundancy in the sense that not all of the possible 2^4 code combinations are used. However, this is inadequate since there is still a chance of undetected errors occurring—for example, in the 5421 code the combination 1001 can, by picking up a digit, become 1101 or 1011. The first is obviously in error as no such number exists in the code, but the second would be treated as a valid combination.

The simplest way of adding redundancy is to insert an extra bit, called a *parity* bit, into each code combination. The value of this, 0 or 1, is such as to make the total number of 1s in each combination either even or odd according to the checking convention adopted. Should an error occur, the sum of the digits will no longer be odd (even), thus indicating the presence of an error in the code. Note, however, that only single errors, or errors resulting in an odd (even) number of 1s, will be detected. Odd parity is most frequently used (for example, in teleprinter codes for computer punched tape input/output), since there is no all-zero combination. Table 1.5 shows examples of typical codes. Another approach is to arrange that an error gives rise to a non-valid combination; an example is the 2-out-of-5 code shown in Table 1.5. It is possible to devise many codes like this (2-out-of-7, 3-out-of-8, or in general *p-out-of-q*). Since if an error occurs the number of 1s will be wrong, all these methods will only detect single errors.

Blocks of information can be checked by arranging the data in the form of a matrix and then making parity checks on the rows and columns, including the extra row formed by the column parity check in the data sent (note that an extra bit is required to check this row). Consider the following example which uses odd parity:

```
9   11001
8   01000
7   00111
3   10011
    _____
    011010   parity check on columns
```

The encoded data would be sent as

```
11001      01000      00111      10011      011010
```

To decode, individual parity checks are made on the rows and columns, and this will pinpoint the incorrect digit in the matrix. Since we are working in a two-valued system the correct digit can be obtained by inversion. This technique will detect multiple errors and also correct single errors,

Table 1.5 Error-detecting codes

Decimal number	Odd parity check pure binary					2-out-of-5 code	Diamond code
	PB	2^3	2^2	2^1	2^0		
0	1	0	0	0	0	01100	00010
1	0	0	0	0	1	10001	00101
2	0	0	0	1	0	10010	01000
3	1	0	0	1	1	00011	01011
4	0	0	1	0	0	10100	01110
5	1	0	1	0	1	00101	10001
6	1	0	1	1	0	00110	10100
7	0	0	1	1	1	11000	10111
8	0	1	0	0	0	01001	11010
9	1	1	0	0	1	01010	11101

becoming more efficient if different checking methods are used for rows and columns.

Hamming[4] has described a single-error detecting and correcting code (which also detects multiple errors) which employs check digits distributed throughout the message group. These check digits provide even parity checks on particular message digit positions in such a way that, if the parity checks are made in order, successful checks being designated by 0 and failure by 1, the resulting binary number gives the position of the incorrect digit. For this to apply, the first parity digit must check those positions (see Table 1.6) which contain a 1 in the 2^0 column (that is, 1, 3, 5, 7, 9, 11, 13 etc.); similarly, the second digit checks those positions which contain a 1 in the 2^1 column (2, 3–6, 7–10, 11–14, 15 etc.); the third check digit checks the positions with a 1 in the 2^2 column (4, 5, 6, 7–12, 13, 14, 15 etc.), and so on. This process may be extended indefinitely for message groups of any length. The amount of redundancy required becomes appreciably less as the message length is increased; for example, a seven-bit message group requires three check bits, but only five check bits are required for a 30-bit group. As an example of its use, suppose the message group 1011 is to be transmitted; the check digits are placed in the 2^0, 2^1, 2^2 etc. positions with data digits taking up the remaining places. Thus the message group would be encoded as 0110011. Should an error occur in the fourth position from the right, giving 0111011, application of the checks would yield: 1st check 0, 2nd check 0, 3rd check 1; taking the first check digit to be the least significant, this gives position 4 as incorrect. Unfortunately, the coding and encoding procedure has been found rather expensive to implement in terms of hardware, and simpler coding methods are generally employed in practical systems. However, as we shall see in Chapter 4, if recourse is made to linear switching circuit theory,[5] and the Hamming code is treated as a *cyclic* or *chain code*, simple logic circuits can be evolved using feedback shift registers.

Diamond[5] has devised a checking code for multiple errors which uses the properties of all numbers which obey the formula $3n + 2$. This BCD code complements on nine, and the sum of two numbers (in the Diamond code) can be obtained by normal binary addition and then subtraction of binary two. The checking process is to subtract 2 from the received combination and reduce it to modulo 3 (the remainder after dividing by binary three). If

Table 1.6 Hamming code

Digit position	Binary equivalent 2^2 2^1 2^0			Digit function	Weight for data bits
1	0	0	1	Check	
2	0	1	0	Check	
3	0	1	1	Data	8
4	1	0	0	Check	
5	1	0	1	Data	4
6	1	1	0	Data	2
7 (l.s.d.)	1	1	1	Data	1

there is no remainder after division, it is a valid number, with a low probability of error. The code may be extended to 8-bit numbers by using the formula $27n + 6$ in a similar way.

1.6   Logic circuits and symbols

So far we have concerned ourselves with the way in which data and information can be represented and encoded in a digital system. The next step is to consider how these binary representations can be manipulated using logic circuits. Before attempting a serious study of switching theory, however, let us first consider, in simple terms, the circuits and symbols used in logic design.

The basic logic circuits are shown in Fig. 1.1 using MILSPEC logic symbols[6] which are now accepted as standard throughout industry.

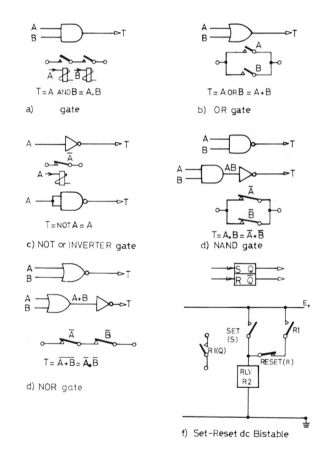

Figure 1.1 Basic logic switching circuits

The AND gate is a circuit that provides an output only when all the input signals are present simultaneously. In the equivalent relay circuit, normally open contacts A and B need to be made (i.e. the appropriate relay coils must be energized) before a signal can be transmitted through the circuit. If we represent putting contacts in series by a dot (.) then the action of the circuit can be expressed as $T = A.B$. In the next chapter a more rigorous mathematical definition will be given to this symbol, and to those that follow below.

The OR circuit (known as the inclusive OR) requires at least one input to be present in order to produce an output. Thus a signal will be transmitted through the relay circuit if A or B or both contacts are made. If in this case we represent the contacts in parallel by a plus (+) sign; we say that $T = A + B$.

The NOT circuit shown in Fig. 1.1(c) produces an output when the input is absent. This is equivalent to a normally closed relay contact; that is the presence of a signal energizes the relay coil and opens the contact. We shall denote a normally closed contact by using a bar over the input signal, thus $\bar{A}$ is the opposite or inverse of A.

The NOR circuit produces an output when all the inputs are absent, the equivalent relay circuit consisting of normally closed contacts in series. A signal will be transmitted through the circuit if contacts $\bar{A}$ and $\bar{B}$ remain closed, i.e. no inputs to the circuit. Note that the same response can be obtained by inverting the output of an OR gate (the OR/NOT function). The logical action of the circuit may be expressed as

$$T = \overline{A + B} = \bar{A}.\bar{B}$$

A similar circuit to the NOR is the NAND gate, which produces an output when at least one input is absent. The corresponding relay circuit consists of normally closed contacts in parallel; thus if any one input is absent, the contact remains closed and a path is established through the circuit. From Fig. 1.1(e), we say that the circuit performs the function AND/NOT and may be expressed logically as

$$T = \overline{A.B} = \bar{A} + \bar{B}$$

The operation of logic gates may be defined functionally in a tabular form by enumerating all possible input values together with the resultant outputs. This is shown in Table 1.7 for two-input gates (two binary variables hence four possible input conditions) where the 1 (TRUE) and 0 (FALSE) entries in the tables represent the presence or absence of an input or output. Note that although we have used two-input gates in our description of logic functions the *Boolean connectives* (i.e. OR, AND etc.) hold for any number of variables.

The two-state logic variables 0 and 1, as we have seen, can be represented by contacts (open or closed). Alternatively, and more usual in electronic circuits, they are represented by voltage levels arising from the output of a transistor inverter amplifier. Typically these voltages would be in the order of 0 V (ground) and 5 V (supply) called the LOW and HIGH values respectively. For a *positive logic* system logic 0 would be represented by a LOW value and logic 1 by a HIGH value; this is shown in Table 1.7(b). The alternative *negative logic* system, shown in Table 1.7(c), represents logic 0 by a HIGH value and logic 1 by LOW.

If Table 1.7(b) and (c) are compared it will be seen that the output of the positive logic AND gate is identical with that of the negative logic OR gate. Similarly the positive OR is identical to the negative AND. Thus if we are working with a positive logic convention and invert the inputs to an AND gate (which will be the same physical circuit whatever the convention) it will function as an OR gate but with the output inverted (i.e. a NOR gate)—this is shown in Fig. 1.2(a). We have also seen, however, that a NAND gate is an AND gate with the output inverted; thus to use a NAND gate to perform the OR function (in positive logic convention) the inputs must first be inverted. A dual situation exists with the OR gate as shown in Fig. 1.2(b), where if the inputs to an OR gate are inverted it will function as a positive logic AND gate with the output inverted, that is a NAND gate. These alternative configurations for the NOR/NAND gates have given rise to its own symbolism, as shown in Fig. 1.2(c); note that by convention the small circle always represents a signal inversion.

As we shall see later this alternative viewpoint of logic gates and their operation is very useful when actually implementing logic circuits, espe-

Table 1.7 Logic gates

INPUTS A B	AND A.B	OR A + B	NAND $\overline{A.B}$	NOR $\overline{A + B}$	EXOR $A \oplus B$	EXNOR $\overline{A \oplus B}$
00	0	0	1	1	0	1
01	0	1	1	0	1	0
10	0	1	1	0	1	0
11	1	1	0	0	0	1

(a) Truth values

INPUTS A B	AND A.B	OR A + B	NAND $\overline{A.B}$	NOR $\overline{A + B}$
L L	L	L	H	H
L H	L	H	H	L
H L	L	H	H	L
H H	H	H	L	L

(b) Voltage levels—positive logic

INPUTS A B	AND A.B	OR A + B	NAND $\overline{A.B}$	NOR $\overline{A + B}$
H H	H	H	L	L
H L	H	L	L	H
L H	H	L	L	H
L L	L	L	H	H

(c) Voltage levels—negative logic

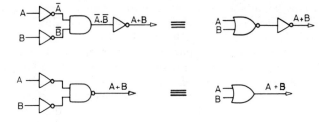

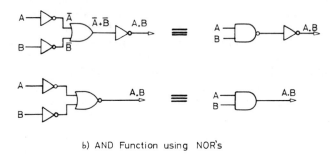

a) OR Function using NAND's

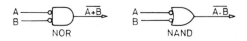

b) AND Function using NOR's

A ———o⟩ $\overline{A+B}$ A ———o⟩ $\overline{A.B}$
B ———o⟩   B ———o⟩
  NOR   NAND

c) Alternative symbols for NOR/NAND

Figure 1.2 Positive/negative logic

cially when mixed logic elements are employed. In this text, however, unless otherwise stated, we shall always use a positive logic convention.

As well as logic gates a digital system must also have some means of storing binary information. To store a single bit of digital information requires a two-state device called a *bistable* or *flip-flop* circuit. A typical example of a single-bit memory, called a *set-reset bistable*, is shown symbolically, together with its relay equivalent, in Fig. 1.1(f). This particular type of bistable is also known as a d.c. bistable or *latch* since it responds directly to voltage level changes on its inputs.

The operation is such that the output Q (the complement or inverse of the output appears at $\bar{Q}$) is set ON to, say, a positive voltage level representing a 1, when the set input receives a 1. The output is reset OFF to a negative voltage level representing a 0, when the reset input receives a 1. Once the output Q is at 1, further inputs to the set input have no effect. Likewise, when the output is at 0, a change to 1 can only be effected by applying an input to the set terminal. A simple equivalent relay circuit may be constructed by using a hold-on contact R_1 in series (a form of feedback) with the reset button or relay contact. To set the bistable the set contact is made; this energizes the relay coil and causes contacts R_1 and $R_1(Q)$ to close; R_1 maintains the current to the coil when the set contact opens. To reset the device, the hold-on circuit must be broken using the reset contact. The output is obtained via the normally open contact $R_1(Q)$; the logical action of the SR bistable will be analysed in greater detail in later chapters on sequential logic.

As we shall see in Chapter 2 a complete digital system can be built up using either AND/NOT, OR/NOT, NOR or NAND gates. However, it is useful in design to have available a much wider range of elements, for example different types of bistable circuits (in particular clocked or edge-triggered devices) for synchronous sequential systems, power amplifiers for buffering signals, equivalence and non-equivalence circuits (exclusive OR gates) etc. In addition, it is also convenient to have larger logic blocks and subsystem modules such as counters, arithmetic units, encoders and decoders, registers (comprising a number of single-bit bistable stages) etc. The list is far from complete and does not include any of the special purpose units available for interfacing and data communications. Figure 1.3 shows the

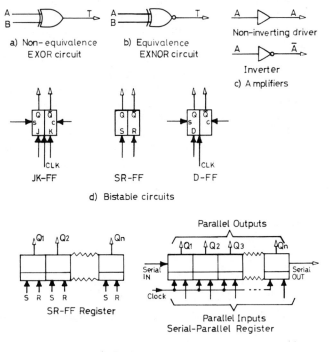

a) Non-equivalence EXOR circuit

b) Equivalence EXNOR circuit

Non-inverting driver

Inverter

c) Amplifiers

JK-FF   SR-FF   D-FF

d) Bistable circuits

SR-FF Register

Parallel Outputs

Parallel Inputs
Serial-Parallel Register

e) Register Modules

Figure 1.3 Logic modules

most commonly used logic units depicted in their corresponding MILSPEC symbols.

1.7 Digital circuits

So far we have only considered the functional characteristics of logic gates and two-state devices omitting, except for the case of relay networks, their physical properties. In the main, logic gates employ electronic circuits which are usually realized with semiconductor materials using integrated circuit techniques. Although a detailed understanding of the circuitry is unnecessary some knowledge is essential in order to choose and exploit a given logic family. As we shall see, to perform the complete range of logical operations requires a minimum of an invertor and either an AND or OR gate; alternatively the NAND/NOR gate may be used. All these basic logic elements are possible and available using electronic circuitry; a set of logic elements with compatible supply voltages and logic 0 and logic 1 values is called a *logic family*.

Transistors are used almost exclusively in these circuits and would normally take the form of either *bipolar* or *metal oxide semiconductor* (MOS) devices. The major difference between the two devices is that whereas bipolar transistors are current controlled, MOS devices (also called field-effect transistors—FETs) are voltage controlled. When used as a switch the bipolar transistor works between the *cut-off* and *saturation regions* of its characteristic curve.[7]

The simplest way to implement AND/OR gates is to use the passive diode–resistor network shown in Fig. 1.4; this form of circuit was used extensively in early logic systems. Figure 1.4(a) is an AND gate for positive logic (OR gate for negative logic) and Fig. 1.4(b) shows the OR gate. If any diode is conducting (i.e. biased in the forward direction with the anode more positive than the cathode) the output takes up the level of the input to that diode (less the voltage drop across the diode). Thus for the AND gate, if all inputs are positive, i.e. logic 1, the diodes are reverse-biased and do not conduct and the output stays at E. If, however, any one of the inputs goes negative, the appropriate diode will conduct, and the output

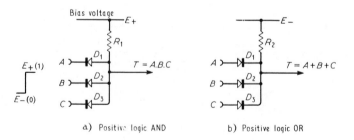

a) Positive logic AND   b) Positive logic OR

Figure 1.4 Diode–resistor logic

will follow and go negative, i.e. logic 0. It will be apparent that if we reverse the logic convention (positive voltage is logic 0) the circuit functions as an OR gate. It is possible to choose the bias voltage to be more positive than logic 1 (generally five times the output swing) and under this condition all diodes will conduct when the inputs are present together, clamping the output to the logic 1 level.

If circuits are cascaded, the input drive current to the circuit must be provided by the preceding stages. In general, this means that the bias resistors must be reduced in value in order to maintain the required drive currents. This is not possible in practice unless different types of AND/OR gates are available in the system, and buffer amplifiers are used instead. Though these circuits are no longer in use the principle is employed in AND/OR matrix arrays (such as the programmable logic array—PLA) using semiconductor devices.

The natural extension of diode logic is to incorporate an invertor amplifier after each diode gate, the result being called *diode-transistor logic—DTL*. Figure 1.5 shows the circuit diagram of this well-tried and reliable logic configuration which was one of the first logic elements to be fabricated as an integrated circuit, since all the circuit design rules had previously been well established using discrete components.

Basically the operation is as follows. Any input going low (equivalent to logic 0) will cause the base of the transistor to go negative with respect to the emitter and hence the transistor will turn OFF going into the cut-off region. This means that no current flows in the collector circuit and the output goes positive, pulled up to V_c equivalent to logic 1. When the inputs are high (logic 1) the base of the transistor will be positive and hence goes into saturation with the collector output approaching the emitter value, i.e. 0 V equivalent to logic 0. As we have seen, changing the logic convention produces the dual function—the NOR operation. When logic circuits are referred to as NAND/NOR it is generally to be assumed that positive logic convention is indicated.

In order to operate the bipolar transistor as a switch it is necessary to

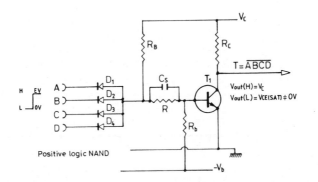

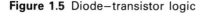

Figure 1.5 Diode–transistor logic

supply sufficient base current to saturate (turn on) the transistor; moreover the charge induced in the base by this current must be removed when the transistor is turned off. In practice due to the variation of h_{FE} (the current gain of the transistor, collector current/base current) it is difficult not to generate more base current than actually required. Since the turn-on time of the transistor depends on generating a large base current and the turn-off time on removing the consequent stored charge the dynamic performance (switching times) of the circuit are critically dependent on these factors. To offset these adverse effects improvements are made to the basic circuit, for example in Fig. 1.5 the capacitor C_s in the base circuit produces a capacitive overdrive current to switch the transistor on without unduly increasing the stored base charge. Again the transistor can be prevented from being driven into hard saturation by inserting a clamping diode (usually a Schottky diode is used in integrated circuits which has zero stored charge) between base and collector. This ensures that the base–collector junction is never fully biased diverting excess current away from the base.

With the introduction of IC technology it quickly became apparent that transistor devices are just as easy to fabricate as diodes and *transistor transistor logic* (TTL) has now become an accepted industrial standard embedded in the SN74 series of logic devices.[8]

A typical circuit, employing a multiple emitter input transistor T_1, is shown in Fig. 1.6. Though the static characteristics of the circuit are similar in operation to DTL the dynamic performance is very much improved. This is due to the current gain of transistor T_1 which ensures that the stored base charge in T_2 is quickly removed when the transistor is being turned off. The circuit differs from the usual gate plus simple invertor structure in that the output transistor is replaced by a push-pull pair, T_3 and T_4, known as a *totem-pole*, driven by a phase-splitter transistor T_2. The operation of the circuit is such that if any of the inputs are held low the transistor T_1 saturates and the base goes low holding off T_2. With T_2 off T_3 is driven on and the output goes high. With all the gate inputs high T_2 and T_3 are both on and the output goes low assuming the saturation voltage of T_3. The diode D_1 is a voltage level shifting diode and ensures that transistor T_3 can be turned off under all operating conditions. Note that T_3 acts as an emitter follower, giving a low output resistance for both logic states.

If R_c, T_3 and D_1 are omitted from the TTL circuit in Fig. 1.6 an *open-collector* output gate is obtained. This is an important circuit since it is possible to perform a *wired-OR* logic function by simply connecting together the outputs of open collector gates and using an external pull-up resistor. (The use of distributed logic NOR and NAND networks using wired-OR will be discussed in a later chapter.) TTL normally requires a +5 V supply with logic levels of 2–5.5 V for logic 1 (assuming positive logic) and 0–0.8 V for logic 0. TTL logic can operate at speeds of 25–50 MHz but has a heavy power consumption.

MOS transistor logic, though slower in operation than bipolar devices, is used extensively in integrated circuit systems mainly because of the ease of fabrication in silicon and the higher packing densities achieved. As we

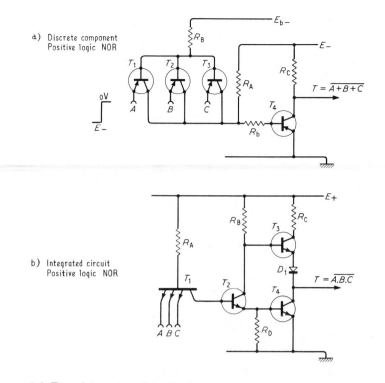

a) Discrete component
Positive logic NOR

$T = \overline{A+B+C}$

b) Integrated circuit
Positive logic NOR

$T = \overline{A.B.C}$

Figure 1.6 Transistor–transistor logic NAND gate

stated earlier the MOS transistor is a voltage controlled device and functions primarily as a switching circuit. The MOS transistor is a three-terminal device (consisting of *gate, drain* and *source*) and operates such that a voltage applied to the gate (greater than some *threshold voltage* V_T, in the order of -2 to 5 V) will cause a current to flow between source and drain. Since the value of the drain current is zero for zero gate voltage and increases for gate voltages greater than V_T, this type of device is called an *enhancement mode MOS*. As with bipolar transistors it is possible to make both n- and p-type MOS devices, though in the main n-channel devices are used.

Figure 1.7(a) shows a typical NMOS invertor circuit where the load R_L is in effect a resistive (passive) pull-up. When V_{IN} is high, at a voltage close to V_D, T_1 is turned hard on and V_{OUT} goes to a low level close to ground. If the input goes low T_1 turns off and the output is pulled high, close to V_D, by R_L. It is possible to replace the pull-up resistor with another MOS transistor, which has the advantage of requiring less chip area than a resistor; this is shown in Fig. 1.7(b). The transistor is used as a two-terminal device, with the gate connected to the drain, and though an active device, it has characteristics similar to those of a resistor.

The enhancement mode transistor, however, has a drawback: V_{OUT} high

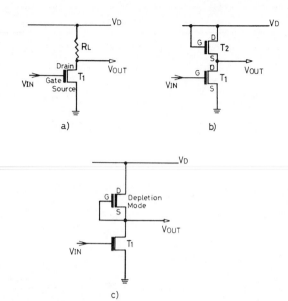

Figure 1.7 MOS invertor circuits

has a value of $V_D - V_T$ rather than V_D due to the voltage drops in the device. To overcome this problem a *depletion mode* transistor is used which allows a current to flow when the gate voltage is zero. When used as an active load the depletion transistor can supply an almost constant current thus allowing V_{OUT} to rise to V_D; the circuit is shown in Fig. 1.7(c). A very useful attribute of MOS transistors is that they may be considered as *transfer gates*, that is as the electronic equivalent of the relay contact allowing a bidirectional signal flow when closed and presenting a high impedance when open. Thus logic functions can be readily realized using MOS devices in series and parallel combinations, analogous to the relay networks considered earlier. For example, in the NAND circuit of Fig. 1.8(a) when A and B are both high the transfer gates close (i.e. conduct) behaving like an ideal switch with near zero resistance and putting V_{OUT} to ground; if any input is low the corresponding gate is effectively open-circuit and V_{OUT} goes high.

Though NMOS circuits have many advantages, bipolar devices are vastly superior in terms of speed. Moreover, the power consumption of both NMOS and bipolar circuits tends to be rather large in the saturated on condition. These problems may be overcome by using complementary n- and p-channel MOS devices (CMOS) in a push-pull configuration; a CMOS inverter circuit is shown in Fig. 1.9(a). With V_{IN} high (negative with respect to V_C) T_1 is turned off and T_2 on simultaneously giving a low V_{OUT}. With V_{IN} low, T_1 is on and T_2 off and V_{OUT} goes high. Note that in the steady state one device is always off; thus the standby dissipation is

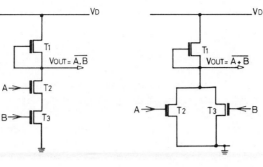

a) NAND gate   b) NOR gate

Figure 1.8 NMOS gates

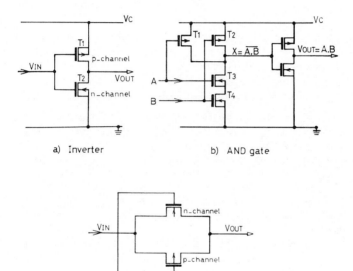

a) Inverter   b) AND gate

c) Transmission gate

Figure 1.9 CMOS circuits

very small. Moreover the push-pull configuration increases the speed of operation, though still not higher than that obtainable with bipolar devices.

A CMOS AND gate is shown in Fig. 1.9(b). The operation of the circuit is such that both inputs must be high to turn on the series pair T_3, T_4 and to turn off both the pull-up transistors T_1 and T_2. This condition produces a low at junction X, the input to the inverter. If any input is low then the serial chain is broken and one of the upper MOS transistors will bring the output high. This NAND function is then inverted to give the AND output (alternatively a double inverter buffer may be used to produce the NAND function).

CMOS devices may also be used in a switch configuration or *transmission gate* circuit as shown in Fig. 1.9(c). The control line C goes high to turn the switch on and establish a conducting path between V_{IN} and V_{OUT}. MOS devices normally require slightly higher power supply voltages than TTL (though they can vary from 3 to 20 V) with corresponding logic levels. CMOS devices are very susceptable to damage from static electricity (due to their high input impedance) and consequently unused inputs should always be connected to an appropriate high or low level source. Special precautions must also be taken in handling the circuits.

1.7.1 Electrical characteristics of logic gates

The properties of logic gates are characterized by their steady state (dc) and transient (ac) parameters: this data are freely available from the published data-sheets for the chosen logic family.

The dc conditions would include the maximum and minimum voltage levels for logic 1 and logic 0, the available output current both as a source and sink, power dissipation in the quiescent state etc. However, if the same logic family is used throughout a design then all inputs and outputs will be compatible and the following simple loading rules can be used:

Fan-out: A number indicating the ability of an output to drive other inputs (for TTL this is in the order of 10). Buffer amplifiers can be used to increase the fan-out if required.

Fan-in: The number of gate inputs; in some cases expander units are provided.

When mixed logic families are used, such as TTL and MOS, and when interfacing logic generally it is essential to consult the electrical characteristics.

Another important parameter is the *dc noise immunity* which is the amount of stray voltage (noise) which can be tolerated on a gate input without adversely affecting the output. Noise margins must be defined for both low and high output conditions. One of the major transient characteristics of a logic gate is the *propagation delay* defined as the delay (normally in nanoseconds) between the signal appearing at the input and the corresponding response at the output. Though average values are normally quoted (typically 15 ns for TTL and 50–100 ns for MOS) the propagation delay can vary considerably for high to low and low to high output transitions. Figure 1.10 shows how the propagation delay is measured for a typical inverter gate. Note that the delays t_1 and t_2 are specified at the 50% voltage levels of the input and output signals (rise- and fall-time values for a waveform are specified between the 10% and 90% points). The propagation delay t_{PD} quoted in the literature is normally an average value given by

$$t_{PD} = \frac{(t_1 + t_2)}{2}$$

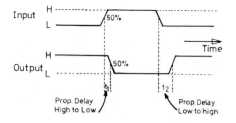

Figure 1.10 Gate propagation delays

The propagation delay will obviously depend on the type of logic gate and its capacitive loading.

1.8 Data representation and transmission

Having looked at the physical form of logic circuits let us now consider how binary information may be represented and transmitted in a digital system. Binary information is normally represented in one of two ways, either in the *serial* or *parallel* modes. In parallel working, the binary digits, represented by high and low voltage values, are each allocated a separate device (such as a bistable element) and/or connecting wire. Thus all digits of an *n*-bit number would appear simultaneously on *n* different wires or bistable outputs.

The binary digits in serial operation are represented by voltage levels on a single wire, but displaced in time. Thus an *n*-bit number would appear on a single wire in sequence, least significant digit first in time, and would require *n*-digit times for the complete number to appear. Figure 1.11 shows a comparison of these two techniques to represent the number 10001. The parallel method requires approximately *n* times as much hardware, but has the advantage of increased speed, and is almost *n* times as fast. Serial

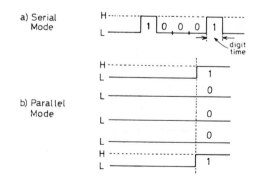

Figure 1.11 Parallel/serial systems

systems operating on a bit by bit process are much slower but require considerably less hardware. It is also possible to have a mixed serial–parallel mode of operation where, for example, the information may be transmitted in parallel mode with the processing being performed in serial.

As we shall see later digital systems require some form of central timing or *clock* to effect control of the logical processes. This is obvious in the case of serial systems where the binary digits occur at a constant rate determined by a clock (normally a square-wave) source. In general, however, any system where the inputs and outputs are periodically sampled at some clock frequency is called *synchronous* in contrast to the free-running or *asynchronous* mode of operation. All these modes of operation can be used to transmit binary information both internally to the system, say between registers and special logic units, and to external peripheral devices such as printers etc. Similar forms of representation would also be used to transmit information over data networks and in digital communication systems.

One common method of exchanging parallel information between units, say in a computer system, is to use a *data bus* where the inputs and outputs of all units are connected to a common set of wires (one for each bit of the data word, see Fig. 1.12). The control of the bus is such that only one unit may place data on the bus at any given time, but all units can simultaneously receive data. There is, however, a circuit problem: it is not possible to connect directly the output of logic gates together since the outputs will always be at either a low or high value. What is needed is some means of disconnecting all the unselected outputs from the data bus. This can be achieved in two ways—either by using open-collector gates (with external pull-up resistors for each line of the bus) or with special *tri-state logic*. Tri-state logic is a misnomer, since it is not a three-valued system but simply a normal logic gate with a third, open-circuit, output state. A separate enable input is required to put the output into the open-circuit state regardless of the input conditions. Both circuits are used in a similar way but tri-state devices are preferred because of their superior performance. The bus operation is as follows: each unit that wants to communicate with the

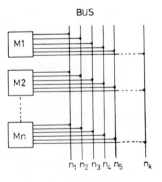

Figure 1.12 Parallel highway or BUS

bus connects to it with, say, an open-collector NAND output. At any instant all units except one will have their normal outputs disabled either by holding one input or the open-collector NAND to low thereby forcing the output high (output transistor off, which will not effect the buslines) or by enabling the third open-circuit input of the tri-state device.

References and bibliography

1. McCarthy, O. J. *MOS Device and Circuit Design*. Wiley, New York, 1982.
2. Barna, A. *High Speed Pulse and Digital Techniques*. Wiley, New York, 1980.
3. Gosling, J. B. *Design of Arithmetic Units for Digital Computers*. Macmillan, London, 1980.
4. Hamming, R. W. Error-detecting and error-correcting codes. *Bell Syst. Tech. J.* **29**, 147–60 (1950).
5. Diamond, J. M. Checking codes for digital computers. *Proc. Inst. Radio Engineers* **43**, 487–8 (1955).
6. Military Standard—Graphical Symbols for Logic Diagrams Mil-STD-806B, 1962, Naval Publication and Form Centre, Philadelphia, USA.
7. Sedra, A. and Smith, K. *Microelectronic Circuits*. Holt, Rinehart & Winston, New York, 1982.
8. *System 74 Designers Manual*. Texas Instruments Ltd, Bedford, 1973.
9. Glaser, A. B. and Subak-Sharpe, G. E. *Integrated Circuit Engineering*. Addison-Wesley, London, 1977.

Tutorial problems

1.1 Convert the following decimal numbers to their pure binary equivalent:
(a) 2397.55   (b) 0.79   (c) −90

1.2 Convert the following binary numbers to their decimal equivalent:
(a) 1011011.101   (b) 10111010111.0   (c) 0.111011

1.3 Calculate the hexadecimal multiplication table and then multiply the following numbers showing full working:
(a) B7 × 53   (b) 9C × CD   (c) FF × FF
Check your calculation by taking the binary equivalents and repeating and then converting to decimal.

1.4 Express the number 149 in the following coded systems:
(a) Binary coded decimal 8421 with odd parity check;
(b) ASCII codes expressed in hexadecimal;
(c) Hamming code;
(d) Diamond code;
(e) 2-out-of-5 code.

1.5 In highly reliable computers *residue codes* are sometimes used to enable checking of the basic arithmetic operations. The residue of a number N is defined by

$$R(N) = N \bmod p$$

where p is called the *modulus*, that is the remainder after dividing a number N

by another number p. Construct the multiplication and addition tables for modulus 3 and modulus 5 arithmetic using binary notation.

1.6 Determine if the following coded data are error free:
   (a) 11010   01011   10111   11100   (five bit code odd parity);
   (b) 0110111 (seven bit Hamming code);
   (c) 1101010 (Diamond $3n + 2$ code).

1.7 Draw an equivalent contact network for the following Boolean functions:
   (a) $F_0 = A\bar{B} + \bar{A}B$ (b) $F_1 = AB\bar{C} + ABC + \bar{A}BC$
   Can any of the circuits be simplified?

2 Principles of switching algebra

2.1 Introduction

In order to describe, analyse and design logical circuitry it is first necessary to become conversant with the underlying basic mathematics of the subject. The majority of books and research papers on switching theory are written in set-theoretic terminology which presents an immediate stumbling block to the uninitiated. It is the object of this chapter to review the relevant modern mathematical theory, including Boolean algebras, in such a way that any engineer may readily appreciate and apply it to the design of digital systems. For a more rigorous treatment of the subject the reader is referred to the bibliography at the end of this chapter.

2.2 Set theory

The concept of a set is already familiar to most people as a collection of objects. The items comprising the set may take any form, but must have at least one property in common, e.g. a set of vintage motor car prints, uniforms of the British army, families with more than three children, integers less than ten but greater than zero, inputs to a logical circuit. A set of sets is called a *class* and a set of classes is called a *family*.

Sets are normally specified by stating the actual members or elements of the set, the order being immaterial, and enclosing them in brackets:

N = Set of integers less than 10 but greater than zero
  = {1, 2, 3, 4, 5, 6, 7, 8, 9}

or

I = Set of inputs to logical circuit = $\{x_1, x_2, x_3, x_5, x_4\}$

Note that capital letters are used to represent sets and small letters or numbers to represent the members. Those sets in which all the members are known and can be described individually are called finite; we need a different notation to describe very large or infinite sets. A set may be defined in terms of some 'property' which all elements of the set are required to possess. Thus the set N above becomes

$N = \{x: 0 < x < 10 \text{ and is an integer}\}$

This is read as 'the set of all x, *such that* x is greater than zero but less than 10 and is an integer'. Again the set I may be defined as

$$I = \{x_i: i = 1, 2, 3, 4, 5\} \qquad \text{or sometimes just} \qquad \{x_i\}$$

It will be obvious that sets with an infinite number of members may be specified in this way, for example:

$$P = \{x: x \text{ is an even number}\}$$

Before we go further we must explain the symbology that is used. Membership of an element in a particular set is represented by the symbol $\in$. Thus $x_1 \in I$ means that x_1 is an element of the set I. Sets may contain single elements or no elements at all; these are called *unit* $\{1\}$ and *null* $\{\phi\}$ sets respectively.

In general, a set is a collection of objects chosen from some larger collection or *universal* set $\{I\}$. For example the set $N = \{1, 2, 3, 4, 5, 6, 7, 8, 9\}$ is contained in the set $X = \{x: x \text{ is a real number}\}$, this may be symbolized as $N \subseteq X$, where N is described as a subset of X. The expression $N \subseteq X$, however, could mean that N contains all members of X, or just a few members; it is in fact a generic term (analogous to the mathematical symbol $\leq$, less than or equal to). To distinguish between these two cases we use the symbol $\subset$ when we mean a subset that does not contain all the members of the parent set, e.g. $N \subset X$, where N is called a *proper subset*. For two sets A and B to be equal $(A = B)$, every element of A must be in B and vice versa, that is, they contain identical elements. If $A = B$ it follows that $A \subseteq B$ and $B \subseteq A$. Consider the set $I = \{a, b, c\}$; the complete list of subsets is

$$\{\phi\}; \quad \{a\}; \quad \{b\}; \quad \{c\}; \quad \{a, b\}; \quad \{a, c\}; \quad \{b, c\}$$

which are proper subsets, and $\{a, b, c\}$ which is a subset, i.e. the universal set.

2.3   Algebra of sets

Sets may be manipulated and combined algebraically using rather special operations which we now define.

  (a) *Intersection*, represented by the symbol $\cap$ (cap). If $A = \{1, 5, 7, 8, 9\}$ and $B = \{4, 5, 6, 9, 12\}$ then $A \cap B = \{5, 9\}$. That is, the elements common to each set are used to form the new set.
  (b) *Union*, represented by the symbol $\cup$ (cup). With the same sets as above, $A \cup B = \{1, 4, 5, 6, 7, 8, 9, 12\}$. In this case the elements in both sets are combined to form the new set; note that elements need occur once only in a set.
  (c) *Complementation*, represented as a bar over a set, e.g. $\bar{A}$, is defined as the set of elements which do not belong to A.

Using these operations we are now in a position to define two important properties:

Covering—A class of sets $A_1, A_2, A_3 \cdots$, such that their union $A_1 \cup A_2 \cup A_3 \cdots$ is the set A, is said to *cover* A.

Partition—A partition of the set A is a covering of A such that the intersection of any pair of the sets in the covering has no members, i.e. is *disjoint*.

The sets forming the partition or covering are called *blocks*. Let $A = \{1, 2, 3, 4, 5, 6, 7, 8, 9\}$; then the class of sets is given by

$$A_1 = \{1, 2, 3, 4\} \qquad A_2 = \{2, 4, 5, 6\}$$
$$A_3 = \{4, 6, 7\} \qquad A_4 = \{1, 8, 9\}$$

To form a covering

$$C = (1, 2, 3, 4) \quad (2, 4, 5, 6) \quad (4, 6, 7) \quad (1, 8, 9)$$

This is not a partition since $A_1 \cap A_2 = \{2, 4\}$, $A_2 \cap A_3 = \{4, 6\}$ etc. The sets $A_1 = \{1, 2\}$, $A_2 = \{3, 4\}$, $A_3 = \{5, 6, 7\}$, $A_4 = \{8, 9\}$ are disjoint and hence form a partition $P = (1, 2) \, (3, 4) \, (5, 6, 7) \, (8, 9)$.

The basic operations obey the following algebraic laws.

1. *Commutative law* $A \cap B = B \cap A; \quad A \cup B = B \cup A$
2. *Associative law* $A \cap (B \cap C) = (A \cap B) \cap C$
   $$A \cup (B \cup C) = (A \cup B) \cup C$$
3. *Distributive law* $A \cap (B \cup C) = (A \cap B) \cup (A \cap C)$
   $$A \cup (B \cap C) = (A \cup B) \cap (A \cup C)$$
4. *Absorption law* $A \cap (A \cup B) = A \cup (A \cap B) = A$
5. *Idempotent law* $A \cap A = A \cup A = A$
6. *Complement law* $A \cup \bar{A} = I \quad A \cap \bar{A} = \phi \quad (\bar{\bar{A}}) = A$
   $$\overline{A \cap B} = \bar{A} \cup \bar{B}; \quad \overline{A \cup B} = \bar{A} \cap \bar{B}$$
   (De Morgan's Theorem)
7. *Universal and* $\phi \cap A = \phi, \quad I \cap A = A$
      null law $\phi \cup A = A, \quad I \cup A = I$

where A, B and C are subsets of some universal set I, and ϕ is the null set.

These algebraic laws, though similar to the laws of normal algebra (the associative and distributive laws are in fact idential), contain surprising differences, e.g. the absorption and idempotent laws which have no equivalent. Although the operation of subtraction as such has not been defined, for certain restricted cases where the sets are comparable, it is possible to say that $\bar{A} = I - A$; however, the operation is best avoided.

As well as the union and intersection operations, there is a further operation which is often used in switching theory and hence worth including. This is set inclusion ($\subseteq$) which can be treated in the same way, but which obeys different basic laws; these are:

1. *Transitive Law*   If $A \subseteq B$ and $B \subseteq C$ then $A \subseteq C$. More strictly the relationship is called *implication*, and is denoted by the symbol $\Rightarrow$, i.e. $A \subseteq B$ and $B \subseteq C \Rightarrow A \subseteq C$.
2. *Reflexive Law* $A \subseteq A$
3. *Anti-symmetric Law* $A \subseteq B$ and $B \subseteq A \Rightarrow A = B$

2.4   Venn diagrams

Though most of the laws can be proved intuitively or by logical induction using the basic postulates, we shall first use a method which will be of value later when considering switching problems. This is the *Venn diagram* approach. The Venn diagram is an aid to intuitive reasoning which represents sets as areas. In Fig. 2.1(a) the universal set, sometimes called the *universe of discourse*, is represented by the large rectangle, and the set A by a circle drawn inside it, thus representing $A \subset I$. Note that the shape of the areas is completely irrelevant. The shaded portion of Fig. 2.1(b) represents the union of the two sets A and B. Similarly in Fig. 2.1(c) the shaded portion represents $A \cap B$.

Using these diagrams it is easy to appreciate the significance of $A \cup \bar{A} = I$ and $A \cap \bar{A} = \phi$ etc. De Morgan's law of complementation can easily be shown, e.g. $\overline{A \cup B} = \bar{A} \cap \bar{B}$, since $\overline{A \cup B}$ means the set of elements which do not belong to $A \cup B$; this is the shaded area in Fig. 2.1(h). For two sets A and B, the Venn diagram can represent all possible combinations of the intersection operations (subsets of A and B) (see Fig. 2.1(i)), and this can be extended to any number of sets, but is limited of course by the difficulty of drawing overlapping areas.

2.5   Example

The Venn diagram is useful in many areas, but particularly so in applying set theory to switching and statistical problems. As an example let us use the technique to solve a simple problem in statistics.

In a sports club of 120 members, 80 people played golf, 50 people played tennis, 60 played squash, 22 played squash and tennis, 28 played tennis and golf, 44 played squash and golf, and 10 people played all sports. Find how many people played tennis only, and the number of non-playing members?

In this problem, the universe of discourse $\{I\}$ is the 120 members of the club, and the number of people playing golf, tennis and squash are the subsets G, T, and S of this universal set. The proportions of games-playing members can be easily visualized from the Venn diagram shown in Fig. 2.2. The diagram is obtained by considering the known information.

$$G \cap T \cap S = 10 \text{ members} \qquad G \cap S = 44 \text{ members}$$
$$T \cap S \quad\;\; = 22 \text{ members} \qquad G = 80 \text{ members}$$
$$T \cap G \quad\;\; = 28 \text{ members} \qquad S = 60 \text{ members}$$
$$\qquad\qquad\qquad\qquad\qquad\qquad\quad T = 50 \text{ members}$$

Then, by deduction, we have

$$T \cap S \cap \bar{G} = 22 - 10 = 12 \text{ members}$$
$$T \cap G \cap \bar{S} = 28 - 10 = 18 \text{ members}$$
$$G \cap S \cap \bar{T} = 44 - 10 = 34 \text{ members}$$
$$G \cap \bar{S} \cap \bar{T} = 80 - (18 + 10 + 34) = 18 \text{ members}$$
$$T \cap \bar{S} \cap \bar{G} = 50 - (18 + 10 + 12) = 10 \text{ members}$$

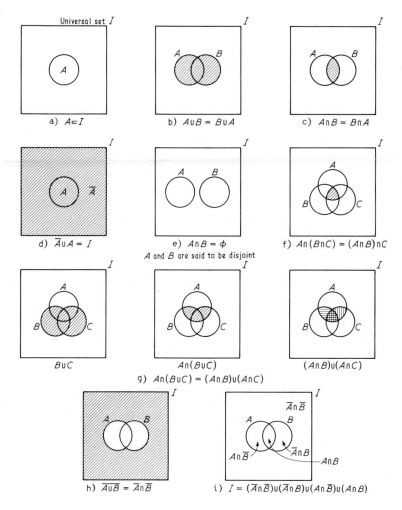

Figure 2.1 Venn diagram proofs

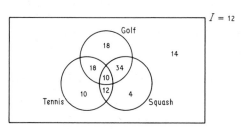

Figure 2.2 Sports club problem

$$S \cap \bar{T} \cap \bar{G} = 60 - (34 + 10 + 12) = 4 \text{ members}$$
$$\bar{S} \cap \bar{T} \cap \bar{G} = 120 - (18 + 18 + 10 + 34 + 10 + 12 + 4)$$
$$= 14 \text{ members}$$

It is clear from the diagram that 10 people play tennis only and there are 14 non-playing members.

2.6 Boolean algebra

A Boolean algebra may be defined for a set $I = \{a, b, c, d \cdots\}$ by assuming the following postulates:

(a) For two binary operations (i.e. between two elements), say $(+)$ and $(.)$, the result must be a unique element in the set I; this is called the *closure property*.

(b) The operations must be commutative, i.e. $a + b = b + a$ and $a.b = b.a$.

(c) Both operations must be distributive, i.e. $a(b + c) = ab + ac$ and $a + bc = (a + b).(a + c)$.

(d) Identity elements must exist such that $0 + a = a$ and $1.a = a$ for all elements of I.

(e) For every element of I there must exist an inverse such that $a + \bar{a} = 1$ and $a.\bar{a} = 0$.

Thus, by comparison, it is clear that under the operations of intersection, union and complementation, the set of all subsets of a universal set I form a Boolean algebra, or as it is sometimes called an *algebra of classes*. Note that there is no restriction on the number of elements contained in the set I, which could be either finite or infinite. For example the set $S = \{a, b\}$ contains the subsets: $\{\phi\} = 0$; $\{a, b\} = 1$; $\{a\} = 2$; and $\{b\} = 3$. These four elements constitute a Boolean algebra with the laws of combination shown in Table 2.1.

Suppose now that the set I contains just one set consisting of the two subsets or elements, say 0 and 1, which could represent the truth (1) or falsity (0) of a proposition or statement. The laws of combination for the operations $(+)$ and $(.)$ (also called inclusive OR and AND) are shown in Table 2.2. These operations, called logical connectives, are represented by $\vee$ (vee) and $\wedge$ (wedge) respectively in this *propositional* or *logical algebra* first developed by Boole.[1] Thus for any given proposition it is also possible to form its complement which is called the denial, indicating that the proposition is false. Using this symbolic notation it is possible to reduce verbal statements (i.e. propositions) to an algebraic form. The algebra may then be used to evaluate the truth or falsity (called the truth value) of a set of propositions either singly or joined together by the basic logical connectives. This may be done by applying the basic postulates of Boolean algebra algebraically or by means of a *truth table*.

To illustrate this let us examine the conditions for employment at a certain company. Applicants must be

Table 2.1 Laws of combination for four-element Boolean algebra

+	0 1 2 3		.	0 1 2 3		x $\bar{x}$
0	0 1 2 3		0	0 0 0 0		0 1
1	1 1 1 1		1	0 1 2 3		1 0
2	2 1 2 1		2	0 2 2 0		2 3
3	3 1 1 3		3	0 3 0 3		3 2
Addition			Multiplication			Complementation

Table 2.2 Laws of combination for two-element Boolean algebra

+	0 1		.	0 1		x $\bar{x}$
0	0 1		0	0 0		1 0
1	1 1		1	0 1		0 1

(1) male, married and under 35, or
(2) male, unmarried and under 35, or
(3) female, married and under 35, or
(4) female, unmarried and over 35.

In order to apply logical algebra to the problem we must first represent each of the basic propositions by a symbol:

Male M Unmarried S Under 35 T Applicants A
Female $\bar{M}$ Married $\bar{S}$ Over 35 $\bar{T}$

We can now combine the basic propositions using the logical connectives AND ($\wedge$) and OR ($\vee$):

$A = M$ and $\bar{S}$ and T, or M and S and T, or $\bar{M}$ and $\bar{S}$ and T, or $\bar{M}$ and S and $\bar{T}$

or

$$A = M \wedge \bar{S} \wedge T \vee M \wedge S \wedge T \vee \bar{M} \wedge \bar{S} \wedge T \vee \bar{M} \wedge S \wedge \bar{T}$$

By applying the idempotent postulate $A \wedge A = A$, we can include an extra redundant term $M \wedge \bar{S} \wedge T$ in the equation, which then becomes

$$A = M \wedge \bar{S} \wedge T \vee M \wedge S \wedge T \vee \bar{M} \wedge \bar{S} \wedge T \vee \bar{M} \wedge S \wedge \\ \bar{T} \vee M \wedge \bar{S} \wedge T$$

If we now apply the distributive law we have

$$A = M \wedge T(S \vee \bar{S}) \vee \bar{S} \wedge T(M \vee \bar{M}) \vee \bar{M} \wedge S \wedge \bar{T}$$

Since $A \vee \bar{A} = 1$ and $A \wedge 1 = A$ from the complement and universal laws respectively, the final and simpler expression becomes

$$A = M \wedge T \vee \bar{S} \wedge T \vee \bar{M} \wedge S \wedge \bar{T}$$

In other words applicants must be

   (1) male and under 35, or
   (2) married and under 35, or
   (3) female, unmarried, and over 35.

Thus by representing the logical statements symbolically and applying the rules of logical algebra, we have arrived at a simpler statement of the original propositions.

   To show that these two equations are equivalent (which does, of course, follow directly from the algebra) we can use the truth table method. We have to prove that $A_1 = A_2$, where

$$A_1 = M \wedge T \vee \bar{S} \wedge T \vee \bar{M} \wedge S \wedge \bar{T}$$
$$A_2 = M \wedge \bar{S} \wedge T \vee M \wedge S \wedge T \vee \bar{M} \wedge \bar{S} \wedge T \vee \bar{M} \wedge S \wedge \bar{T}$$

The first step in the process is to form a table containing all possible combinations of the truth values (either 0 or 1) of the basic propositions M, S and T. It may be seen from Table 2.3 that these are the normal combinations formed by three two-valued binary variables. These values are then substituted into the equations for A_1 and A_2, and the resulting truth value of the equations entered in the table. For example, consider the combination $M S \bar{T}$, i.e. 110, which is the symbolic representation of male, unmarried and over 35. If we substitute these values into the equations for A_1 and A_2, we have

$$A_1 = 1 \wedge 0 \vee 0 \wedge 0 \vee 0 \wedge 1 \wedge 1 = 0$$
$$A_2 = 1 \wedge 0 \wedge 0 \vee 1 \wedge 1 \wedge 0 \vee 0 \wedge 0 \wedge 0 \vee 0 \wedge 1 \wedge 1 = 0$$

The results follow from the table of combinations shown in Table 2.2. If this procedure is repeated for all combinations, we eventually arrive at the full truth table shown in Table 2.3. It will be seen that the results for the two equations are the same, thereby proving that they are identical expressions. This method of proving theorems or establishing equality is called *perfect induction*.

   As we have seen propositional logic can be used to analyse arguments that involve relationships among propositions (declarative sentences) which have fixed truth values (either true or false). Unfortunately in many

Table 2.3 Truth table for employment conditions

M	S	T	A_1	A_2
0	0	0	0	0
0	0	1	1	1
0	1	0	1	1
0	1	1	0	0
1	0	0	0	0
1	0	1	1	1
1	1	0	0	0
1	1	1	1	1

cases (for instance, in the formal verification of software systems) we would like to examine the elements of the propositions themselves, to see why it is true or false or to attempt to establish relationships between propositions. For example, the fact that 'all ducks in this village, that are branded B, belong to Mrs Bond' can be represented by propositional variables but there is no way of representing Mrs Bond or the ducks!

Propositional logic can be extended to allow the representation of *individual entities* (people, numbers, things etc.) and *functions* that describe the translation of one individual entity into another. In this *predicate logic*, a predicate is a function whose value is a truth value, true or false, depending on which individual it is applied to. For example, if f is the 'branded by B' function then $f(y)$ represents the branding of whatever individual is represented by y. Again, if G is the predicate that specifices Mrs Bond's ownership than $G(x)$ is true whenever x is owned by Mrs Bond and false otherwise. Then Mrs Bond's ownership of branded ducks can be expressed by the predicate

$$G(f(\text{duck}))$$

which is asserted to be true.

As with propositional variables these 'atomic formulas' can be combined together to form complete 'sentences' using the propositional connectives AND, OR and invert. A full treatment of predicate calculus is beyond the scope of this book; however, the calculus is becoming an important tool in digital system design since it can be used to express almost any computational or mathematical concept.

At this stage it might be rather difficult for the reader to fully understand or see the significance of set theory and Boolean algebras, and to appreciate their use in designing switching circuits. We shall now consider this particular application in more detail.

2.7 Switching algebra

Any switching circuit or network can be considered as a two-valued system, that is, at any given instant of time either an open or closed transmission path is established between its terminals. Furthermore, as we have seen, the binary method of counting with the use of only two symbols (0 and 1) is used extensively in digital systems to codify and manipulate data. Thus it is not surprising that the mathematical theory of switching circuits, first postulated by Shannon,[2] is based on a two-valued Boolean algebra closely related to the logical algebra described in the last section.

In switching algebra, the operations are called AND (.), OR (+) and NOT (−) (also known as inversion and complement). These operations are identical to those defined in Table 2.2, and also correspond to the basic switching circuit functions discussed in Chapter 1. *Transmission* through a switching circuit is represented by the value 1, and corresponds to a short-circuit; the *hindrance*, corresponding to an open-circuit, is represented by the value 0. Each binary signal or variable is represented by a symbol, which can have a value of either 0 or 1, and these variables can be combined together under the AND/OR operations to form a switching function. Variables may be in an uncomplemented or complemented form in these equations; when a variable appears, in either form, it is called a *literal*. Note that the algebraic definition is independent of the particular type of switching circuit under discussion—it could, for example, represent relay, electronic, or even hydraulic logic.

Using the basic postulates discussed earlier for a two-valued Boolean algebra, numerous theorems may be deduced. These theorems apply equally well to the propositional logic but are stated here in switching terms; Table 2.4 shows a list of the most important theorems. The switching variables are represented by the symbols A, B, C etc., and may assume a value of either 0 or 1. It follows from the definition of a two-state system, i.e. binary variable, that if $A = 0$, $\bar{A} = 1$, and if $A = 1$, $\bar{A} = 0$. It is interesting to note that a duality principle exists for the basic laws of combination (Table 2.2): if the digits 0 and 1 and the operations $(+)$ and $(.)$ are interchanged, the alternative operation is obtained. This duality is expressed in general form by theorems 17 and 18 in Table 2.4, which we have met before in set theory, and is known as De Morgan's theorem. The theorem may be expressed in words: the complement of any function is obtained by replacing each variable by its complement and, at the same time, interchanging the symbol for the AND/OR operations.

The theorems may be proved algebraically or by using the truth table method (sometimes called a conditions-function table in switching algebra). Before we proceed with this, however, let us try to interpret these theorems physically in terms of simple electrical contact circuits. Figure 2.3 shows the contact representation of the more important theorems; the contacts are assumed ideal in that common contacts open and close together, with no bounce or delay. The AND operation represents contacts in series, each contact being a binary variable; similarly the OR operation represents contacts in parallel. In these circuits, variables are represented by normally open contacts (the presence of a signal closes the contact, e.g. current through a relay coil), and complemented variables are represented by normally closed contacts. This representation is also applicable to gating circuits using active (as distinct from path-closing) devices such as transistors, etc. In this case the binary variables are the two-state voltage levels

Table 2.4 Theorems of switching algebra

(1)	$A + 0 = A$	(10)	$A + B = B + A$
(2)	$A \cdot 1 = A$	(11)	$A \cdot B = B \cdot A$
(3)	$A + 1 = 1$	(12)	$A + AB = A$
(4)	$A \cdot 0 = 0$	(13)	$A + \bar{A}B = A + B$
(5)	$A + A = A$	(14)	$AB + AC = A(B + C)$
(6)	$A \cdot A = A$	(15)	$(A + B)(A + C) = A + BC$
(7)	$(\bar{\bar{A}}) = A$	(16)	$A + B + C = (A + B) + C = A + (B + C)$
(8)	$A + \bar{A} = 1$	(17)	$\overline{A + B} = \bar{A} \cdot \bar{B}$
(9)	$A \cdot \bar{A} = 0$	(18)	$\overline{A \cdot B} = \bar{A} + \bar{B}$

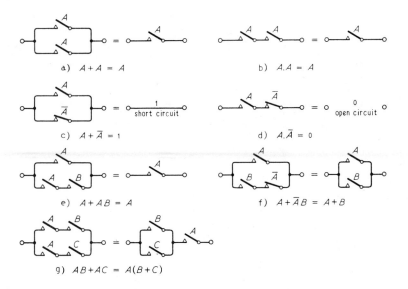

Figure 2.3 Physical interpretation of the algebraic theorems

(described as high or low, up or down, positive or negative) representing the values 0 or 1, which are applied to the input of the gates.

The techniques of theorem-proving using algebra or truth tables is identical with that described above for propositional logic. Let us consider one more example, however. Suppose we have to prove the relationship

$$T = (A + C)(\bar{A} + B) = AB + \bar{A}C$$

Algebraically, we have

$T = A\bar{A} + AB + \bar{A}C + BC$	(Theorem 14)
$= AB + \bar{A}C + BC$	(Theorem 9)
$= AB + \bar{A}C + BC(A + \bar{A})$	(Theorem 8)
$= AB + \bar{A}C + BCA + BC\bar{A}$	
$= AB(1 + C) + \bar{A}C(1 + B)$	(Theorem 12)

Hence

$$T = AB + \bar{A}C$$

The truth table for the two functions is shown in Table 2.5. Again the method of perfect induction is used, whereby all possible values of the binary variables are substituted into the functions under comparison.

Many different switching functions can be obtained from the basic binary variables. In general, for n variables we have 2^n possible combinations of these variables and hence 2^{2^n} different switching functions. For example, if we have two binary variables A and B, we can obtain 16 unique functions; these are shown in Table 2.6. Note that these include many functions we

Table 2.5 Theorem proving using perfect induction

A	B	C	$(A + C)(\bar{A} + B)$	$(AB + \bar{A}C)$
0	0	0	0	0
0	0	1	1	1
0	1	0	0	0
0	1	1	1	1
1	0	0	0	0
1	0	1	0	0
1	1	0	1	1
1	1	1	1	1

Table 2.6 Functions of two variables

A	B	F_0	F_1 AND	F_2	F_3	F_4	F_5	F_6 EXOR	F_7 OR	F_8 NOR	F_9 EXNOR	F_{10}	F_{11}	F_{12}	F_{13}	F_{14} NAND	F_{15}
0	0	0	0	0	0	0	0	0	0	1	1	1	1	1	1	1	1
0	1	0	0	0	0	1	1	1	1	0	0	0	0	1	1	1	1
1	0	0	0	1	1	0	0	1	1	0	0	1	1	0	0	1	1
1	1	0	1	0	1	0	1	0	1	0	1	0	1	0	1	0	1

have already encountered: F_1 is the AND, F_7 the inclusive OR, F_8 the NOR, F_{14} the NAND, etc.

Any logical function can be expressed in terms of one of the basic operations (AND or OR) and the NOT operation. This follows from the duality of the basic postulates and can be seen by applying De Morgan's theorem. For example, the sum-of-products expression

$$T = \bar{a}bc + a\bar{b}\bar{c} + ad + bcd$$

may be expressed in terms of AND/NOT, i.e. NAND, as

$$T = \overline{(\overline{\bar{a}bc})(\overline{a\bar{b}\bar{c}})(\overline{ad})(\overline{bcd})}$$

and using OR/NOT, i.e. NOR, as

$$\bar{T} = \overline{(a + \bar{b} + \bar{c}) + (\bar{a} + b + c) + (\bar{a} + d) + (\bar{b} + \bar{c} + \bar{d})}$$

The implication of this is that the hardware for a switching system need only consist of either NAND/NOR logic, or basic AND/OR gates plus invertor amplifiers.

2.8 Derivation and classification of switching functions

Logical design problems are usually presented in the form of oral or written requirements that dictate the terminal behaviour of the required circuit. From this specification a mathematical statement for the switching circuit must be formulated. Consider the requirements for a hall lighting system that enables the hall light to be switched 'on' or 'off' from either one of two switches, situated downstairs and upstairs. This is a system found in most houses in some form or other. The first step is to form a truth table (Table 2.7) in which all possible input combinations of the two binary variables are examined. The variables A and B represent the two switches, which may be either on (A, B) or off $(\bar{A}, \bar{B})$. If the particular AB combination selected results in the hall light coming on, it is represented by a 1 in column T, otherwise by a 0. The truth table may be expressed as a switching or transmission function by extracting from the table those combinations of AB that produce an output, i.e. cause the hall light to come on. Thus

$$T = A\bar{B} + \bar{A}B \tag{2.1}$$

This particular equation represents an important logical function in its own right (F_6 in Table 2.6): it is called the *exclusive OR* or *non-equivalence* circuit and is defined as a circuit that produces an output only when the two inputs are dissimilar. Furthermore, the equation also represents the result of adding two binary digits together (compare with the binary addition table given in the last chapter).

The form of equation we derived from the truth table, eqn (2.1), is called a *sum-of-products*. The *dual* equation may be obtianed by extracting from the table those combinations that *do not* produce an output, i.e. the hindrance terms:

$$H = \bar{T} = \bar{A}\bar{B} + AB \tag{2.2}$$

Then by applying De Morgan's theorem we have

$$T = (\overline{\bar{A}\bar{B} + AB}) = (A + B)(\bar{A} + \bar{B})$$
$$= A\bar{A} + A\bar{B} + \bar{A}B + B\bar{B}$$
$$T = A\bar{B} + \bar{A}B$$

Therefore the equation

$$T = (A + B)(\bar{A} + \bar{B}) \tag{2.3}$$

is an equivalent and alternative form of eqn (2.1) and is called the *product-of-sums* form. When implemented in terms of switching circuits (Fig. 2.4), the two forms of the function T given above have identical switching

Table 2.7 Truth table for exclusive OR

A	B	T
0	0	0
0	1	1
1	0	1
1	1	0

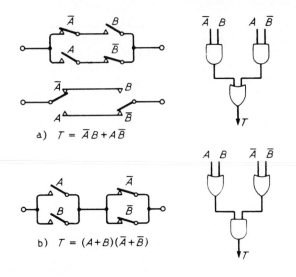

a) $T = \bar{A}B + A\bar{B}$

b) $T = (A + B)(\bar{A} + \bar{B})$

Figure 2.4 Switching networks

characteristics. Note, however, that the actual logic gates required to implement these functions are different: the sum-of-products requires two AND gates and one OR, while the product-of-sums needs two OR gates and one AND gate.

In these two alternative forms of switching function, all the binary variables or their complements (literals) appear once, and only once, in each term or factor. For this reason they are called the *canonical* (i.e. standard or normal) forms of the switching function. The sum-of-products is also called the normal *minterm* form, and the product-of-sums is called the normal *maxterm* form, when referring to switching circuits. The minterms are all possible combinations of the binary variables joined together by the AND operation; similarly the maxterms are expressed in the OR relationship. Table 2.8 shows all possible minterms and maxterms for the three binary variables A, B, and C.

Table 2.8 Minterm and maxterm forms

A	B	C	Minterm	Maxterm
0	0	0	$M_0 = \bar{A}\bar{B}\bar{C}$	$M_0 = A + B + C$
0	0	1	$M_1 = \bar{A}\bar{B}C$	$M_1 = A + B + \bar{C}$
0	1	0	$M_2 = \bar{A}B\bar{C}$	$M_2 = A + \bar{B} + C$
0	1	1	$M_3 = \bar{A}BC$	$M_3 = A + \bar{B} + \bar{C}$
1	0	0	$M_4 = A\bar{B}\bar{C}$	$M_4 = \bar{A} + B + C$
1	0	1	$M_5 = A\bar{B}C$	$M_5 = \bar{A} + B + \bar{C}$
1	1	0	$M_6 = AB\bar{C}$	$M_6 = \bar{A} + \bar{B} + C$
1	1	1	$M_7 = ABC$	$M_7 = \bar{A} + \bar{B} + \bar{C}$

The following theorems apply to minterm/maxterm relationships.

(1) For n binary variables there are exactly 2^n minterms and 2^n maxterms.
(2) The union (i.e. logical sum or OR) of all minterms of n binary variables is equal to 1. The converse is also true—the intersection (i.e. logical product or AND) of all maxterms of n binary variables is equal to 0. If we consider the binary input variables as three sets A, B and C and let $T = \{A, B, C\}$, then the union of all minterms (i.e. subsets of T) forms a covering of T. This covering is also a partition since the intersection of any pair of minterms is disjoint (Section 2.3). The sets A, B and C are referred to as the *generating sets* for the partition.
(3) The complement of any minterm is a maxterm and vice versa.

Canonical expressions are useful in comparing and simplifying switching functions. For example, if we manipulate the functions to be examined to the same basic canonical form, they may then be compared term by term. Expansion formulae exist, due to Shannon, which may be used to expand non-canonical functions to the canonical form. These are

Minterm form $f(A, B) = \bar{A}\bar{B}.f(0, 0) + \bar{A}B.f(0, 1) + A\bar{B}.f(1, 0)$
$+ AB.f(1, 1)$

Maxterm form $f(A, B) = [\bar{A} + \bar{B} + f(1, 1)][\bar{A} + B + f(1, 0)]$
$[A + \bar{B} + f(0, 1)][A + B + f(0, 0)]$

where f is a Boolean function of the binary variables A and B; the formulae hold for any number of variables. The formula for the minterm expansion is applied by considering all possible values of the binary variables in turn, and then logically multiplying them by the truth value of the actual function, for these values. For example, let us expand the function $T = AB + B\bar{C}$ into its canonical forms. There are three variables, so we have

$T = AB + B\bar{C}$
$= \bar{A}\bar{B}\bar{C}.0 + \bar{A}\bar{B}C.0 + \bar{A}B\bar{C}.1 + \bar{A}BC.0 + A\bar{B}\bar{C}.0$
$+ A\bar{B}C.0 + AB\bar{C}.1 + ABC.1$
$T = \bar{A}B\bar{C} + AB\bar{C} + ABC$

Again

$T = [\bar{A} + \bar{B} + \bar{C} + 1][\bar{A} + \bar{B} + C + 1][\bar{A} + B + \bar{C} + 0]$
$[\bar{A} + B + C + 0][A + \bar{B} + \bar{C} + 0][A + \bar{B} + C + 1]$
$[A + B + \bar{C} + 0][A + B + C + 0]$
$T = (\bar{A} + B + \bar{C})(\bar{A} + B + C)(A + \bar{B} + \bar{C})(A + \bar{B} + C)$
$(A + B + C)$

An alternative and perhaps easier algebraic method of expansion for minterms is to multiply logically each term by the absent variables expressed in the form $A + \bar{A}$. For example

$T = AC + B$
$= AC(B + \bar{B}) + B(A + \bar{A})(C + \bar{C})$

$$= AC B + AC\bar{B} + ABC + AB\bar{C} + \bar{A}BC + \bar{A}B\bar{C}$$
$$T = ABC + A\bar{B}C + AB\bar{C} + \bar{A}BC + \bar{A}B\bar{C}$$

This procedure is very easy to apply using a truth table or the Karnaugh map technique which will be described in the next chapter.

Another useful way of representing switching functions expressed as a sum-of-products is to replace each combination of binary variables by its direct decimal equivalent, the operation of logical addition being symbolized by a summation sign. Thus

$$T = \bar{A}\bar{B}C + ABC + A\bar{B}\bar{C} + A\bar{B}C$$

becomes

$$T = \Sigma(1, 7, 4, 5)$$

2.9 Cubic notation for Boolean functions

An alternative representation for Boolean functions is based on a geometrical method whereby switching variables can be plotted as a series of points in space. This topological representation due to Roth[3] is called the *cubic notation* and is based on the methods used to specify error-correcting codes in which points in *n-dimensional space* are used to represent all possible binary codes, or *n-tuples*, and the *distance* between them.

What do we mean by *n*-dimensional space? This is best explained by example: a single binary variable has two values, 0 and 1, and hence can be represented by a single line or *edge* in space with the two values plotted at the two ends. To plot all values of two binary variables requires 2^2, that is four points, and this can be accommodated by using the four vertices of a square, called a *plane*. With three variables and eight combinations we must go to three dimensions (3-space), that is the cube, where again each vertex of the cube represents a minterm. Higher orders can be represented in the same way but the *n*-cubes or *hypercubes* that are formed in *n*-dimensional space are difficult to visualize! The binary codes, equivalent to the minterms of the function, are called *n-tuples*, a term which can be used to describe any sequence of *n*-binary digits; for example, the binary sequence 0111 would be called a 4-tuple.

Using the cubic notation a Boolean function in three variables, a, b and c for instance, may be represented as a 3-dimensional unit cube as shown in Fig. 2.5(a). Note that each canonical product term (minterm) of the function is associated with a unique point (or vertex) of the cube. Note also that each vertex is *adjacent* in that they differ in one variable only, for example, moving left to right along the top front edge we go from 010 to 110 with only the a variable changing. (Note from our Boolean algebra that $\bar{a}bc + abc = bc(\bar{a} + a) = bc$, that is if these two minterms are present than a can be eliminated.)

The *distance* between *n*-tuples is the number of variables that must change value in going along the edges from one *n*-tuple to another (in the case of the 3-cube, the terms on opposite vertices would be termed distance-

one apart). The *Hamming distance*, to give it its full name, is as we shall see
later an important characteristic of error detecting and correcting codes.

Using this property it is possible to represent non-canonical product
terms by *planes* and *edges* of the cube, as shown in Fig. 2.5(b), (c), (d) and
(e). The minterms of a function are referred to as *0-cubes* and represent the
canonical form with all variables present. Thus the 3-cube may be considered as comprising eight vertices (0-cubes), six planes (2-cubes, two variables eliminated) and twelve edges (1-cubes, one variable eliminated). It
will also be seen that the general term k-cube can be used to describe the
number of literals in a product term where for *n* variables the number of
literals is $(n - k)$. In Fig. 2.5(b), for example, the 2-cubes $1XX$ and $0XX$
are represented by the shaded *a* planes and in Fig. 2.5(e) the 1-cubes $X11$,
$X10$, $X00$ and $X01$ by the blocked *ab* edges.

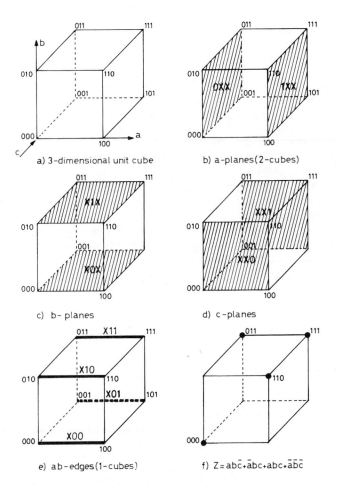

Figure 2.5 Cubic notation for Boolean functions

The cubic method of representing Boolean functions provides an alternative approach to the truth table for defining switching circuits; consider for example the function

$$Z = ab\bar{c} + \bar{a}bc + abc + \bar{a}b\bar{c}$$

which can be represented in the cubic notation as shown in Fig. 2.5(f). Note that only those vertices corresponding to the product terms present in the function are indicated (by dots) on the diagrams. The set of 3-tuples (or 0-cubes) comprising the function, that is those terms which give $Z = 1$, is called the *ON array*. The set of vertices not picked out correspond to $Z = 0$ and is called the *OFF array*. Together the ON and OFF arrays give the same amount of information as the truth table; thus we may specify the function as:

$$\text{ON-array} \begin{cases} \begin{bmatrix} a & b & c \\ 0 & 0 & 0 \\ 0 & 1 & 1 \\ 1 & 1 & 0 \\ 1 & 1 & 1 \end{bmatrix} \end{cases} \quad \text{OFF-array} \begin{cases} \begin{bmatrix} a & b & c \\ 0 & 0 & 1 \\ 0 & 1 & 0 \\ 1 & 0 & 0 \\ 1 & 0 & 1 \end{bmatrix} \end{cases}$$

As we shall see later in some functions a number of the output values may be unspecified (for example, it may be of no consequence whether the output goes to 0 or 1 for a particular input combination). These *'don't care'* terms can be represented by a corresponding *DC array* or obtained by default from the ON and OFF arrays.

Using the cubic notation it is possible to specify and manipulate switching functions expressed as arrays of n-tuples and a complete algebra may be established with defined operations on arrays of cubes.[4]

The use of switching algebra, then, allows switching circuits to be described symbolically, thereby providing a formal statement of the problem. Moreover, in constructing the truth table for the circuit, the designer is forced to consider all possible inputs, thereby eliminating any errors due to certain input conditions being overlooked. Furthermore, by using the postulates and theorems deduced above it is possible to analyse and synthesize this class of circuit. The resulting switching equations may then be simplified, or clarified, before being translated directly into equivalent hardware devices.

References and bibliography

1. Boole, G. *An Investigation of the Laws of Thought*. Dover Publications, New York, 1954 (reprint of original publication in 1854)
2. Shannon, C. E. A symbolic analysis of relay and switching circuits. *Trans. Am. Inst. Elect. Engrs.* **57**, 713–23 (1938).
3. Roth, J. P. Algebraic topological methods for the synthesis of switching systems. I. *Trans. Am. Math. Soc.* **88**, 301–26 (1958).
4. Tilson, P. Generalization of concensus theory and application to the minimization of Boolean functions. *IEEE Trans. Computers* **EC16**, 446–56 (1967).

5. Levy, L. S. *Discrete Structures of Computer Science*. Wiley, New York, 1980.
6. Raphael, B. *The Thinking Computer*. W. H. Freeman, San Francisco, 1976.
7. Hohn, F. R. *Applied Boolean Algebra*. Macmillan, New York, 1966.
8. Korfhage, R. *Logic and Algorithms*. Wiley, New York, 1966.
9. Johnsonbaugh, R. *Discrete Mathematics*. Macmillan, New York, 1984.

Tutorial problems

2.1 Prove the following identities algebraically and then by perfect induction using a truth table.

(a) $B + \bar{A}C = (A + B + C)(\bar{A} + B + C)(\bar{A} + B + \bar{C})$

(b) $\bar{A}D + \bar{C}D + A\bar{B} = \bar{A}\bar{C}D + A\bar{C}D + A\bar{B}\bar{C} + A\bar{B}C + \bar{A}CD$

(c) $D(\bar{A} + B + C + \bar{D})(A + B + \bar{C} + \bar{D})$
$$= (D + A\bar{C} + \bar{A}C)(\bar{A}\bar{C} + BD + AC)$$

2.2 Reduce the following functions by taking complements.

(a) $T = \overline{[\overline{(\overline{ab})}.a].[\overline{(\overline{ab})}.b]}$

(b) $T = \overline{\overline{(a + b + \bar{c})(\overline{ab} + cd)} + \overline{(bcd)}}$

(c) $T = \overline{(abc + b\bar{c}d) + (\overline{acd} + \bar{b}\bar{c}d + bc\bar{d})}$

2.3 Write down the switching function representing the circuits shown in Fig. 2.6. Then, for each circuit:

(a) expand the function into the canonical sum-of-products and product-of-sums forms;

(b) simplify the original function algebraically and draw the resulting circuit.

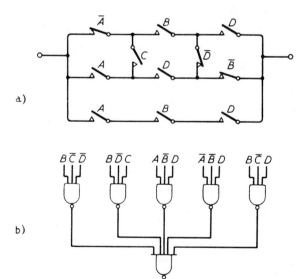

Figure 2.6 Problem 2.3

2.4 Write down all the subsets of the set $A = \{x_1, x_2, x_3, x_4\}$. Draw a Venn diagram showing these subsets. Consider if this representation could be used to aid the simplification of Boolean functions.

2.5 It is required to design a lighting circuit for a warehouse such that the lights may be switched on or off from any one of three switch points. Set up the truth table for the problem and derive the corresponding switching equation. Simplify this equation if possible and draw the resulting contact and logic gate circuit.

2.6 Four people, members of a television panel game, each have an on/off button that is used to record their opinion of a certain pop record. Instead of recording individual scores, some data processing is required such that the scoreboard shows a 'hit' when the majority vote is in favour and a 'miss' when against. Provision must also be made to indicate a 'tie'. From the verbal statement:
   (a) derive a truth table for the problem;
   (b) extract the sum-of-products and product-of-sums equations;
   (c) simplify the equations and suggest a suitable circuit.

2.7 Consider the following logical statements.
   (a) Hubert never drinks.
   (b) Joe drinks if, and only if, Hubert and Donald are present.
   (c) Sidney drinks under all conditions—even by himself!
   (d) Donald drinks if, and only if, Hubert is not present.
   If A represents Hubert's presence in the bar, B Joe's presence, C Sidney's and D Donald's, determine the function representing a state of no drinking taking place in the bar. Express this function as a word statement.

2.8 Show how a four-variable Boolean function can be represented in cubical notation as points in n-space. Plot the function

$$f(a, b, c, d) = \Sigma(1, 5, 9, 11, 12, 15)$$

and determine the ON and OFF arrays.

3 Design of combinational switching circuits

3.1 Combinational switching circuits

This class of switching circuit is perhaps the most important of all, as a thorough understanding of the principles of combinational design is an essential prerequisite to the study of more complex systems. A *combinational* logic circuit is one in which the *output* (or outputs) obtained from the circuit is *solely dependent on the present state of the inputs*. Combinational switching networks normally take one of two forms, distinguished by whether they have a *single-terminal* or a *multi-terminal* output circuit; this is shown in Fig. 3.1(a) and (b). Note that the single-terminal circuit is a special case of the more general multi-terminal model. Multi-terminal combinational circuits are often used, and described, as *decoders*; that is, a circuit which selects a unique output line for each value of a set of binary inputs (*addresses*) (see Fig. 3.1(e)). A special form of the single-terminal circuit is the *multiplexer*, shown in Fig. 3.1(d), which routes one of n possible input lines to a selected single output line. Both decoders and multiplexers are available as standard MSI modules.

The classical objective of combinational design is to produce a circuit having the required switching characteristics but utilizing the minimum

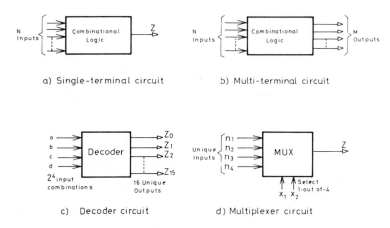

Figure 3.1 Combinational switching circuits

number of components, either in terms of relay contacts, basic gating
elements, or gate inputs. Though this objective still applies in the case of
discrete circuits the availability of MSI/LSI modules (such as multiplexers)
and the requirement to design logic for direct realization as an LSI/VLSI
circuit has radically changed the criteria for minimization. The need now is
to conserve module types and their interconnections and, in the case of
LSI, silicon area rather than gates or gate inputs. Nevertheless, the theory
is still highly relevant and, with a change of criteria, directly applicable to
the design of integrated circuit systems.

Switching problems are usually presented to the designer in the form of
an oral or written requirement specifying the logical behaviour of the
circuit. From this specification a mathematical statement of the problem
can be formulated by the construction of a truth table, or conditions-
function table. An algebraic representation, in the sum-of-products (or
product-of-sums) form, can then be derived from the table and simplified
where possible. These simplified equations may then be directly related to
a hardware diagram, in terms of actual circuits such as relay contacts,
NAND/NOR gates etc. Let us now consolidate these ideas by considering a
design example in more detail.

3.2   Design example: binary full-adder circuit

Suppose it is required to design a switching circuit that will add two binary
digits plus a carry digit. In other words, the circuit has three inputs x, y and
c, and two outputs, the sum S and the next carry C_+. This is the operation
required in adding together two binary numbers; if parallel representation
is used, one such full-adder stage per bit will be required. We begin the
problem by constructing a truth table which shows all the possible input
conditions and the resulting outputs. This is done with reference to the
binary addition rules; the complete table is shown in Table 3.1. Note that
exactly the same procedure is followed as described in the previous
chapter; with three inputs we have 2^3 possible input conditions and all
these must be enumerated in the table. Writing down those input conditions which produce an output, we obtain the normal sum-of-products
expressions for the sum and next carry:

$$S = \bar{x}\bar{y}c + \bar{x}y\bar{c} + x\bar{y}\bar{c} + xyc$$
$$C_+ = \bar{x}yc + x\bar{y}c + xy\bar{c} + xyc$$

The next stage in the process is to reduce and simplify the equations, if
possible; S will not reduce further, but C_+ can be expressed as

$$C_+ = \bar{x}yc + xyc + x\bar{y}c + xyc + xy\bar{c} + xyc$$
$$= yc(\bar{x} + x) + xc(\bar{y} + y) + xy(\bar{c} + c)$$
$$= yc + xc + xy$$

Note that the term xyc has been included three times in the equation; in
other words, we have added redundant terms to effect the reduction. That
this is possible and does not affect the equations follows from the Boolean
algebra relationship $A + A = A$. The equations for S and C_+ may now be
implemented in hardware to give the required switching circuit.

Table 3.1 Truth table
for binary full adder

x	y	c	S	C_+
0	0	0	0	0
0	0	1	1	0
0	1	0	1	0
0	1	1	0	1
1	0	0	1	0
1	0	1	0	1
1	1	0	0	1
1	1	1	1	1

It is, however, possible to manipulate the equations in a different way by factorizing the canonical expressions:

$$S = c(\bar{x}\bar{y} + xy) + \bar{c}(\bar{x}y + x\bar{y})$$

Let $z = \bar{x}y + x\bar{y}$; then we have

$$S = c\bar{z} + \bar{c}z$$

since

$$(\overline{\bar{x}\bar{y} + xy}) = \bar{x}y + x\bar{y}$$

This is the exclusive OR function described in the last chapter, and often referred to as a *modulo 2 adder* circuit, i.e. it gives no carry output. Again, for the carry equation, we can say

$$C_+ = c(\bar{x}y + x\bar{y}) + xy(\bar{c} + c)$$
$$C_+ = cz + xy$$

These equations may now be implemented using the exclusive OR gate, which is frequently obtained as a basic logic module. Figure 3.2 shows both forms of the required switching circuit. It is interesting to note that though circuit (b) uses one more element than circuit (a), only two-input gates are required in (b). Clearly, then, the appropriate simplification of canonical switching equations depends on hardware constraints, such as the type of logic module available, and the number of inputs to a gate (fan-in factor).

3.3 Minimization of switching equations

We have seen that the formal statement of a combinational problem leads to a Boolean canonical function which must, in the majority of cases, be simplified in some way to provide an economical hardware solution. Though it is possible to simplify algebraically, if the laws of switching algebra are applied haphazardly to, say, a function with more than five variables, the problem becomes prohibitively difficult. We now consider procedures which have been evolved to facilitate the reduction of switching equations to some minimal form (in the sense that there is no unique minimum). These methods, both graphical and algorithmic, do not represent a departure from algebraic principles since they are still based on the fundamental Boolean laws.

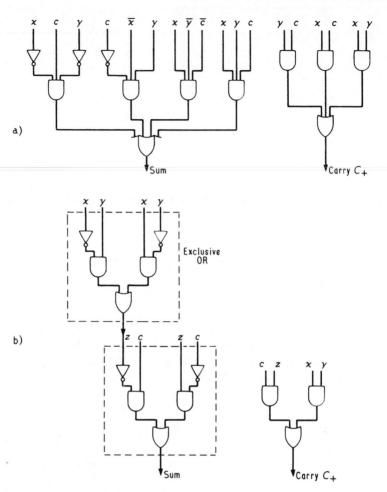

Figure 3.2 Full-adder circuits

There are three main criteria which may be used to determine a minimal sum-of-products (or product-of-sums) expression:

(1) The expression with the fewest literals; a literal is defined as either a complemented or uncomplemented binary variable.
(2) The expression with the fewest terms.
(3) The expression which requires the least number of logical units in its circuit implementation.

The last criterion is the one which must be met in practice, since this will affect the economics of the project. This means that the characteristics of the logic modules, such as type of logic function, number of inputs, speed of operation etc. must be taken into account and used as constraints upon the design procedure. Most of the techniques we will discuss lead to a minimal result in terms of the fewest number of literals, and the least number of terms. However, this is a very good starting point since the simplification process yields the essential terms of the switching function and allows the implementation to proceed in an optimum manner.

Moreover, in designing MSI/LSI circuits an important cost factor is the number of input/output connections (pins) and the total surface area of the chip. Thus minimizing the number of input variables and gates can still be an important consideration.

As an illustration of the type of problem involved, consider the minimal switching equation

$$Z = \bar{a}\bar{b}\bar{c}\bar{d}\bar{e} + \bar{a}\bar{b}\bar{c}d + ab\bar{c}d + a\bar{b}cde$$

Direct implementation would require five-input AND gates and a four-input OR gate, with an overall total of five basic modules, assuming that the complemented variables are already available. But suppose the logic modules to be used are three-input AND/OR gates; then the equation must be factorized accordingly:

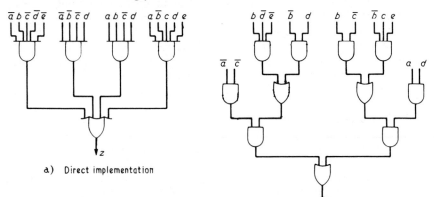

a) Direct implementation

b) Factorized implementation

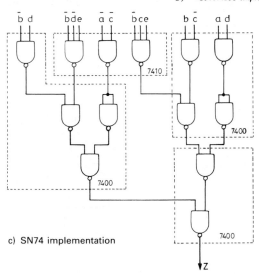

c) SN74 implementation

Figure 3.3 Circuit implementation

$$Z = \bar{a}\bar{c}(b\bar{d}\bar{e} + \bar{b}d) + ad(b\bar{c} + \bar{b}ce)$$

Note, however, that this now uses a total of nine basic modules and in-volves many more interconnections. Furthermore, as can be seen from Fig. 3.3(b), the factorized version has more stages and hence the switching delays through the circuit (propagation delays) will be greater, which could adversely affect the speed of operation. In high-speed logic systems, particularly those using fast integrated circuits, the minimization of inter-connection paths could easily become the major design problem.

Again the requirement may well be to realize the circuit using a specific and minimum number of SSI units, for example SN7400 and SN7410 TTL units containing respectively four two-input NAND gates and three three-input NAND gates per package. In this case the factorized circuit would require one SN7410 and three SN7400 packages as shown in Fig. 3.3(c).

Thus the reduction of a switching equation to a form containing the least number of literals and (or) terms is not the complete answer to logical circuit design and the equations must be manipulated further to realize an optimum design in terms of an actual hardware realization. These prob-lems will be discussed in more detail in later chapters, but first we must consider the basic techniques of minimization.

3.4   Karnaugh map method

This graphical technique is based on the Venn diagram and was originally proposed by Veitch[1] but later improved by Karnaugh.[2] The starting point is really the Venn diagram which we have seen can be used to represent all the possible combinations of n switching variables as areas on a plane; Fig. 3.4 shows the Venn diagram for four variables. The diagram shows that in moving from one sub-area to another in a horizontal or vertical direction a

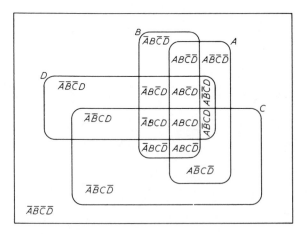

Figure 3.4 Four-variable Venn diagram

change in one variable only is involved, e.g. $\bar{A}\bar{B}\bar{C}\bar{D}-\bar{A}\bar{B}C\bar{D}-A\bar{B}\bar{C}\bar{D}-$
$A\bar{B}C\bar{D}$. Thus all *adjacent states* (that is, combinations which differ in only
one variable and which can be related according to the Boolean law ($BA +$
$B\bar{A} = B$) are depicted on the Venn diagram.

The technique is applied by plotting on the Venn diagram those variable
combinations which occur in the basic canonical switching equation. For
example, consider the equations obtained in the full-adder circuit:

$$C_+ = \bar{x}yc + x\bar{y}c + xy\bar{c} + xyc$$

This is shown in Fig. 3.5 with the appropriate terms shaded. Clearly, the
terms can be combined to give the reduced form obtained previously, i.e.
$C_+ = xy + xc + yc$. However, it is very difficult to draw the Venn diagram
for more than four variables, and furthermore it is not organized for convenient plotting and extraction of the switching terms.

In the Karnaugh map (hereafter abbreviated to K-map) or matrix
method, every possible combination of the binary input variables is represented on the map by a square (or cell); thus, for n variables we have 2^n
squares. The squares in the matrix are generally coded using the reflected
binary notation for columns and rows, which ensures that there is a change
in one variable only between adjacent vertical or horizontal squares. In this
way it becomes immediately obvious by inspection which terms can be
combined and simplified using the relationship $BA + B\bar{A} = B$. K-maps for
two, three, and four variables are shown in Table 3.2. Note that cells in
adjacent rows differ by one variable, as do cells in adjacent columns.
Moreover, there is a correspondence between top and bottom rows, and
between extreme left- and right-hand columns.

To plot a canonical sum-of-products function on the K-map, we enter a 1
in each square of the map corresponding to a term in the function; thus the
map completely defines the switching function. To save time, the K-map
may be plotted directly from the truth table without extracting the
canonical equations. For example, consider the function

$$T = \bar{A}\bar{B}\bar{C}\bar{D} + \bar{A}\bar{B}C\bar{D} + \bar{A}BCD + \bar{A}BCD$$
$$+ A\bar{B}\bar{C}\bar{D} + A\bar{B}C\bar{D} + A\bar{B}CD + ABCD$$

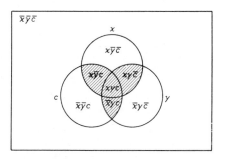

Figure 3.5 Full-adder carry equations

Table 3.2 Karnaugh maps

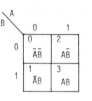

(a) Two-variable map

(b) Three-variable map

(c) Four-variable map

or the numerical representation of it

$$T = \Sigma(0, 2, 3, 7, 8, 10, 11, 15)$$

If the numerical notation is used, it is an easy matter to plot the function—simply place a 1 in each square of the map corresponding to a number in the function (Table 3.3).

The inverse product of sums form is, of course, given by the remaining squares, and these may be indicated by a 0:

$$T = (A + \bar{B} + C + D)(\bar{A} + \bar{B} + C + D)(A + B + C + \bar{D})$$
$$(A + \bar{B} + C + D)(\bar{A} + \bar{B} + C + D)(\bar{A} + B + C + \bar{D})$$
$$(A + \bar{B} + \bar{C} + D)(\bar{A} + \bar{B} + \bar{C} + D)$$

Note that in this case the inverse of the terms must be used to plot the function on the map. For example, $(A + \bar{B} + C + \bar{D})$ would be mentally inverted to $\bar{A}B\bar{C}D$ and plotted by placing a 0 in the appropriate cell.

It will be seen from Table 3.3 that sets of adjacent 1s have been enclosed; these correspond to terms in the function that can be combined (and reduced) using the expression $AB + \bar{A}B = B$. Thus, in moving from one square to another adjacent square, a variable will pass through both its possible values, i.e. 0 and 1; this variable must then be redundant. In Table

Table 3.3 Function plotting with K-maps

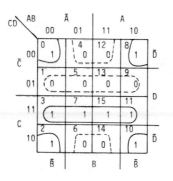

$$T = \bar{B}\bar{D} + CD = (\bar{B} + D)(C + \bar{D})$$

3.3 we have enclosed the terms $\bar{A}\bar{B}CD$, $\bar{A}BCD$, $ABCD$ and $A\bar{B}CD$; these combine to form $BCD(A + \bar{A})$ and $\bar{B}CD(A + \bar{A})$, the results of which may then be combined to give $CD(B + \bar{B}) = CD$. Note that the loop formed by the enclosed 1s extends over both states of the two variables A and B. The final result is identical to that obtained by algebraic manipulation, i.e.

$$T = \bar{A}\bar{B}\bar{C}\bar{D} + \bar{A}\bar{B}C\bar{D} + \bar{A}BCD + \bar{A}BCD + A\bar{B}\bar{C}\bar{D} + A\bar{B}C\bar{D}$$
$$+ A\bar{B}CD + ABCD$$
$$= \bar{A}\bar{B}\bar{D}(\bar{C} + C) + \bar{A}CD(\bar{B} + B) + A\bar{B}\bar{D}(\bar{C} + C) + ACD(\bar{B} + B)$$
$$= \bar{B}\bar{D}(\bar{A} + A) + CD(\bar{A} + A)$$
$$T = \bar{B}\bar{D} + CD$$

The cells containing 0s may be combined in the same way, except that in reading the result, the inverses of the variables must be used (this again can be done mentally) and combined in the product-of-sums form:

$$T = (\bar{B} + D)(C + \bar{D})$$

It is apparent, then, that the K-map is best suited to sum-of-products working, and for this reason it is often referred to as a *minterm* map.

Note that if the switching function is not in the canonical form it must first be expanded. This can be done by using the techniques described in Chapter 2, or it can be performed mentally during the plotting procedure. The latter approach is easier than it sounds because of the matrix organization of the map. For example, suppose we wished to plot the function

$$T = \bar{B}\bar{D} + CD$$

Referring to Table 3.3, we insert 1s in those squares where the $\bar{B}$ and $\bar{D}$ areas intersect; likewise we put 1s in the C and D intersecting areas; in this way the expansion is obtained automatically.

The looped terms which appear on the map and in the final expression are called *prime implicants* (they are called prime *implicates* in the case of product-of-sums). In this particular example, since each prime implicant contains original switching terms (represented by the cells containing 1) which are not involved in any other, they are called *essential prime implicants*. More precisely, an essential prime implicant is one which contains a 1 in a cell which cannot be included in any other prime implicant. That is, they must be included in the final minimal expression, and in this case the result obtained is a unique minimal function. In general, however, the method does not result in a unique solution since there are many possible ways of combining the cells; nevertheless all reduced expressions are valid solutions of the original switching problem. Table 3.4 shows all possible groupings for a particular switching problem; the minimal solutions are

$$T = \bar{A}\bar{D} + AD + BD + A\bar{B}\bar{C}$$
$$T = \bar{A}\bar{D} + AD + \bar{A}B + A\bar{B}\bar{C}$$
$$T = \bar{A}\bar{D} + AD + BD + \bar{B}\bar{C}\bar{D}$$
$$T = \bar{A}\bar{D} + AD + \bar{A}B + \bar{B}\bar{C}\bar{D}$$

Table 3.4 Prime implicants

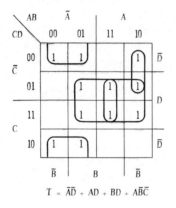

$$T = \bar{A}\bar{D} + AD + BD + A\bar{B}\bar{C}$$

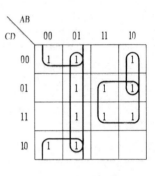

$$T = \bar{A}\bar{D} + AD + \bar{A}B + A\bar{B}\bar{C}$$

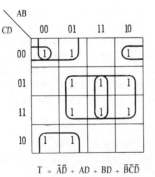

$$T = \bar{A}\bar{D} + AD + BD + \bar{B}\bar{C}\bar{D}$$

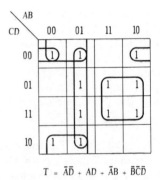

$$T = \bar{A}\bar{D} + AD + \bar{A}B + \bar{B}\bar{C}\bar{D}$$

Prime implicant set: $\bar{A}\bar{D}$, BD, AD, $A\bar{B}\bar{C}$, $\bar{A}B$, $\bar{B}\bar{C}\bar{D}$

Essential prime implicants: $\bar{A}\bar{D}$, AD

Note that the essential prime implicants are $\bar{A}\bar{D}$ and AD, and the minimal solution is obtained by selecting a minimal subset from the complete set of prime implicants.

It will be clear that, unless some systematic procedure is followed, haphazard grouping of the cells will result in an expression which is not necessarily in the minimal form. In combining squares, the following general rules should be obeyed:

(a) Every cell containing a 1 must be included at least once.
(b) The largest possible group of squares (powers of 2 only) must be formed; these are the prime implicants. This ensures that the maximum number of variables is eliminated, i.e. groups formed of 2^n cells eliminate n variables.
(c) The 1s must be contained in the minimum number of groups to avoid duplication.

To obtain a minimal result we first select the essential prime implicants, and then those additional prime implicants necessary to completely cover the original function.

The K-map method may be extended to the solution of five- and six-variable problems, but above this number the method becomes ungainly and the adjacencies become difficult to recognize. The four-variable map is symmetrical in shape, having top-to-bottom and side-to-side adjacencies, as well as row-to-row and column-to-column adjacencies. In fact, the four-variable map can be likened to a motor car tyre inner-tube, or, in topological terms, a torus. For this reason it is used as the basis of five- and six-variable maps, as shown in Table 3.5. The five-variable map for $VWXYZ$ consists of two four-variable maps, each representing all possible combinations of the variables $WXYZ$, for the conditions $V = 0$ and $V = 1$. Thus, the maps are effectively grouped to correspond to the binary powers. The normal adjacencies apply to the four-variable maps, with a one-to-one correspondence between them, rather as if the two four-variable maps were superimposed. To use the maps, we first have to recognize and group the terms on the four-variable maps, and then search for correspondence between individual maps. An alternative method of drawing high order K-maps is to label the columns and rows according to the reflected Gray code. This ensures that there is a change of only one variable between each row and column, but there are also other column and row adjacencies which are spread throughout the map, thus making it difficult to recognize the adjacent groups.

It will be obvious that there is a direct correspondence between the K-map and the cubical representation discussed earlier in Chapter 2. The 0-cubes correspond to the individual cells on the K-map, the 1-cubes (edges) to pairs of adjacent cells, 2-cubes (planes) to groupings of four cells etc.

When designing a switching circuit it often happens in practice that it is not possible, or even desirable, to specify the output conditions arising from all possible input values. This is not necessarily through ignorance but because particular input conditions just 'can't-happen'; for example, some

Table 3.5 Five- and six-variable K-maps

(a) Five-variable map

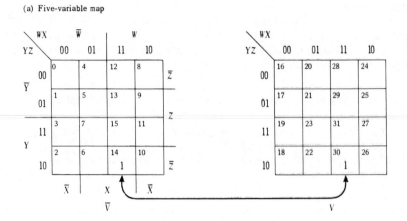

(b) Six-variable map

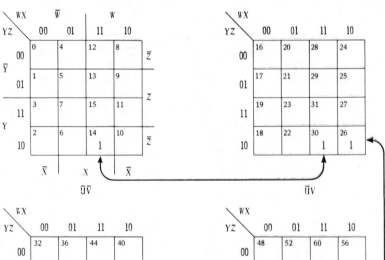

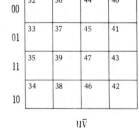

input variables may be mutually exclusive. Again in some cases if an output does occur it may be of no consequence and can be ignored, that is we 'don't-care' whether the output is 0 or 1. These *incompletely specified functions* are generally more amenable to minimization than fully specified functions since the 'don't-care' output values can be assigned either 0 or 1; this condition is normally indicated by an X (or a dash '–') in the truth table. For example, if we were designing a combinational logic circuit to encode a four-bit binary coded decimal to Gray code, 6 out of the possible 16 four-bit combinations would not be used since only the decimal numbers 0–9 have to be represented (Table 3.6). Thus, since these combinations will not occur in normal usage, the outputs are immaterial and can be assigned either 0 or 1. However, it should be noted that an output could occur due to an error in the input and in practice the outputs from the unused input conditions might be used to generate an error signal. Table 3.7 shows the K-maps for this coding problem, and it is immediately obvious which is the best way to include the 'don't care' conditions, which take the value 1 within the loops, and 0 elsewhere. Note that in all cases the 'don't care' conditions have been included in such a way as to complete a maximum sized group.

3.5 Reduced expressions

So far we have only considered canonical truth tables in which all possible values of the input variable are tabulated. In practice, and especially when handling large networks (a 10 variable circuit would require a 1024 row truth table!) a reduced form of table is often used since the designer is primarily interested in the output that must be generated when a particular

Table 3.6 BCD to Gray code don't-care conditions

	B_8	B_4	B_2	B_1		G_4	G_3	G_2	G_1
0	0	0	0	0		0	0	0	0
1	0	0	0	1		0	0	0	1
2	0	0	1	0		0	0	1	1
3	0	0	1	1		0	0	1	0
4	0	1	0	0		0	1	1	0
5	0	1	0	1		0	1	1	1
6	0	1	1	0		0	1	0	1
7	0	1	1	1		0	1	0	0
8	1	0	0	0		1	1	0	0
9	1	0	0	1		1	1	0	1
'Don't care' conditions	1	0	1	0		X	X	X	X
	1	0	1	1		X	X	X	X
	1	1	0	0		X	X	X	X
	1	1	0	1		X	X	X	X
	1	1	1	0		X	X	X	X
	1	1	1	1		X	X	X	X

Table 3.7 K-map BCD to Gray code with don't-care conditions

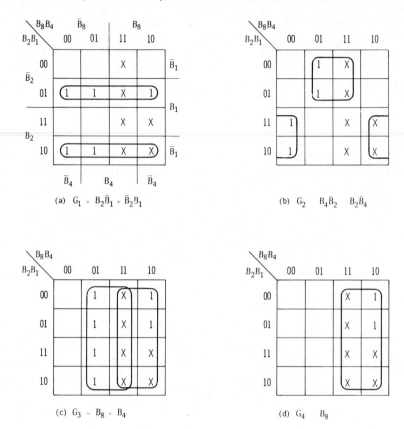

(a) $G_1 = B_2\bar{B}_1 + \bar{B}_2 B_1$

(b) $G_2 \quad B_4\bar{B}_2 \quad B_2\bar{B}_4$

(c) $G_3 = B_8 + B_4$

(d) $G_4 \quad B_8$

input variable or combination of variables occurs. Note that this approach is responsible for many of the 'don't care' conditions (and errors!) that are generated in a design, since the undefined input values can of course still give rise to an output which may or may not be relevant. Note also that some minimization procedures require canonical functions as their starting point; this in essence is true for the K-map but the expression is automatically expanded on plotting. In practice the *collapsed* form of truth table is often the only way of efficiently specifying a large switching network. As an example of this approach consider the carry logic discussed earlier and given by the equation:

$$C_+ = yc + xc + xy$$

This would be represented in a collapsed truth table form as shown in Table 3.8. The X entries in this case signify that an input variable has been eliminated and can be treated as a 'don't care' state.

Table 3.8 Collapsed
truth table

x	y	C	C_+
0	0	X	0
0	X	0	0
X	0	0	0
X	1	1	1
1	X	1	1
1	1	X	1

3.6 Design example: 9s complement converter for 2-out-of-5 BCD

Let us consider another design example to illustrate the treatment of 'don't care' conditions. Suppose a switching circuit is required to convert binary coded decimals in the 2-out-of-5 code to their corresponding 9s complement, as used to represent negative numbers. Thus when, for example, the number 5, coded as 00101, appears as an input to the circuit, the output must go to $9 - 5 = 4$, coded as 10100. The truth table is shown in Table 3.9; since we are only concerned with the decimal numbers 0–9, the 'don't care' conditions are

$$D = (0, 1, 2, 4, 7, 8, 11, 13, 14, 15, 16, 19, 21, 22,$$
$$23, 25, 26, 27, 28, 29, 30, 31)$$

If we insert these 'don't care' conditions in the five-variable K-maps, together with the actual $VWXYZ$ terms required to give an output for $ABCDE$, we obtain the set of maps shown in Table 3.10. The first step in minimization is to form the largest groups, including the 'don't care' conditions, on the four-variable maps, noting at the same time if a corresponding group occurs in the other map. For example, in Table 3.10(a) the

Table 3.9 Truth table: 9s complement for 2-out-of-5 BCD

Decimal	V	W	X	Y	Z	A	B	C	D	E	Decimal equivalent of 2-out-of-5 code combination
0	0	1	1	0	0	0	1	0	1	0	12
1	1	0	0	0	1	0	1	0	0	1	17
2	1	0	0	1	0	1	1	0	0	0	18
3	0	0	0	1	1	0	0	1	1	0	3
4	1	0	1	0	0	0	0	1	0	1	20
5	0	0	1	0	1	1	0	1	0	0	5
6	0	0	1	1	0	0	0	0	1	1	6
7	1	1	0	0	0	1	0	0	1	0	24
8	0	1	0	0	1	1	0	0	0	1	9
9	0	1	0	1	0	0	1	1	0	0	10

Remaining combinations are 'don't-care'

Table 3.10 K-maps: 9s complement for 2-out-of-5 BCD

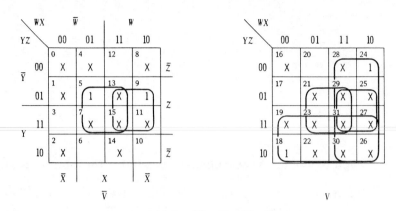

(a) $A = VW + VY + XZ + WZ$

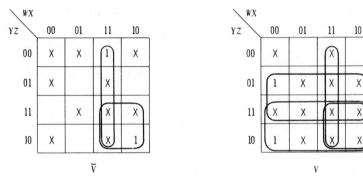

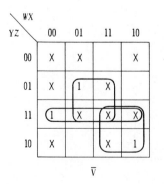

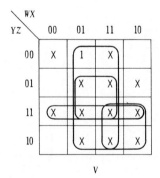

(b) $B = WX + WY + VZ + VY$

(c) $C = ZX + YZ + VX + WY$

Table 3.10 continued

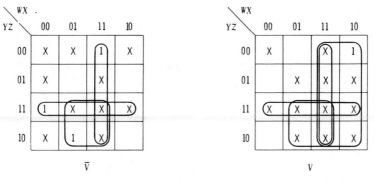

(d) $D = YZ + WV + YX + WX$

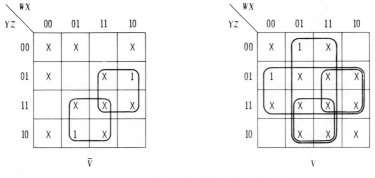

(e) $E = XY + WZ + VX + VZ$

group ZX appears on both V and $\bar{V}$ maps; thus the variable V may be eliminated; the same applies to the group WZ. Proceeding in this way we eventually obtain the output equations

$$A = VW + VY + XZ + WZ$$
$$B = WX + WY + VZ + VY$$
$$C = ZX + YZ + VX + WY$$
$$D = YZ + WV + YX + WX$$
$$E = XY + WZ + VX + VZ$$

These equations are shown implemented in AND/OR logic in Fig. 3.6. This problem is really an example of a multi-terminal circuit (to be dealt with in later sections), since it has more than one output. In this type of circuit a greater degree of minimization may be achieved if prime implicants are selected which are common to more than one output function. In the equations above, all five output functions can be formed using the prime implicants VW, VY, XZ, WZ, WX, WY, XY, VZ, VX and YZ. Conse-

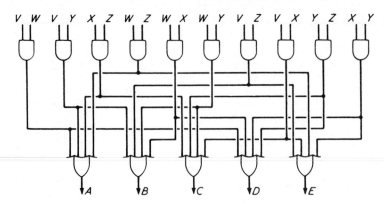

Figure 3.6 9s complement 2-out-of-5 BCD

quently, only one gate is necessary to form each prime implicant, with the output being distributed to the appropriate output gates. This means that, for multi-terminal circuits, forming the largest group for individual output maps does not necessarily lead to the minimal result.

3.7   Tabular methods of minimization

For problems with a large number of switching variables the K-map approach breaks down. Up to six variables can be handled conveniently, especially if printed sheets of four-variable maps are available as standard stationery, but above this number the technique becomes too complicated. For problems with a large number of variables a tabular or algorithmic method due to McCluskey,[3] based on an original technique due to Quine,[4] is used. The advantage of this method is that it does not depend on pattern recognition procedures but operates directly on the actual switching terms. It may be used as a hand computation technique or, better still (see later chapter), programmed for a digital computer; in addition it can be extended to handle any number of variables. The basic idea is to examine each term, and its reduced derivatives, exhaustively and systematically to see if the Boolean theorem $AB + A\bar{B} = A$ can be applied. This results in a complete list of all the prime implicants for the function concerned. The procedure is best illustrated by means of an example.

Suppose we wish to minimize the expression expressed in the numerical sum-of-products form as

$$T = \Sigma(1, 2, 3, 4, 5, 6, 7, 10, 14, 20, 22, 28)$$

The first step is to tabulate the terms of the switching function into groups according to the number of 1s contained in each term. In Table 3.11(a) the first group consists of the terms (1, 2, 4) all of which contain one binary digit; the second group consists of the terms (3, 5, 6, 10, 20) which contain two binary digits; etc. We then compare each term with the terms in the group below it, looking for entries that differ by one variable only and which can be combined according to the theorem $AB + A\bar{B} = A$. For example, 00001 is compared with 00011 and found to differ by one variable; the term 000–1 (the dash represents the eliminated variable) is used to start a new group in the next listing, List 2. Both combining terms are ticked off on the original list and the comparison continued until no more combinations can be formed. The comparison is exhaustive and it would compare, in the case of the example, terms $(1, 3)(1, 5)(1, 6)(1, 10)$ $(1, 20)(2, 3)(2, 5)(2, 6)(2, 10)(2, 20)(4, 3)(4, 5)(4, 6)(4, 10)(4, 20)$, then $(3, 7)(3, 14)(3, 22)(3, 28)(5, 7)(5, 14)\cdots(20, 28)$. Terms need only be compared with those in the group immediately below (and numerically greater), as these are the only ones that can differ by one variable. The process is continued by comparing terms in the derived lists, Lists 2 and 3, in the same way, except that this time the 'dashed' variables must also correspond. It is also necessary to keep a check of the actual combining terms as they will be required later; this is done by noting at the side of each list entry the actual terms of the original switching function included in the reduced expression. The unticked entries (terms that cannot be combined further) are the prime implicants of the switching function. From Table 3.11 these are

Prime implicants, PI $= (VX\bar{Y}\bar{Z}, \bar{V}\bar{W}Z, \bar{V}\bar{W}Y, \bar{V}Y\bar{Z}, \bar{V}\bar{W}X, \bar{W}X\bar{Z})$

That these are, in fact, the complete set can easily be ascertained from the K-map in Table 3.11(b).

During the comparison process it is possible that the same reduced term may be formed in more than one way, for example in List 3 of Table 3.11(a), prime implicant B may be formed by combining 1, 3/5, 7 and 1, 5/3, 7. The repeated terms are best ignored and not included in succeeding new lists, though they are retained in the examples that follow to ensure a better appreciation of the technique.

3.8   Selecting a minimal PI subset: prime implicant tables

Having ascertained the complete PI set, it is now necessary to choose a minimal subset that will include, or *cover*, all the terms in the original switching expression. Each product term of the function must be included in at least one of the prime implicant terms in the minimal expression. The relationship between the prime implicants and the switching terms can best be seen by means of a *prime implicant table*; this is shown in Table 3.12 for the example above. The table takes the form of a matrix with the original product terms as columns and the prime implicants as rows. For each prime implicant row, a cross is placed in those columns that contain a term of the original switching function; for example, prime implicant row A, comprising terms 20, 28, would have crosses in columns 20 and 28. To choose an optimum set of prime implicants we first examine the table for any columns with only one cross. The corresponding rows are called *basis rows*

Table 3.11 Tabular minimization

(a) Determination of prime implicants

	V W X Y Z			V W X Y Z			V W X Y Z
1	0 0 0 0 1 √	1,3	0 0 0 - 1 √	1,3/5,7	0 0 - - 1 B
2	0 0 0 1 0 √	1,5	0 0 - 0 1 √	1,5/3,7	0 0 - - 1
4	0 0 1 0 0 √	2,3	0 0 0 1 - √	2,3/6,7	0 0 - 1 - C
3	0 0 0 1 1 √	2,6	0 0 - 1 0 √	2,6/3,7	0 0 - 1 -
5	0 0 1 0 1 √	2,10	0 - 0 1 0 √	2,6/10,14	0 - - 1 0 D
6	0 0 1 1 0 √	4,5	0 0 1 0 - √	2,10/6,14	0 - - 1 0
10	0 1 0 1 0 √	4,6	0 0 1 - 0 √	4,5/6,7	0 0 1 - - E
20	1 0 1 0 0 √	4,20	- 0 1 0 0 √	4,6/5,7	0 0 1 - -
7	0 0 1 1 1 √	3,7	0 0 - 1 1 √	4,6/20,22	- 0 1 - 0 F
14	0 1 1 1 0 √	5,7	0 0 1 - 1 √	4,20/6,22	- 0 1 - 0
22	1 0 1 1 0 √	6,7	0 0 1 1 - √		
28	1 1 1 0 0 √	6,14	0 - 1 1 0 √		List 3
	List 1	6,22	- 0 1 1 0 √		
		10,14	0 1 - 1 0 √		
		20,22	1 0 1 - 0 √		
		20,28	1 - 1 0 0 A		
			List 2		

(b) K-maps

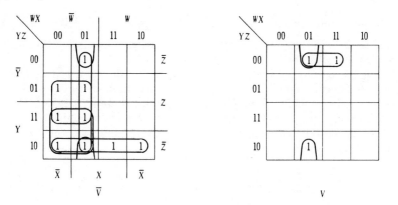

and represent, of course, the essential prime implicants; these are marked in Table 3.12 by an asterisk, with the included column terms ticked off.

When this is done, a check is made to see if any column terms are uncovered by these basis rows, i.e. check for any terms of the original switching function not included in the essential prime implicants. In this example, the basis rows (essential prime implicants) include all the switching terms, so that no other rows (prime implicants) are required. The minimal switching equation is given by the logical sum of the prime implicants, i.e.

$$T = A + B + D + F$$
$$= VX\bar{Y}\bar{Z} + \bar{V}\bar{W}Z + \bar{V}Y\bar{Z} + \bar{W}X\bar{Z}$$

Table 3.12 Extraction of minimal PI subset

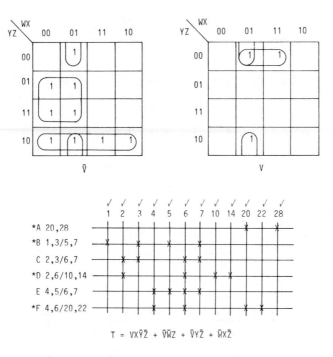

$$T = V X \bar{Y} \bar{Z} + \bar{V} \bar{W} Z + \bar{V} Y \bar{Z} + \bar{W} X \bar{Z}$$

If this had not been the case, however, it would have been necessary to select other rows to complete the covering. Thus, after selecting the basis rows, one would proceed by inspecting the table for those rows which can cover the remaining column terms. If more than one row exists, a selection is made in favour of the row which can cover the maximum number of remaining terms, or which contains the most crosses (and hence the fewest literals) in the prime implicant term. If two rows can cover the same terms, the row with the fewest literals is chosen. The process is repeated until all the rows are covered.

Consider the example shown in Table 3.13. The basis rows are G, H and I, and these cover the column terms 1, 2, 5, 6, 13, 17, 18, 21, 22, 29. Now we examine the table for rows which can cover the remaining terms 7, 8, 9, 10. Each of the remaining terms can be covered by two possible rows, that is, term 7 is covered by D and E; term 8 by B and C; term 9 by B and F; and term 10 by A and C. We shall choose F rather than B to cover 9 since it contains the fewest literals, even though prime implicant B does cover two of the required terms, 8 and 9. The next choice is C since this covers terms 8 and 10; then we take either D or E as there is nothing to choose between them. The minimal solution, then, is

$$T = G + H + I + F + C + D \text{ (or } E)$$
$$= \bar{W} \bar{Y} Z + \bar{W} Y \bar{Z} + X \bar{Y} Z + \bar{V} \bar{Y} Z + \bar{V} W \bar{X} \bar{Z} + \bar{V} \bar{W} X Z \text{ (or } \bar{V} \bar{W} X Y)$$

Table 3.13 Prime implicant table

Function $T = \Sigma(1, 2, 5, 6, 7, 8, 9, 10, 13, 17, 18, 21, 22, 29)$

	V W X Y Z			V W X Y Z			V W X Y Z
1	0 0 0 0 1 √		1,5	0 0 – 0 1 √		1,5/9,13	0 – – 0 1 F
2	0 0 0 1 0 √		1,9	0 – 0 0 1 √		1,5/17,21	– 0 – 0 1 G
8	0 1 0 0 0 √		1,17	– 0 0 0 1 √		1,9/5,13	0 – – 0 1
			2,6	0 0 – 1 0 √		1,17/5,21	– 0 – 0 1
5	0 0 1 0 1 √		2,10	0 – 0 1 0 A		2,6/18,22	– 0 – 1 0 H
6	0 0 1 1 0 √		2,18	– 0 0 1 0 √		2,18/6,22	– 0 – 1 0
9	0 1 0 0 1 √		8,9	0 1 0 0 – B			
10	0 1 0 1 0 √		8,10	0 1 0 – 0 C		5,21/13,29	– – 1 0 1 I
17	1 0 0 0 1 √					5,13/21,29	– – 1 0 1
18	1 0 0 1 0 √		5,7	0 0 1 – 1 D			
			5,13	0 – 1 0 1 √			
7	0 0 1 1 1 √		5,21	– 0 1 0 1 √			
13	0 1 1 0 1 √		6,7	0 0 1 1 – E			
21	1 0 1 0 1 √		6,22	– 0 1 1 0 √			
22	1 0 1 1 0 √		9,13	0 1 – 0 1 √			
			17,21	1 0 – 0 1 √			
29	1 1 1 0 1 √		18,22	1 0 – 1 0 √			
			13,29	– 1 1 0 1 √			
			21,29	1 – 1 0 1 √			

Coverage chart (columns: $1^{√}$ $2^{√}$ $5^{√}$ $6^{√}$ 7 8 9 10 $13^{√}$ $17^{√}$ $18^{√}$ $21^{√}$ $22^{√}$ $29^{√}$):

	1	2	5	6	7	8	9	10	13	17	18	21	22	29
A		✕						✕						
B						✕	✕							
** C						✕		✕						
*** D			✕		✕									
E				✕	✕									
** F	✕		✕				✕		✕					
* G	✕		✕							✕		✕		
* H		✕		✕							✕		✕	
* I			✕						✕			✕		✕

$$T = G + H + I + F + C + D \text{ (or } E)$$
$$= \overline{W}\overline{Y}Z + \overline{W}Y\overline{Z} + X\overline{Y}Z + \overline{V}\overline{Y}Z + \overline{V}W\overline{X}\overline{Z} + \overline{V}\overline{W}XZ \text{ (or } \overline{V}\overline{W}XY)$$

This result is just one of the many possible solutions, giving the minimum number of terms, though this particular one is truely minimal in the sense of both literals and terms.

For some functions it may happen that, after the basis rows have been selected, the remainder of the chart is such that there is no obvious way to choose the next rows; this structure is said to be *cyclic*. In other words, there are no unique rows which can be chosen so that other rows may be covered. Table 3.14 shows an example of a cyclic prime implicant table. After the basis row A has been choosen, there are several alternative ways of selecting rows to cover the remaining column terms. Since all terms contain the same number of literals and cover the same number of column

Table 3.14 Cyclic prime implicant table

Function $T = \Sigma(1, 3, 4, 5, 6, 8, 9, 10, 14)$

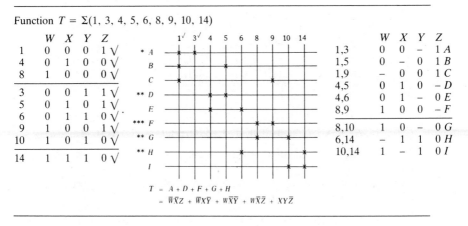

	W	X	Y	Z	
1	0	0	0	1	✓
4	0	1	0	0	✓
8	1	0	0	0	✓
3	0	0	1	1	✓
5	0	1	0	1	✓
6	0	1	1	0	✓
9	1	0	0	1	✓
10	1	0	1	0	✓
14	1	1	1	0	✓

	W	X	Y	Z	
1,3	0	0	–	1	A
1,5	0	–	0	1	B
1,9	–	0	0	1	C
4,5	0	1	0	–	D
4,6	0	1	–	0	E
8,9	1	0	0	–	F
8,10	1	0	–	0	G
6,14	–	1	1	0	H
10,14	1	–	1	0	I

$$T = A + D + F + G + H$$
$$= \bar{W}\bar{X}Z + \bar{W}X\bar{Y} + W\bar{X}\bar{Y} + W\bar{X}\bar{Z} + XY\bar{Z}$$

terms there is no obvious starting point. We must proceed, therefore, by selecting an arbitrary row, say D, and continuing until all column terms are covered. In this case a possible minimal selection is

$$T = A + D + F + G + H$$
$$= \bar{W}\bar{X}Z + \bar{W}X\bar{Y} + W\bar{X}\bar{Y} + W\bar{X}\bar{Z} + WY\bar{Z}$$

To ascertain if this is a unique minimal result, all possible alternative prime implicant sets must be obtained and examined in detail. This is essentially a trial and error process, the complexity of which rapidly increases with problem size.

3.9 Algebraic solution of prime implicant tables

Prime implicant tables, including cyclic ones, can be 'solved' by using an algebraic approach, due to Petrick[5] and later modified by Pyne and McCluskey,[6] which produces all the irredundant sums, that is, solutions from which no prime implicant may be removed if all the terms are to be accounted for. The minimal result can be obtained by direct examination and comparison of the alternative irredundant sum-of-products expressions. In this method, each row of the prime implicant table is considered as a binary variable, and a product-of-sums expression is derived for the complete table. This function is called a prime implicant function (P), and each variable corresponds to a prime implicant of the original switching function. From Table 3.14 we can see that, to account for all the terms in the table, the P function must contain A or B or C from column 1, A from column 3, and D or E from column 4, etc.

Combining these statements we have $(A + B + C)(A)(D + E)$ etc., and continuing in this way for the entire table we have

$$P = (A + B + C)(A)(D + E)(B + D)(E + H)$$
$$(F + G)(C + F)(G + I)(H + I)$$

Since this is a Boolean expression, we may reduce it in the normal way:

$$P = (A)(D + BE)(E + H)(F + GC)(I + GH)$$
$$= (A)(FD + BEF + DGC + BEGC)(EI + HG + HI)$$
$$= AFDEI + AFDHG + AFDHI + ABEFI + ABEFHG$$
$$+ ABEFHI + ADGCEI + ADGCH + ADGCHI$$
$$+ ABEGCI + ABEGCH + ABEGCHI$$

Thus,

$$P = AFDEI + AFDHG + AFDHI + ABEFI + ABEFHG$$
$$+ ADGCEI + ADGCH + ABEGCI + ABEGCH$$

There are, then, five minimal solutions each containing five prime implicant terms.

Again, let us consider the prime implicant table shown in Table 3.13; this may be expressed in the product-of-sums form:

$$P = (F + G)(A + H)(D + F + G + I)(E + H)(D + E)(B + C)$$
$$(B + F)(A + C)(F + I)(G)(H)(G + I)(H)(I)$$
$$= (G)(H)(I)(D + E)(B + C)(B + F)(A + C)$$
$$= (G)(H)(I)(C + AB)(B + F)(D + E)$$
$$= (GHI)(BC + AB + FC + FAB)(D + E)$$
$$= (GHI)(BCD + ABD + FCD + FABD + BCE$$
$$+ ABE + FCE + FABE)$$

Thus,

$$P = GHIBCD + GHIABD + GHIFCD + GHIBCE$$
$$+ GHIABE + GHIFCE$$

Now since $FGHI$ are the prime implicant terms with the fewest literals, there are two minimal solutions:

$$T = G + H + I + F + C + D$$

and

$$T = G + H + I + F + C + E$$

which is the same result that we obtained using the prime implicant chart. This technique is the only way to manipulate complex tables, and moreover it is a very convenient algorithm for machine computation.

3.10 'Don't care' conditions

So far in our discussions of the tabulation method we have not mentioned how incompletely specified functions are treated. In fact the procedure is very nearly identical. The switching terms, including 'don't cares', are tabulated and examined as before to produce the complete set of prime implicants. The prime implicant table is then constructed in the usual way, but the 'don't care' conditions are ignored. This is feasible since the inclusion, or exclusion, of 'don't care' terms in the final expression is immaterial. Suppose, for example, we wish to minimize the function

$$T = \Sigma(5, 15, 20, 29, 41, 42, 45, 47, 53, 58, 61, 63)$$

with the 'don't care' conditions

$$D = (7, 9, 11, 13, 21, 25, 27, 31, 37, 39, 40, 43, 56, 57, 59)$$

The complete set of terms, including 'don't cares', are shown in Table 3.15.

Proceeding with the minimization routine, we establish a set of prime implicants:

$$PI = (A, B, C, D, E) = (\bar{U}V\bar{W}X\bar{Y}, \bar{V}XZ, X\bar{Y}Z, UW\bar{X}, WZ)$$

If these are now plotted on a prime implicant chart (Table 3.16) containing the original transmission terms only, we find that A, C, D and E are essential prime implicants which give a unique minimal function:

Table 3.15 Minimization with don't-care condition

	U	V	W	X	Y	Z	
5	0	0	0	1	0	1	√
9	0	0	1	0	0	1	√
20	0	1	0	1	0	0	√
40	1	0	1	0	0	0	√
7	0	0	0	1	1	1	√
11	0	0	1	0	1	1	√
13	0	0	1	1	0	1	√
21	0	1	0	1	0	1	√
25	0	1	1	0	0	1	√
37	1	0	0	1	0	1	√
41	1	0	1	0	0	1	√
42	1	0	1	0	1	0	√
56	1	1	1	0	0	0	√
15	0	0	1	1	1	1	√
27	0	1	1	0	1	1	√
29	0	1	1	1	0	1	√
39	1	0	0	1	1	1	√
43	1	0	1	0	1	1	√
45	1	0	1	1	0	1	√
53	1	1	0	1	0	1	√
57	1	1	1	0	0	1	√
58	1	1	1	0	1	0	√
31	0	1	1	1	1	1	√
47	1	0	1	1	1	1	√
59	1	1	1	0	1	1	√
61	1	1	1	1	0	1	√
63	1	1	1	1	1	1	√

List 1

	U	V	W	X	Y	Z	
5,7	0	0	0	1	-	1	√
5,13	0	0	-	1	0	1	√
5,21	0	-	0	1	0	1	√
5,37	-	0	0	1	0	1	√
9,11	0	0	1	0	-	1	√
9,13	0	0	1	-	0	1	√
9,25	0	-	1	0	0	1	√
9,41	-	0	1	0	0	1	√
20,21	0	1	0	1	0	-	A
40,41	1	0	1	0	0	-	√
40,42	1	0	1	0	-	0	√
40,56	1	-	1	0	0	0	√
7,15	0	0	-	1	1	1	√
7,39	-	0	0	1	1	1	√
11,15	0	0	1	-	1	1	√
11,27	0	-	1	0	1	1	√
11,43	-	0	1	0	1	1	√
13,15	0	0	1	1	-	1	√
13,29	0	-	1	1	0	1	√
13,45	-	0	1	1	0	1	√
21,29	0	1	-	1	0	1	√
21,53	-	1	0	1	0	1	√
25,27	0	1	1	0	-	1	√
25,29	0	1	1	-	0	1	√
25,57	-	1	1	0	0	1	√
37,39	1	0	0	1	-	1	√
37,45	1	0	-	1	0	1	√
37,53	1	-	0	1	0	1	√
41,43	1	0	1	0	-	1	√

List 2

	U	V	W	X	Y	Z	
41,45	1	0	1	-	0	1	√
41,57	1	-	1	0	0	1	√
42,43	1	0	1	0	1	-	√
42,58	1	-	1	0	1	0	√
56,57	1	1	1	0	0	-	√
56,58	1	1	1	0	-	0	√
15,31	0	-	1	1	1	1	√
15,47	-	0	1	1	1	1	√
27,31	0	1	1	-	1	1	√
27,59	-	1	1	0	1	1	√
29,31	0	1	1	1	-	1	√
29,61	-	1	1	1	0	1	√
39,47	1	0	-	1	1	1	√
43,47	1	0	1	-	1	1	√
43,59	1	-	1	0	1	1	√
45,47	1	0	1	1	-	1	√
45,61	1	-	1	1	0	1	√
53,61	1	1	-	1	0	1	√
57,59	1	1	1	0	-	1	√
57,61	1	1	1	-	0	1	√
58,59	1	1	1	0	1	-	√
31,63	-	1	1	1	1	1	√
47,63	1	-	1	1	1	1	√
59,63	1	1	1	-	1	1	√
61,63	1	1	1	1	-	1	√

Table 3.15 continued

	U	V	W	X	Y	Z	
5,7/13,15	0	0	–	1	–	1	√
5,7/37,39	–	0	0	1	–	1	√
5,13/7,15	0	0	–	1	–	1	√
5,13/21,29	0	–	–	1	0	1	√
5,13/37,45	–	0	–	1	0	1	√
5,21/13,29	0	–	–	1	0	1	√
5,21/37,53	±	–	0	1	0	1	√
5,37/7,39	–	0	0	1	–	1	√
5,37/13,45	–	0	–	1	0	1	√
5,37/21,51	–	–	0	1	0	1	√
9,11/13,15	0	0	1	–	–	1	√
9,11/25,27	0	–	1	0	–	1	√
9,13/11,15	0	0	1	–	–	1	√
9,13/25,29	0	–	1	–	0	1	√
9,13/41,45	–	0	1	–	0	1	√
9,25/11,27	0	–	1	0	–	1	√
9,25/13,29	0	–	1	–	0	1	√
9,25/41,57	–	–	1	0	0	1	√
9,41/11,43	–	0	1	0	–	1	√
9,41/13,45	–	0	1	–	0	1	√
9,41/25,27	–	–	1	0	0	1	√
40,41/42,43	1	0	1	0	–	–	√
40,41/56,57	1	–	1	0	0	–	√
40,42/41,43	1	0	1	0	–	–	√
40,56/41,57	1	–	1	0	0	–	√
7,15/39,47	–	0	–	1	1	1	√
7,39/15,47	–	0	–	1	1	1	√
11,15/27,31	0	–	1	–	1	1	√
11,15/43,47	–	0	1	–	1	1	√
11,27/15,31	0	–	1	–	1	1	√
11,43/15,47	–	0	1	–	1	1	√
11,43/27,59	–	–	1	0	1	1	√
13,15/29,31	0	–	1	1	–	1	√
13,15/45,47	–	0	1	1	–	1	√
13,29/15,31	0	–	1	1	–	1	√
13,29/45,61	–	–	1	1	0	1	√
13,45/15,47	–	0	1	1	–	1	√

	U	V	W	X	Y	Z	
13,45/29,61	–	–	1	1	0	1	√
21,53/29,61	–	1	–	1	0	1	√
25,29/27,31	0	1	1	–	–	1	√
25,29/57,61	–	1	1	–	0	1	√
25,57/27,59	–	1	1	0	–	1	√
25,57/29,61	–	1	1	–	0	1	√
37,39/45,47	1	0	–	1	–	1	√
37,45/39,47	1	0	–	1	–	1	√
37,45/53,61	1	–	–	1	0	1	√
37,53/45,61	1	–	–	1	0	1	√
41,43/45,47	1	0	1	–	–	1	√
41,45/43,47	1	0	1	–	–	1	√
41,45/57,61	1	–	1	–	0	1	√
41,57/43,59	1	–	1	0	–	1	√
41,57/45,61	1	–	1	–	0	1	√
42,43/58,59	1	–	1	0	1	–	√
42,58/43,59	1	–	1	0	1	–	√
42,58/45,61	1	–	1	–	0	1	√
56,57/58,59	1	1	1	0	–	–	√
56,58/57,59	1	1	1	0	–	–	√
15,31/47,63	–	–	1	1	1	1	√
15,47/31,63	–	–	1	1	1	1	√
27,31/59,63	–	1	1	–	1	1	√
27,59/31,63	–	1	1	–	1	1	√
29,31/61,63	–	1	1	1	–	1	√
29,61/31,63	–	1	1	1	–	1	√
43,47/59,63	1	–	1	–	1	1	√
43,59/47,63	1	–	1	–	1	1	√
45,47/61,63	1	–	1	1	–	1	√
45,61/47,63	1	–	1	1	–	1	√
57,59/61,63	1	1	1	–	–	1	√
57,61/59,63	1	1	1	–	–	1	√

List 3

	U	V	W	X	Y	Z	
5,7/13,15/37,39/45,47	–	0	–	1	–	1	B
5,7/37,39/13,15/45,47	–	0	–	1	–	1	
5,13/7,15/37,39/45,47	–	0	–	1	–	1	
5,13/21,29/37,45/53,61	–	–	–	1	0	1	C
5,13/37,45/7,15/39,47	–	0	–	1	–	1	
5,13/37,45/21,53/29,61	–	–	–	1	0	1	
5,21/13,29/37,45/53,61	–	–	–	1	0	1	
5,21/37,53/13,29/45,61	–	–	–	1	0	1	
5,37/7,39/13,15/45,47	–	0	–	1	–	1	
9,11/13,15/25,29/27,31	0	–	1	–	–	1	√
9,11/13,15/41,43/45,47	–	0	1	–	–	1	√
9,11/25,27/13,15/29,31	0	–	1	–	–	1	√
9,4/25,27/41,57/43,59	–	–	1	0	–	1	√
9,13/11,15/25,29/27,31	0	–	1	–	–	1	√

Table 3.15 continued

	U	V	W	X	Y	Z	
9,13/11,15/41,43/45,47	–	0	1	–	–	1	√
9,13/25,29/11,15/27,31	0	–	1	–	–	1	√
9,13/25,29/41,45/57,61	–	–	1	–	0	1	√
9,13/41,45/11,15/43,47	–	0	1	–	–	1	√
9,13/41,45/25,29/57,61	–	–	1	–	0	1	√
9,25/11,27/13,15/29,31	0	–	1	–	–	1	√
9,25/41,57/11,43/27,59	–	–	1	0	–	1	√
9,25/41,57/13,29/45,61	–	–	1	–	0	1	√
9,41/11,43/13,15/45,47	–	0	1	–	–	1	√
9,41/11,43/25,57/27,29	–	–	1	0	–	1	√
40,41/42,43/56,57/58,59	1	–	1	0	–	–	D
40,41/56,57/42,43/58,59	1	–	1	0	–	–	
11,15/27,31/43,47/59,63	–	–	1	–	1	1	√
11,15/43,47/27,31/59,63	–	–	1	–	1	1	√
11,43/27,59/15,31/47,63	–	–	1	–	1	1	√
13,15/29,31/45,47/61,63	–	–	1	1	–	1	√
13,15/45,47/29,31/61,63	–	–	1	1	–	1	√
13,29/45,61/15,31/47,63	–	–	1	1	–	1	√
25,29/27,31/57,59/61,63	–	1	1	–	–	1	√
25,29/57,61/27,31/59,63	–	1	1	–	–	1	√
25,57/27,29/29,31/61,63	–	1	1	–	–	1	√
41,43/45,47/57,59/61,63	1	–	1	–	–	1	√
41,45/57,61/43,47/59,63	1	–	1	–	–	1	√
41,57/43,59/45,47/61,63	1	–	1	–	–	1	√

List 4

	U	V	W	X	Y	Z	
9,11/13,15/25,29/27,31 41,43/45,47/57,59/61,63	–	–	1	–	–	1	E
9,11/13,15/41,43/45,47 25,29/27,31/57,59/61,63	–	–	1	–	–	1	
9,4/25,27/41,57/43,59 13,15/29,31/45,47/61,63	–	–	1	–	–	1	
9,13/25,29/41,45/57,61 11,15/27,31/43,47/59,63	–	–	1	–	–	1	

List 5

Prime implicant set = $\bar{U}V\bar{W}X\bar{Y}$, $\bar{V}XZ$, $X\bar{Y}Z$, $UW\bar{X}$, WZ

Table 3.16 Prime implicant table with 'don't cares'

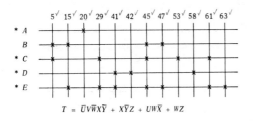

$$T = \bar{U}V\bar{W}X\bar{Y} + X\bar{Y}Z + UW\bar{X} + WZ$$

$$T = A + C + D + E$$
$$= \bar{U}V\bar{W}X\bar{Y} + X\bar{Y}Z + UW\bar{X} + WZ$$

If the reader draws the K-map for this problem it will be clear that the map method involves far less computation, and the use of 'don't care' conditions is immediately apparent. However, the strength of the tabular method lies in its easy extension to larger problems and suitability for both hand and machine computation.

3.11 Minimization of multiple output circuits

In general, the design of multi-terminal circuits follows very closely the procedures already described for single-terminal networks. By deriving from the truth table separate minimal equations for each output, in terms of the input variables, it is possible to arrive at a perfectly viable solution. However, in so doing it is very likely that redundancies may occur which could have been avoided if all the output functions were minimized collectively. This may be achieved by comparing K-maps, or by a modification to the McCluskey tabular method.

The best way of explaining the procedure is by means of an actual example. Suppose we wish to design an encoding circuit to convert pure binary numbers into a binary coded decimal number with the weights 5421. The truth table is shown in Table 3.17. Note that there are six unused combinations which can be considered as 'don't care' terms for each output function.

From the table, the equations for each output condition may be derived; they are

Table 3.17 Encoder for 5421 code

a 8	b 4	c 2	d 1		w 5	x 4	y 2	z 1
0	0	0	0		0	0	0	0
0	0	0	1		0	0	0	1
0	0	1	0		0	0	1	0
0	0	1	1		0	0	1	1
0	1	0	0		0	1	0	0
0	1	0	1		1	0	0	0
0	1	1	0		1	0	0	1
0	1	1	1		1	0	1	0
1	0	0	0		1	0	1	1
1	0	0	1		1	1	0	0
1	0	1	0	'Don't care' terms				
1	0	1	1					
1	1	0	0					
1	1	0	1					
1	1	1	0					
1	1	1	1					

$$Z_w = \bar{a}b\bar{c}d + \bar{a}bcd + \bar{a}bcd + a\bar{b}\bar{c}d + a\bar{b}\bar{c}d$$
$$Z_x = \bar{a}b\bar{c}d + a\bar{b}\bar{c}d$$
$$Z_y = \bar{a}b\bar{c}d + \bar{a}bcd + \bar{a}bcd + a\bar{b}\bar{c}d$$
$$Z_z = \bar{a}b\bar{c}d + \bar{a}bcd + \bar{a}bcd + a\bar{b}\bar{c}d$$

'Don't care' conditions are

$$Z'_w, Z'_x, Z'_y, Z'_z = a\bar{b}c\bar{d} + a\bar{b}cd + ab\bar{c}\bar{d} + ab\bar{c}d + abc\bar{d} + abcd$$

If we minimize each function separately using K-maps (see Table 3.18) we get the minimal forms:

$$Z_w = a + bd + bc$$
$$Z_x = ad + b\bar{c}\bar{d}$$
$$Z_y = cd + \bar{b}c + a\bar{d}$$
$$Z_z = a\bar{d} + \bar{a}\bar{b}d + bc\bar{d}$$

These equations when implemented directly in AND/OR invertor logic require 17 basic elements and 32 inputs.

However, since this is a multiple output circuit it would seem profitable to identify common product terms and attempt to share these between the output functions, thus reducing the total number of gates required. This may be done using K-maps as shown in Table 3.19. Here maps have been

Table 3.18 K-maps for 5421 encoder

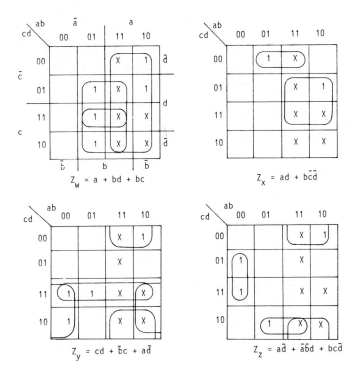

$$Z_w = a + bd + bc$$

$$Z_x = ad + b\bar{c}\bar{d}$$

$$Z_y = cd + \bar{b}c + a\bar{d}$$

$$Z_z = a\bar{d} + \bar{a}\bar{b}d + bc\bar{d}$$

Table 3.19 Multiple output minimization

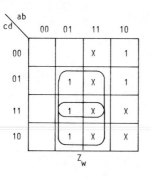

Z_w

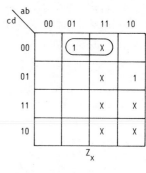

Z_x

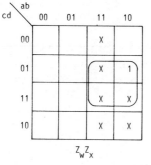

Z_wZ_x

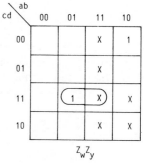

Z_wZ_y

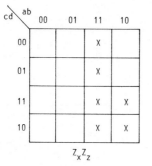

Z_xZ_z

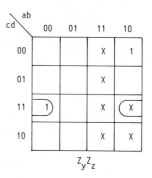

Z_yZ_z

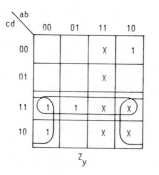

Z_y

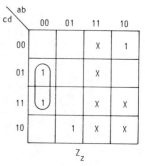

Z_z

Table 3.19 continued

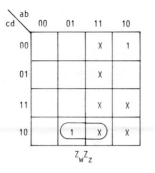

$Z_w \overline{Z}_z$

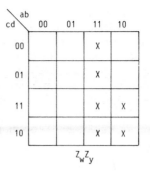

$Z_w Z_y$

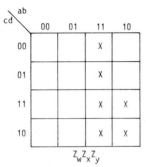

$Z_w \overline{Z}_x Z_y$

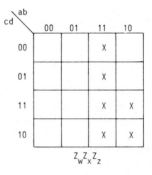

$Z_w \overline{Z}_x Z_z$

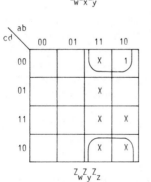

$Z_w Z_y Z_z$

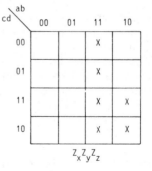

$Z_x Z_y Z_z$

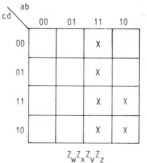

$Z_w Z_x Z_y Z_z$

drawn for all four functions followed by the maps of all the intersections, or products, of the four functions taken in pairs, then triples and finally altogether. Note that the intersection of two or more K-maps is simply the set of common cells. The *multiple output prime implicants* (MOPI) are now extracted starting with the product of all the functions, $Z_w Z_x Z_y Z_z$, followed by the triples etc. The first MOPI is $a\bar{d}$ on map $Z_w Z_y Z_z$ and then $\bar{b}cd$ on map $Z_y Z_z$. We continue marking the multiple output PIs in this way but ignoring those cells which are already covered.

The next step is to select a suitable set of prime implicants, which can be done either directly from the maps or using a PI chart as described earlier. The set of equations so obtained is as follows:

$$Z_w = a\bar{d} + bd + bc\bar{d} + ad$$
$$Z_x = b\bar{c}\bar{d} + ad$$
$$Z_y = cd + a\bar{d} + \bar{b}c$$
$$Z_z = a\bar{d} + bc\bar{d} + \bar{a}\bar{b}d$$

Note that a more minimal expression for Z_w (found earlier) is

$$Z_w = a + bd + bc$$

but since the terms $bc\bar{d}$, $a\bar{d}$ and ad are necessary for the other functions it is generally better to represent the function as shown above.

If the equations above are implemented directly in AND/OR invertor logic only 16 units are required, which gives a saving of one unit compared with the original design.

For large networks a modified Quine–McCluskey tabular approach can be used. In this case the terms are tabulated, as usual, in groups according to the number of 1s contained in each term, but identifying the originating function with each term (Table 3.20). Note that the term $a\bar{b}\bar{c}\bar{d}$ contains one 1 in its binary equivalent 1000, and appears in the output functions $Z_w Z_y Z_z$; thus, this term is placed in group 1 and identified accordingly.

The comparison routine must include the function tags, as well as the eliminated variables (represented by a dash), in its search for changes in one variable only. Referring to Table 3.20 the term $Z_z 0001$ in group 1 is compared with the first term in group 2, $Z_y Z_z 0011$, and is found to contain the common function Z_z and to differ in one variable only; the term $Z_z 00\!-\!1$ is formed and used to start a new group in the next listing. The variable c has been eliminated in the process; the common function Z_z is indicated by a tick in the first list. This procedure is repeated until no more terms combine; the multiple output prime implicant (MOPI) terms are those which contain an unticked function and are identified by a capital letter. It will be seen from Table 3.20 that there are 14 prime implicants. For example, the expression $Z_w Z_y Z_z 1\!-\!-\!0$ means that $a\bar{d}$ is a prime implicant of Z_w, Z_y and Z_z.

The next step in the process is to select the minimal subset. This is done using a PI chart as usual, but the four functions $Z_w Z_x Z_y Z_z$ are plotted side by side (Table 3.21). For each prime implicant row, a cross is placed in those columns that contain a term of the original switching function; 'don't care' conditions are ignored. To choose an optimum set, the table is first

Table 3.20 McCluskey tables for 5421 encoder

Decimal term	Function	Binary term
1	$Z_z^\vee$	0001
2	$Z_y^\vee$	0010
4	$Z_x^\vee$	0100
8	$Z_w^\vee Z_y^\vee Z_z^\vee$	1000
3	$Z_y^\vee Z_z^\vee$	0011
5	$Z_w^\vee$	0101
6	$Z_w^\vee Z_z^\vee$	0110
9	$Z_w^\vee Z_x^\vee$	1001
10	$Z_w^\vee Z_x^\vee Z_y^\vee Z_z^\vee$	1010
12	$Z_w^\vee Z_x^\vee Z_y^\vee Z_z^\vee$	1100
7	$Z_w^\vee Z_y^\vee$	0111
11	$Z_w^\vee Z_x^\vee Z_y^\vee Z_z^\vee$	1011
13	$Z_w^\vee Z_x^\vee Z_y^\vee Z_z^\vee$	1101
14	$Z_w^\vee Z_x^\vee Z_y^\vee Z_z^\vee$	1110
15	$Z_w^\vee Z_x^\vee Z_y^\vee Z_z^\vee$	1111

List 1

Decimal term	Function	Binary term
1,3	Z_z	00-1 A
2,3	$Z_y^\vee$	001-
2,10	$Z_y^\vee$	-010
4,12	Z_x	-100 B
8,9	$Z_y^\vee$	100-
8,10	$Z_w^\vee Z_y^\vee Z_z^\vee$	10-0
8,12	$Z_w^\vee Z_y^\vee Z_z^\vee$	1-00
3,11	$Z_y^\vee Z_z$	-011 C
3,7	$Z_y^\vee$	0-11
5,7	$Z_w^\vee$	01-1
5,13	$Z_w^\vee$	-101
6,7	$Z_w^\vee$	011-
6,14	$Z_x^\vee Z_z$	-110 D
9,11	$Z_w^\vee Z_x^\vee$	10-1
9,13	$Z_w^\vee Z_x^\vee$	1-01
10,11	$Z_w^\vee Z_x^\vee Z_y^\vee Z_z^\vee$	101-
10,14	$Z_w^\vee Z_x^\vee Z_y^\vee Z_z^\vee$	1-10
12,13	$Z_w^\vee Z_x^\vee Z_y^\vee Z_z^\vee$	110-
12,14	$Z_w^\vee Z_x^\vee Z_y^\vee Z_z^\vee$	11-0
7,15	$Z_w^\vee Z_y^\vee$	-111
11,15	$Z_w^\vee Z_x^\vee Z_y^\vee Z_x^\vee$	1-11
13,15	$Z_w^\vee Z_x^\vee Z_y^\vee Z_z^\vee$	11-1
14,15	$Z_w^\vee Z_x^\vee Z_y^\vee Z_z^\vee$	111-

List 2

Decimal term	Function	Binary term
2,3/10,11	Z_y	-01- E
2,10/3,11	Z_y	-01-
8,9/10,11	$Z_w^\vee$	10--
8,9/12,13	$Z_w^\vee$	1-0-
8,10/9,11	$Z_w^\vee$	10--
8,10/12,14	$Z_w^\vee Z_y Z_z$	1--0 F
8,12/9,13	$Z_w^\vee$	1-0-
8,12/10,14	$Z_w^\vee Z_y Z_z$	1--0
3,11/7,15	Z_y	--11 G
3,7/11,15	Z_y	--11
5,7/13,15	Z_w	-1-1 H
5,13/7,15	Z_w	-1-1
6,7/14,15	Z_w	-11- I
6,14/7,15	Z_w	-11-
9,11/13,15	$Z_w^\vee Z_x$	1--1 J
9,13/11,15	$Z_w^\vee Z_x$	1--1
10,11/14,15	$Z_w^\vee Z_x Z_y Z_z$	1-1- K
10,14/11,15	$Z_w^\vee Z_x Z_y Z_z$	1-1-
12,13/14,15	$Z_w^\vee Z_x Z_y Z_z$	11-- L
11,14/13,15	$Z_w^\vee Z_x Z_y Z_z$	11--

List 3

Decimal term	Function	Binary term
8,9,10,11/12,13,14,15	Z_w	1---M
8,9,10,11/12,14,13,15	Z_w	1---
8,9,12,13/10,14,11,15	Z_w	1---
8,9,12,13/10,11,14,15	Z_w	1---
8,10,9,11/12,13,14,15	Z_w	1---
8,10,9,11/12,14,13,15	Z_w	1---
8,10,12,14/9,11,13,15	Z_w	1---
8,10,12,14/9,13,11,15	Z_w	1---
8,12,9,13/10,11,14,15	Z_w	1---
8,12,9,13/10,14,11,15	Z_w	1---
8,12,10,14/9,11,13,15	Z_w	1---
8,12,10,14/9,13,11,15	Z_w	1---

List 4

Table 3.21 Prime implicant chart

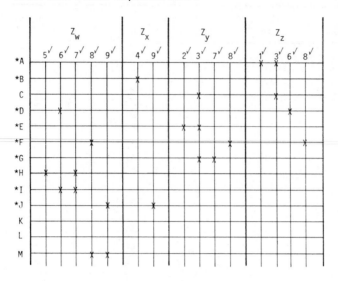

examined for any column with one entry, marked by a cross, which is an essential PI and must be included in the final result for the particular function concerned.

From Table 3.21 it will be seen that prime implicants A, B, D, E, F, G, H and J are all essential and the rows are marked appropriately. The terms covered by these PIs are then ticked off on the chart, when it becomes obvious that the functions are completely specified by the essential prime implicants. The minimal equations for the output functions Z_w, Z_x, Z_y and Z_x are given by

$$
\begin{aligned}
Z_w &= D + F + H + J = bc\bar{d} + a\bar{d} + bd + ad \\
Z_x &= B + J &= b\bar{c}\bar{d} + ad \\
Z_y &= E + F + G &= \bar{b}c + a\bar{d} + cd \\
Z_z &= A + D + F &= \bar{a}\bar{b}d + bc\bar{d} + a\bar{d}
\end{aligned}
$$

which will be seen to be identical with those obtained earlier using K-maps.

The final choice of logic circuit, however, is influenced by many factors; for example, the type of logic system used, the fan-in factor, the possibility of factoring the equations and the availability of logic signals within the system. These points will be considered later in more detail.

3.12　Minimization using topological methods[7,8]

For large switching circuits with more than six variables the K-map is unsuitable and recourse must be made to tabular methods, like Quine–McCluskey, which can be programmed for a digital computer. The Quine–McCluskey algorithm, however, suffers from the major disadvantage that the canonical (minterm) form of the Boolean function must be used as the initial starting point, necessitating an expansion of the function if the circuit has been specified in a reduced form. If we note that a 20-variable function may have up to 10^6 minterms it is clear that the algorithm is impracticable, requiring a very large amount of memory space and extremely long computation times. Using the cubic array notation minimization techniques have been developed which do not require a minterm (0-cube) specification. These algorithms are based on algebraic operations (such as the *sharp* and *star* operators) performed on the ON, OFF and DC arrays of n-tuples (which may or may not be sets of 0-cubes) and as such are ideally suited for computer realization.

The sharp function (denoted by the hash (#) sign) is a cubical method of extracting the prime implicants of a function; it is based on De Morgan's theory and could be considered as a process of subtraction. For example, consider the function

$$Z = \bar{a}b + bc$$

Complementing and expanding we obtain

$$\bar{Z} = \overline{\bar{a}b + bc} = (a + \bar{b})(\bar{b} + \bar{c}) = a\bar{c} + \bar{b}$$

Note that each expression is a prime implicant of $\bar{Z}$ and there are no others; this is shown in Table 3.22(a) using K-maps. In effect we have 'subtracted' the function Z from the unit cube (containing all possible 0-cubes) and obtained the prime implicants of the residual function.

The sharp product $a \# b$, where a and b are two sets, is defined as containing all elements of a not in b. The sharp product between two arrays

Table 3.22 K-mapping of sharp products

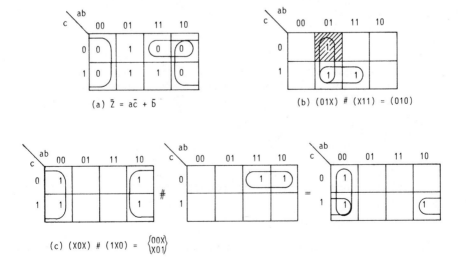

(a) $\bar{Z} = a\bar{c} + \bar{b}$

(b) (01X) # (X11) = (010)

(c) (X0X) # (1X0) = $\left\{ \begin{matrix} 00X \\ X01 \end{matrix} \right\}$

of cubes A and B is the set of cubes such that $P(A \# B)$ is the set of all prime implicants of the function defined by $P(A)\overline{P(B)}$. Thus, $A \# B$ is the set of the largest cubes containing those 0-cubes (minterms) in A but not in B.

To facilitate the manipulation of arrays of cubes it is more convenient to define the sharp product algebraically by means of a Coordinate table, as shown in Table 3.23, together with the following rules:

$$a \# b = a \quad \text{if } a_i \# b_i = \phi \quad \text{for any } i$$
$$a \# b = \psi \quad \text{if } a_i \# b_i = Z \quad \text{for all } i$$
$$a \# b = U_i(a_1, a_2, \cdots, a_i, \cdots, a_n) \quad \text{for all } i \text{ for which}$$
$$a_i \# b_i = a_i \in \{0, 1\}$$

where ψ is the null cube and ϕ the empty set; note that the entire n-dimensional unit cube is denoted by $U_n = XX \cdots X$. This procedure is best illustrated by way of an example. For instance, to compute the sharp product of the cubes $01X$ and $X11$ the cubes are combined according to the coordinate table as follows:

$$
\begin{array}{ccc}
 & 1 & 2 & 3 \\
a & 0 & 1 & X \\
\# \, b & X & 1 & 1 \\
\hline
 & Z & Z & 0 \\
\end{array}
\qquad (01X) \# (X11) = 010
$$

Note that the order is important when looking up the table (a is being 'sharped' by b); $a_1 \# b_1$ gives Z as does $a_2 \# b_2$, which indicates that no cubes can be formed and the only cube obtainable is $a_1a_20 = 010$. This is shown on the K-map in Table 3.22(b) where it is clear that the cube 010 is the only element in a which is not contained in b. Again, considering the sharp product of the cubes $X0X$ and $1X0$, we have

$$
\begin{array}{ccc}
X & 0 & X \\
\# \, 1 & X & 0 \\
\hline
0 & Z & 1 \\
\end{array}
\qquad (X0X) \# (1X0) = \begin{Bmatrix} 00X \\ X01 \end{Bmatrix}
$$

which is given by $0a_2a_3 = 00X$ and $a_1a_21 = X01$ as shown in Table 3.22(c). Note that the rules equally apply to sharping two 0-cubes but, unless they are equal, will always yield an empty set. An interesting case is the unit cube sharped by a 0-cube:

$$
\begin{array}{ccc}
X & X & X \\
\# \, 1 & 1 & 0 \\
\hline
0 & 0 & 1 \\
\end{array}
\qquad (XXX) \# (110) = \begin{Bmatrix} 0XX \\ X0X \\ XX1 \end{Bmatrix}
$$

When sharping a cubic array A by a cube b, that is $A \# b$, the product is given by

$$A \# b = [(a_1 \# b) \, U \, (a_2 \# b) \, U \cdots]$$

where the operator U is called the *cube-union* and is the *cover* of the union

Table 3.23 Coordinate sharp product

$a_i \# b_i$	b_i		
	0	1	X
a_i 0	Z	ϕ	Z
1	ϕ	Z	Z
X	1	0	Z

of the individual cubes. A cube b is said to cover another cube a if it has a 0 (or a 1) in all coordinate positions in which a has a 0 (or a 1); those positions of b corresponding to positions in which $a_1 = X$ may be filled with either value and vice versa. In this way the absorption law of Boolean algebra $(A + AB = A(1 + B) = A)$ is employed to remove redundant cubes of smaller dimensions.

The sharp product of $a \# B$ is given by

$$a \# B = [(a \# b_1) \# b_2] \# b_3 \cdots$$

When the sharp product of two arrays is required either procedure may be used.

We saw in Chapter 2 how Boolean functions may be specified in terms of ON, OFF and DC arrays of input n-tuples for which the function $Z = 1, 0$ and X (don't-care) respectively. Only two of these three arrays are needed to completely specify a function. Let U_n be the n-tuple of Xs (the unit cube); then

$$U_n = \text{ON } U \text{ OFF } U \text{ DC}$$

and

$$\text{ON} = U_n \# (\text{OFF } U \text{ DC})$$
$$\text{OFF} = U_n \# (\text{ON } U \text{ DC})$$
$$\text{DC} = U_n \# (\text{ON } U \text{ OFF})$$

For example, consider the function $Z = \bar{a}b + bc$; the ON array is given by

$$\text{ON} = \begin{Bmatrix} 01X \\ X11 \end{Bmatrix}$$

OFF $= U_n \#$ ON (since there are no don't-care terms)

$$= XXX \# \begin{Bmatrix} 01X \\ X11 \end{Bmatrix}$$

$$= (XXX \# 01X) \# X11 = \begin{Bmatrix} 1XX \\ X0X \end{Bmatrix} \# X11$$

$$= (1XX \# X11) \, U \, (X0X \# X11)$$

$$= \begin{Bmatrix} 10X \\ 1X0 \end{Bmatrix} U \, X0X = \begin{Bmatrix} X0X \\ 1X0 \end{Bmatrix} \text{ (note } X0X \text{ covers } 10X)$$

which is the OFF array established earlier and equivalent to $\bar{Z} = a\bar{c} + \bar{b}$.

It can be shown[7] that the set of all prime implicants of a function Z are given by

$$PI(Z) = U_n \# (U_n \# \text{ON})$$

or when don't-cares are present

$$PI(Z) = U_n \# (U_n \# (\text{ON} \cup \text{DC}))$$

Note that canonical terms are not required. In practice the function would be described in terms of its OFF and DC arrays and computed using the expression

$$PI(Z) = U_n \# \text{OFF}$$

The sharp product generates the complete set of prime implicants but does not in general give a minimum cover which must be found using other methods. Consider the function Z whose OFF matrix is given by

$$\text{OFF} = \begin{Bmatrix} XX00 \\ X00X \\ 111X \end{Bmatrix}$$

then

$$PI(Z) = XXXX \# \begin{Bmatrix} XX00 \\ X00X \\ 111X \end{Bmatrix}$$

$$= ((XXXX \# XX00) \# X00X) \# 111X$$

$$= \left[\begin{Bmatrix} XX1X \\ XXX1 \end{Bmatrix} \# X00X \right] \# 111X$$

$$= ((XX1X \# X00X) \cup (XXX1 \# X00X)) \# 111X$$

$$= \begin{Bmatrix} XX1X \\ X1X1 \end{Bmatrix} \# 111X$$

$$= (XX1X \# 111X) \cup (X1X1 \# 111X) \cup (XX11 \# 111X)$$

$$= \begin{Bmatrix} 0X1X \\ X01X \end{Bmatrix} \cup \begin{Bmatrix} 01X1 \\ X101 \end{Bmatrix} \cup \begin{Bmatrix} 0X11 \\ X011 \end{Bmatrix}$$

Thus

$$PI(Z) = \begin{Bmatrix} 01X1 \\ X101 \\ 0X1X \\ X01X \end{Bmatrix} \qquad \begin{array}{l} \text{Note: } 0X1X \text{ covers } 0X11 \\ \text{and} \quad X01X \text{ covers } X011 \end{array}$$

A minimal cover may be found using prime implicant charts in the usual way but it would seem pointless to expand the ON array into canonical form when the entire point of using the cubical method is to extra $PI(Z)$ using non-canonical terms. There is, however, an alternative algebraic method which operates directly on the PI array.

Let $PI = \{P_1, P_2, P_3, \cdots, P_n\}$ be the set of cubes of PIs and also let $(PI - p_i)$ denote the PI array less cube p_i. Now if we define $A = p_i \# (PI - p_i)$ for $DC = 0$, where cube p_i covers some members of the complete set of 0-cubes for the function Z and $(PI - p_i)$ the rest. Now if $A = \phi$ then all the

0-cubes covered by p_i are also covered by other prime implicants and p_i is not an essential prime implicant; if $A \neq \phi$ then p_i is an essential prime implicant.

If the original function contained don't-cares then the 0-cubes covered by p_i may be don't-care terms; this possibility may be eliminated by computing the expression

$$A = (p_i \mathbin{\#} (\text{PI} - p_i)) \mathbin{\#} \text{DC}$$

As an example consider the PI array obtained above:

$$\text{PI} = \begin{Bmatrix} 01X1 \\ X101 \\ 0X1X \\ X01X \end{Bmatrix}$$

and

$$01X1 \mathbin{\#} \begin{Bmatrix} X101 \\ 0X1X \\ X01X \end{Bmatrix} = (((01X1 \mathbin{\#} X101) \mathbin{\#} 0X1X) \mathbin{\#} X01X)$$

$$= \phi$$

Again

$$X101 \mathbin{\#} \begin{Bmatrix} 01X1 \\ 0X1X \\ X01X \end{Bmatrix} = (((X101 \mathbin{\#} 01X1) \mathbin{\#} 0X1X) \mathbin{\#} X01X)$$

$$= 1101$$

$$0X1X \mathbin{\#} \begin{Bmatrix} 01X1 \\ X101 \\ X01X \end{Bmatrix} = (((0X1X \mathbin{\#} 01X1) \mathbin{\#} X101) \mathbin{\#} X01X)$$

$$= ((001X \mathbin{\#} X101) \, U \, (0X10 \mathbin{\#} X101)) \mathbin{\#} X01X$$
$$= (001X \mathbin{\#} X01X) \, U \, (0X10 \mathbin{\#} X01X)$$
$$= 0110$$

Finally

$$X01X \mathbin{\#} \begin{Bmatrix} 01X1 \\ X101 \\ 0X1X \end{Bmatrix} = (((X01X \mathbin{\#} 01X1) \mathbin{\#} X101) \mathbin{\#} 0X1X)$$

$$= 101X$$

Thus the essential prime implicants are $X101$, $0X1X$ and $X01X$; note that the non-empty result of a cubical expansion, for example 1101, is the minterm(s) which is covered solely by the essential prime implicant. It is necessary but not sufficient simply to extract essential prime implicants since, as we have seen, a minimal (or near-minimum) subset of prime implicants is required to cover the switching function.

An irredundant cover may be generated using the following algorithm:

(1) Compute PI array
(2) For each p_i in turn

(3) Replace PI by $(PI - p_i)$
compute $A = p_i \mathbin{\#} (PI - p_i)$
(4) If $A = \phi$ then discard p_i else add p_i back to PI
(5) Increment i; repeat 3.

As an example consider the PI array

$$PI = \begin{Bmatrix} 00X1 \\ X011 \\ 0X0X \\ X10X \\ 1X1X \\ 11XX \end{Bmatrix}$$

Pass 1

$$00X1 \mathbin{\#} \begin{Bmatrix} X011 \\ 0X0X \\ X10X \\ 1X1X \\ 11XX \end{Bmatrix}$$
$$= (((((00X1 \mathbin{\#} X011) \mathbin{\#} 0X0X) \mathbin{\#} X10X) \mathbin{\#} 1X1X) \mathbin{\#} 11XX$$
$$= \phi$$

Pass 2

$$X011 \mathbin{\#} \begin{Bmatrix} 0X0X \\ X10X \\ 1X1X \\ 11XX \end{Bmatrix}$$
$$= (((X011 \mathbin{\#} 0X0X) \mathbin{\#} X10X) \mathbin{\#} 1X1X) \mathbin{\#} 11XX$$
$$= 0011$$

Pass 3

$$0X0X \mathbin{\#} \begin{Bmatrix} X011 \\ X10X \\ 1X1X \\ 11XX \end{Bmatrix}$$
$$= (((0X0X \mathbin{\#} X011) \mathbin{\#} X10X) \mathbin{\#} 1X1X) \mathbin{\#} 11XX$$
$$= 00XX$$

Pass 4

$$X10X \mathbin{\#} \begin{Bmatrix} X011 \\ 0X0X \\ 1X1X \\ 11XX \end{Bmatrix}$$
$$= (((X10X \mathbin{\#} X011) \mathbin{\#} 0X0X) \mathbin{\#} 1X1X) \mathbin{\#} 11XX$$
$$= \phi$$

Pass 5

$$1X1X \text{ \# } \begin{Bmatrix} X011 \\ 0X0X \\ 11XX \end{Bmatrix}$$

$= ((1X1X \text{ \# } X011) \text{ \# } 0X0X) \text{ \# } 11XX$
$= ((111X \text{ \# } 0X0X) \text{ U } (1X10 \text{ \# } 0X0X)) \text{ \# } 11XX$
$= (111X \text{ \# } 11XX) \text{ U } (1X10 \text{ \# } 11XX)$
$= 1010$

Pass 6

$$11XX \text{ \# } \begin{Bmatrix} X011 \\ 0X0X \\ 1X1X \end{Bmatrix}$$

$= ((11XX \text{ \# } X011) \text{ \# } 0X0X) \text{ \# } 1X1X$
$= 110X$

Thus the irredundant cover is given by

$$\text{cov} = \begin{Bmatrix} X011 \\ 0X0X \\ 1X1X \\ 11XX \end{Bmatrix}$$

A similar procedure for finding prime implicants from a set of reduced terms is based on a method first described by Quine[10] called *consensus*. The consensus of two product terms A and B is the largest product P such that P does not imply either A or B, but P implies $A + B$. Thus A is the consensus of AB and $A\bar{B}$ and BC is the consensus of AB and $\bar{A}C$. Roth[8] has proposed an equivalent cubical operation called the *star* (*) product which is defined algebraically in a similar way to the sharp function. The cubical concept of consensus has been generalized by Tilson[11] who has also evolved efficient algorithms for generating prime implicants.

Both the sharp and consensus algorithms can be adopted for use with multiple output circuits using some form of *tagging* (in principle similar to that used in Quine–McCluskey) to identify the outputs associated with each cube.

Most of these methods have been programmed for a digital computer but the most successful application of the cubic methods is embodied in the MINI system[12] where the authors claim they are able to minimize multiple output functions with 20–30 effective inputs.

3.13 Criteria for circuit minimization

In minimizing a logic function due consideration must be given to the criteria used for optimization, particularly with regard to cost and performance.

The performance of a combinational network is primarily dependant on the propagation delay through the total circuit and hence is directly related to the number of logic levels the signals have to pass through. It follows that two-level AND/OR networks (or their NAND/NAND etc. equivalent) will always give the minimum delay and hence the fastest circuit. Note, however, that in most cases invertors must also be used to generate the complemented variables which in practice necessitates a third logic level (unless *dual-polarity* systems are employed when both complemented and uncomplemented signals are distributed). Though in general this rule will still hold true for LSI realizations there are other considerations such as the physical structure of the circuit and the signal path connections.

The cost factors for a logic circuit can vary considerably depending very much on the particular form the realization will take. For example, the following parameters can all be used to determine a minimal cost circuit:

(a) The *gate cost* depends on deriving the expression with the minimum number of product terms.
(b) The *input cost* is determined by the number of literals or the actual number of inputs all assumed to be in uncomplemented form.
(c) The *chip cost* criterion according to which a minimal expression is one that utilizes the least number of standard IC modules.
(d) The *interconnection cost* is determined by the number of interconnections in a circuit or between standard IC modules.
(e) The *silicon area* cost (in the case of integrated circuits) which is dependent on the total area as opposed to the number of devices enclosed within that area.

The most common criterion used for the minimization of discrete gate networks is a combination of gate and input cost, choosing the minimal sum-of-products expression with the minimum number of literals. In general, and providing we are satisfied with a two-level realization, the determination of prime implicants constitutes the first step in any minimization procedure. As we have seen the next stage is to determine a minimum irredundant set of prime implicants which cover the original function or functions in the case of multiple output circuits; it is here that the cost criterion is employed.

For instance, in solving single output prime implicant tables using the minimum literal criteria each PI would be scored by a weight depending on the number of literals (and hence inputs); the gate cost is obtained simply by summing the PIs. After extracting the essential prime implicants selection would proceed on the basis of choosing PIs with the minimal number of literals to cover the function. It is possible in some cases that the minimum literal solution does not yield the minimum number of gates: in practice the minimal sum of products is chosen with the least number of literals.

Another important consideration is the number of inverted inputs to the first level gates (which will require extra invertor gates) and some routines incorporate this in the cost factor. In practice it involves putting an additional (lower) cost on PIs depending on the number of uninverted literals. Also, in practice, account should be taken of the fan-in factors of the gates available for the actual implementation—extra gates will be required if the fan-in is exceeded.

The multiple-output case is more complicated in that it is necessary to consider the total circuit, that is on *both levels*, taking account of the shared terms.

On method of costing an overall realization is as follows: Consider the multiple output functions shown in Table 3.21:

$$Z_w = H + D + F + J$$
$$Z_x = B + J$$
$$Z_y = E + G + F$$
$$Z_z = A + D + F$$

Then representing the total circuit by the expression

$$(HDFJ)(BJ)(EGF)(ADF)$$

and multiplying out the PI terms algebraically we have

$$H_2^1 \ D_3^2 \ F_2^3 \ J_2^2 \ B_3^1 \ G_2^1 \ E_2^1 \ A_3^1 \quad \text{literals}$$

where each letter represents a PI (gate) at the first level and the power the number of second level gates fed by that PI. The total cost of the circuit is the sum of the first level inputs (the sum of the literals—19) and second level inputs (the sum of the powers—12). Thus the circuit requires 31 inputs and 12 gates.

The classical method of using cost factors when minimizing multiple-output circuits is illustrated in Table 3.24. In this chart a separate input cost C_D and gate cost C_G is associated with each prime implicant. Note that the input cost is composed of the number of literals in the PI term (the inputs at the first level) and an additional input to account for the second level gate. The gate cost for each PI is 1, which will reduce to 0 if selected.

The initial step in generating a minimal cover is to select essential prime implicant terms. Note that a PI may be essential for one function but not another; the actual functions for which particular PIs are essential are listed in the table. Having determined the essential PIs the terms covered are ticked off in the usual way but only for the functions concerned. For instance, the terms covered by the essential prime implicant E are only included in $Z2$ and not $Z1$.

The procedure leads to the reduced chart shown in Table 3.24(b). In this table all the PIs are again included, except G which only covers terms in $Z2$, together with the outstanding (uncovered) terms. Note also that the costs are now re-evaluated as a result of selecting certain PIs. For example, the gate costs for A, B, D and E are reduced to 0 and the input cost to 1, that is the input required for the second level only.

It is possible at this stage to reduce the table further by using the idea of *dominance*. Thus H can be eliminated since it is dominated by D (that is it contains all the terms covered by H) and moreover costs more to implement. Similarly C can be eliminated since it is dominated by F and has a higher cost. Although I dominates both A and E it nevertheless has a

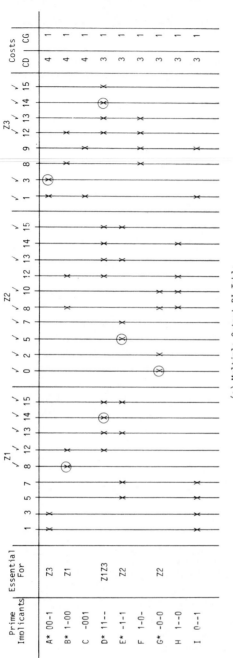

(a) Multiple Output PI Table

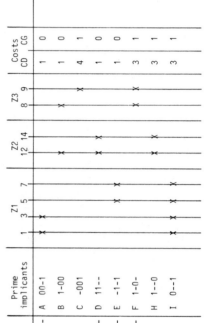

(b) Reduced Table

Table 3.24 Costing MOPI tables

higher cost factor than the two together and hence A and E are to be preferred.

Thus the final equations are given by

$$Z1 = B + D + A + E$$
$$Z2 = E + G + D$$
$$Z3 = A + D + F$$

As we shall see in later chapters this method of minimization using optimization criteria based on gate and input costs is inappropriate when dealing with LSI circuits and modules and must be modified to suit parti-cular realizations. However, the concept of attributing a weighting factor to each term which is related in some way to cost and then selecting such as to minimize this weight is a general one and has many applications.

References and bibliography

1. Veitch, E. W. A chart method for simplifying truth functions. *Proc. Ass. comput. Mach.* May, 127–33 (1952).
2. Karnaugh, M. The map method for synthesis of combinational logic circuits. *Trans. Am. Inst. elect. Engrs Comm. Electron.* **72**, 593–9 (1953).
3. McCluskey, E. Minimization of Boolean functions. *Bell Syst. tech. J.* **35**, 1417–44 (1956).
4. Quine, W. V. The problem of simplifying truth functions. *Am. math. Mon.* **59**, 521–31 (1952).
5. Petrick, S. R. *A direct determination of the irredundant forms of a Boolean function from the set of prime implicants.* Air Force Cambridge Research Center Report, Bedford, Mass. (1956). AFCRC-TR-56-110.
6. Pyne, I. B. and McCluskey, E. The reduction of redundancy in solving prime implicant tables. *IRE Trans. electron. comput.* **EC11**, 473–82 (1962).
7. Dietmeyer, D. L. *Logical Design of Digital Systems.* Allyn & Bacon, 1971, Chapter 3.
8. Roth, J. P. Algebraic topological method for the synthesis of switching systems. I. *Trans. Am. Math. Soc.* July, 301–6 (1958).
9. Roth, P. *Computer Logic, Testing and Verification.* Computer Science Press, Maryland, 1980, Chapter 1.
10. Quine, W. V. A way to simplify truth functions. *Am. Math. Mon.* Nov. 627–31 (1955).
11. Tilson, P. Generalization of concensus theory and application to the minimization of Boolean functions. *IEEE Trans. Computers* **EC16**, 446–56 (1967).
12. Hong, S. J., Cain, R. G. and Ostapko, D. L. MINI: A heuristic approach for logic minimization. *IBM J. Res. Dev.* Sept. 443–58 (1974).

Tutorial problems

3.1 Design a combinational switching circuit that will perform the functions of both binary addition and subtraction. The circuit has three inputs, x and y (the digits to be added or subtracted), and a carry (or borrow) b/c; the outputs required are the sum (or difference) s/d, and the next carry (or borrow) b_+/c_+.

A control waveform M determines the mode of operation, i.e. when $M = 1$ the circuit adds, when $M = 0$ it subtracts.

3.2 Convert the relay circuit shown in Fig. 3.7, simplifying if necessary, into a NAND circuit.

3.3 Minimize the following switching functions using K-maps:
   (a) $T(ABCD) = \Sigma(0, 1, 5, 7, 8, 9, 12, 14, 15)$
       'Don't cares' $D = (3, 11, 13)$
   (b) $T(ABCDE) = (1, 2, 4, 5, 9, 10, 11, 15, 16, 18, 19, 21, 22, 25, 27, 31)$
       'Don't cares' $D = (0, 8, 13, 17, 24, 26, 29)$
   (c) $T(ABCDE) = \Pi(5, 6, 8, 11, 12, 19, 21, 22, 23, 24, 25, 28)$
       'Don't cares' $D = (2, 10, 13, 14, 20, 26, 29)$

3.4 Minimize the following switching function using the McCluskey tabular technique, and algebraic extraction of the prime implicant set.

$$T(ABCDEF) = \Sigma(4, 12, 13, 15, 21, 23, 24, 29, 31, 36, 37, 44, 45, 51,$$
$$52, 53, 56, 58, 59, 60, 63)$$
$$\text{Don't-cares } D = (2, 5, 14, 20, 28, 34, 49, 61)$$

3.5 In a digital servo system an error comparator is required which will compare two three-bit binary numbers, A and B, and give separate outputs for the conditions $A = B$, $A > B$ and $A < B$. Design a combinational logic circuit that will perform this function, and implement the design in terms of NAND logic.

3.6 Minimize the following switching function represented in the cubical notation using the sharp algorithm:

$$\text{ON} = \begin{Bmatrix} 0XX1 \\ 10X0 \\ 1100 \end{Bmatrix} \qquad \text{DC} = \begin{Bmatrix} 111X \\ 1101 \\ 0000 \end{Bmatrix}$$

Comment on the efficacy of the method.

3.7 Design a two-input mod-3 adder network, that is a circuit that will accept digits 0, 1 and 2, represented in binary, and produce the sum and carry. Implement the circuit in either NOR or NAND gates and then show how the element could be used to form a full parallel adder for n-digit mod-3 numbers.

3.8 Design a minimal logic gate circuit to translate binary-coded decimal numbers in the 8421 code into excess-three code.

3.9 Derive a logic gate circuit to convert pure five-bit binary numbers to Gray code, i.e. reflected binary notation.

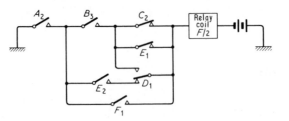

Figure 3.7

3.10 Repeat question 3.9 but this time convert from Gray code to pure binary and implement the circuit in NAND logic.

3.11 A typical seven-segment display as used in digital equipment is shown in Fig. 3.8. Numbers are displayed by selecting and energizing the appropriate sectors *a* to *f*. Design a logic circuit that will accept the numbers 0–9 in pure binary and output the appropriate signals, *a* to *f*, to energize the display unit. Attempt to produce a minimal circuit using NAND gates.

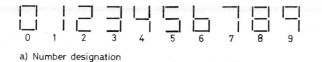

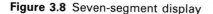

a) Number designation

    b) Segment identification

Figure 3.8 Seven-segment display

4 Implementation of combinational networks

4.1 Introduction

Most of the design algorithms we have encountered so far have been essentially technology independent in that the sum-of-products or product-of-sums forms represent AND/OR or OR/AND two-level networks. In practice there are a large number of ways of realizing logic functions using available technology. These range from NAND/NOR gates in the form of SSI units such as the SN74 series through medium and large scale integrated circuit modules, such as read-only memory (ROM) and programmable logic arrays (PLA) to semi-custom and custom LSI circuits. Moreover, by designing for a particular technology it is often possible to produce a more viable and economic circuit. In this chapter we shall consider the various methods of implementation that are currently available and, where one exists, the attendant design philosophy. We shall also confine ourselves to realization using SSI and MSI/LSI modules, the problems of designing custom LSI/VLSI circuits being left to a later chapter.

4.2 NOR/NAND implementation

The design problem here is one of translating Boolean equations into a suitable form for realization as NOR/NAND logic, called the *Pierce* or *Dagger* function and *Sheffer Stroke* function respectively in the literature. There are basically two methods used: these are the *factoring* and *mixed mode logic* approaches.

4.2.1 Algebraic and K-map factoring techniques[1-3]

The NAND/OR equivalents of the AND and OR logic functions are shown in Fig. 4.1(a) and (b); the single input NOR/NAND unit acts as a simple invertor. Direct implementation of the function $T = AB + C$, assuming positive logic convention throughout, in NAND/NOR produces the circuits shown in Fig. 4.1(c). Note that, for the NAND element, odd levels function as OR gates with the variables complemented, and even levels as AND gates; level 1 is assumed to be the output gate. The same circuit configuration (Fig. 4.1(d)), using NOR gates and complementing the inputs to odd levels, gives the function $T = (A + B)C$. Thus, for the NOR unit, odd levels act as

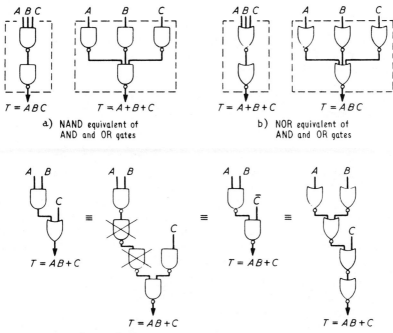

a) NAND equivalent of AND and OR gates

b) NOR equivalent of AND and OR gates

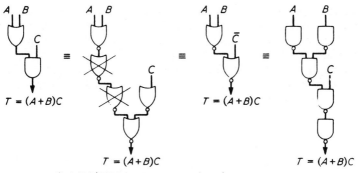

c) NAND/NOR equivalent of $T = AB + C$

d) NAND/NOR equivalent of $T = (A+B)C$

Figure 4.1 NOR/NAND implementation

AND gates, and even levels as OR gates. For both types of circuit, when counting the levels, which is a necessary operation in any transform process, single input invertors are not counted as levels. Thus, in order to implement NOR/NAND circuits, the sum-of-products equations (AND/OR form, with OR as the final output gate) must be used for NAND logic, and product-of-sums (OR/AND form) for NOR logic. For two-level circuits, the direct transform is very simple and, assuming that both the variables and their complements are available ('double rail' working) is the most economic form. Furthermore, the minimality of the original equations will always be retained in two-level NOR/NAND equivalent circuits, but this does not always follow for factored forms of the equation.

Now consider the exclusive OR function:

$$T = A\bar{B} + \bar{A}B$$

or in the product-of-sums form

$$T = (\bar{A} + \bar{B})(A + B)$$

Both these functions may be implemented directly in two-level circuits (see Fig. 4.2) and five NAND/NOR elements will be required if variable complements are not available. Is it possible, though, by suitable factoring of the circuit to produce a more economical result, which does not rely on complements always being available? In general it is possible if the equations can be manipulated so that the complemented variables occur on odd levels, and the uncomplemented on even levels. Collecting all these facts together we may state the following transform rules:

(1) NAND implementation: factor the equations to an AND/OR−AND/OR form (always with the OR as the final output), with the complemented variables on odd levels and the uncomplemented variables on even levels. More than two levels will be required in general.

(2) NOR implementation: as above, except that final output must be an AND, and the equations should have the form OR/AND−OR/AND.

Once the equations have been factored in this way, the circuit may be drawn in NAND/NOR logic, but the gates should be laid out as if they were in AND/OR logic following the factored equations, and the variables entering the logic at odd levels should be complemented. This transformation process, however, is not easy, since it involves considerable algebraic manipulation and most of the laws and tricks of Boolean algebra are called for.

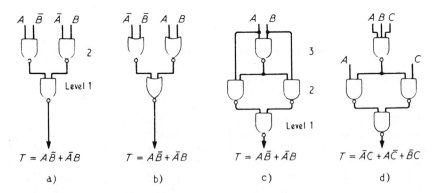

$$T = A\bar{B} + \bar{A}B$$ $$T = A\bar{B} + \bar{A}B$$ $$T = A\bar{B} + \bar{A}B$$ $$T = \bar{A}C + A\bar{C} + \bar{B}C$$

a)   b)   c)   d)

Figure 4.2 NOR/NAND implementation

Nevertheless, for simpler equations the results can be very effective. Consider our exclusive OR circuit:

$$T = (\bar{A} + \bar{B})(A + B)$$

Using partial multiplication of the terms, we have

$$T = A(\bar{A} + \bar{B}) + B(\bar{A} + \bar{B})$$

This is a three-level circuit with the complemented terms occurring on the third and odd level. Direct implementation gives the four-element circuit shown in Fig. 4.2(c). Again, consider the function

$$T = \bar{A}C + A\bar{C} + \bar{B}C$$

This may be factorized as

$$T = (A + C)(\bar{A} + \bar{B} + \bar{C})$$

Note that the complemented and uncomplemented terms are kept together. Expanding the expression gives

$$T = A(\bar{A} + \bar{B} + \bar{C}) + C(\bar{A} + \bar{B} + \bar{C})$$

the circuit for which is shown in Fig. 4.2(d). Note that the equation for T is redundant in the sense that all the prime implicant terms are represented. In many cases this redundancy is essential to achieve the factored form of the equation. The factorization can be performed either algebraically or using a mapping technique due to Caldwell. In the latter case the function is plotted on two K-maps placed side by side (Table 4.1); 1s are then inserted in both maps with the objective of producing maximal groupings, but each time a 1 is inserted, a 0 must be placed in the corresponding position on the other map. The product of the terms on each map is the required factorized function.

An alternative method, for a small number of variables, is to derive the factored form directly from a K-map. This is preferable to the algebraic method since the technique is easier to apply and, as before, can be used to implement both NAND and NOR elements. Consider the function above, $T = \bar{A}C + A\bar{C} + \bar{B}C$; this is shown plotted on a K-map in Table 4.2. We can form two loops—A but not ABC, and C but not ABC—from the map; the loop ABC is known as the inhibiting loop. Hence we can represent the function as

Table 4.1 Factorization using K-maps

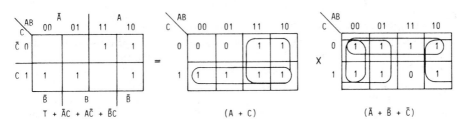

$T + \bar{A}C + A\bar{C} + \bar{B}C$ = $(A + C)$ $(\bar{A} + \bar{B} + \bar{C})$

Table 4.2 NAND implementation of $T = \bar{A}C + A\bar{C} + \bar{B}C$

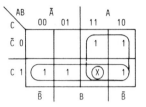

Inhibiting loops are marked with a $\times$

$$T = A(\overline{ABC}) + C(\overline{ABC})$$
$$= A(\bar{A} + \bar{B} + \bar{C}) + C(\bar{A} + \bar{B} + \bar{C})$$

which is identical to the NAND function we arrived at earlier.

Thus we can interpret the main loops (those containing 1s) on the K-map as representing the inputs to NAND elements at the even circuit levels, and the inhibiting loops as inputs to NAND elements at the odd levels. The choice of loops, however, is restricted to those representing combinations of *uncomplemented* variables, i.e. for three binary variables we have A, B, C, AB, AC, BC and ABC. As a loop of 1s can be inhibited by a loop of 0s, so the reverse applies—a loop of 0s can be inhibited by a loop of 1s. In practice we can start with loops of either 1s or 0s, but the process must continue taking alternative loops in order to account for the sequence of odd and even levels. In this way multi-level circuits may be designed which contain many levels of factored gating. Quite often in complex circuits, the first loops to be formed are those whose outputs will be needed later, as inhibitors, in order to produce the required factored circuit.

In the example above we started with a 1s loop and inhibited with a 0s loop, which gave all the 1 entries. We must now account for the 0 entries, since all entries, both 1 and 0, must be included in the final circuit. Thus if we take the unity loop (the loop comprising the whole map) and inhibit this with the outputs representing the 1 entries (which were obtained earlier) we shall get the final result. This simply means putting all the 1s outputs as inputs to a single gate, representing in this case the final OR gate.

To illustrate these ideas further let us consider a more complicated function:

$$Z = \bar{A}\bar{B}C + ABC + A\bar{B}\bar{C}$$

To implement this function in straightforward two-level logic would require, with invertors, seven NAND elements. Table 4.3 shows the K-maps for the factoring process. The first step is to form loops which may be useful later in obtaining the final output function. In this case we take the loops BC and AC and inhibit with loop AB; this gives the outputs ② and ③ (see Table 4.3(b)) representing the 0s $A\bar{B}C$ and $\bar{A}BC$ respectively. We next take the loop C and inhibit this with the outputs ② and ③ (Table

Table 4.3 NAND implementation of $T = \bar{A}\bar{B}C + ABC + A\bar{B}\bar{C}$

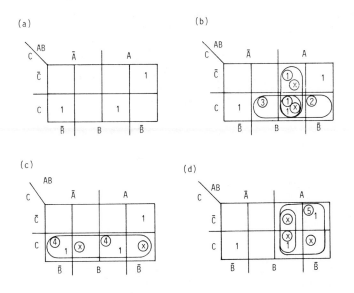

4.3(c)) which yields the 1s, output ④, $\bar{A}\bar{B}C$ and ABC; we now have to find the remaining 1, $A\bar{B}\bar{C}$. We do this by taking loop A and inhibiting with AB and output ②, to give $A\bar{B}\bar{C}$. We now have all the 1s represented by outputs ④ and ⑤, and to obtain the final output we have only to account for the 0s in the final gate. The complete circuit is shown in Fig. 4.3; it uses a total of six NAND elements, a saving of one unit.

Both the algebraic and map techniques need considerable practise before one can manipulate the switching equations into the appropriate form with anything like the required skill and ease. Even then, the methods are only practicable for small-variable problems. In point of fact there are several disadvantages to using this factoring technique.

(1) The method of factoring means, in most cases, that different length signal paths exist for the same variable because of increased redundancy. Figure 4.3 is an example of this, where the logic signal A comes in at three different levels. As we shall see later this can give rise to dynamic circuit hazards in both combinational and sequential circuits.

(2) The number of interconnecting paths is increased, and in high speed systems this can cause cross-talk problems, as well as an increase in the cost of system wiring.

(3) In a number of cases the saving in logic units is not great, and in any case the cost of logic units is rapidly becoming of secondary importance in integrated circuit systems.

(4) In many practical systems the complement of the variables will be readily available, as alternative outputs from bistable registers and

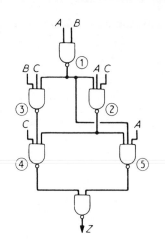

Figure 4.3 Factored circuits

counters etc., making the normal two-level circuit a much better proposition.

(5) The logical function of the circuit is often masked by the factoring, and this can be a handicap to the service technician. In fact it has often been suggested that logic design should be kept simple to allow for easy and rapid fault-finding. Furthermore, the logic diagram and connection diagrams are quite different, bearing little relationship to one another. This again can lead to production and commissioning difficulties.

Thus it would seem wiser to keep to two-level logic circuits where possible, depending on the availability of complemented variables and large enough fan-in factors. Even so there are some difficulties in implementation, and logic circuit drawing and comprehension.

4.2.2   Mixed mode logic[4,5]

As we have seen, the physical transformations necessary to realise AND/OR/NOT logic equations in terms of NAND/NOR can be awkward and, moreover, often lead to a circuit diagram which bears little relationship to the conceptual logic diagram. Again the freedom to mix together NOR/NAND in the same circuit can lead to a more compact design but one that is difficult to interpret and prone to errors. The mixed mode logic technique overcomes many of these difficulties.

Using this technique the inversion property of the NOR/NAND element is dissociated from the logical function. In other words, we treat the units as performing the normal OR/AND functions, but we assume a change of signal polarity (i.e. a change in logic convention) rather than the logical NOT function. Using this approach, the basic logical design is executed in terms of two-level (or otherwise) logic using AND/OR/NOT elements. Then the final logic *and* wiring diagram is drawn using NAND/NOR elements bearing in mind that

(i) the NAND element acts as an AND gate for positive logic and as an OR gate for negative logic, with a change of logic convention at the output in both cases;

(ii) the NOR element acts as an AND gate for negative logic and as an OR gate for positive logic, again with a change of logic convention at the output.

It will be apparent that we will require a notation to indicate the logic convention, either positive or negative, of the logic signals. Mixed mode logic distinguishes negative logic by a small circle on the corresponding terminal of the logic symbol. Note that the circles *do not change the logic function*. It is also convenient to indicate the convention with the symbol itself by appending, say, an 'L' or 'H' character to the variable name. These conventions are shown in Fig. 4.4. In particular note that a change of logic convention is effected (see Fig. 4.4(d)) using a phase reversing amplifier (an invertor circuit in practice) but that the logic variable is unaffected (there is no *logical* inversion). Inversion of logic variables is obtained simply by redefining the logic convention of the variable, indicated by a slash on the signal line as shown in Fig. 4.4(e). Note, however, that there must also be a change of logic convention.

The technique is illustrated in Fig. 4.5, which shows the implementation of the function $Z = \bar{A}\bar{B}C + ABC + A\bar{B}\bar{C}$ in terms of two-input NOR/NAND gates with positive logic inputs. Note that the use of this method allows logic diagrams to be used directly as wiring diagrams with complete understanding of the signal polarities at any point in the circuit. Furthermore, theoretical circuits may be directly converted to NAND/OR circuits, and the problem of complemented inputs is automatically accounted for in the process.

Before we conclude the discussion on NAND/NOR logic systems, it is worth considering the implementation of one further class of circuit. In general, the NAND/NOR unit is at its worst when performing AND−AND or OR−OR operations, because of the need for inversion of the input signals to the second stages. Surprisingly, however, when implementing an electronic tree-circuit, which requires AND−AND operations, the equivalent NAND circuit is fairly simple. Consider the circuit shown in Fig. 4.6(a); this is a NAND tree for generating all possible combinations of two variables. It is very similar to the exclusive OR circuit which gives three of the four combinations $(\overline{AB}, A\bar{B}$ and $\bar{A}B)$; the fourth is obtained by taking these three to an output gate which gives $\overline{AB}$. Since the circuit has a systematic structure (easily apparent on a close examination) it may be extended to any number of variables. Figure 4.6(b) shows the three-variable tree, and an algorithm for generating *n*-variable tree-circuits is described by Maley and Earle.[3]

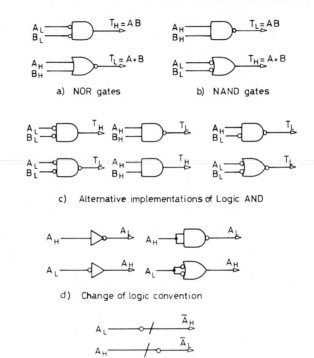

a) NOR gates   b) NAND gates

c)   Alternative implementations of Logic AND

d)   Change of logic convention

e)   Performing the NOT function

Figure 4.4 Mixed mode logic notation

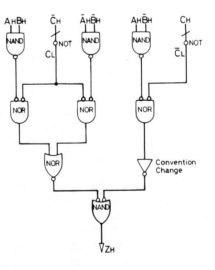

Figure 4.5 Mixed mode logic using NAND/NOR

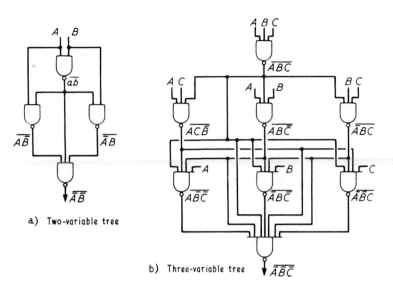

a) Two-variable tree

b) Three-variable tree

Figure 4.6 NAND tree circuits

4.3 Implementation using MSI modules

All the design techniques described so far have tacitly assumed that the final logic network will be realized using basic NAND/NOR gates. This has resulted in an emphasis on algorithms for economical realization in terms of the number of gates and inputs required to implement the circuit. Though this approach is still fundamentally sound and, as we shall see later, with modification is appropriate for designing IC logic, the availability of complex MSI logic elements has opened up a whole new range of possibilities.

MSI modules providing compound logic circuits such as counters, shift registers, decoders and encoders, multiplexers, ROMs, PLAs etc. are now freely available as small packaged units which can be connected directly to a PC board. These devices enable many logic circuits, such as the encoding for an LED display unit, to be implemented directly using a single MSI module rather than built up from individual gate packages.

In many cases it is better design practice to utilize standard MSI units, even if this introduces redundant, or unused gates, than develop optimized logic for implementation at the gate level. Thus the criteria for economical realization have been raised to a higher, modular level, and must now take into account the number of MSI packages and the cost of printed circuit boards and back wiring.

Consequently, the design of logic circuits has been elevated to a systems activity where complex MSI/LSI modules rather than gates are interconnected to give the required system functions. Though, as we shall see in the following sections, some design rules do exist, in the main there is as yet no established theory for design at the systems level. Design procedures are in general based on partitioning the required logic specification into a number of functional blocks which may then be realized using MSI modules, dropping down to SSI gates where necessary to interface the blocks together. Alternatively the more usual formal design techniques may be adopted but implementing the logic equations in terms of MSI modules such as PLAs multiplexers etc. Although the systems approach is adopted initially, a mixture of the two techniques is used. Due to the lack of any overall formal theory the approach is inevitably intuitive based on a mixture of experience and what formal theory seems applicable.

There are a considerable number of commercially available MSI units and Table 4.4 shows a typical selection. The circuits normally employ Schottky, bipolar (TTL) NMOS or CMOS technology and would generally be supplied in flat pack or dual-in-line packages (approximately on average $1\frac{1}{2}$ in. $\times$ $\frac{1}{2}$ in. in size) with up to 64 pins. In many cases the devices would also be directly compatiable with TTL logic.

Table 4.4 Typical MSI/LSI circuits

Module	Comment
Decade counters 4–14 stage binary counters BCD counters Johnson counters Programmable divide by n	Synchronous and asynchronous versions, also up/down counters
4–64-bit shift registers	Parallel in/parallel out Parallel in/serial out Serial in/parallel out and Serial in/serial out Static and dynamic versions
BCD to binary convertor Binary to BCD convertor BCD to decimal decoder BCD to 7 segment LED decoder 4 line to 16 line decoder 4, 8, 16-bit data selectors	Also called a demultiplexer Selects 1 out of n lines, also called a multiplexer (MUX)
2–4-bit binary full-adders Look ahead generator BCD adders 8-bit odd/even parity generation/checker 4-bit magnitude comparitor 4-bit arithmetic logic unit (ALU)	Fast carry propagation Performs $A > B, A < B, A = B$ Performs 16 binary arithmetic and logic functions
64–64Kbit random access memory (RAM) 64–32Kbit read only memory (ROM)	Read–write and read only stores: organized as 1, 4, 8 or 16-bit words
Programmable logic array (PLA)	AND/OR gate array typically 16 inputs, 48 products, 8 outputs
Universal logic array (ULA)	Equivalent to a 100–2000 2-input gate array

From the viewpoint of implementing combinational switching circuits the most important devices are the read only memory (ROM), the programmable logic array (PLA), the data selector or multiplexer circuit (MUX) and the universal logic array (ULA).

4.3.1   Read only memory (ROM)

The ROM is a permanent (in the sense that loss of power will not destroy its contents) bit storage device that accepts an n-bit binary *address* at its input terminals and delivers a predetermined m-bit *word* at its output terminals for each of the input address combinations. Thus the contents are arranged as m-bit words which are accessed by selecting one word out of 2^n. The words are either prewritten at time of manufacture from a user specification (usually a truth table) or written directly by the user into a programmable ROM or PROM. Thus the functional relationship between input and output (called the ROM's *personality*) is fixed and cannot be overwritten.

The time delay between supplying the address input and the appearance of the output word is called the *access* or *read* time and is in the order of 30–300 ns depending on the size and the technology used. Both MOS and bipolar technology is used for ROMs but when very large ROMs are required MOS technology (though much slower) is generally prefered. ROMs behave effectively like a combinational circuit and as such are used particularly for code conversion and general function generation using table-look-up techniques. Figure 4.7(a) shows how a ROM could be constructed from an address decoder, decoding a 4-bit address into 1 of 16 outputs (in this case a multiplexer is used) in conjunction with a diode–resistor matrix. The outputs are determined by the presence or absence of a diode connected to the bit-lines.

Figure 4.7(b) shows the block diagram for an IC ROM module; the inputs A, B and C are used to select 1 out of 8 input lines into the memory array, while D and E select 1 out of 4 lines; note that there are 8 identical Y decoders, one for each bit of the 8-bit word, and that they are accessed simultaneously. The read-out of a particular bit location in the array is accomplished by sensing for current flow at the intersection of the selected X and Y lines.

The basic storage element of the MOS read only memory is the NMOS transistor, which is used in an analogous way to the diode–resistor combination in the diode matrix. Thus, a logical 1 is represented in the memory array by the physical presence of a transistor in that position; to specify a logical zero, the transistor is omitted. Insertion or deletion of a transistor in the matrix can be accomplished by changing a simple photomask in the MOS production process. Alternatively, each bit may be represented by a bipolar (or MOS) transistor and the coding performed by connecting or disconnecting the emitter lead (gate input) of the transistor as shown in Fig. 4.7(c). In order to keep the number of circuit connections to a minimum the address decoding logic and sense amplifiers are produced on the same substrate as the memory matrix.

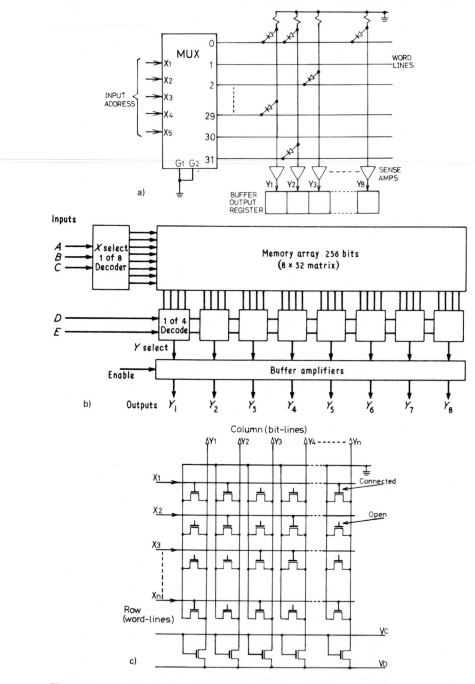

Figure 4.7 Read only memory. Top: ROM using diode matrix; middle: block diagram of a 32 × 8-bit word ROM; bottom: NMOS memory array

PROMs can be electrically programmed by the user and are normally supplied with all the storage bits set to zero. A logic 1 can be programmed at any of the bit locations by selecting, and then physically severing, a fusible metal connection using a high current source to melt the fuse link. An alternative approach is to create a short circuit, using a method known as avalanche induced migration, which involves electrically 'blowing' a diode junction into a short circuit. Note that a zero error can always be rectified but not vice versa!

Using MOS devices a reprogrammable ROM or EPROM is also possible. In this case an electric field or electromagnetic radiation is used to inject charge into an oxide trap above the gate region of a connecting MOS transistor: note that the MOS transistor memory cell has essentially an insulated 'floating gate'. When the gate is charged a conducting path between source and drain is induced which allows the representation of a logic 1. The *entire* contents of the PROM may be erased by irradiating the chip with ultraviolet radiation through a quartz window in the package. EPROMs are normally relatively slow (by a factor of ten) compared to fusable link PROMs.

The random access memory (RAM) is a similar device except that in this case the stored words may be addressed and written directly using logical inputs, as well as being read. Standard static bistable circuits are usually employed in this form of memory, again being implemented in either MOS or bipolar technology. The writing operation is accomplished by applying the word to be written to separate data inputs, addressing the store in the usual way, and then enabling the write control waveform. Note that the RAM, though still possessing the characteristics of a combinational circuit, is a true storage device in that it possesses a clocked delay time determined by the write enable waveform. Like most bistable devices the RAM is *volatile*, in the sense that the stored data are lost when the power supply is removed; this does not, of course, apply to the read only memory. For many applications this can be a severe disadvantage; however, it is possible to make special provisions (such as providing a separate battery power supply) which render the memory effectively non-volatile.

Using ROMs, the design and implementation of combinational and sequential switching circuits becomes a rather trivial operation.[6,7] There is no need to employ any of the conventional minimization techniques (indeed to do so could lead to pitfalls in the design) and the designer can work directly from a truth or assigned state table (see later).

As an illustration of the technique, Table 4.5 shows the layout of a ROM (containing 32 words each of 8 bits) to perform the switching function $T = \Sigma(1, 2, 3, 4, 5, 6, 7, 10, 14, 20, 22, 28)$. The format is identical to the initial truth table; the five input variables (minterms) correspond to the ROM address, and the contents of the words (one bit in each) to the output function. Thus to implement the switching circuit the designer simply specifies the position in the ROM of the 1s in the required output function. To use the ROM, the unit is addressed with the input variables which causes the corresponding word containing the required output bit to be read down.

Note that only one bit is used in each output word of the ROM; with this

Table 4.5 Read only memory formats

Address input variables					Output word output functions							
A_0	A_1	A_2	A_3	A_4	Z_0	Z_1	Z_2	Z_3	Z_4	Z_5	Z_6	Z_7
0	0	0	0	0	0	0	0	0	0	0	0	0
0	0	0	0	1	1	0	0	0	0	0	0	0
0	0	0	1	0	1	0	0	0	0	0	0	0
0	0	0	1	1	1	0	0	0	0	0	0	0
0	0	1	0	0	1	0	0	0	0	0	0	0
0	0	1	0	1	1	0	0	0	0	0	0	0
0	0	1	1	0	1	0	0	0	0	0	0	0
0	0	1	1	1	1	0	0	0	0	0	0	0
0	1	0	0	0	0	0	0	0	0	0	0	0
0	1	0	0	1	0	0	0	0	0	0	0	0
0	1	0	1	0	1	0	0	0	0	0	0	0
0	1	0	1	1	0	0	0	0	0	0	0	0
0	1	1	0	0	0	0	0	0	0	0	0	0
0	1	1	0	1	0	0	0	0	0	0	0	0
0	1	1	1	0	1	0	0	0	0	0	0	0
0	1	1	1	1	0	0	0	0	0	0	0	0
1	0	0	0	0	0	0	0	0	0	0	0	0
1	0	0	0	1	0	0	0	0	0	0	0	0
1	0	0	1	0	0	0	0	0	0	0	0	0
1	0	0	1	1	0	0	0	0	0	0	0	0
1	0	1	0	0	1	0	0	0	0	0	0	0
1	0	1	0	1	0	0	0	0	0	0	0	0
1	0	1	1	0	1	0	0	0	0	0	0	0
1	0	1	1	1	0	0	0	0	0	0	0	0
1	1	0	0	0	0	0	0	0	0	0	0	0
1	1	0	0	1	0	0	0	0	0	0	0	0
1	1	0	1	0	0	0	0	0	0	0	0	0
1	1	0	1	1	0	0	0	0	0	0	0	0
1	1	1	0	0	1	0	0	0	0	0	0	0
1	1	1	0	1	0	0	0	0	0	0	0	0
1	1	1	1	0	0	0	0	0	0	0	0	0
1	1	1	1	1	0	0	0	0	0	0	0	0

vertical layout the bits in a word may be totally unrelated for a given input address. Consequently, multiple output switching functions can easily be programmed, and in fact the ROM shown in Table 4.5 represents what is essentially a multiple output circuit with up to 8 output functions, one for each bit of the word.

When designing with ROMs there is no need to consider 'don't-care' conditions since the devices are only available in standard modules and hence no savings are possible. The choice of ROM is determined solely by the number of variables involved and the size of ROM which will accommodate them; in our example we have 5 variables; therefore a ROM with at least 2^5 words will be required. If the number of minterms for a given function of n variables is greater than 2^{n-1} it is sometimes more convenient to program the complement of the function (equivalent to using maxterms) and invert the output of the ROM. One obvious disadvantage is that it is necessary to use the canonical or expanded form of the switching functions.

Contrasting this form of implementation with conventional NAND gate logic, it is obviously uneconomic for a single-output circuit, since the cost of a ROM package is some ten times greater. However, when used as a multiple output circuit, for example when translating from one code to another, the economic gains are considerable.

The major advantages over the hardwired approach are:

(a) The total chip count is reduced since one ROM may replace a number of SSI gate units.
(b) With fewer chips and a more regular wiring pattern less time is required to carry out a design.
(c) Circuit testing is simplified.
(d) Circuit modification and error correction can be accomplished by changing the ROM pattern rather than rewiring the system.

However, for large-variable problems direct implementation using a single ROM soon becomes impractical since every additional switching variable doubles the number of words required in the memory. This limitation may be overcome in the majority of cases by employing smaller ROMs in cascaded or multi-level circuits. Since it is always possible to connect ROMs together to produce a larger sized store, for example two $32 \times$ 8-bit word modules can be connected together to give a $64 \times$ 8-bit word store, the techniques of cascading are effectively those of minimization at the subsystem (i.e. ROM) level.

We will now consider a simple example of cascaded ROM circuits using the switching function shown in Table 4.6, that is:

$$T = \Sigma(4, 5, 15, 20, 29, 41, 42, 45, 47, 53, 58, 61, 63)$$

and

$$DC = \Sigma(7, 9, 10, 13, 21, 25, 26, 31, 36, 37, 39, 40, 52, 56, 57)$$

The circuit can of course be implemented directly using a ROM with $2^6 =$ 64 words which would be obtained by connecting together four $16 \times$ 4-bit word stores. However, we shall show that it is possible to implement this function using two ROMs connected in cascade with the output of one feeding directly into the other.

In the cascaded technique[8] the variables are partitioned and recoded in order to achieve data compression. This is possible since, in general, most switching functions contain terms with common variables or minterms, for example the terms $\bar{A}\bar{B}CDEF$ and $ABCDEF$ share the variables $CDEF$. Table 4.7(a) shows the ON terms listing for the original function (which has been partitioned into blocks containing variables AB and $CDEF$) and the shared terms in $CDEF$. Any other permutation or combination of variables may be used, but the best choice is that set (or sets) of variables which gives rise to the smallest number of shared terms. From Table 4.7(b) it is apparent that 3 bits are necessary to code the shared $CDEF$ terms; thus a ROM with 16 words of 3 bits will be required to generate the coded

Table 4.6 Function for ROM implementation

	Input variables $A^* B^* C\ D\ E\ F$	Output Z		Input variables $A^* B^* C\ D\ E\ F$	Output Z
0	0 0 0 0 0 0	0	32	1 0 0 0 0 0	0
1	0 0 0 0 0 1	0	33	1 0 0 0 0 1	0
2	0 0 0 0 1 0	0	34	1 0 0 0 1 0	0
3	0 0 0 0 1 1	0	35	1 0 0 0 1 1	0
4	0 0 0 1 0 0	1	36	1 0 0 1 0 0	X
5	0 0 0 1 0 1	1	37	1 0 0 1 0 1	X
6	0 0 0 1 1 0	0	38	1 0 0 1 1 0	0
7	0 0 0 1 1 1	X	39	1 0 0 1 1 1	X
8	0 0 1 0 0 0	0	40	1 0 1 0 0 0	X
9	0 0 1 0 0 1	X	41	1 0 1 0 0 1	1
10	0 0 1 0 1 0	X	42	1 0 1 0 1 0	1
11	0 0 1 0 1 1	0	43	1 0 1 0 1 1	0
12	0 0 1 1 0 0	0	44	1 0 1 1 0 0	0
13	0 0 1 1 0 1	X	45	1 0 1 1 0 1	1
14	0 0 1 1 1 0	0	46	1 0 1 1 1 0	0
15	0 0 1 1 1 1	1	47	1 0 1 1 1 1	1
16	0 1 0 0 0 0	0	48	1 1 0 0 0 0	0
17	0 1 0 0 0 1	0	49	1 1 0 0 0 1	0
18	0 1 0 0 1 0	0	50	1 1 0 0 1 0	0
19	0 1 0 0 1 1	0	51	1 1 0 0 1 1	0
20	0 1 0 1 0 0	1	52	1 1 0 1 0 0	X
21	0 1 0 1 0 1	X	53	1 1 0 1 0 1	1
22	0 1 0 1 1 0	0	54	1 1 0 1 1 0	0
23	0 1 0 1 1 1	0	55	1 1 0 1 1 1	0
24	0 1 1 0 0 0	0	56	1 1 1 0 0 0	X
25	0 1 1 0 0 1	X	57	1 1 1 0 0 1	X
26	0 1 1 0 1 0	X	58	1 1 1 0 1 0	1
27	0 1 1 0 1 1	0	59	1 1 1 0 1 1	0
28	0 1 1 1 0 0	0	60	1 1 1 1 0 0	0
29	0 1 1 1 0 1	1	61	1 1 1 1 0 1	1
30	0 1 1 1 1 0	0	62	1 1 1 1 1 0	0
31	0 1 1 1 1 1	X	63	1 1 1 1 1 1	1

outputs; variables AB assume all possible values and a more economic coding is impossible. The cascaded circuit takes the form shown in Fig. 4.8, in which the variables $CDEF$ go to ROM 2, which generates the coded output $Z_1 Z_2 Z_3$, and this output together with the variables AB go to ROM 1 which generates the final switching function. The layout of ROM 1 is shown in Table 4.7(c); note that only those input terms which generate an output are shown; all the other words in the ROM will contain zeros.

Thus, using cascaded implementation the number of 16 4-bit word ROMs has been reduced from four to three, assuming that ROM 1 consists of two such ROMs connected together. The technique produces even greater savings when large multi-output functions are to be implemented. However, there is a penalty to be paid in terms of speed since, as with all cascaded networks, the propagation delay is significantly increased.

A much greater saving is possible if the number of input variables can be reduced since eliminating one variable will immediately halve the ROM

Table 4.7 Cascaded ROMs

Decimal form	Variables A B	C D E F
4	0 0	0 1 0 0
5	0 0	0 1 0 1
15	0 0	1 1 1 1
20	0 1	0 1 0 0
29	0 1	1 1 0 1
41	1 0	1 0 0 1
42	1 0	1 0 1 0
45	1 0	1 1 0 1
47	1 0	1 1 1 1
53	1 1	0 1 0 1
58	1 1	1 0 1 0
61	1 1	1 1 0 1
63	1 1	1 1 1 1

(a) ON terms listing

Variables C D E F	Coded form Z_1 Z_2 Z_3
0 1 0 0	0 0 0
0 1 0 1	0 0 1
1 0 0 1	0 1 0
1 0 1 0	0 1 1
1 1 0 1	1 0 0
1 1 1 1	1 0 1

(b) Shared terms and coding

Input A B	Variables Z_1 Z_2 Z_3	Outputs T_1 T_2 T_3 T_4
0 0	0 0 0	1 0 0 0
0 0	0 0 1	1 0 0 0
0 0	1 0 1	1 0 0 0
0 1	0 0 0	1 0 0 0
0 1	1 0 0	1 0 0 0
1 0	0 1 0	1 0 0 0
1 0	0 1 1	1 0 0 0
1 0	1 0 0	1 0 0 0
1 0	1 0 1	1 0 0 0
1 1	0 0 1	1 0 0 0
1 1	0 1 1	1 0 0 0
1 1	1 0 0	1 0 0 0
1 1	1 0 1	1 0 0 0

(c) Layout of first-level ROM

Variables C D E F	Coded T_1 T_2 T_3 T_4
0 0 0 0	0 0 0 0
0 0 0 1	0 0 0 0
0 0 1 0	0 0 0 0
0 0 1 1	0 0 0 0
0 1 0 0	1 0 0 0
0 1 0 1	1 0 0 0
0 1 1 0	0 0 0 0
0 1 1 1	0 0 0 0
1 0 0 0	0 0 0 0
1 0 0 1	1 0 0 0
1 0 1 0	1 0 0 0
1 0 1 1	0 0 0 0
1 1 0 0	0 0 0 0
1 1 0 1	1 0 0 0
1 1 1 0	0 0 0 0
1 1 1 1	1 0 0 0

(d) Layout with eliminated variables

size. In many practical cases there are more variables used than the minimum required to generate the outputs. For example, in the truth table shown in Table 4.6, 6 variables are used to define 13 ON terms whereas in the limit only four are required.

A procedure for determining redundant input variables[9] is to delete each variable in turn and then test if there are any *contradictions*, that is when an input condition gives rise to conflicting 0 and 1 output values. Don't-care terms may be used as appropriate to resolve contradicting output values. For example, in Table 4.6 if we remove variable *F* we see immediately that

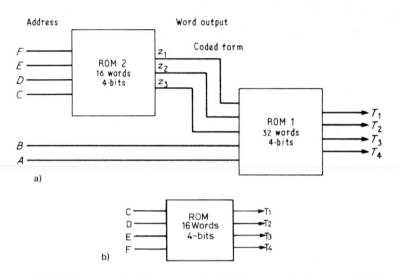

Figure 4.8 ROM networks. Top: cascaded ROM network; bottom: ROM with eliminated inputs

a contradiction arises between inputs 14 and 15 where the reduced term 00111 now has both 0 and 1 outputs. Thus in this case F is an essential input variable and cannot be eliminated. However, if we consider variables A and B and choosing the don't-cares appropriately we find that both variables may be removed giving the ROM layout shown in Table 4.7(d) and the single 16 × 4-bit ROM circuit shown in Figure 4.8(b).

An alternative method for reducing ROM inputs is to attempt to perform simple logic operations such as AND/OR, NAND/NOR on the input variables. The possibility of applying any of these methods will, of course, depend on the particular structure of the function and the number and disposition of don't-cares (as with all minimization). Unfortunately no formal design algorithms exist for implementing circuits in terms of ROM modules. Some work has been done using computers to determine the best partitioning and coding for cascaded networks but these have been developed intuitively.

4.3.2 Multiplexer circuits (MUX)

Another MSI circuit which has general application in implementing combinational logic is the *data selector* or *multiplexer* circuit shown in Fig. 4.9. This is a combinational switching circuit which generates the function

$$Z = \bar{x}\bar{y}\bar{z}A + \bar{x}\bar{y}zB + \bar{x}y\bar{z}C + \bar{x}yzD + x\bar{y}\bar{z}E + x\bar{y}zF + xy\bar{z}G + xyzH$$

where x, y and z are the control waveforms and A, B, C, D, E, F, G and H the input lines which are required to be switched to the output. Thus, the logical operation of the circuit is to select 1 out of 8 possible inputs and

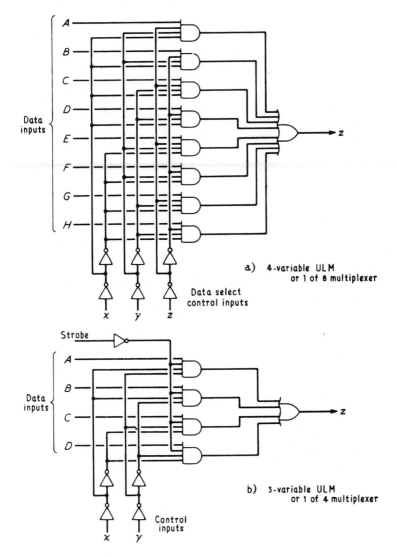

a) 4-variable ULM
or 1 of 8 multiplexer

Data select
control inputs

b) 3-variable ULM
or 1 of 4 multiplexer

Control
inputs

Figure 4.9 Multiplexer circuits

present it at the output. Other versions of this circuit are also available, for instance a dual 1-out-of-4 data selector, implemented in one package, with two common control inputs. It is also possible to obtain the converse of this function, that is, a demultiplexer circuit which performs the operation of decoding a binary input code to separate (and mutually exclusive) output lines, for example, a 4-line to 16-line decoder.

Multiplexer circuits are capable of generating Boolean equations and may consequently be used to implement combinational switching circuits.[10] For example, a four-way multiplexer with four data input lines (A, B, C and D) and two control lines x and y can produce any Boolean function of three variables. This can be achieved by connecting two of the binary variables to the control inputs and the other variable (or variables) to the data input lines. Now, using each of the four possible combinations of the control lines, the required output terms may be generated by applying either logic 1, logic 0, the variable or the inverse of the variable to the input lines.

The technique is illustrated in Fig. 4.10(a) and (b), where the output functions

$$Z_1 = \bar{x}_1\bar{x}_2y_1 + \bar{x}_1x_2 + x_1\bar{x}_2\bar{y}_1$$

and

$$Z_2 = \bar{x}_1\bar{x}_2y_3 + x_1x_2y_4$$

are generated using a four-way multiplexer unit. The operation of the circuit shown in Fig. 4.10(a) is such that when $\bar{x}_1\bar{x}_2$ is 1 the output signal should be 1 or 0 according to whether y_1 is 1 or 0, corresponding to the term $\bar{x}_1\bar{x}_2y_1$; therefore y_1 is connected to the A input line. Similarly, since the output should always be 1 when $\bar{x}_1x_2$ is 1, the B input is connected permanently to logic 1. In the case of the term $x_1\bar{x}_2\bar{y}_1$, the inverse of y_1 is connected to the C input line. Finally, since there are no terms which require x_1x_2, input D is put to logic 0.

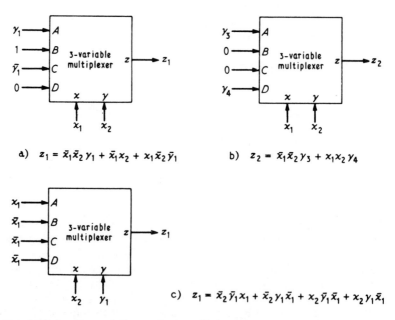

a) $z_1 = \bar{x}_1\bar{x}_2 y_1 + \bar{x}_1 x_2 + x_1\bar{x}_2 \bar{y}_1$

b) $z_2 = \bar{x}_1\bar{x}_2 y_3 + x_1 x_2 y_4$

c) $z_1 = \bar{x}_2 \bar{y}_1 x_1 + \bar{x}_2 y_1 \bar{x}_1 + x_2 \bar{y}_1 \bar{x}_1 + x_2 y_1 \bar{x}_1$

Figure 4.10 Implementation using multiplexers

Note that in the above examples the variables x_1 and x_2 appear in each term of the required expressions and are therefore the obvious candidates for connection to the control lines of the multiplexer. However, the choice of variables for the control inputs is generally not unique, though in many cases it can be rather critical. Consider implementing function Z_1 above, with the variables $x_2 y_1$ going to the control input. The first step is to expand the function to its full canonical form, giving the expression

$$Z_1 = \bar{x}_1 \bar{x}_2 y_1 + \bar{x}_1 x_2 \bar{y}_1 + \bar{x}_1 x_2 y_1 + x_1 \bar{x}_2 \bar{y}_1$$

which can then be implemented directly as shown in Fig. 4.10(c). Note, however, that if we had chosen $x_1 y_1$ as the control inputs it would have been impossible to generate the output equations.

In the general case, any one of the four signals 0, 1, y_1 and $\bar{y}_1$ (shown in Fig. 4.10(a)) could have been connected to any one of the four multiplexed input lines, giving $4^4 = 256$ different input combinations. Since this corresponds exactly to the number of different Boolean functions of three variables (that is 2^{2^n} where n is the number of variables), it follows that multiplexer circuits can be used to generate any 'random' switching function. In fact the multiplexer modules described above are identical to the circuits proposed by Yau and Tang[11,12] for use as universal logic modules (ULM).

The action of the multiplexer circuit may be expressed more formally by noting that any logic function $f(x_1, x_2, \cdots x_n)$ of n variables, where $n \geq 3$, can be expanded to the form

$$
\begin{aligned}
f(x_1, x_2, \cdots x_n) = &\; \bar{x}_1 \bar{x}_2 f(0, 0, x_3, \cdots x_n) \\
&+ \bar{x}_1 x_2 f(0, 1, x_3, \cdots x_n) \\
&+ x_1 \bar{x}_2 (1, 0, x_3, \cdots x_n) \\
&+ x_1 x_2 f(1, 1, x_3, \cdots x_n)
\end{aligned}
\tag{4.1}
$$

For example, in the case of a function of three variables, i.e. $f(x, y, z)$, we have, by expanding with respect to the variables x and y, the following equation:

$$f(x, y) = \bar{x}\bar{y}f(0, 0) + \bar{x}yf(0, 1) + x\bar{y}f(1, 0) + xyf(1, 1)$$

where the *residue functions*, $f(0, 0)$, $f(0, 1)$, $f(1, 0)$ and $f(1, 1)$ are functions of z only, and each of these functions assumes one of the four values 0, 1, z or $\bar{z}$. Note that this equation describes the 1-of-4 data selector described above, where x and y are the control lines. Moreover, it follows that Boolean functions may also be expanded with respect to any number of variables, for example expanding with respect to four variables results in the 1-to-16 data selector which enables all Boolean functions of five variables to be generated. It is also possible to expand about a single variable; for instance:

$$f(x_1, x_2, \cdots x_n) = \bar{x}_1 f(0, x_2, \cdots x_n) + x_1 f(1, x_2, \cdots x_n)$$

which can be realized using a three-variable multiplexer by connecting both control lines together and only using the A and D input lines, i.e. the input lines corresponding to the control terms $\bar{x}\bar{y}$ and xy.

Though it will be obvious from above that ULMs handling any number of variables can be produced, the complexity of the circuits increases rapidly and from economical and maintenance considerations it is better to implement logic functions using a number of identical small variable ULMs. This is possible by connecting multiplexers in arrays of two or more levels, corresponding to repeated expansion of the residue functions in eqn (4.1) above until they are functions of the variable x_n only.

As an example of this process we shall consider the implementation of the five-variable switching function discussed in Section 3.7 using only three-variable multiplexer circuits. The canonical form of the function is given by

$$T = \bar{V}\bar{W}\bar{X}\bar{Y}Z + \bar{V}\bar{W}\bar{X}Y\bar{Z} + \bar{V}\bar{W}\bar{X}YZ + \bar{V}\bar{W}X\bar{Y}\bar{Z} + \bar{V}\bar{W}X\bar{Y}Z$$
$$+ \bar{V}\bar{W}XY\bar{Z} + \bar{V}\bar{W}XYZ + \bar{V}W\bar{X}Y\bar{Z} + \bar{V}WXY\bar{Z} + V\bar{W}X\bar{Y}\bar{Z}$$
$$+ V\bar{W}XY\bar{Z} + VWX\bar{Y}\bar{Z}$$

which can be implemented directly using a five-variable ULM, that is, a 1-to-16 multiplexer unit. To do this the variable $WXYZ$ would be used as the control inputs and the data inputs would be $\bar{V}$, V, 0 or 1. Note that the terms $V\bar{W}X\bar{Y}\bar{Z}$ and $V\bar{W}XY\bar{Z}$ must be accommodated by simplifying with $\bar{V}\bar{W}X\bar{Y}\bar{Z}$ and $\bar{V}\bar{W}XY\bar{Z}$ respectively and applying logic 1 to the corresponding inputs of the multiplexer. To implement using three-variable multiplexers the process is similar. First, we expand the function about the variables Y and Z, simplifying where necessary, i.e.

$$T = \bar{Y}\bar{Z}(\bar{W}X + VWX) + \bar{Y}Z(\bar{V}\bar{W}\bar{X} + \bar{V}\bar{W}X) + Y\bar{Z}(\bar{V}\bar{W}\bar{X}$$
$$+ \bar{W}X + \bar{V}W\bar{X} + \bar{V}WX) + YZ(\bar{V}\bar{W}\bar{X} + \bar{V}\bar{W}X)$$

This gives the first level of implementation. The residue terms can now be expanded further to obtain the second level; this may be done by inspection using WX as the control inputs. From the equations above it is obvious that the residue terms for $\bar{Y}Z$ and YZ are identical and consequently only one multiplexer is required to generate the output function; the output can of course be shared at the input to the first level multiplexer. The final circuit is shown implemented using three-variable multiplexers in Fig. 4.11(b).

An alternative, and perhaps simpler, design technique involves the use of K-maps to determine the input variables. After deciding the type of multiplexer to be used (i.e. three- or four-variable ULM) and the control input variables, an ON term listing is made for the switching function (see Table 4.8(a)). In this table only those input combinations which generate an output are listed and the ordering of the table is given by the choice of control variables. Note that the same control variables have been chosen as before (YZ) and that the entries in the input column refer to the first level multiplexer. The next step in the procedure is to plot the input and control variables (VWX) on K-maps for each of the three-variable multiplexers in the second level; this is shown in Table 4.8(b). Note that it is again obvious that the inputs B' and D' to the first level multiplexer are identical and D' can thus be ignored.

The K-maps are interpreted in terms of the control inputs WX and the

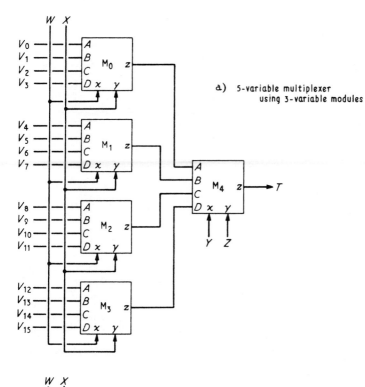

a) 5-variable multiplexer
   using 3-variable modules

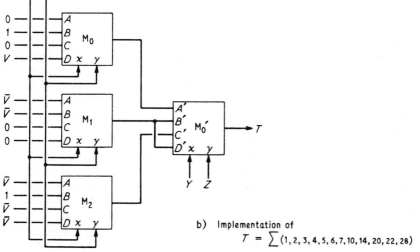

b) Implementation of
   $T = \sum(1, 2, 3, 4, 5, 6, 7, 10, 14, 20, 22, 28)$

Figure 4.11 Multilevel implementation using multiplexers

Table 4.8 Multiplexer circuit design using K-maps

V W X	Input	Y Z
0 0 1		0 0
1 0 1	A'	0 0
1 1 1		0 0
0 0 0	B'	0 1
0 0 1		0 1
0 0 0		1 0
0 0 1		1 0
0 1 0	C'	1 0
0 1 1		1 0
1 0 1		1 0
0 0 0	D'	1 1
0 0 1		1 1

(a) ON terms listing

M_0:

WX / V	00	01	11	10
0		1		
1		1	1	
	A	B	D	C

$A = 0$
$B = 1$
$C = 0$
$D = V$

M_1:

WX / V	00	01	11	10
0	1	1		
1				
	A	B	D	C

$A = \bar{V}$
$B = V$
$C = 0$
$D = 0$

M_2:

WX / V	00	01	11	10
0	1	1	1	1
1		1		
	A	B	D	C

$A = \bar{V}$
$B = 1$
$C = \bar{V}$
$D = \bar{V}$

(b) K-maps for second level multiplexers

data inputs $ABCD$; for instance, the column $\bar{W}\bar{X}$ corresponds to input A, $\bar{W}X$ to input B, and so on. To determine the value of the input variable V we note that if a column has two zeros the data input must be $V = 0$; similarly for two ones, $V = 1$. The other values are obtained by noting the position of the minterm and reading the corresponding value of the input variable V. For instance, from Table 4.8(b) for M_0 we have $A = 0$ (all zeros), $B = 1$ (all ones), $C = 0$ (all zeros) and $D = V$. Note that the results obtained are identical with those found using the algebraic technique, but in this case automatic minimization has been performed on the map (for example, multiplexer M_2 term). The mapping technique can be used for single level circuits and may easily be extended to four- and five-variable ULMs.

Another approach to designing multiplexer circuits is based on a matrix notation.[13] For example, a Boolean function can be expressed as

$$[f] = [I][S]$$

where $[I]$ is the input matrix and represents the input values to the multiplexer and $[S]$ is the selector matrix. Thus $[S]$ for the three-variable 1-out-of-4 multiplexer is expressed as

$$[S] = \begin{bmatrix} x \\ y \end{bmatrix} \begin{bmatrix} 0 & 0 & 1 & 1 \\ 0 & 1 & 0 & 1 \end{bmatrix}$$

and for 1-out-of-8

$$[S] = \begin{bmatrix} x \\ y \\ z \end{bmatrix} \begin{bmatrix} 0 & 0 & 0 & 0 & 1 & 1 & 1 & 1 \\ 0 & 0 & 1 & 1 & 0 & 0 & 1 & 1 \\ 0 & 1 & 0 & 1 & 0 & 1 & 0 & 1 \end{bmatrix} = [0\ 1\ 2\ 3\ 4\ 5\ 6\ 7]$$

The output $[f]$ for any given selector input can be obtained by a simple logic 'multiplication' of $[S]$ and $[I]$, equivalent to an addressing function. Thus $[f]$ is made equal to the input (column) selected by the value of $[S]$. From Fig. 4.10(b), if $[I] = [ABCD] = [y_3 00 y_4]$ and $[S] = [0123]$ then for $[S] = 3$, $[F] = D = y_4$. The notation would be employed in conjunction with the K-map method described above to effect an implementation and as such adds nothing novel; however, the notation is extremely useful when realizing sequential circuits using MUXs.

It is illuminating to contrast the implementation of combinational logic using multiplexers with the more conventional NOR/NAND realizations. Taking the example shown in Fig. 4.11(b) (that of Section 3.7), the reduced form of the equation was found to be

$$T = VX\bar{Y}\bar{Z} + \bar{V}\bar{W}Z + \bar{V}Y\bar{Z} + \bar{W}X\bar{Z}$$

which requires two packages for its implementation, i.e. one dual 4-input NAND unit and a triple 3-input NAND unit. If a 1-of-16 multiplexer had been used only one package would have been required: using 1-of-4 multiplexers four packages (all identical) are needed. This could be reduced to three if a dual 1-to-4 package with common control lines is used.

Though the cost of multiplexer units at the present time is over double that of basic gate packages there is nevertheless considerable advantage to be gained from this method of implementation. The obvious gain if one 1-of-16 multiplexer is used is the reduced cost of wiring and layout of printed circuit boards. Other considerations include the reduced number of spare packages that will be required and the ease of testing and maintenance. An obvious disadvantage in the case of cascaded modules is the increased propagation time through the circuits. Multiplexer units are, on average, some two or three times slower than corresponding TTL NAND/NOR units. This difference balances itself out with single package multiplexer implementation since most conventional logic circuits involve at least two gate levels.

It will be apparent that switching functions can be implemented using multiplexers in a variety of ways; the example given above is by no means a unique solution. Note, moreover, that in this case the canonical equation was used as the starting point for the factorization procedure. This in general seems to be a better approach; using the reduced form of equation can often lead to difficulties—the reader is invited to try this for himself! In fact this is an important practical advantage of the technique, in that it is no longer necessary to find the minimal sum or product of a function in order to effect an economical realization.

In multi-level implementation the selection of suitable control inputs at the first level is very important, since this can affect the number of multiplexers required in the second and subsequent levels. If possible the choice must be made so as to optimize the number of 0, 1, and common inputs (that is, inputs which can be shared at the data inputs of the multiplexer see Fig. 4.11(b)). An alternative approach is to select variables for higher order levels that are either identical or the inverse of one another; this latter characteristic is only applicable when the multiplexer has both true and complemented outputs. In all cases the objective is to reduce the number of multiplexers required in the higher order levels. The problem is further complicated since there is no reason why the control inputs at higher order levels should all be the same (though they often are); the use of individual control inputs can often lead to a reduction in the number of modules required in the preceding levels.

Unfortunately there is no formal design theory, as yet, which considers implementation at the subsystem level, for example the systematic minimization of ULMs. Note also that only single-output switching circuits have been discussed. Multiple-output networks present yet another problem; they can, of course, be designed as separate single-output circuits. These aspects of logic design are ideal topics for further research but ones which seems to have claimed little attention, perhaps because it is difficult to decide on the ideal form of ULM. However, the multiplexer circuit seems able to perform this function and, since it is commercially available, work could profitably proceed using this device as the basic element or cell in LSI arrays.

Before concluding this section it is worth mentioning the *demultiplexer* (DMUX), also available as a MSI package, whose function is effectively the inverse of the MUX. A demultiplexer can be considered as an n-to-2^n decoder, where n is the number of inputs, which uses its enable control as a data input line which can be routed to a selected output; Fig. 4.12 shows a circuit for the device. Note that the data input is routed to output f_0, f_1, f_2 or f_3 depending on the setting of the selector inputs xy. The main use of DMUXs (and MUXs) is in data handling applications such as routing data in digital communication systems.

4.3.3 *Programmable logic arrays (PLAs)*[14,15]

The PLA is similar to the ROM in that it has a fixed addressable memory but differs in that the address section itself can be programmed. Thus whereas the ROM must utilize all possible input combinations due to its fixed address decoder (and consequently generating the canonical minterm form of the function), the PLA allows the input addresses to be specified as reduced product terms. This overcomes one of the major disadvantages of the ROM, that is the number of empty words obtained when realizing a sparse truth table containing a small number of ON terms. This is a particular problem when the ROM is implemented as part of an LSI/VLSI circuit where silicon area is at a premium.

Logically the PLA functions as a two-level AND/OR array which gener-

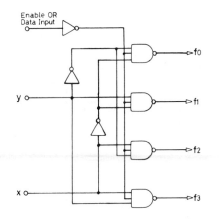

Figure 4.12 Demultiplexer circuit

ates sum-of-product outputs; the structure in diode/resistor form is shown in Fig. 4.13(a). PLAs can be either mask or field programmable as in the case of ROM; in the fusible link case (as shown) the device is supplied with all the diodes intact and the user must disconnect those not required according to the particular product terms and output functions to be realized. PLAs are normally available with 16 inputs, 48 product gates and 8 outputs; the number of product terms and output functions may be expanded by paralleling the appropriate input–output connections as shown in Fig. 4.13(b) and (c).

Because of its regular structure the PLA is the preferred method of implementing random logic in LSI/VLSI circuits; a MOS version of the PLA is shown in Fig. 4.14. Note that the circuit is effectively a positive logic NOR/NOR network. From Fig. 4.14, if any of the inputs A, B or $\bar{C}$ is high (turning on the gate) then the product line R_3 will be pulled down to earth and thus will be low. In this case $R_3 = \overline{(A + B + \bar{C})} = \bar{A}\bar{B}C$ and similarly $R_4 = \overline{(A + \bar{B} + C)} = \bar{A}B\bar{C}$. For the OR (NOR) array output, if R_3 or R_4 is high then $\bar{Z}_4$ is low, thus $\bar{Z}_4 = \overline{(\bar{A}\bar{B}C + \bar{A}B\bar{C})}$ which is inverted to give Z_4. In a digital system the inputs and outputs to the PLA (as with most combinational circuits) would be clocked out of and into a register to avoid timing problems. This is provided for in the PLA by using MOS pass transistor gates controlled by two separate and out of phase clock inputs.

To implement a logic function using PLAs it is first necessary to derive a minimal sum of products expression, which may be done in the usual way using K-maps etc., except that the minimization criterion is quite different. For example, if the network can be realized within the capabilities of a single PLA package there is no reason why minimization should even be attempted. Thus the minimization routine should terminate when the total number of product terms is less than or equal to the maximum number of available AND gates provided by the PLA chip (or a multiple of this figure if product expansion is necessitated). In essence, then, the criterion is based

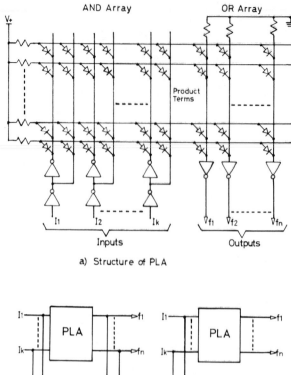

a) Structure of PLA

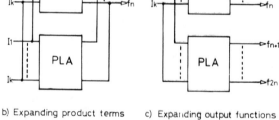

b) Expanding product terms   c) Expanding output functions

Figure 4.13 Programmable logic arrays

on obtaining the minimum number of PLA *packages* (and hence connections) to implement the function.

The minimization problem is similar to the classic two-level multiple output case but with the essential difference that the cost is simply determined by the number of AND gates at the first level. The number of inputs, that is the size of the product terms and/or the number of inputs to the second level OR gates are of no consequence. Note also that a product consisting of only one literal, a 'feed through' in the normal case, must still use an AND gate prior to the OR gate. Moreover if an AND gate (PI) is essential for the realization of any one of the functions the cost for covering other functions is nil. Thus for PLAs the covering problem, although multiple output, can be effectively treated as a single output problem.[16]

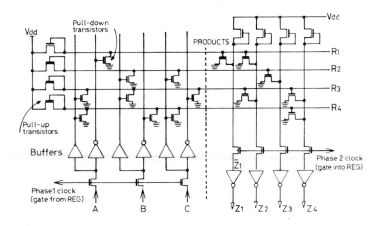

Figure 4.14 PLA circuit using NMOS transistors

As an example consider the MOPI chart shown earlier in Table 3.24. Treating this as a single output problem and ignoring gate costs we obtain Table 4.9. If we now choose essential prime implicants we find that all the functions are covered except Z_3; if we now arbitrarily choose C to cover term 9 we have

$$Z_1 = A + B + D + E$$
$$Z_2 = B + D + E + G$$
$$Z_3 = A + B + C + D$$

which gives a total of 6 gates at the first level. Note that compared to the minimal cost solution we obtained earlier we require an extra 3 inputs but the amount of computation has been drastically reduced—a considerable boon when devising algorithms to be implemented on a digital computer.

Table 4.9 PLA minimization

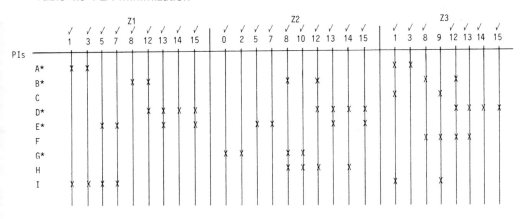

	Z1									Z2										Z3							
PIs	1	3	5	7	8	12	13	14	15	0	2	5	7	8	10	12	13	14	15	1	3	8	9	12	13	14	15
A*	X	X																		X	X						
B*					X	X								X	X							X	X				
C														X						X							
D*						X	X	X	X							X	X	X	X					X	X	X	X
E*			X	X			X		X			X	X														
F																								X	X	X	X
G*										X	X			X	X												
H														X	X	X		X									
I	X	X	X	X																X			X				

Though we have said that the number of inputs is of no consequence there is an exception to this. That is when the number of *external* inputs exceed the number of available inputs to the PLA. As we have seen, inputs may be expanded using another module but it is good practice to attempt to reduce the inputs as described in Section 4.3.1 for ROM implementations.

Another exception that can arise is when a physical contact needs to be made to connect the diodes (MOS transistors) in the array. Then if the number of interconnections can be reduced overall the likelihood of faults due to bad connections of the MOS gates would be minimized as well as reducing the overall power consumption. This would entail reverting to the normal minimization criteria, for multiple-output circuits.

An alternative approach to minimizing the number of AND gates, i.e. product terms, is to precode the input variables prior to connecting them to the PLA.[17] This allows the ANDing together of binary functions, rather than literals, in the generation of the output and hence requires fewer AND gates. In practice the input variables would be partitioned, for ease of calculation, into disjoint groups of two variables each with its own decoder. To implement this method the functions must first be factorized into an appropriate form, which will determine the choice of variables in the partitioning.

As an example consider the truth table shown in Table 4.10; the reduced expressions are

$$Z_1 = AC + ABD + BCD$$
$$Z_2 = \bar{A}\bar{B}C + A\bar{B}\bar{C} + \bar{A}C\bar{D} + A\bar{C}\bar{D} + \bar{A}B\bar{C}D + ABCD$$
$$Z_3 = B\bar{D} + \bar{B}D = (B + D)(\bar{B} + \bar{D})$$

Factorizing Z_1 and Z_2 gives

Table 4.10

A	B	C	D	Z_1	Z_2	Z_3
0	0	0	0	0	0	0
0	0	0	1	0	0	1
0	0	1	0	0	1	0
0	0	1	1	0	1	1
0	1	0	0	0	0	1
0	1	0	1	0	1	0
0	1	1	0	0	1	1
0	1	1	1	1	0	0
1	0	0	0	0	1	0
1	0	0	1	0	1	1
1	0	1	0	1	0	0
1	0	1	1	1	0	1
1	1	0	0	0	1	1
1	1	0	1	1	0	0
1	1	1	0	1	0	1
1	1	1	1	1	1	0

$$Z_1 = AC + (A + C)BD$$

and

$$Z_2 = (A\bar{C} + \bar{A}C)(\bar{B} + \bar{D}) + (\bar{A}\bar{C} + AC)BD$$
$$= (\bar{A} + \bar{C})(A + C)(\bar{B} + \bar{D}) + (\bar{A} + C)(A + \bar{C})BD$$

which requires the input variables to be partitioned into the groups AC and BD. A schematic diagram for the PLA is shown in Fig. 4.15; note that only five AND gates are required compared to eleven for the direct implementation. The decoder could of course be a basic 2–4 line demultiplexer unit. This method can be further extended by using two-bit decoders which only generate the maxterms. For example, the partition $X_1 = (AB)$ of function $Z = f(A, B, C, D)$ would have the maxterms $(A + B)$, $(A + \bar{B})$, $(\bar{A} + B)$ and $(\bar{A} + \bar{B})$.

If the decoders are included as an integral part of the PLA structure, as would be the case in an LSI realization, we obtain a three level OR-AND-OR network. Sasao has shown[18] that on average the use of two-bit decoders will lead to a smaller overall area for the PLA—as we shall see later this is an important consideration in the design of VLSI circuits.

To realize a function using an OR-AND-OR network it is advantageous to represent it as a *generalized Boolean function*, as described by Sasao. For example, function Z_2 in Table 4.10 would be expressed in this form as

$$Z_2 = A^0B^0C^1D^0 + A^0B^0C^1D^1 + A^0B^1C^0D^1 + A^0B^1C^1D^0$$
$$+ A^1B^0C^0D^0 + A^1B^0C^0D^1 + A^1B^1C^0D^0 + A^1B^1C^1D^1$$

where the indices represent the value of the binary variables: X^s is of course a literal. In the usual way a product of distinct literals forms a term and a sum of terms is a sum-of-products expression. Now if we represent a

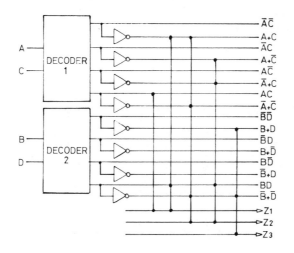

Figure 4.15 Decoded PLA

partition of the variables A, B, C, D by $X_1 = (AC)$ and $X_2 = (BD)$ we may write

$$Z_2 = X_1^{01}X_2^{00} + X_1^{01}X_2^{01} + X_1^{00}X_2^{11} + X_1^{01}X_2^{10}$$
$$+ X_1^{10}X_2^{00} + X_1^{10}X_2^{01} + X_1^{10}X_2^{10} + X_1^{11}X_2^{11}$$

The indices may be combined to affect a reduction of terms using the relationships

$$X^{s_1} \cdot X^{s_2} = X^{s_1 \cap s_2}$$
$$X^{s_1} + X^{s_2} = X^{s_1 \cup s_2}$$
$$\overline{X^{s_1}} = X^{I-s_1}; \ X^I = 1 \text{ and } X^\phi = 0$$

where I is the universal set and consists of all possible combinations of the variables comprising s, and ϕ is the null set. For instance, in function Z_2 above with a two variable partition $I = \{00, 01, 10, 11\}$.

As an example consider the terms $X_1^{01}X_2^{00}$ and $X_1^{10}X_2^{01}$; these may be combined as $X_1^{0110}X_2^{0001}$. Note that multiplying out the index terms gives 0100, 0101, 1000 and 1001 which are all minterms of Z_2 (this is an essential condition, otherwise the function is not covered). Thus, though combining $X_1^{01}X_2^{00}$ and $X_1^{11}X_2^{11}$ gives $X_1^{0111}X_2^{0011}$ it is invalid since the terms 0111 and 1100 are not minterms of the function.

Each term in the function must be compared with every other term (similar to Boolean minimization), rejecting those combinations which generate invalid minterms. Proceeding in this way for the functions Z_1, Z_2 and Z_3 given in Table 4.10 we obtain the equations

$$Z_1 = X_1^{0110}X_2^{11} + X_1^{11}X_2^{0001} + X_1^{11}X_2^{1011}$$
$$Z_2 = X_1^{0011}X_2^{11} + X_1^{0110}X_2^{0110} + X_1^{0110}X_2^{1000}$$
$$Z_3 = X_1^{0010}X_2^{0110} + X_1^{0111}X_2^{0110}$$

These equations may be translated directly into a product-of-sums expressions by taking the inverse, i.e. $\overline{X_1^{0110}} = X_1^{I-(0110)} = X_1^{1100}$ and expressing in maxterm form as $(\bar{A} + \bar{B})(A + B)$. Note that this procedure is identical to reading a product-of-sums function directly from a K-map using the OFF terms.

Thus the equations for Z_1, Z_2 and Z_3 may be expressed as

$$Z_1 = (A + C)(\bar{A} + \bar{C})(\bar{B} + D)(B + \bar{D})(B + D)$$
$$+ (\bar{A} + C)(A + \bar{C})(A + C)(\bar{B} + \bar{D})(\bar{B} + D)$$
$$+ (\bar{A} + C)(A + \bar{C})(A + C)(B + \bar{D})(B + D)$$
$$Z_2 = (\bar{A} + C)(A + \bar{C})(\bar{B} + D)(B + \bar{D})(B + D)$$
$$+ (A + C)(\bar{A} + \bar{C})(B + D)(\bar{B} + \bar{D})$$
$$+ (A + C)(\bar{A} + \bar{C})(B + \bar{D})(\bar{B} + \bar{D})$$
$$Z_3 = (\bar{A} + \bar{C})(A + \bar{C})(B + D)(\bar{B} + \bar{D})$$
$$+ (\bar{A} + C)(A + C)(B + D)(\bar{B} + \bar{D})$$

Note that the equations have a sum-of-product-of-sums form which may be implemented directly using either separate 2-bit decoders or as an integral OR-AND-OR array as shown in Fig. 4.16.

Note that the size of the PLA is influenced by the way the input variables

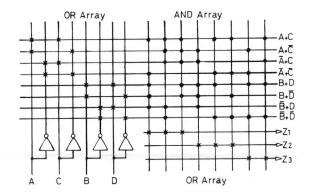

Figure 4.16 Three-level PLA realization

are assigned to the decoders, that is the particular partitioning of the variables. It would also be possible to use decoders with a larger number of input variables and unequal numbers in the partitions (including shared variables), but then the problem of finding the best partition becomes extremely difficult.

4.4 Uncommitted logic arrays (UCLA)[19]

The UCLA concept is based on the principle of producing integrated circuits (normally CMOS or TTL) consisting of an array (matrix) of identical basic logic cells (for instance, a 2-input NAND gate) which has been processed up to but excluding the final interconnection stage. The user would derive the specific interconnections necessary to realize his circuit and then this pattern would be supplied to the component manufacturer who would produce a *metallization mask* to enable the LSI chip to be fabricated. Typical gate arrays would comprise some 500–2000 gates with input/output buffers in a 14–64 pin package.

The gate array, which can be regarded as a *mask programmable* component, enables the benefits of a dedicated custom designed LSI chip to be obtained without being directly concerned with LSI fabrication technology; for this reason the method is also known as *semi-custom IC design*.

The designer may use conventional methods of design, based for example on SN74 series logic, to realize his circuit. Consequently many of the techniques described earlier would be applicable. For large systems the logic diagram would need to be partitioned into modules containing input and output connections equal to or less than the number of pins available on the particular gate array chip.

The designer is, however, more involved with problems of layout and interconnections. This can be performed manually, using techniques similar to printed circuit board layout, or, as is more usual, using computer aided design (CAD) methods.

As we shall see later, UCLAs can be used to develop entire logic systems, including counters, shift registers etc., and in general require sophisticated CAD tools for layout, interconnections, simulation and testing to ensure that the design is functionally and physically correct. The UCLA has been used extensively by computer manufacturers such as IBM and DEC to implement computer mainframe systems[20] and is also used in developing personal microcomputer systems. The concept can also be extended further by considering the use of more complex cells, such as the ULM discussed earlier, to give increased efficiency.

4.5 Circuit hazards

In our discussion of combinational logic we have assumed throughout that logic variables change instantaneously from one logic state to another. However, we have also seen that a finite delay (the propagation time) is encountered when a signal is transmitted through an active gate. Delays in the response of a logic network to a changing signal can cause *hazards*, defined as an actual or potential malfunction of the circuit as a result of signals encountering delays in the paths of the network.

In many cases hazards are generated because of the need to generate the complement of a variable or the reliance on the Boolean theorem $A\bar{A} = 0$ and $A + \bar{A} = 1$. That these theorems cannot hold in practice is obvious when we consider that the inverse of a variable is normally produced by using an inverter element which inserts a delay in the signal path (see Fig. 4.17).

If using these theorems gives rise to hazards how can we rely so much on them in our design work? The answer lies in the temporal nature of the hazard; if we are prepared to wait for the spurious signals to settle out then the correct output will always result. (This is, as we shall see, the reason for using clocked synchronous systems—the inputs and outputs to a circuit are sampled at specific instants of time, after the delays have settled).

The characteristics of a *static hazard* is that it will give rise to a transition in an output (glitch) which is required to remain constant at one value, during a variable change. Static hazards may be caused by single variable changes (the usual case) or multivariable changes such as when two variables are required to change state simultaneously.

Static hazards may easily be detected (predicted) by plotting the function on a K-map and looking for looped variables (those combined together to effect minimization) which share common boundaries. For example, Table 4.11 shows the K-map for a combinational circuit given by the function

$$Z = A\bar{B} + BD$$

Now if $A = D = 1$ the output Z should be 1 irrespective of the value of B, but due to the inherent delay $\bar{B} + B \neq 1$ and therefore a hazard is possible. Note that on the K-map this is predicted by the adjacent loops and that a change in B will cause a transversal.

The prediction of a hazard does not necessarily mean that it will occur in

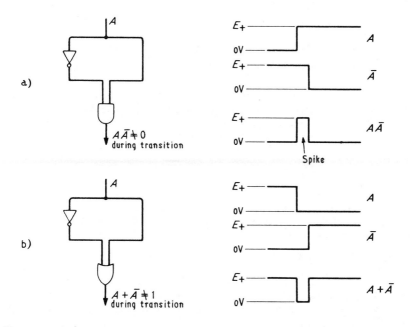

Figure 4.17 Static hazards

practice or, if it does, that it will necessarily have adverse effects. It will depend on the response time and noise immunity of the actual logic circuits used in the network.

Static hazards may be eliminated logically[21] or by the inclusion of additional delays in the network. In the case of our example the hazard may be logically eliminated by including an extra *overlapping* loop AD which gives the function

$$Z = A\bar{B} + BD + AD$$

Now when $A = D = 1$ the output will hold solid for any changes in B. Note that we have added redundancy to the circuit, contrary to our need for minimization!

Minimization procedures quite often lead to hazardous conditions. It can be shown[22] that if a circuit is realized with the complete set of prime implicants no hazards will occur—but redundancy will, of course. Since we always endeavour to find the optimal set of PIs to cover a function it follows that hazards could occur. In effect what is required is a set of PIs which covers the function without generating hazardous conditions; this would mean choosing PIs such that adjacent minterms (on the K-maps) are covered by at least one PI. A tabular method for obtaining minimal hazard-free networks has been described by McCluskey.[23] Note that no change is required in the generation of prime implements but the method of selecting PIs using the PI chart must be modified. Moreover, even if we have a hazard-free circuit such as $Z = A\bar{B} + BD + AD$, it can by factoring

Table 4.11 Static hazards

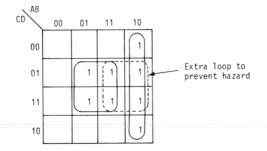

become hazardous again. For example, if the equation was factored for a NOR implementation as below:

$$Z = (\bar{B} + D)(A + B)$$

and the circuit implemented as shown in Fig. 4.18, it is obvious that a hazard will arise when $A = D = 0$ since a change in B from 0 to 1 will cause the output to go momentarily to 1 (when it should stay at zero) due to the delay through the inverter.

The dynamic hazard, which also occurs in combinational circuits, causes an output to change three or more times instead of only once, and so produce sporadic outputs because of a single-variable change. Thus an output required to change from 1 to 0 would, due to a dynamic hazard, change $1 \rightarrow 0 \rightarrow 1 \rightarrow 0$ (note the minimum three changes of output). Should the duration of the output transition be long enough to cause any following circuits to switch over, for example if connected in a feedback loop, the circuit could malfunction.

Dynamic hazards are caused by the existence in the circuit of three or more different signal paths for the same variable, each with differing delay times. This is generally the result of factoring or using long lead connections in fast logic circuits; in relay circuits, contacts with different closure times can produce the same effect. NOR/NAND logic in particular gives rise to this problem, since each unit includes an invertor amplifier with a significant propagation time. Consider the Y-map shown in Table 4.12; there are no static hazards since all prime implicants are present and the excitation equation is given by

$$Y_1 = \bar{x}_2 y_2 + x_1 y_1 + x_1 y_2$$

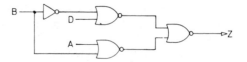

Figure 4.18 Hazard caused by factoring

Table 4.12 Dynamic hazards

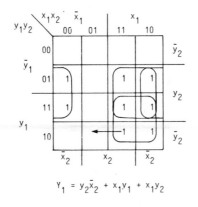

$$Y_1 = y_2\bar{x}_2 + x_1 y_1 + x_1 y_2$$

Now if this equation was implemented directly using three-input NAND gates, the circuit would function perfectly correctly. However, suppose it is necessary to factorize the equations, so that existing gates (part of a much larger logic system) can be used, then a dynamic hazard could arise. Figure 4.19 shows one possible way of factorizing the circuit, assuming gate C giving $\bar{y}_2 \bar{x}_2$ is already available in the system. This is not the best way of implementing the circuit, but since in practice dynamic hazards arise mainly from careless, and sometimes unnecessary, factorization it provides a good example. Analysing the circuit to assure ourselves that it faithfully reproduces the switching equation, we have

$$
\begin{aligned}
Y_1 &= \{[\overline{\overline{(\bar{y}_2 x_1)}(x_2)][(\bar{y}_2 x_1)][(\bar{y}_2 \bar{x}_2)]}}\} \{[\overline{(\bar{y}_2 \bar{y}_1)}(x_1)]\} \\
&= [\overline{(\bar{y}_2 x_1)}(x_2)][(\bar{y}_2 x_1)][(\bar{y}_2 \bar{x}_2)] + [\overline{(\bar{y}_2 \bar{y}_1)}(x_1)] \\
&= [(\bar{y}_2 x_1) + \bar{x}_2][y_2 + \bar{x}_1][y_2 + x_2] + x_1 y_2 + x_1 y_1 \\
&= (\bar{y}_2 y_2 x_1 + \bar{y}_2 x_1 \bar{x}_1 + y_2 \bar{x}_2 + \bar{x}_1 \bar{x}_2)(y_2 + x_2) + x_1 y_2 + x_1 y_1 \\
&= y_2 \bar{x}_2 + y_2 \bar{x}_2 x_2 + \bar{x}_1 \bar{x}_2 y_2 + \bar{x}_1 \bar{x}_2 x_2 + x_1 y_2 + x_1 y_1
\end{aligned}
$$

Hence

$$Y_1 = y_2 \bar{x}_2 + x_1 y_2 + x_1 y_1$$

Now suppose the circuit is in the stable condition $x_1 = x_2 = y_1 = 1$ and $y_2 = 0$, and let x_1 change to 0. The output Y_1 is initially at 1 and when $x_1 \rightarrow 0$ it should go to 0 and stay there. However, from Fig. 4.19 there are three paths, each of different length, from the input signals x_1 to the output gate:

(1) via gates F, G
(2) via gates A, E, G
(3) via gates A, B, E, G.

Thus we can expect dynamic hazards to occur. A convenient way of analysing and appreciating the operation of NAND circuits is to invoke the basic logic properties of the gate. That is, the output will only be logical 0

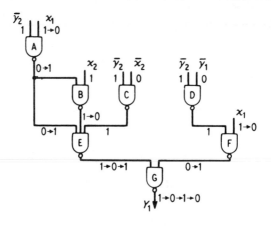

Figure 4.19 Dynamic hazard caused by factoring

when all its inputs are logical 1, thus a 0 on any input will always cause the output to go to 1. Now, if we insert the initial truth values of the variables ($x_1 = x_2 = y_1 = 1$, $y_2 = 0$) on the logic diagram, and then let $x_1 \rightarrow 0$, we can easily see the circuit operation. Assuming equal delays through the NAND elements, when $x_1 \rightarrow 0$ at gate F the output will go to 1; the output of gate E has not yet changed, and consequently the output of gate G will go from 1 to 0. Meanwhile, $x_1 \rightarrow 0$ at gate A causes its output to go to 1; this in turn causes the output of gate E to go to 0 because so far the output of B is unchanged. This causes the output of gate G to change once again to 1; thus so far it has changed $1 \rightarrow 0 \rightarrow 1$. Finally gate B will respond, its output going to 0, which in turn causes the output of gate E to go to 1, giving a final change at the output of G; thus the output Y_1 has changed $1 \rightarrow 0 \rightarrow 1 \rightarrow 0$.

This mechanism is typical of a dynamic hazard and is primarily caused by inept factorization. The hazard cannot be overcome by basic logical design, since the fault arises from the circuit structure; the only remedy is to refactor the circuit. It is good practice always to avoid dynamic hazards where possible because, even if the output transitions are of very short duration, they still have the effect of introducing 'glitches' or noise in the system which is very undesirable in low-level logic circuits.

As we shall see later one way of checking out hazards in a logic circuit is to use a *logic simulator* which effectively goes through the logical evaluation described above. However, it is still necessary to take care that the hazards reported do in fact occur in practice.

References and bibliography

1. Grisamore, N. T., Rotolo, L. S. and Uyehara, G. Y. Logical design using the Stroke function. *IRE Trans Electronic Comput.* **EC7**, 181–3 (1958).

2. Earle, J. Synthesizing minimal Stroke and Dagger functions. *IRE Trans Circuit Theory* **CT7**, 144–54 (1960).
3. Maley, G. A. and Earle, J. *The Logic Design of Transitor Digital Computers*. Prentice Hall, Englewood Cliffs, 1963, Chapter 6.
4. Kintner, P. M. Dual polarity logic as a design tool. *IRE Trans Electron Comput.* **EC8**, 227–8 (1959).
5. Prosser, F. and Winkel, D. Mixed logic leads to maximum clarity with minimum hardware. *Computer Design* **16**(5), 111–7 (1977).
6. Wyland, D. C. Using p/ROMs as logic elements. *Computer Design* **13**(9), 98–100 (1974).
7. Fletcher, W. I. and Despain, A. M. Simplify combination logic circuits. *Electronic Design* **13**, 72–3 (1971).
8. Kramme, F. Standard read-only memories simplify complex logic design. *Electronics* **43**, 89–95 (1970).
9. Dietmeyer, D. L. *Logic Design of Digital Systems* 2nd edition. Allyn & Baker, Boston, Mass., 1979, Chapter 11.
10. Anderson, J. L. Multiplexers double as logic circuits. *Electronics* **42**, 100–5 (1969).
11. Yau, S. S. and Tang, C. K. Universal logic circuits and their modular realisations. *AFIPS Proc. SJCC* **32**, 297–305 (1968).
12. Yau, S. S. and Tang, C. K. Universal logic modules and their applications. *IEEE Trans. Comput.* **C19**, 141–9 (1970).
13. Hope, G. *Integrated Devices in Digital Circuit Design*. Wiley, New York, 1981, Chapter 2.
14. Reyling, G. PLAs enhance digital processor speed and cut component count. *Electronics* Aug. 109–114 (1974).
15. Mitchell, T. Programmable logic arrays. *Electronic Design* **19**(15), 95–101 (1976).
16. Kobylarz, T. and Al-Najjar, A. An examination of the cost functions for programmable logic arrays. *IEEE Trans. Comput.* **C28**, 586–90 (1979).
17. Fleisher, H. and Maissel, L. I. An introduction to array logic. *IBM J. Res. Dev.* Mar. 98–109 (1975).
18. Sasao, T. Multiple-valued decomposition of generalised Boolean functions and the complexity of programmable logic arrays. *IEEE Trans. Comput.* **C30**(9), 635–43 (1981).
19. Tobias, J. R. LSI/VLSI Building Blocks. *IEEE Computer* **14**(8), 83–101 (1981).
20. Cane, D. Semicustom technology derives minicomputer architecture. *Computer Design* Dec. 103–108 (1980).
21. Huffman, D. A. The design and use of hazard-free switching networks. *J. Ass. Comput. Mach.* **4**, 47–62 (1957).
22. Eichelberger, E. B. Hazard detection in combinational and sequential switching circuits. *IBM. J. Res. Dev.* **9**, Jan. 90–9 (1965).
23. McCluskey, E. J. *Introduction to the Theory of Switching Circuits*. McGraw Hill, New York, 1965, Chapter 7.

Tutorial problems

4.1 Implement the logic function $M = \bar{A}\bar{B}C + \bar{D}E + FG$ using 2-input NOR/NAND gates and the mixed logic method of design. All inputs are uncomplemented and in negative logic convention; the output M is required in positive logic.

4.2 Design a convertor circuit for seven-segment LED code to pure binary using ROMs to realize the network. The seven segment code is given in problem 3.11.

4.3 Implement the following logic functions using 3-variable, that is 1-out-of-4, multiplexer units:
(a) the binary full-adder sum logic $S = A\bar{B}\bar{C} + \bar{A}B\bar{C} + \bar{A}\bar{B}C + ABC$
(b) $Z = AB\bar{C}D + B\bar{C}DE + BC\bar{D}\bar{E} + \bar{A}BC\bar{D}$
(c) a 5-bit odd parity bit generator given by
$Z = \Sigma(0, 3, 5, 6, 9, 10, 12, 15, 17, 18, 20, 23, 24, 27, 29, 30)$

4.4 Implement the seven segment decoder described in problem 3.11 using PLAs.

4.5 Implement the function shown in the truth table in Table 4.13 using a decoded PLA with 2-bit decoders.

4.6 Determine whether the circuit shown in Fig. 4.20 will give rise to static hazards and if so show how they may be eliminated.

Table 4.13 Problem 4.5

X_1	X_2	X_3	X_4	F
0	0	0	0	1
0	0	0	1	1
0	0	1	0	1
0	0	1	1	0
0	1	0	0	1
0	1	0	1	1
0	1	1	0	0
0	1	1	1	1
1	0	0	0	1
1	0	0	1	0
1	0	1	0	0
1	0	1	1	1
1	1	0	0	0
1	1	0	1	1
1	1	1	0	1
1	1	1	1	0

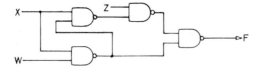

Figure 4.20 Problem 4.6

5 Modular approach to combinational logic

5.1 Introduction

So far our design techniques for combinational logic have been primarily based on a truth table specification and realization in a minterm or sum-of-products/product-of-sums form. In many cases, however, this technique of expressing oral statements for a logic circuit in the form of a truth table is inadequate. For a simple network a terminal description will often suffice but for more complex circuits the truth table approach can often lead to a laborious and inelegant solution.

In particular, when designing logic networks which are to be realized as an integrated circuit it is most important that regular structures are employed in order to conserve physical space on the chip. For instance, functions derived from a truth table and implemented using NOR/NAND logic (in its equivalent MOS transistor circuit form) would be very unstructured and hence inefficient in terms of silicon area utilization and the number of interconnections required. For this reason we tend to use the term *random logic* when referring to circuits of this type. Ideally what is required is a cellular or modular structure which can either be used by itself (such as the PLA) or be repeated (cascaded) to achieve the required logic functions.

The use of *iterative networks* was current practice in the early days of switching systems when relay contacts were the major means of realizing logic circuits. This design methodology, together with other switching contact techniques,[1] fell into disuse when electronic gates became widely available. However, the developments in integrated circuit technology, particularly the use of MOS transfer gates (also called a *pass transistor*) has reawakened interest in these techniques. The NMOS transistor as we have seen in Section 1.7 functions in essence as a simple make or break contact with control being effected via the gate terminal. Since once the gate is high, current and hence logic signals can flow in either direction, the device is bilateral (as is the relay contact) in contrast to electronic gates which only permit a unilateral flow of current. There is, however, a major difference and that is the delay encountered when pass transistors are connected in series. The delay causes a degradation of the signal level which must be reinstated at some stage using an invertor amplifier. In the case of the relay contact, which is essentially a passive switch, the degradation is negligible.

The delay through n pass transistors can be shown to be proportional to n^2 and for NMOS transistors the maximum number of pass transistors that can be connected in series is four;[2] this number can be increased when CMOS technology is used.

In this chapter we shall present some of the fundamental ideas of contact switching circuits and attempt to show their relevance to the logic design of MOS circuits in silicon.

5.2 Iterative circuits

Suppose, for example, a logic system could be decomposed into a number of identical subsystems; then if we could produce a design for the subsystem, or *cell*, the complete system could be synthesized by cascading these cells in series. The problem has now been reduced to that of specifying and designing the cell, rather than the complete system. In Fig. 5.1 the outputs of one cell form the inputs to the next one in the chain and so on, and thus each cell is identical except for the first one (and frequently the last) whose cell inputs must be deduced from the initial conditions. Each cell has external inputs as well as inputs from the preceding cell, which are distinguished by defining the outputs of a cell as its *state*.

We now describe the design of a switching circuit using these ideas, and in so doing show how a typical cell may be specified. Suppose we wish to design a logic network that will detect the occurrence of an error in a five-bit parallel binary number which includes an odd-parity check digit; this is commonly called a parity check circuit. The truth table is shown in Table 5.1 and it is clear on drawing the K-maps that no simplification can be effected using standard techniques. A straightforward approach would mean implementing the equations

$$\begin{aligned}
T = &\ \bar{x}_1\bar{x}_2\bar{x}_3\bar{x}_4\bar{x}_5 + \bar{x}_1\bar{x}_2\bar{x}_3x_4x_5 + \bar{x}_1\bar{x}_2x_3\bar{x}_4x_5 + \bar{x}_1\bar{x}_2x_3x_4\bar{x}_5 \\
&+ \bar{x}_1x_2\bar{x}_3\bar{x}_4x_5 + \bar{x}_1x_2\bar{x}_3x_4\bar{x}_5 + \bar{x}_1x_2x_3\bar{x}_4\bar{x}_5 + \bar{x}_1x_2x_3x_4x_5 \\
&+ x_1\bar{x}_2\bar{x}_3\bar{x}_4x_5 + x_1\bar{x}_2\bar{x}_3x_4\bar{x}_5 + x_1\bar{x}_2x_3\bar{x}_4\bar{x}_5 + x_1\bar{x}_2x_3x_4x_5 \\
&+ x_1x_2\bar{x}_3\bar{x}_4\bar{x}_5 + x_1x_2\bar{x}_3x_4x_5 + x_1x_2x_3\bar{x}_4x_5 + x_1x_2x_3x_4\bar{x}_5
\end{aligned}$$

either as a two-level circuit or using ROM, PLA or multiplexer units (see problem 4.3).

How, then, do we decompose this circuit into an iterative cell configuration? The first step is to decide on the number of state variables and exter-

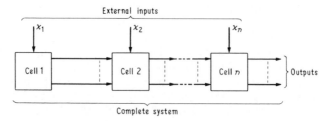

Figure 5.1 Iterative switching system

Table 5.1 Truth table
for parity check circuit

x_1	x_2	x_3	x_4	x_5	T
0	0	0	0	0	1
0	0	0	0	1	0
0	0	0	1	0	0
0	0	0	1	1	1
0	0	1	0	0	0
0	0	1	0	1	1
0	0	1	1	0	1
0	0	1	1	1	0
0	1	0	0	0	0
0	1	0	0	1	1
0	1	0	1	0	1
0	1	0	1	1	0
0	1	1	0	0	1
0	1	1	0	1	0
0	1	1	1	0	0
0	1	1	1	1	1
1	0	0	0	0	0
1	0	0	0	1	1
1	0	0	1	0	1
1	0	0	1	1	0
1	0	1	0	0	1
1	0	1	0	1	0
1	0	1	1	0	0
1	0	1	1	1	1
1	1	0	0	0	1
1	1	0	0	1	0
1	1	0	1	0	0
1	1	0	1	1	1
1	1	1	0	0	0
1	1	1	0	1	1
1	1	1	1	0	1
1	1	1	1	1	0

nal inputs required. In this case we shall choose one single-bit external input per cell; we could equally well have chosen two bits or even more— there is no absolute rule, except the requirement to keep the number as small as possible compared with the total number of system inputs. The choice of one bit per cell simplifies the design of the cell and produces optimum results. Since we have to distinguish between an odd or even number of 1s in the binary word, a single-bit state variable S is sufficient, and we shall call its two values odd and even; the cell may now be represented as in Fig. 5.2. We now have to express the next output state S_n^+ as a function of S_{n-1} and x_n, and this may be done using a *state transfer table* (Table 5.2(a)) where the entries are the resulting (next) output states. From the table it may be seen that if the input state variable indicates that so far the number of 1s is odd ($\bar{S}$) and if the external input is 0, then the condition is unchanged and the output state of the cell must still indicate an odd number of 1s, i.e. $\bar{S}^+$. Thus we may write

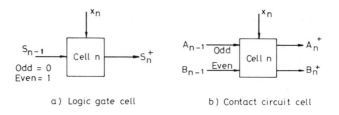

a) Logic gate cell   b) Contact circuit cell

Figure 5.2 Typical cells for parity check circuit

$$S_n^+ = x_n \bar{S}_{n-1} + \bar{x}_n S_{n-1}$$

This equation is the well-known exclusive OR logic function, or half-adder, which is available as a basic element in most logic systems. Thus the parity check circuit can be implemented by cascading the cell circuit, i.e. exclusive OR as shown in Fig. 5.3(b). Because the first cell has no state variable input, the external input alone determines the output state; therefore this may be used as the input to the second cell, that is, external inputs $x_1 x_2$ go to the second cell. This circuit could also have been arrived at by algebraic manipulation of the basic switching equation, but in many cases of this type the algebra involved is tedious.

If the design is to be executed in terms of contact logic, the approach must be slightly modified. Again, the choice of state and external inputs is the preliminary step but the representation of these variables, particularly the output states, must be such as to allow a path-closing contact circuit which gives separate voltage outputs. This means in practice that the odd and even states must be on two separate lines, both connected to a voltage source. This can be represented in the state transfer table (Table 5.2(b)) in the same way as above except that A is the odd line and B is the even line. It is now necessary to derive separate equations for A and B:

$$A_n^+ = \bar{x}_n A_{n-1} + x_n B_{n-1}$$
$$B_n^+ = x_n A_{n-1} + \bar{x}_n B_{n-1}$$

The relay contact circuit is shown in Fig. 5.4; note that the external inputs energize a relay coil with two change-over contacts. As before, the first cell requires only the external input switching line B, as it is assumed that

Table 5.2 State transfer tables for parity check circuit

Input state variable S	External input x_n		Input state variable	External input x_n	
	0	1		0	1
$\bar{S}_{n-1}(0)$	$\bar{S}_n^+(0)$	$S_n^+(1)$	A_{n-1}	A_{n+}	B_{n+}
$S_{n-1}(1)$	$S_n^+(1)$	$\bar{S}_n^+(0)$	B_{n-1}	B_{n+}	A_{n+}
(a) Logic gate implementation			(b) Contact circuit implementation		

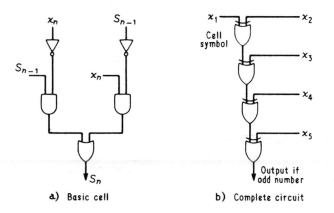

a) Basic cell   b) Complete circuit

Figure 5.3 Parity check circuit using logic gates

previous digits are even, i.e. all zeros. The final output is determined by the presence of a voltage level on either the A or the B line.

The contact network may to all intents and purposes be translated directly into an NMOS pass transistor network as shown in Fig. 5.5. Note the need for a depletion mode transistor to drive the network and the use of invertor amplifiers to provide the control signals.

Iterative cell techniques are particularly well suited to pattern recognition and encoding and decoding circuits with a large number of parallel inputs. Furthermore, circuit specification is simplified and large variable problems reduced to a more tractable size. The method is also directly applicable to the design of LSI circuits and has the advantage of producing a modular structure based on a standard cell which may be optimized independently in terms of layout etc. Circuits containing any number of input

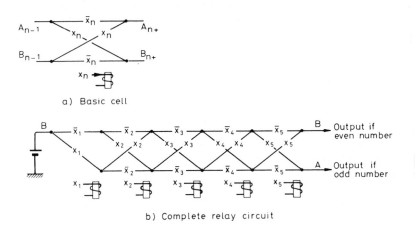

a) Basic cell

b) Complete relay circuit

Figure 5.4 Parity check circuit using relays

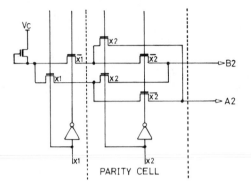

Figure 5.5 NMOS pass transistor version of parity circuit

variables can easily be constructed by simply extending the network with more cells.

The parity check circuit, for example, was designed for a 5-bit word, but should the requirement change and a 15-bit message be desired, then the circuit could easily be modified by the addition of the appropriate number of exclusive OR cells. With a conventionally designed system using random logic a modification would necessitate a major redesign exercise.

However, iterative circuits do have the disadvantage of reducing the speed of a system because of the time required for the signals to propagate through the network; the number of interconnections is also considerably increased. With conventional implementations these factors would be considered major deterrents to their use. With LSI circuits, however, the physical length of the signal paths are orders of magnitude smaller, hence negating to a large extent the problems of overall propagation speed. Again though the interconnections are increased they form a regular pattern which allows an economical (in terms of silicon area) layout.

One disadvantage is the need to insert invertors to resuscitate the signals when using pass transistors. However, the clustering of logic in iterative networks makes it fairly easy to insert interstage buffers as part of the cell. Note also that logic performed in this way, by steering signals with pass transistors, does not require static power dissipation, a considerable advantage with LSI circuits.

5.2.1 Design for an iterative decoder

To conclude our discussion of iterative networks let us apply the technique to a more complicated circuit, and implement the design in both contact and LSI logic. In many data transmission systems a coding method is employed which defines a codeword as having a certain number of 1s, e.g. 2-out-of-5 code discussed earlier; errors due to digits being dropped or inserted would corrupt this pattern. We will now design a circuit to detect errors occuring in an eight-bit parallel word coded in the 3-out-of-8 code. The standard approach to this problem would lead to a truth table with 256 combinations, of which 56 (normal combination $_8C_3$) would need to be recognized as codewords. Using NAND/NOR logic with a fan-in of eight, the circuit implementation would require 72 basic units; considerably more would be required with a smaller fan-in factor due to the necessity for branching. Realizing the circuit directly in LSI modules would require a 64×1 bit ROM or a PLA with 8 inputs, 56 product terms and 1 output. Either would be an effective implementation at the subsystem level if the word length was known to be fixed at 8 bits.

In the iterative design, we shall again choose a one-bit external variable, but in this case we have a larger number of state variables. It is necessary to know whether the preceding digits sum to zero, one, two, three, or greater than three digits. The state variable indicating three digits is used to signify a correct codeword. The state transfer table is shown in Table 5.3(b). Since we have five states A, B, C, D, E, we will need three bits (using gated logic) to represent them: these are x_1, x_2 and x_3; the external input is designated y_n. Note that in assigning the state variables, we have chosen 100 to indicate the correct codeword, i.e. three digits only; this allows us to economize in the final cell, as only x_1 need be examined.

The output state equations are obtained by inspecting the transfer table for the conditions that cause the output variables to go to 1. For example x_{1+}, the next output state of x_1, goes to 1 when input states x_1, x_2, x_3 are equal to 010, and the external input goes to 1, i.e. $\bar{x}_1 x_2 \bar{x}_3 y_n$. Similarly, x_{1+} goes to 1 when $x_1 \bar{x}_2 \bar{x}_3 \bar{y}_n$ occurs. Thus we may write

Table 5.3 State transfer tables for 3-out-of 8 circuit

Input state variables	External input y_n 0	1
A Sum zero	A_+	B_+
B Sum one	B_+	C_+
C Sum two	C_+	E_+
D Sum > three	D_+	D_+
E Sum three	E_+	D_+

(a) General table, used for contact circuit.

| | Input state variables | | | External input y_n | |
	x_1	x_2	x_3	0	1
A	0	0	0	0 0 0	0 0 1
B	0	0	1	0 0 1	0 1 0
C	0	1	0	0 1 0	1 0 0
D	0	1	1	0 1 1	0 1 1
E	1	0	0	1 0 0	0 1 1
Don't-care terms {	1	0	1		
	1	1	0		
	1	1	1		

(b) Logic gate table

$$x_{1+} = x_1\bar{x}_2\bar{x}_3\bar{y}_n + \bar{x}_1x_2\bar{x}_3y_n$$
$$x_{2+} = \bar{x}_1\bar{x}_2x_3y_n + \bar{x}_1x_2\bar{x}_3\bar{y}_n + x_1\bar{x}_2\bar{x}_3y_n + \bar{x}_1x_2x_3\bar{y}_n + \bar{x}_1x_2x_3y_n$$
$$x_{3+} = \bar{x}_1\bar{x}_2\bar{x}_3y_n + \bar{x}_1\bar{x}_2x_3y_n + x_1\bar{x}_2\bar{x}_3y_n + \bar{x}_1x_2x_3y_n + \bar{x}_1x_2x_3\bar{y}_n$$

These equations may be minimized using standard techniques, in this case K-maps (Table 5.4). It should be pointed out that the way in which the assignment of state variables is made determines to what extent the equations can be minimized. In the example above, the assignment has been a straightforward allocation of ascending pure binary, but it is conceivable that a different assignment could give a more minimal final set of equations. This problem is identical to that of assigning internal states in a sequential logic circuit, and will be discussed in detail in later chapters on sequential logic. From the K-maps the minimal expressions are

$$x_{1+} = x_1\bar{y}_n + x_2\bar{x}_3y_n$$
$$x_{2+} = x_2\bar{y}_n + x_3y_n + x_1y_n$$
$$x_{3+} = x_2x_3 + x_3\bar{y}_n + \bar{x}_2\bar{x}_3y_n$$

Table 5.4 K-maps for 3-out-of-8 iterative circuit

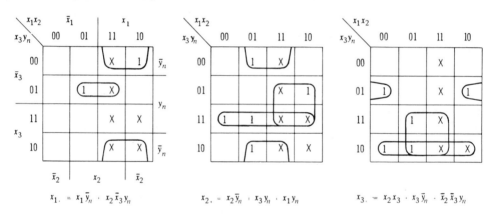

These equations could either be implemented in NOR/NAND logic or preferably, and applicable to both MSI and LSI realization, a PLA or multiplexer circuit. The first cell can be simplified since the initial number of digits must be zero and the output states will depend only on the value of y_1. The network is shown implemented with PLAs in Fig. 5.6; note that each PLA is identical comprising 4 inputs, 8 product terms and 3 outputs.

Using path closing logic the procedure is modified, since we require one line per state variable; hence the following state variable equations are obtained directly from Table 5.3(a):

$$A_+ = A\bar{y}_n$$
$$B_+ = Ay_n + B\bar{y}_n$$
$$C_+ = C\bar{y}_n + By_n$$
$$D_+ = Dy_n + D\bar{y}_n + Ey_n = D + Ey_n$$
$$E_+ = E\bar{y}_n + Cy_n$$

As a correct word will be indicated by the presence of a voltage on line E, variable D is redundant since it is not necessary to indicate directly when more than three digits occur. This can be ascertained by line E being un-energized; thus we can say that no output is required for the conditions Ey_n or D. This differs from the design using logic gates in which each input state variable must result in a definite output state variable. The contact circuit for the cell is shown in Fig. 5.7(a). The first cell can be simplified since the only input variable is A (sum zero); similarly, the second cell requires only A and B (sum one), and the third requires only A, B and C (sum two). Furthermore, the last cells in the system can also be simplified, since it is only necessary to retain the final output line E. Thus the last three cells can be contracted as shown in the complete contact circuit shown in Fig. 5.7(b). The circuit may also be converted directly into NMOS pass transistors as shown earlier; note once again the ease of in-serting interstage buffer amplifiers.

5.3 Symmetric functions

From the last section it will be clear that there are considerable advantages in designing iterative networks for realization using path closing logic such

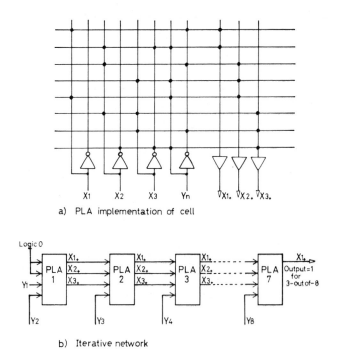

a) PLA implementation of cell

b) Iterative network

Figure 5.6 3-out-of-8 decoder circuit using PLAs

as contacts or pass transistors. Moreover, the final circuit is in the form of a bridge circuit rather than the more normal series–parallel arrangement (SOP or POS) obtained by truth table methods. Bridge networks and non-planar circuits (circuits that cannot be drawn without crossing lines) require far less devices and hence are more economical and reliable in use.

However, the design of such circuits can lead to difficulties in cell speci-fication and to problems in determining the conditions that govern the simplification of the initial and final cells. For certain types of circuit (those which can be represented by a *symmetric logic function*[3]) the design can be greatly simplified.

A symmetric function is a logic function which has the property of remaining unaltered when any two at least of its variables (called the *variables of symmetry*) are interchanged. For example, the function

$$T = \bar{x}\bar{y}z + \bar{x}y\bar{z} + xy\bar{z}$$

is symmetric, since if the variables x and y are interchanged (i.e. replace all x's with y's and all $\bar{x}$'s with $\bar{y}$'s, and vice versa) we obtain

$$T = \bar{y}\bar{x}z + \bar{y}x\bar{z} + yx\bar{z}$$

which is identical to the original function. Note that all terms in the func-tion are prime implicants and all three are required in the minimum sum;

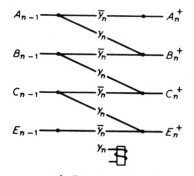

a) Typical contact cell

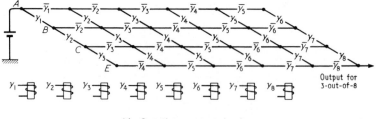

b) Complete contact circuit

Figure 5.7 3-out-of-8 decoder circuit using relays

this is normally the case with symmetric functions. It is also worth noting that the exclusive OR function is, of course, symmetric. The variables of symmetry can also take the complemented form, e.g.

$$T = \bar{x}\bar{y}z + x\bar{y}\bar{z} + xyz$$

is symmetric with the variables x and $\bar{y}$. In this case, we replace x by $\bar{y}$, $\bar{x}$ by y, y and $\bar{x}$, obtaining the identical function

$$T = yxz + \bar{y}x\bar{z} + \bar{y}\bar{x}z$$

Symmetric functions with uncomplemented variables of symmetry are called *n-out-of-m functions*, that is, the logic function equals 1 when precisely n variables out of the total m are equal to 1. For example, the decoder circuit designed in Section 5.2 could be described as a symmetric 3-out-of-8 function. The equation

$$T = \bar{x}\bar{y}z + \bar{x}y\bar{z} + x\bar{y}\bar{z}$$

represents a 1-out-of-3 function and can be symbolized as $S_1^3(xyz)$; the decoder circuit would be represented as $S_3^8(y_1y_2y_3y_4y_5y_6y_7y_8)$. The number of variables which must be one for the function to be true is normally called the *a-number* after Shannon.[3]

The symbology can be extended to functions which equal 1 when, say, two or three of the variables equal 1, i.e. $S_{23}^4(ABCD)$. The parity checking circuit discussed earlier could be described as an $S_{024}^5(ABCDE)$ function. Furthermore, it can be shown that symmetric functions can be manipulated algebraically, e.g. cascaded circuits, equivalent to logical AND, would be combined:

$$[S_{23}^5(ABCD)][S_{12}^5(ABCD)] = S_2^5(ABCD)$$

and in parallel (logical OR):

$$[S_{23}^5(ABCD)] + [S_{12}^5(ABCD)] = S_{123}^5(ABCD)$$

One of the difficulties of using symmetric functions is their recognition.

This is fairly simple if we have an *n-out-of-m* description but otherwise we need to resort to specific identification procedures. Techniques have been described by both Caldwell[1] and McClusky;[4] we shall use the latter's approach.

Consider the function

$$Z = \Sigma(7, 11, 13, 14, 19, 21, 22, 25, 26, 28)$$

To determine if the function is symmetric it is first expressed in binary form and arranged in columns, as shown in Table 5.5. Now for each column we sum the number of 1s and express it as a ratio to the number of 0s. If the function is symmetric and the variables are not mixed this ratio will be the same for all columns. If the variables are mixed then the reciprocal ratio will be obtained for those columns which represent the complementary variables. The number of 1s in each row of the table is also determined which should again be the same for all rows if there are no mixed variables and a single *a*-number.

Table 5.5 Identification of symmetric functions

	A	B	C	D	E	Number of ones
	0	0	1	1	1	3
	0	1	0	1	1	3
	0	1	1	0	1	3
	0	1	1	1	0	3
	1	0	0	1	1	3
	1	0	1	0	1	3
	1	0	1	1	0	3
	1	1	0	0	1	3
	1	1	0	1	0	3
	1	1	1	0	0	3
Ratio	$\frac{6}{4}$	$\frac{6}{4}$	$\frac{6}{4}$	$\frac{6}{4}$	$\frac{6}{4}$	

In the case of our example we find that all columns have the ratio 6/4 and all rows contain three 1s. We must, however, check that all possible combinations are present; in this case the number of combinations of 5 things taken 3 at a time is given by

$$_5C_3 = \frac{5!}{3!(5-3)!} = 10$$

Since all combinations are present the function is symmetric and can be represented by $S_3^5(ABCDE)$. Again, consider the function shown in Table 5.6(a). This time two of the column ratios (for variables A and C) are the inverse of the others indicating that they could be complemented variables. Note that if there are more than two different column ratios the function cannot be symmetric.[5]

Complementing the variables A and C we obtain the table shown in Table 5.6(b). Note that all the ratios are now the same and the row sums indicate that the a-numbers are 2 and 4—this is confirmed by checking that all combinations are indeed present. The function can now be identified as $S_{2,4}^5(\bar{A}B\bar{C}DE)$.

Note that three conditions must be satisfied for the function to be symmetric:

(a) The ratio of 1s to 0s must be the same for all columns.
(b) The number of ones in each term must be the same for all terms representing a particular a-number.
(c) The number of terms of the same a-number must be as given by the combinatorial formula.

In some cases it can happen that all the column ratios are the same but the number of row occurances for a particular a-number is incomplete (see Table 5.7(a)). If the function is symmetric the number of 1s in each column (which will be identical) will be equal to half the number of rows in the

Table 5.6 Identification of complemented variables

A	B	C	D	E	Number of ones		$\bar{A}$	B	$\bar{C}$	D	E	Number of ones
0	0	0	0	0	0		0	0	0	1	1	2
0	0	0	1	1	2		0	0	1	0	1	2
0	0	1	0	1	2		0	0	1	1	0	2
0	0	1	1	0	2		0	1	0	0	1	2
0	1	0	0	1	2		0	1	0	1	0	2
0	1	0	1	0	2		0	1	1	0	0	2
0	1	1	0	0	2		0	1	1	1	1	4
0	1	1	1	1	4		1	0	0	0	1	2
1	0	0	0	1	2		1	0	0	1	0	2
1	0	0	1	0	2		1	0	1	0	0	2
1	0	1	1	1	4		1	0	1	1	1	4
1	1	0	0	0	2		1	1	0	0	0	2
1	1	0	1	1	4		1	1	0	1	1	4
1	1	1	0	1	4		1	1	1	0	1	4
1	1	1	1	0	4		1	1	1	1	0	4

Ratio $\quad \frac{7}{8} \quad \frac{8}{7} \quad \frac{7}{8} \quad \frac{8}{7} \quad \frac{8}{7}$ $\qquad$ Ratio $\quad \frac{8}{7} \quad \frac{8}{7} \quad \frac{8}{7} \quad \frac{8}{7} \quad \frac{8}{7}$

(a) $\qquad\qquad\qquad\qquad\qquad$ (b)

table.[5] In this case there are six rows and the column sum of 1s is three; hence the function is symmetric.

To determine the actual function it is necessary to partition about any of its variables and perform a partial summation but ignoring the partitioning variable. In this case we have chosen to partition about $A = 0$ and $A = 1$ to give Table 5.7(b). It will be apparent from the table that variable B requires to be inverted which then yields Table 5.7(c) and the conclusion that the function is $S_2^4(A\bar{B}CD)$.

5.3.1 Design of symmetric circuits

How, then, can the recognition and use of symmetric functions assist us in the design of switching circuits? Symmetric functions can be represented by a basic contact (path closing) network which has one input and which branches out to give $m + 1$ outputs, where m is the number of variables (Fig. 5.8). Tracing through the network it will be seen that all possibilities are covered for the variables $(ABCD)$ in this topological representation; the pattern can of course be enlarged to cover any number of variables.

Symmetric circuits are of considerable importance in LSI/VLSI design (called *tally circuits* by Mead and Conway[2]) since the contact network may be mapped directly into NMOS circuitry—see Fig. 5.9. The circuit is such that a logic 1 (high) signal will propagate through the network from the pull-up transistor to an output, with the particular path being defined by the states of the input variables (a high signal effectively closes the path); logic 0 signals will propagate from ground to all other outputs (note the need to insert additional pass transistors).

Table 5.7 Special case of symmetric identification

A	B	C	D	Number of ones
0	0	0	1	1
0	0	1	0	1
0	1	1	1	3
1	0	0	0	1
1	1	0	1	3
1	1	1	0	3
$\frac{3}{3}$	$\frac{3}{3}$	$\frac{3}{3}$	$\frac{3}{3}$	

(a)

A	B	C	D	A	B	C	D
0	0	0	1	1	0	0	0
0	0	1	0	1	1	0	1
0	1	1	1	1	1	1	0
$\frac{1}{2}$	$\frac{2}{1}$	$\frac{2}{1}$		$\frac{2}{1}$	$\frac{1}{2}$	$\frac{1}{2}$	

(b)

A	$\bar{B}$	C	D	Number of ones
0	1	0	1	2
0	1	1	0	2
0	0	1	1	2
1	1	0	0	2
1	0	0	1	2
1	0	1	0	2
$\frac{3}{3}$	$\frac{3}{3}$	$\frac{3}{3}$	$\frac{3}{3}$	

(c)

As before, a larger number of input variables can be handled by simply extending the array. However, because of the delay through the pass transistors it may be necessary to insert level restoration circuits between stages. The design of the 3-out-of-8 circuit now becomes very simple: we merely draw the appropriate symmetric contact circuit for eight variables, but only include the contacts necessary to give an output when $n = 3$, ignoring all other outputs. If we compare the resultant circuit (Fig. 5.10) with that obtained earlier, we see that they are identical. The cell structure can easily be seen, and the contraction of the initial and final cells is obtained automatically. The same approach may be used to design sym-

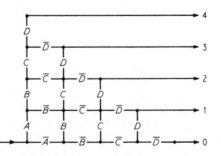

Figure 5.8 Basic symmetric contact circuit

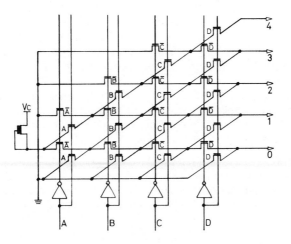

Figure 5.9 NMOS implementation of symmetric network

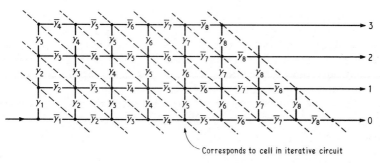

Corresponds to cell in iterative circuit

a) Basic circuit with 0-,1-,2- and 3-out-of-8 outputs

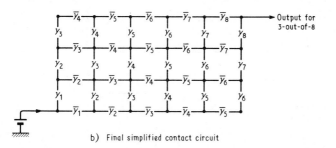

b) Final simplified contact circuit

Figure 5.10 Symmetric contact circuit for 3-out-of-8 decoder

metric circuits with multiple outputs, i.e. circuits represented by functions of the type $S_{024}^5(ABCDE)$. In this case, the circuit for m variables is drawn for the required outputs n_1, n_2, etc., which are then simply joined together. Simplification may be effected by applying the following rules.[11]

(a) If the difference between the subscript is 1 (that is, adjacent outputs) when the outputs are combined, we can apply the $A + \bar{A} = 1$ rule. For example, in Fig. 5.11(a) when we combine outputs 1 and 2 they are connected to point X by the contacts $D + \bar{D} = 1$, which can be eliminated.

(b) If the difference between the subscripts is greater than 1 and they form an arithmetic progression, the network may be 'folded over'— the next term in the progression must be greater than the number of variables.

Suppose we were to implement $S_{13}^4(ABCD)$. We first draw a circuit for 1-out-of-4 (the lowest subscript); then instead of drawing 3-out-of-4 in the normal way to complete the circuit (Fig. 5.11(b)), we 'fold' the circuit over and utilize the common set of contacts in the 1-out-of-4 circuit to get the circuit shown in Fig. 5.11(c).

5.4   Realization of non-series–parallel circuits

The design of combinational circuits using iterative or symmetric techniques leads directly to non-series–parallel circuits. Is it possible, however, to devise bridge networks, starting from the transmission function, for circuits derived from a truth table approach? In fact several methods have been evolved[4,6] but the technique we shall discuss here is one that employs Boolean matrices and is due to Hohn and Schissler.[7]

The advantage of using non-series–parallel circuits is that they offer the ultimate in contact (and hence MOS transistor) economy. Care must be

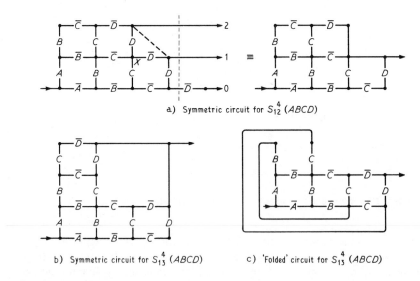

a) Symmetric circuit for S_{12}^4 $(ABCD)$

b) Symmetric circuit for S_{13}^4 $(ABCD)$ c) 'Folded' circuit for S_{13}^4 $(ABCD)$

Figure 5.11 Multiple-output symmetric circuit

taken, however, to ensure that 'sneak paths' do not occur in the network due to the bilateral nature of the devices. In such cases an undesirable path is established which can have the effect of introducing unwanted switching terms.

Two-terminal contact circuits can be precisely described using a Boolean matrix. Consider the circuit shown in Fig. 5.12: the *primitive connection* matrix alongside it specifies the identify and location of each contact.

Note that the entries are the actual single contacts (in some cases contacts in parallel) between the pairs of nodes specified and *not* the transmission between nodes. Thus we enter a 1 for a connection between a node and itself, which forms the principal diagonal, and a 0 for no connection; furthermore, the matrix is symmetrical about its principal diagonal.

The primitive connection matrix may be reduced by a process known as *node removal*. For a matrix of n columns and m rows with entries Y_{nm}, to remove a node each Y_{nm} is replaced by

$$Y_{nm} = Y_{nm} + (Y_{nx} . Y_{xm})$$

where x is the removed node. To remove node 4 in the matrix and by so doing forming a new matrix with four columns and rows, we compute the new entries:

$$\begin{aligned} Y_{15} &= Y_{15} + (Y_{14} . Y_{45}) = C + B.0 = C \\ Y_{13} &= Y_{13} + (Y_{14} . Y_{43}) = A + B\bar{C} \\ Y_{12} &= Y_{12} + (Y_{14} . Y_{42}) = 0 + B\bar{D} = B\bar{D} \\ Y_{11} &= Y_{11} + (Y_{14} . Y_{41}) = 1 + B.B = 1 \end{aligned}$$

When all the entries have been modified in this fashion, the reduced matrix is

$$\begin{array}{c} \\ 1 \\ 2 \\ 3 \\ 5 \end{array} \begin{array}{c} \begin{array}{cccc} 1 & 2 & 3 & 5 \end{array} \\ \begin{bmatrix} 1 & B\bar{D} & A + B\bar{C} & C \\ B\bar{D} & 1 & \bar{B} + \bar{C}\bar{D} & D \\ A + B\bar{C} & \bar{B} + \bar{C}\bar{D} & 1 & 0 \\ C & D & 0 & 1 \end{bmatrix} \end{array}$$

If this process is repeated by removing node 3, and then node 5, we finally arrive at the matrix:

$$\begin{bmatrix} 1 & B\bar{D} + (A + B\bar{C})(\bar{B} + C\bar{D}) + CD \\ B\bar{D} + (A + B\bar{C})(\bar{B} + \bar{C}\bar{D}) + CD & 1 \end{bmatrix}$$

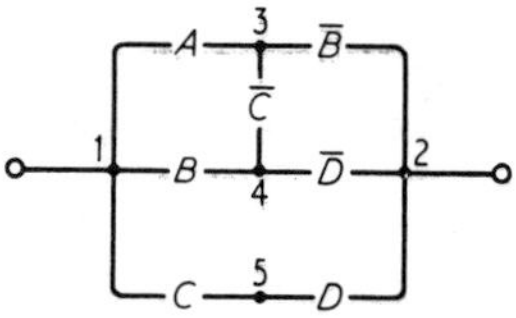

$$\begin{array}{c} \text{Node} \\ 1 \\ 2 \\ 3 \\ 4 \\ 5 \end{array} \begin{array}{c} \begin{array}{ccccc} 1 & 2 & 3 & 4 & 5 \end{array} \\ \begin{bmatrix} 1 & 0 & A & B & C \\ 0 & 1 & \bar{B} & \bar{D} & D \\ A & \bar{B} & 1 & \bar{C} & 0 \\ B & \bar{D} & \bar{C} & 1 & 0 \\ C & D & 0 & 0 & 1 \end{bmatrix} \end{array}$$

Figure 5.12 Typical contact network

Now the expression $B\bar{D} + CD (A + B\bar{C})(\bar{B} + \bar{C}\bar{D})$ simplifies to $B\bar{D} + CD + A\bar{B} + A\bar{C}\bar{D}$ which is the transmission between nodes 1 and 2. Thus the output matrix of a two-terminal circuit may be written

$$\begin{bmatrix} 1 & T \\ T & 1 \end{bmatrix}$$

where T is the transmission function of the entire network.

Thus if the transmission function for a required circuit is known, or can be deduced from the truth table, it can be represented in matrix form and expanded into a primitive connection matrix; in this way a non-series–parallel circuit may be synthesized. Unfortunately, the expansion of an output matrix to a primitive connection matrix does not yield a unique solution, since the result depends on the expansion method and the order employed; furthermore, the resulting circuit is not necessarily minimal. For some circuits it is not possible, or even desirable, to expand to a full primitive matrix. When this happens a *connection matrix* results with a reduced number of nodes, and contacts consequently appear in series.

5.4.1 Synthesis of bridge networks

To illustrate the arguments above we will synthesize a non-series–parallel circuit starting with the transmission function

$$T = A\bar{C}\bar{D} + ACD + \bar{A}C\bar{D}$$

Before we represent the function in matrix form it is preferable, and sometimes essential, to factorize the equation. This may be done using the mapping technique described in Section 4.2.1; the process is shown in Table 5.8. The resulting factored form of the function is

$$T = (\bar{A} + \bar{C}\bar{D} + CD)(A + C\bar{D})$$

This may be expressed in the matrix form

$$\begin{bmatrix} 1 & 0 & \bar{A} + \bar{C}\bar{D} + CD \\ 0 & 1 & A + C\bar{D} \\ \bar{A} + \bar{C}\bar{D} + CD & A + C\bar{D} & 1 \end{bmatrix}$$

Observing that $\bar{D}$ is common to both $\bar{A} + CD + \bar{C}\bar{D}$ and $A + C\bar{D}$ we can expand using this factor:

$$\begin{bmatrix} 1 & 0 & \bar{A} + CD & \bar{C} \\ 0 & 1 & A & C \\ \bar{A} + CD & A & 1 & \bar{D} \\ \bar{C} & C & \bar{D} & 1 \end{bmatrix}$$

Finally, we arrive at the primitive connection matrix:

$$\begin{array}{c} \\ 1 \\ 2 \\ 3 \\ 4 \\ 5 \end{array} \begin{array}{ccccc} 1 & 2 & 3 & 4 & 5 \\ \begin{bmatrix} 1 & 0 & \bar{A} & \bar{C} & C \\ 0 & 1 & A & C & 0 \\ \bar{A} & A & 1 & \bar{D} & D \\ \bar{C} & C & \bar{D} & 1 & 0 \\ C & 0 & D & 0 & 1 \end{bmatrix} \end{array}$$

Table 5.8 K-map factorization

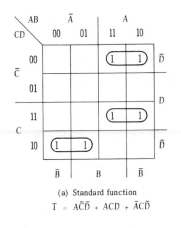

(a) Standard function

$$T = A\bar{C}\bar{D} + ACD + \bar{A}C\bar{D}$$

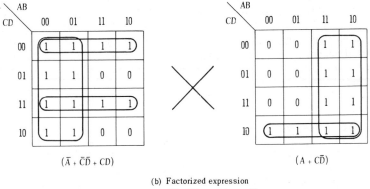

$(\bar{A} + \bar{C}\bar{D} + CD)$ $(A + C\bar{D})$

(b) Factorized expression

$$T = (\bar{A} + \bar{C}\bar{D} + CD)(A + C\bar{D})$$

Note that the method of expansion consists of picking out common factors and applying the reduction formula in reverse. In order to retain the zero entries, it is often necessary to insert other zeros in the appropriate position to ensure cancellation when the matrix is reduced. The final result must be checked to ensure that the original transmission function can be obtained by reducing the matrix. The network obtained by implementing the primitive matrix is shown in Fig. 5.13; the total number of contacts required has been reduced by two to seven.

The matrix method has possibilities of being a very powerful tool for contact network synthesis. As yet, however, very little is known about formal minimizing procedures or ways in which near minimal circuits may be obtained or how they can be applied to VLSI. Since a considerable amount of work is involved in manipulating matrices, even for a small number of variables, digital computer methods must be used to handle large switching matrices.

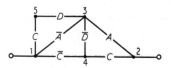

Figure 5.13 Matrix synthesized contact network

5.5 Switching tree circuits

Suppose we need to design a multi-output decoder circuit in which each input combination has a unique output associated with it, but did not wish to use a conventional two-level network—in this case we would use a *tree circuit*.

For example, consider a circuit for decoding four-bit pure binary code to ten output lines each representing one of the ten equivalent decimal numbers 0–9. This could be done quite easily by using normal truth table methods, and the result would be an array of ten four-input AND gates, one for each combination. However, suppose only two-input logic modules were available; this would mean splitting the input combinations, and the modified circuit would require three two-input AND gates per combination, giving a total of 30 gates. Is there any other way of producing a more economic circuit?

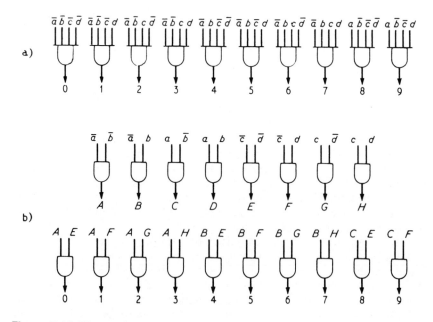

Figure 5.14 Electronic tree circuit

If we take the total number of input variables and divide them as evenly as possible into two integral numbers, continuing until we reach 2 or 3, we obtain

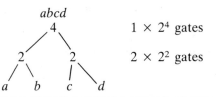

1×2^4 gates

2×2^2 gates

Each number in the flow diagram represents the total number of variables being switched in all possible combinations at that point. Thus, each 2 represents two variables, say a and b, switched in all four combinations $\bar{a}\bar{b}$, $\bar{a}b$, $a\bar{b}$ and ab. Similarly, 4 represents all possible combinations of the four variables a, b, c and d. This means that for each number, n, on the diagram, the total number of AND gates required at that point is 2^n; also the lines give the total number of gate inputs. The diagram may be directly related to a logic circuit (except the logic diagram is drawn reverse-way up) as shown in Fig. 5.14(b). The total number of two-input AND gates required has been reduced to 18, a saving of 12 gates. Note that if all the output combinations had to be decoded, 24 two-input AND gates would be needed as against 16 four-input AND gates, and the total number of gate inputs would remain constant at 48.

Let us now implement the circuit in terms of contact, path closing logic; the full tree for four variables is shown in Fig. 5.15 and may be extended to any number of variables. The contacts within the dotted line are unnecessary in the case of the four-bit decoder, as only the decimal outputs 0–9 are required. In fact, if only these combinations can occur, the others being 'can't happen' conditions, we can also eliminate contacts $\bar{b}$ and $\bar{c}$. Marcus[8]

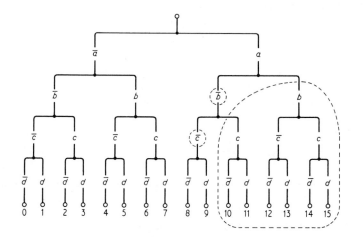

Figure 5.15 Contact tree network

has described a map technique for designing and minimizing partial trees (as in the case above) to obtain an optimum contact utilization. Briefly, this consists of plotting the required combinations on a K-map and subdividing the map in such a way that the *unused combinations* form a minimum number of sub-groups. The subdivision is commenced (see Table 5.9(b)) by separating the combinations into two groups each dependent on the alternative states of a variable, in this case *a*, the appropriate literal being inserted in each square. The subdivision is continued, ignoring unused combinations, with variable *b*, followed by *c* and *d*; note that the subdivision is such that the unused combinations form the largest possible groups. In general, there are many alternative ways of subdividing, each yielding a feasible tree network though not necessarily an optimum one. The order of subdivision, recorded in each square by the literal term (see Table 5.9(e)) gives the contact path directly; this is identical to Fig. 5.15.

If the combinations required are the only ones that can occur, as could be the case in the four-bit decoder, the procedure can be simplified. The division process is repeated, but only to separate the required combinations; for instance, in Table 5.9(f), the map is subdivided as before for $\bar{a}$ and *a*. The $\bar{a}$ group is again divided by *b*, *c* and *d*, but the *a* group needs only to be separated by variable *d*, to yield the final circuit.

5.6 Cellular arrays

A different approach to the implementation of combinational (and sequential) logic systems is the use of *cellular arrays*. This type of circuit, because of its inherent modularity, leading to simple and regular structures and the promise of concurrent (parallel) processing, has attracted considerable interest for VLSI realizations. We have already encountered one simple type of array circuit, the iterative network, in which identical cells generating the same function are cascaded in series.

Arrays may be classified in the following manner:

(a) according to the type of cell, whether all cells generate the same logic function or are programmable to give different functions;
(b) the physical dimensions of the array, whether it is linear (*1-dimensional*) or rectangular (*2-dimensional*);
(c) the direction in which signals can flow through the array, whether in one direction only (*unilateral*) or in two normally opposite directions at the same time (*bilateral*).

For example, the iterative network discussed earlier would be classified as a 1-dimensional unilateral array.

Some examples of array structures are shown in Fig. 5.16. Note that it is necessary to specify the primary inputs and outputs for the array (depending on the application and logical function of the cell) and the values of any unspecified signals entering the edges of the array (called the *boundary conditions*). This latter process is identical to determining the initial input values for the first cell in an iterative network. Note also the intercell con-

Table 5.9 Design of tree networks

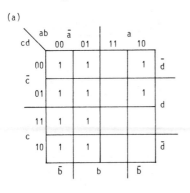

(a)

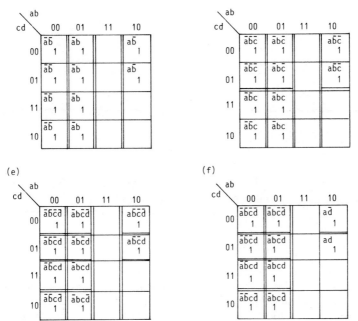

(b) (c) (d) (e) (f)

nections (the output(s) of one cell acting as the input(s) for other cells) and the possibility of bilateral signal flow.

5.6.1 One-dimensional linear arrays

As an example of a simple array let us examine a linear cascade of 3-input 2-output cells where each cell is assumed to be capable of realizing any pair of combinational functions of its three input variables.[9] Since the two

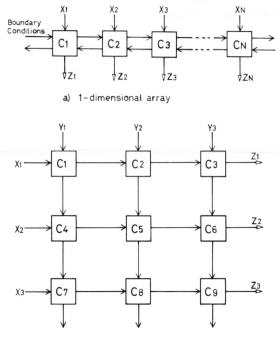

a) 1-dimensional array

b) 2-dimensional array

Figure 5.16 Array structures

outputs of a cell go to the inputs of the following cell in the cascade this
form of array is also known as a *two-rail* cascade.

Now, from Fig. 5.17(a), if the following output functions are selected:

$$f_1 = xy_1; f_2 = \bar{x}y_1; f_3 = 1$$
$$g_1 = y_2 + xy_1; g_2 = y_2 + \bar{x}y_1; g_3 = y_2$$

then we have a total of nine different cell types which may be used to
generate any combinational function in sum of products form. This is
achieved by forming the product terms, on the z_1 output leads of cells
performing the functions $z_1 = f_1; z_2 = g_3$ or $z_1 = f_2; z_2 = g_3$ and summing
the product terms so formed on the z_2 leads of cells with the functions
$(f_3 g_1)$ or $(f_3 g_2)$. This is shown in Fig. 5.17(b) where the function $z = x_1 \bar{x}_2 \bar{x}_3$
$+ \bar{x}_1 x_2 \bar{x}_3$ is to be realized; note the boundary conditions $y_1 = 1$ and $y_2 = 0$
going to cell C1. Note also that the array requires two cells each of functional types $(f_1 g_3)$, $(f_2 g_3)$ and $(f_3 g_2)$ giving a total of six cells.

The operation of the circuit is such that for cell C1 we have

$$z_1 = f_1 = xy_1 = x_1 \qquad \text{and} \qquad z_2 = g_3 = y_2 = 0$$

which forms the input to cell C2 giving

$$z_1 = f_2 = \bar{x}y_1 = x_1 \bar{x}_2 \qquad \text{and} \qquad z_2 = g_3 = y_2 = 0$$

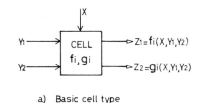

a) Basic cell type

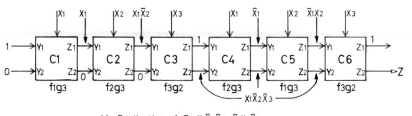

b) Realisation of $Z = X_1\bar{X}_2\bar{X}_3 + \bar{X}_1X_2\bar{X}_3$

Figure 5.17 Two-rail cascaded array

Again for cell C3 we have

$$z_1 = f_3 = 1 \qquad \text{and} \qquad z_2 = g_2 = y_2 + \bar{x}y_1 = x_1\bar{x}_2\bar{x}_3$$

which is the first product term. The process is repeated to obtain the complete combinational function.

This form of circuit is obviously not very efficient, either in terms of speed or the amount of logic required. It can be shown that the upper bound on the number of cells required to realize a function of n variables is $n.2^n$ which assumes all minterms are present.

5.6.2 Two-dimensional arrays

Rectangular arrays are of much greater application since in general they utilize relatively simple logic cells and result in a good utilization of chip area when fabricated in integrated circuit form. Moreover, they form the basis for semi-custom integrated circuits in the form of *gate arrays* using basic NAND/NOR cells. Whether to use multifunction or single function logic cells in logic arrays is still a matter for debate—there are of course advantages and disadvantages to both approaches. As an example let us consider the 2-dimensional array shown in Fig. 5.18. The basic cell in this case is the familiar binary half-adder unit giving the sum and carry outputs for the two inputs x and y. Note that this would be equivalent to using an exclusive OR and an AND gate per cell. The cells may be configured into an orthogonal array of n rows and 2^n columns as shown in Fig. 5.18(b) and used to generate any arbitrary n-variable combinational function.

To achieve this the input variables are applied to the left-hand boundary of the array (the x inputs) and the y inputs of all cells in the top row set to logic 1. It can easily be shown (see Fig. 5.19) that the carry output of the

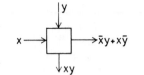

a) Basic half-adder cell

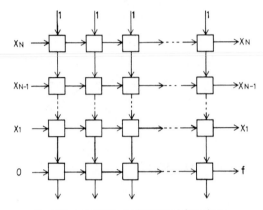

b) Array to realise combinational functions

Figure 5.18 Two-dimensional array using half-adders

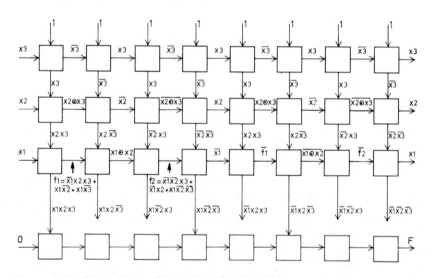

Figure 5.19 Generation of minterms for 3-variables

final cell in each column of the array generates a specific minterm of the n-variable inputs. To realize a particular function, say $F = x_1x_2x_3 + x_1\bar{x}_2\bar{x}_3 + \bar{x}_1x_2\bar{x}_3$, the required minterms are simply connected to a collector row of cells (with initial left-hand boundary value set to logic 0). Note that the size of an array required to realize an n-variable function is given by $(n + 1)2^n$.

An array which may be used to realize any combinational function and using an externally controlled multifunction cell has been described by Akers.[10] The basic cell, shown in Fig. 5.20, has three inputs x, y and z and two identical outputs X and Y. The (x, y) and (X, Y) leads are used to form the actual array connections with input z being externally controlled.

The truth table for the cell is given in Table 5.10. Note that the outputs for $x = 0$ and $y = 1$ are unspecified and as a consequence the boundary conditions to the array must be defined in such a way that these two inputs can never occur. Thus each cell in the array must satisfy the constraint $y \leq x$.

With this constraint and depending on the choice of values for the don't-care conditions, four possible logic functions are obtainable for the cell:

$$f_1 = xy + zy + xz; \qquad f_2 = xy + xz$$
$$f_3 = xz + y\bar{z}; \qquad f_4 = y + xz$$

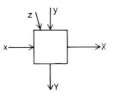

a) Basic cell

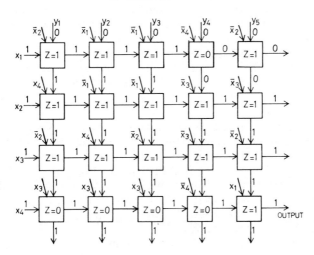

b) Array configuration for combinational logic

Figure 5.20 Aker's array structure

Table 5.10 Truth table—Aker's cell

x	y	z	$X = Y = f$
0	0	0	0
0	0	1	0
0	1	0	X
0	1	1	X
1	0	0	0
1	0	1	1
1	1	0	1
1	1	1	1

It can be shown that if all the y inputs to the top row of the array (see Fig. 5.20(b)) are set to logic 0 and all the x inputs to the left-hand column are set to logic 1 then the inputs and outputs of every cell will satisfy the condition $y \le X = Y \le x$ independant of z. Moreover, this condition will apply irrespective of the choice of logic function for the cell.

Since the x and y inputs on the periphery of the array must always be at the same values the external inputs to the array can only be connected to the z terminal; the output is taken from the bottom right-hand cell. Note that this configuration allows the main array to be fabricated with a fixed interconnection pattern and the input connections to be derived and processed separately at a later time. In other words we have the basis for a semi-custom integrated circuit system where the main array is prefabricated as a standard device and the input interconnections determined by the user for a particular application.

The array functions on the basis of establishing a path (called a 1-path) through the cell network to the output cell for the ON terms of any combinational function (as determined by the z inputs). For example, from Fig. 5.20(b), if the value of z for the top left-hand cell of the array is logic 1 then from the truth table for the cell (Table 5.10) input $x\bar{y}z$ gives the output $Y = X = 1$. It will be seen that all cells in the first column will now have outputs of 1 (due to boundary conditions $X = 1$) and thus can be ignored. Similarly we find that the second and third columns can be cancelled out in the same way.

In column four we find that with $z = 0$ the outputs will be $Y = X = 0$ which means that all the cells in the rest of row one will also have outputs of logic 0 and thus can be ignored. Proceeding in this way until we reach the bottom right-hand cell we find that the final output is 1. Note that in effect we have established a 1-path through the array to the output cell; an identical situation will result if the OFF terms of a function are applied to the z inputs, in this case a 0-path will be established.

To program the array for any combinational function it is first necessary to derive both the sum-of-products and product-of-sums form of the function. A matrix is then formed with the product terms as rows and the sum terms as columns; Table 5.11 shows the matrix for the function

Table 5.11 Realizing combinational functions

	$\bar{x}_2 + x_3 + x_4$	$\bar{x}_1 + x_3 + x_4$	$\bar{x}_1 + \bar{x}_2 + x_3$	$\bar{x}_3 + \bar{x}_4$	$x_1 + \bar{x}_2 + \bar{x}_3$
$\bar{x}_1\bar{x}_2\bar{x}_4$	$\bar{x}_2$	$\bar{x}_1$	$\bar{x}_1\bar{x}_2$	$\bar{x}_4$	$\bar{x}_2$
$\bar{x}_1\bar{x}_3x_4$	x_4	$\bar{x}_1x_4$	$\bar{x}_1$	$\bar{x}_3$	$\bar{x}_3$
$\bar{x}_2\bar{x}_3x_4$	$\bar{x}_2x_4$	x_4	$\bar{x}_2$	$\bar{x}_3$	$\bar{x}_2\bar{x}_3$
$x_1x_3\bar{x}_4$	x_3	x_3	x_3	$\bar{x}_4$	x_1

$$F = \Sigma(0, 1, 2, 5, 9, 10, 14)$$

The entries in the matrix are obtained by simply extracting the common literals which are common to both the product and sum terms. The literals so obtained are used as the z inputs to the cells in the array in direct correspondence to their position in the matrix; note that the array requires 20 cells and is not necessarily minimal. In the case of more than one literal being available any one may be used but the choice could effect the interconnection pattern.

In generating combinational functions using lateral 2-dimensional arrays the minterms are normally produced separately and then 'collected' or summed in a dedicated row or rows of the array (contrast also the AND/OR structure of the PLA as described in the last chapter). However, if a more complex cell is employed it is possible to combine the two requirements into a single bilateral array.

Consider the cell shown in Fig. 5.21(a); the outputs are given by

$$z_1 = w_1x_1y_1 + w_1\bar{x}_1\bar{y}_1$$
$$z_2 = w_2$$
$$z_3 = x_1$$
$$z_4 = x_2 + y_2w_2$$

where y_1 and y_2 are external control inputs to the array and provide the means of programming specific functions. Note that since in the array z_2 outputs will be connected to the w_2 inputs and z_3 to the x_1 inputs these outputs represent direct through connections.

An array of these cells, as shown in Fig. 5.21(b), can be used to generate combinational functions. Minterms are formed using the functions z_1 and z_3 and the control input y_1, where

$$z_1 = w_1x_1 \quad \text{for} \quad y_1 = 1$$
$$z_1 = w_1\bar{x}_1 \quad \text{for} \quad y_1 = 0$$

Similarly the required minterms are summed together using functions z_2, z_4 and y_2 where

$$z_4 = x_2 + w_2 \quad \text{for} \quad y_2 = 1$$
$$z_4 = x_2 \quad \text{for} \quad y_2 = 0$$

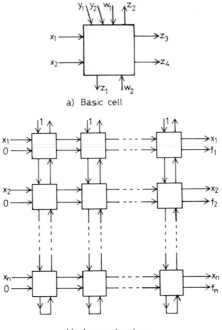

a) Basic cell

b) Array structure

Figure 5.21 Bilateral array

Note that the array is bilateral in the vertical direction with signals flowing from top to bottom and bottom to top of the array; the outputs are taken from the right-hand column of cells.

This procedure is illustrated in Fig. 5.22 which shows an array to generate the functions

$$f_1 = ab + \bar{a}b + a\bar{b} + \bar{a}\bar{b}$$
$$f_2 = ab + \bar{a}\bar{b}$$

Note that to collect, a particular minterm y_2 is set to 1; otherwise $y_2 = 0$.

The applicability of array structures to the design of LSI/VLSI circuits will be obvious. Moreover, arrays are not restricted to combinational logic but, as we shall see later, can also be used to implement sequential systems. The concept of decomposing a logic system into a regular array of simple and ideally identical substructures (but in practice as few different cell types as possible) is one of the major goals of the VLSI designer. In this context *systolic arrays*[11] are attracting considerable interest.

A systolic array consists of a small set of cells, each capable of performing some simple operation, interconnected to form an array or tree structure; the arrays are typically bilateral in operation. Information in a systolic array flows between cells in a pipe-lined fashion, processing stage by stage as on a production line. Since they are invariably sequential in

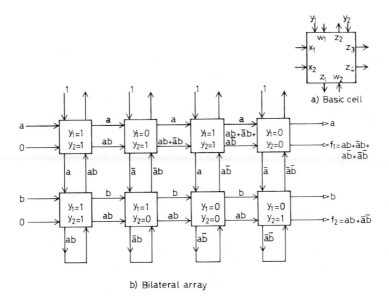

b) Bilateral array

Figure 5.22 Generating 2-variable functions with bilateral array

operation a memory (binary storage device) is used to 'pump' information (that is the inputs) into the array. Note that this implies that only those cells on the array boundaries may be used as input/output ports. Special purpose systolic array structures have been proposed for a large number of applications including matrix arithmetic, signal processing and non-numerical procedures such as sorting, searching etc.

A full treatment of array structures is unfortunately outside the scope of this text, and considerable work is still in progress. In particular the complexity of the cell required in a general purpose array is still very controversial. Also, the need for a design methodology and good CAD algorithms both for synthesis and evaluation is a vital consideration in developing array architectures.

References and bibliography

1. Caldwell, S. *Switching Circuits and Logical Design*. Wiley, New York, 1959.
2. Mead, C. and Conway, L. *Introduction to VLSI Systems*. Addison Wesley, Mass., 1980.
3. Shannon, C. E. A symbolic analysis of relay and switching circuits. *Trans. Am. Inst. Elect. Engrs.* **57**, 713–23 (1938).
4. McCluskey, E. J. Algebraic minimization and the design of two-terminal networks. Ph.D. thesis, Dept. of E.E., MIT, June 1956.
5. Marcus, M. P. The detection and identification of symmetric switching functions and the use of tables of combination. *IRE Trans. Elect. Comput.* **EC5**(4), 237–9 (1956).

6. Shannon, C. E. The synthesis of two terminal switching circuits. *Bell Syst Tech. J.* **28**, 59–98 (1949).

7. Hohn, F. E. and Schissler, L. R. Boolean matrices and the design of combinational relay switching circuits. *Bell Syst Tech. J.* **34**, 177–202 (1955).

8. Marcus, M. P. Minimization of partially-developed transfer tree. *IRE Trans. Electron. Comput.* **EC6**, 92–5 (1957).

9. Short, R. A. Two-rail cellular arrays. *AFIPS Conf. Proc. Pt. 1*, 355–69 (1965).

10. Akers, S. B. A rectangular logic array. *IEEE Trans. Comp.* **C21**, 848–57 (1972).

11. Kung, H. T. Why systolic architectures? *IEE Computer* **15**(1), 37–46 (1982).

12. Hennie, F. C. *Iterative Arrays of Logical Circuits*. MIT Press, Cambridge, Mass., 1961.

Tutorial problems

5.1 Design a circuit using the iterative method that will recognize the occurrence of three consecutive 1s in a ten-bit parallel message. Implement the design in both NAND and contact logic. Can this be designed using symmetric functions?

5.2 Repeat problem 1 using different state assignments and then using two external inputs per cell.

5.3 Redesign the parity-check circuit discussed in Section 5.2 using symmetric functions, and compare the resulting contact circuit with the one found previously.

5.4 Devise a circuit that will detect whenever the number of 1s contained in a nine-bit parallel message is equal to six or eight.

5.5 Using Boolean matrices synthesize a bridge circuit starting from the transmission function $T = A\bar{C}\bar{D}\bar{E} + \bar{A}\bar{B}C\bar{E} + \bar{A}\bar{C}\bar{D}E + ABCE$.

5.6 Design an electronic and relay tree circuit to decode the following combinations:

$$C = (5, 7, 8, 9, 11, 12, 13, 15, 25, 27, 29, 31)$$

5.7 Realize the function

$$Z = \Sigma(2, 3, 4, 5, 8, 9, 14, 15)$$

using Aker's array. Check that the circuit provides both 0- and 1-paths through the network.

6 Sequential switching circuits

6.1 Introduction

Up to now we have dealt with the problems of designing combinational logic circuits, that is, circuits in which the output is a function of the *present inputs* only. However, in a practical system we are concerned with another, more general, type of logic circuit, where the output is a function of both present and *past* inputs. These circuits, variously called *sequential machines* or *finite state machines*, are embodied in most digital systems as counters, pattern generators, sequence detectors, etc. A simple example of a sequential machine is the combination lock, which can only be opened if the correct coded sequence of numbers or letters is set up. Perhaps a more familiar example is the telephone system in which a connection is made by dialling coded symbols in sequence.

As the output response of a sequential machine is dependent on its present internal state (determined by past inputs) and its present input conditions, it should be apparent that the concepts can be used to describe any dynamic system with finite states. Thus a business organization, a game, or a living organism may be defined in these terms.

Sequential systems are basically combinational circuits with the additional properties of *storage* (to remember past inputs) and *feedback* (Fig. 6.1); note that the output z is a function of inputs x and internal states y. There are two main classes of sequential circuit, *synchronous* (clocked or strobed) and *asynchronous* (unclocked or free-running). In the former type the input, output and internal states are sampled at definite intervals of time, controlled by the fundamental *clock* frequency of the system. Since the clock is generally some form of square wave, synchronous circuits are often referred to as pulse circuits, the timing being done by incorporating a clock input to each basic switching module, or as is more usual the storage elements. This type of logic circuit is readily applied to the processing of serial information.

Asynchronous circuits (also called self-timing circuits) proceed at their own speed regardless of any basic timing, the outputs of one circuit immediately becoming the inputs to the next. In contrast to pulse signals, active asynchronous logic uses dc levels, and is generally used in the processing of parallel data. Because of differences in the inherent delays through signal paths, such circuits can give rise to random operation, requiring special treatment in their design, as we shall see in later chapters.

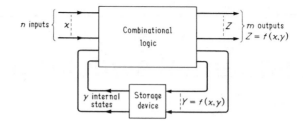

Figure 6.1 Finite state machine—Mealey model

The basic difference between synchronous and asynchronous circuits may be summed up by considering static inputs to each circuit in turn. For the synchronous circuit, if the input combination never changes it will be interpreted as m repetitions (one at each clock pulse) of the input combination. To an asynchronous circuit, however, continued application of a particular input combination appears as a single input.

The rest of this chapter will be devoted to synchronous circuits, and attempts to give a fundamental understanding of the problem, before proceeding to a more formal analysis. In particular, we shall consider the design of counter circuits.

6.2 Bistable circuits

It can be seen from Fig. 6.1 that sequential circuits can be represented by a combinational circuit in conjunction with some form of storage or memory element. Before we can begin to design sequential logic circuits, we must examine the properties of these storage devices and derive *characteristic equations* defining their operation.

There is no restriction on the type of storage that can be employed in a sequential system and RAM, ROM, shift registers etc. have all been used. In general, however, *bistable or flip-flop circuits* are extensively used, their two-state properties providing a single bit store. Several types of bistable circuits are available and are classified according to the input conditions available to cause the device to change state. There is much loose terminology bandied about in connection with bistables, and the only reliable way to specify the logical operation of the device is by means of a truth table. The more familiar types are the set–reset bistable (SR-FF), the dc version of which is also called a *latch*; the trigger bistable (T-FF), or divide-by-two circuit; the JK bistable (JK-FF); and the D-type bistable (D-FF) used as a delay element.

Table 6.1(a) shows a truth table for the *set-reset bistable* giving the next output (or state) Q_+ in terms of the present output (or state) Q and the inputs S and R. The truth table is constructed by considering the physical action of the circuit shown in Fig. 6.2(a). The operation is such that an input $S = 1$ on the set terminal causes an output $Q = 1$, and further inputs

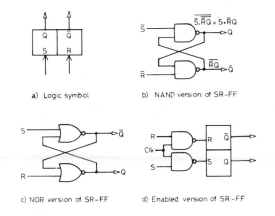

a) Logic symbol

b) NAND version of SR-FF

c) NOR version of SR-FF

d) Enabled version of SR-FF

Figure 6.2 Set–reset bistable circuits

have no effect. Similarly, an input $R = 1$ on the reset terminal causes an output $\bar{Q} = 1$; these results are entered in the truth table. The entries marked with a cross (X) correspond to the 'not allowed' or 'don't care' inputs since, under these conditions, when both R and S are present simultaneously, the operation of the circuit becomes uncertain. Note that Q and Q_+ occur in different time intervals, Q_+ occurring *after* Q. That is to say, Q_+ is a delayed version of Q, the delay being caused by the inherent operation time of the bistable store, as the circuit cannot change state instantaneously. This delay is essential in the operation of sequential circuits, and in some cases it is necessary to add further delay to achieve correct logical operation. In a synchronous circuit, Q_+ would be the output in the next sampling interval, or clock pulse. Though Q_+ and Q occur at different times the switching algebra is not affected as they can be treated as two distinct variables.

Thus we can write the *difference equation* as the combinational expression

$$Q_+ = \bar{Q}S\bar{R} + Q\bar{S}\bar{R} + QS\bar{R}$$

The K-map, shown in Table 6.1(b), gives the minimal *characteristic equation* for the SR-FF element:

$$Q_+ = S + \bar{R}Q$$

Figure 6.2(b) and (c) shows the circuit implemented in terms of NAND and NOR elements. Note the inherent feedback loops in the circuit and that a direct AND/OR realization would not work since there must be a power gain round the feedback loop. Note also that the characteristic equation for the NOR circuit is given in product of sums form, that is

$$Q_+ = (S + Q)\bar{R} = S\bar{R} + Q\bar{R}$$

The circuits described above are essentially dc operated, that is the circuit will respond directly to a change of voltage level. However, since

Table 6.1 Design tables for set—reset bistables

Q	S	R	Q_+
0	0	0	0
0	0	1	0
0	1	0	1
0	1	1	X
1	0	0	1
1	0	1	0
1	1	0	1
1	1	1	X

Q	SR 00	01	11	10
0			X	1
1	1		X	1

(a) Truth table　　　　　　　　　(b) K-map

most digital systems are synchronous it is necessary to have bistable circuits which will only change state when a clock signal is present.

It is possible to synchronize the operation of the dc bistable to an external clock or enabling signal as shown in Fig. 6.2(d). Note that whilst the enable signal is present the circuit will function as a normal SR-FF. Unfortunately the operation of the circuit depends critically on the duration of the clock signal which can give rise to erroneous outputs. The dc bistable takes a finite time to change state due to the need for the internal feedback signals to propagate through the circuit and settle out to a stable condition. If the duration of the clock signal is longer than the propagation delays the circuit could respond to these unstable internal changes. It follows from this that the outputs from dc bistables cannot be used to control their own inputs and any feedback loops of this type would constitute a serious design error. One solution to the problem would be to use very narrow clock pulses but these would be very difficult to generate and distribute around the system.

The ideal solution is to use bistable circuits which only allow a change of state to occur on a voltage transition or *edge* rather than in response to a change in level. Thus most bistable circuits are designed to be triggered by the leading or trailing edge (positive or negative edge triggering) of a clock pulse input; other input signals can of course be dc levels.

One method of achieving this is to use a master—slave arrangement based on the dc SR bistable as shown in Fig. 6.3. In this circuit, which still responds to dc levels, the two latches are isolated by the inverted clock input which also effectively delays the response of the circuit. The master bistable will respond to inputs S and R as long as the clock is present (remains high) but its outputs cannot affect the slave bistable due to the inverted clock input which will be low. When the external clock goes low the master is isolated and the slave, now enabled, will change accordingly. The overall effect is that the circuit output (from the slave) changes only on the negative-going clock edge.

This technique, however, though effective, increases the overall propagation delay of the circuit and as a result adversely affects its speed of operation. Consequently most synchronous edge triggered bistables em-

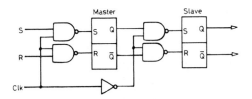

Figure 6.3 Master–slave clocked SRFF

ploy special circuitry[1] to achieve the required characteristics and performance. Hereafter, unless otherwise stated, we shall assume the use of
negative edge triggered bistables.

The *D-type* or *delay bistable*, shown in Fig. 6.4(a) and Table 6.2, has the
property of transferring the logic value on input D to the output Q
whenever a clock pulse is present. Thus the characteristic equation for the
device is given by

$$Q_+ = D$$

Note that the next state is independent of the present state. This element is
equivalent to a 1-bit delay unit, where the bit-time is determined by the
clock rate.

The truth table for the *trigger bistable* is shown in Table 6.3 and its logic
symbol in Fig. 6.4(b). Here the state of the circuit changes each time an
input pulse is received. The characteristic equation in this case is the
familiar exclusive OR relationship:

$$Q_+ = \bar{Q}T + Q\bar{T}$$

Note that when $T = 1$, $Q_+ = \bar{Q}$.

In practice the trigger bistable, also known as a *toggle* or *divide-by-two*
circuit, would not normally be available as a device in its own right but
would be constructed using other bistables as shown in Fig. 6.4(c) and (d).

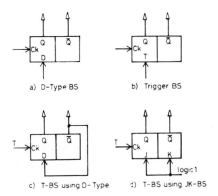

a) D-Type BS

b) Trigger BS

c) T-BS using D-Type

d) T-BS using JK-BS

Figure 6.4 Synchronous bistables—logic symbols

Table 6.2 D-type bistable

Q	D	Q_+
0	0	0
0	1	1
1	0	0
1	1	1

(a) Truth table   (b) K-map

Table 6.3 Design tables for trigger bistable

Q	T	Q_+
0	0	0
0	1	1
1	0	1
1	1	0

(a) Truth table   (b) K-map

By far the most important member of the bistable family is the *JK-bistable* shown in Table 6.4 and Fig. 6.5. From the K-map the characteristic equation is given by

$$Q_+ = J\bar{Q} + \bar{K}Q$$

which may be seen to combine the characteristics of both the SR-FF and T-FF bistables. In other words it behaves like a normal set–reset bistable with the terminals J and K functioning as set–reset inputs, except when J and K are both equal to one when the bistable changes state. That is for $J = K = 1$ the characteristic equation becomes

$$Q_+ = \bar{Q}$$

Table 6.4 Design tables for JK-bistable

Q	J	K	Q_+
0	0	0	0
0	0	1	0
0	1	0	1
0	1	1	1
1	0	0	1
1	0	1	0
1	1	0	1
1	1	1	0

(a) Truth table   (b) K-map

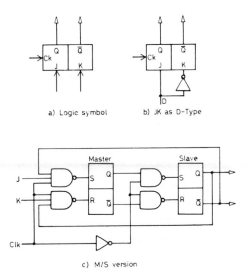

a) Logic symbol   b) JK as D-Type

c) M/S version

Figure 6.5 JK-bistables

i.e. the characteristics of the trigger bistable. Again, if we invert the J input and apply it to the K terminal as shown in Fig. 6.5(b) we obtain the characteristics of the D-type bistable, that is $Q_+ = J$.

A further advantage is that since the uncertainty inherent in the SR-bistable when both $S = R = 1$ has been resolved the JK-bistable, as we shall see later, generally leads to a more economic realization.

In practice all bistables would have additional set and clear inputs. These would be asynchronous in operation, that is independent of the clock, and used to set the bistable to some initial start state.

The choice of storage device or particular type of bistable for a sequential system depends entirely on application, availability of devices, cost and reliability. In particular the choice of bistable type can considerably affect the amount of combinational logic required for the realization.

6.3 Counter design

Counters are special purpose sequential systems designed to count (in some specified number system) the number of transitions, either $0 \rightarrow 1$ or $1 \rightarrow 0$, at its input terminal.[2] Counters can be classified into two main types, *asynchronous* or ripple through counters and *synchronous* or coherent counters.

A 3-bit binary ripple counter is shown in Fig. 6.6(a), realized using JK bistables connected as trigger bistables. In effect we have three cascaded divide-by-two stages giving a binary count of $0 \rightarrow 7$. Note that the input goes direct to the clock terminal of the least significant stage (bistable A) and that it is the negative going transitions of the input that are counted.

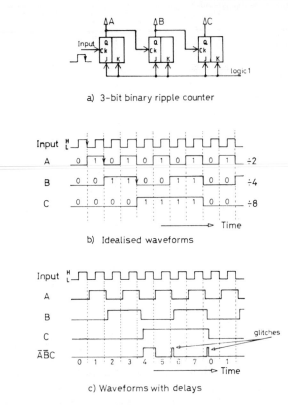

a) 3-bit binary ripple counter

b) Idealised waveforms

c) Waveforms with delays

Figure 6.6 Asynchronous counters

The idealized waveforms for the circuits are shown in Fig. 6.6(b) but, in practice, because of the propagation delays in the bistables, the actual waveforms exhibit edge displacement brought about by the effect of accumulative delays as shown in Fig. 6.6(c). This is because one bistable must change before the next one in line can change, that is, an output change must ripple through *all* lower order stages before it can affect a change in a higher order bit stage. For example, if the counter is in state 110 $(AB\bar{C})$ and it receives another input pulse all the lower significant stages must change in succession before the counter can change to the final state 001 $(\bar{A}\bar{B}C)$.

The effect of this is two-fold; first, it limits the maximum counting frequency the period of which must not exceed the total delay through the counter (called the *resolution time*). Second, the delays will produce *spikes* or *glitches* when the waveforms are decoded. One of the main applications of a counter is to enable timing control waveforms to be generated by decoding the outputs of a counter. For example, suppose for our 3-bit counter we wanted to detect when the state $\bar{A}\bar{B}C$ occurred by gating the relevant outputs from the counter. Because of the overlapping non-coincident edges, as well as obtaining the required output we would also get two spurious pulses (see Fig. 6.6(c)). These pulses could, if at a sufficiently high level, cause malfunctioning in any following circuits.

Ripple counters may be designed to operate in bases other than two by using feedback to eliminate redundant, unwanted, states. For example, to design a decade counter a four-bit binary counter with states A, B, C and D would be used as the starting point. Since only ten ($0000 \rightarrow 1001$) of the possible sixteen states are required, states $1010 \rightarrow 1111$ must be eliminated. This is achieved by detecting the occurrence of state $\bar{A}B\bar{C}D$, using an AND gate, and applying the output to the clear terminals of the bistables, thus resetting the counter to zero.

Though ripple counters are simple to design and relatively cheap to realize the problems associated with their asynchronous nature gives rise to many problems. Propagation delay problems can, however, be drastically reduced by using a synchronous counter in which all the bistables change state at the same time. Synchronous counters are a typical example of a sequential switching system and as such afford a good introduction to the synthesis of synchronous circuits.

6.3.1 Design of synchronous counters

Suppose a counter is required to count incoming pulses up to a maximum of 15, afterwards resetting to zero. The reflected binary system is to be used, each intermediate count being displayed as the count proceeds. This may be considered as a synchronous sequential circuit with the pulse input itself acting as the clock, and gating each stage of the counter. A four-bit store will be required, since there are 16 states, and we shall use SR-FF devices, together with the necessary combinational logic to set and reset the stores. The task of determining the *input equations* for the storage elements is a major part of the logic design.

Table 6.5 shows the truth table, or more correctly the *transition* or *state-table* for the counter. In practice, this is a five-variable problem, the variables being present states A, B, C and D, and the input x. However, since we are concerned with changes from one state to another, which only occur when $x = 1$, we can ignore x. Note that each state may be identified by its four-bit code, determined in this case by the choice of reflected binary. From the table, we could write down equations for A_+, B_+, etc. in terms of A, B, C and D. These *application equations*, together with the characteristic input equations for the storage device, form a set of simultaneous Boolean equations:

$$A_+ = f_1(A, B, C, D)$$
$$A_+ = S_A + \bar{R}_A A$$
$$B_+ = f_2(A, B, C, D)$$
$$B_+ = S_B + \bar{R}_B B$$
$$\dots\dots\dots\dots\dots\dots$$

We have now to solve for S_A, R_A, S_B, R_B, etc. in terms of (A, B, C, D) and thus obtain the input equations for the relevant SR bistables. This may be

Table 6.5 Transition table for reflected binary counter

Present states Time n				Next states Time $(n+1)$			
A	B	C	D	A_+	B_+	C_+	D_+
0	0	0	0	0	0	0	1
0	0	0	1	0	0	1	1
0	0	1	1	0	0	1	0
0	0	1	0	0	1	1	0
0	1	1	0	0	1	1	1
0	1	1	1	0	1	0	1
0	1	0	1	0	1	0	0
0	1	0	0	1	1	0	0
1	1	0	0	1	1	0	1
1	1	0	1	1	1	1	1
1	1	1	1	1	1	1	0
1	1	1	0	1	0	1	0
1	0	1	0	1	0	1	1
1	0	1	1	1	0	0	1
1	0	0	1	1	0	0	0
1	0	0	0	0	0	0	0

done algebraically or by using truth table methods;[3] we shall use a simpler approach, however, and deduce the input switching functions directly from the state table, using the transition table for the SR-FF shown in Table 6.6.

To find S_A, the switching function for setting bistable A, we compare columns A and A_+ in Table 6.5, noting the values of the present state variables for the condition when $A = 0$ and $A_+ = 1$. This value is

$$S_A = \bar{A}B\bar{C}\bar{D}$$

There are also don't-care conditions, when no changes are required to take place, which should, if possible, be included in the simplification process. They occur when $A = 1$ and $A_+ = 1$; thus

$$S'_A(\text{don't-care}) = AB\bar{C}\bar{D} + AB\bar{C}D + ABCD + ABC\bar{D}$$
$$+ A\bar{B}C\bar{D} + A\bar{B}CD + A\bar{B}\bar{C}D$$

The corresponding reset conditions occur when $A = 1$ and $A_+ = 0$, and for the don't-cares when $A = 0$ and $A_+ = 0$:

$$R_A = A\bar{B}\bar{C}\bar{D}$$
$$R'_A = \bar{A}\bar{B}\bar{C}\bar{D} + \bar{A}\bar{B}\bar{C}D + \bar{A}\bar{B}CD + \bar{A}\bar{B}C\bar{D} + \bar{A}BC\bar{D} + \bar{A}BCD$$
$$+ \bar{A}B\bar{C}D$$

Similarly, comparing columns B and B_+, we have

$$S_B = \bar{A}\bar{B}C\bar{D}$$

and

$$S'_B = \bar{A}B\bar{C}\bar{D} + \bar{A}B\bar{C}D + \bar{A}BCD + \bar{A}BC\bar{D} + AB\bar{C}\bar{D} + AB\bar{C}D$$
$$+ ABCD$$

Table 6.6 Transition table
for SR-FF

Q	Q_+	S	R
$0 \rightarrow 0$		0	X
$0 \rightarrow 1$		1	0
$1 \rightarrow 0$		0	1
$1 \rightarrow 1$		X	0

Also

$$R_B = ABC\bar{D}$$

and

$$R'_B = \bar{A}\bar{B}C\bar{D} + \bar{A}\bar{B}CD + \bar{A}BCD + A\bar{B}C\bar{D} + A\bar{B}CD + A\bar{B}CD + A\bar{B}C\bar{D}$$

Continuing in this manner for the other input switching functions, we arrive at the complete solution, shown plotted on K-maps in Table 6.7. The maps lead to the reduced set of input equations for S and R shown below.

$$S_A = B\bar{C}\bar{D} \qquad\qquad R_A = \bar{B}\bar{C}\bar{D}$$
$$S_B = \bar{A}C\bar{D} \qquad\qquad R_B = AC\bar{D}$$
$$S_C = D(\bar{A}\bar{B} + AB) \qquad R_C = D(\bar{A}B + A\bar{B})$$
$$S_D = \bar{C}(\bar{A}\bar{B} + AB) + C(\bar{A}B + A\bar{B}) \quad R_D = \bar{C}(\bar{A}B + A\bar{B}) + C(\bar{A}\bar{B} + AB)$$

Note that $(\bar{A}B + A\bar{B}) = \overline{(\bar{A}\bar{B} + AB)}$; in fact this is the exclusive OR function discussed in a previous chapter. The switching functions are shown implemented in terms of AND/OR logic in Fig. 6.7. The input line acts as the clock, and is taken directly to the clock inputs; this is equivalent to logically multiplying all the equations by the input variable.

Note that all the bistables will change state, simultaneously, triggered by the negative-going edge of the input pulse. Sufficient time, however, must

Table 6.7 K-maps for reflected binary counter

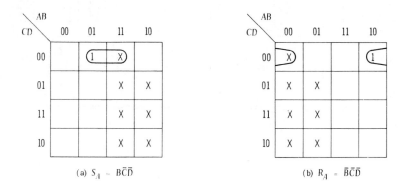

(a) $S_A = B\bar{C}\bar{D}$ (b) $R_A = \bar{B}\bar{C}\bar{D}$

Table 6.7 Continued

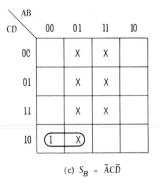

(c) $S_B = \bar{A}C\bar{D}$

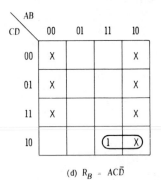

(d) $R_B = AC\bar{D}$

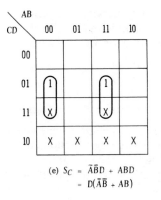

(e) $S_C = \bar{A}\bar{B}D + ABD$
$= D(\bar{A}\bar{B} + AB)$

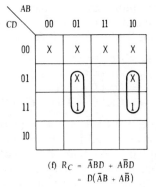

(f) $R_C = \bar{A}BD + A\bar{B}D$
$= D(\bar{A}B + A\bar{B})$

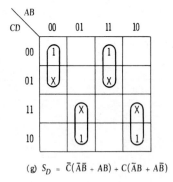

(g) $S_D = \bar{C}(\bar{A}\bar{B} + AB) + C(\bar{A}B + A\bar{B})$

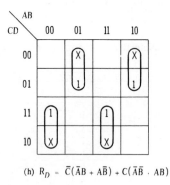

(h) $R_D = \bar{C}(\bar{A}B + A\bar{B}) + C(\bar{A}\bar{B} \cdot AB)$

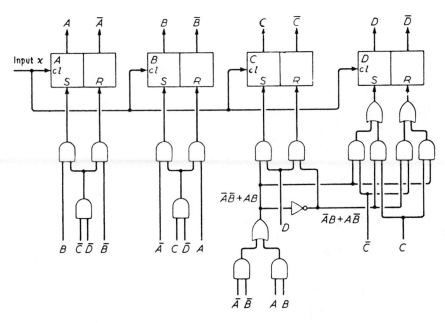

Figure 6.7 Logic diagram for reflected binary counter using SRFF

be allowed for the gates to settle out and establish the proper input levels to the S and R terminals before the next input pulse is applied. The maximum counting rate will be determined by the propagation time for the bistable device plus the decoding time.

From Fig. 6.7 it is easy to identify the inherent structure of a sequential circuit; the division between combinational logic and memory and the essential feedback loops will be obvious.

Any type of bistable may be used for the storage requirement. For example let us redesign the reflected binary counter using D-type bistables. Since this type of bistable essentially gives a delay of one clock period it is possible to use the application equations directly as the input to the device. Thus from Table 6.5 we have:

$$A_+ = \bar{A}B\bar{C}\bar{D} + AB\bar{C}\bar{D} + AB\bar{C}D + ABCD + ABC\bar{D} + A\bar{B}CD$$
$$+ A\bar{B}CD + A\bar{B}\bar{C}D$$
$$B_+ = \bar{A}\bar{B}C\bar{D} + \bar{A}BC\bar{D} + \bar{A}BCD + \bar{A}B\bar{C}D + \bar{A}B\bar{C}\bar{D} + AB\bar{C}\bar{D}$$
$$+ AB\bar{C}D + ABCD$$
$$C_+ = \bar{A}\bar{B}\bar{C}D + \bar{A}\bar{B}CD + \bar{A}B\bar{C}\bar{D} + \bar{A}BC\bar{D} + AB\bar{C}D + ABCD$$
$$+ ABC\bar{D} + A\bar{B}C\bar{D}$$
$$D_+ = \bar{A}\bar{B}\bar{C}\bar{D} + \bar{A}\bar{B}C\bar{D} + \bar{A}BCD + \bar{A}BCD + AB\bar{C}\bar{D} + AB\bar{C}D$$
$$+ A\bar{B}C\bar{D} + A\bar{B}CD$$

Plotting these equations on a K-map (Table 6.8) gives the following minimal input equations:

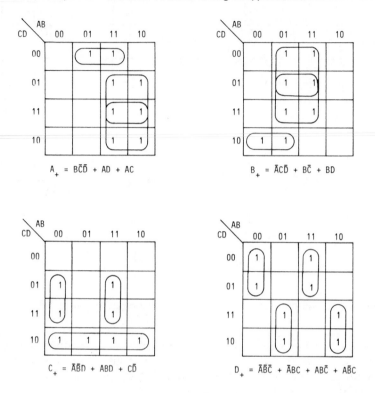

Table 6.8 K-maps for reflected counter using D-type bistables

$A_+ = B\bar{C}\bar{D} + AD + AC$

$B_+ = \bar{A}C\bar{D} + B\bar{C} + BD$

$C_+ = \bar{A}\bar{B}D + ABD + C\bar{D}$

$D_+ = \bar{A}\bar{B}\bar{C} + \bar{A}BC + AB\bar{C} + A\bar{B}C$

$$A_+ = B\bar{C}\bar{D} + AD + AC$$
$$B_+ = \bar{A}C\bar{D} + B\bar{C} + BD$$
$$C_+ = \bar{A}\bar{B}D + ABD + C\bar{D}$$
$$D_+ = \bar{A}\bar{B}\bar{C} + \bar{A}BC + AB\bar{C} + A\bar{B}C$$

These equations are shown implemented using NAND logic in Fig. 6.8. It will be apparent that the amount of combinational logic required is dependent on the type of bistable used. D-type bistables provide the simplest and most direct method of realizing a sequential circuit (the sequential structure is very obvious) and are extensively employed, in conjunction with PLAs to generate the combinational logic, in LSI/VLSI implementations.

Let us consider another example, that of a binary-coded decimal counter, counting from 0 to 9 in the excess-three code; the transition table is shown in Table 6.9. Again a four-bit memory will be required, but in this case only ten states are used. As the remaining six states will never occur in normal operation it is unnecessary to specify their next states. Thus they are don't-care conditions and can be used as such in the minimization process.

The design procedure is as before but this time we shall use the JK bistable for the memory devices; the transition table is given in Table 6.10.

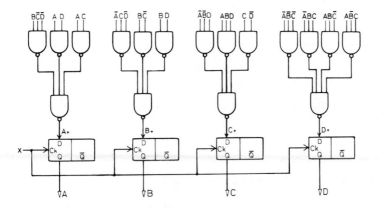

Figure 6.8 Logic diagram for reflected binary counter using D-type BS

Note that there are two extra don't-care conditions compared to the SR-FF, brought about by the toggling function when $J = K = 1$; otherwise the JK bistable behaves in the same way as the SR-FF. The input conditions for the JK bistables are plotted directly on K-maps as shown in Table 6.11 (there is little point in extracting the equations first). Note the use of both types of don't-care terms—those for the bistable and those generated by the application. The minimized input equations are given by

$$J_A = BCD \qquad K_A = B$$
$$J_B = CD \qquad K_B = A + CD$$
$$J_C = D + AB \qquad K_C = D$$
$$J_D = 1 \qquad K_D = 1$$

Table 6.9 Transition table for excess-three counter

A	B	C	D	A_+	B_+	C_+	D_+
0	0	1	1	0	1	0	0
0	1	0	0	0	1	0	1
0	1	0	1	0	1	1	0
0	1	1	0	0	1	1	1
0	1	1	1	1	0	0	0
1	0	0	0	1	0	0	1
1	0	0	1	1	0	1	0
1	0	1	0	1	0	1	1
1	0	1	1	1	1	0	0
1	1	0	0	0	0	1	1
0	0	0	0				
0	0	0	1				
0	0	1	0		Don't-care		
1	1	0	1		terms		
1	1	1	0				
1	1	1	1				

Table 6.10 Transition
table, JK-bistable

Q	Q_+	J	K
$0 \to 0$		0	X
$0 \to 1$		1	X
$1 \to 0$		X	1
$1 \to 1$		X	0

Table 6.11 K-maps for excess-three counter

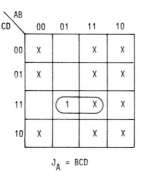

$$J_A = BCD$$

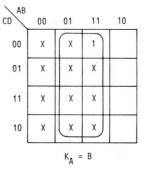

$$K_A = B$$

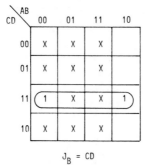

$$J_B = CD$$

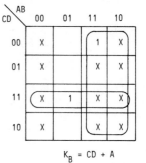

$$K_B = CD + A$$

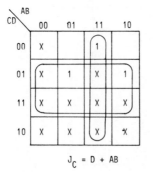

$$J_C = D + AB$$

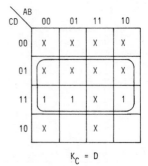

$$K_C = D$$

Table 6.11 Continued

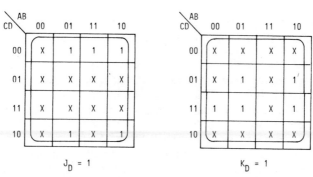

$$J_D = 1 \qquad\qquad K_D = 1$$

which are shown implemented in Fig. 6.9.

Should the number of variables in a design problem exceed five, a convenient number for map manipulation, it will be necessary to use tabular methods of minimization as described in Chapter 3.

Using the procedures described above synchronous counters operating in any sequence or radix may easily be designed in two basic steps. First, the required state transition table is generated and, second, having selected an appropriate bistable element, the relevant input equations are derived and minimized using standard techniques. We shall see in later sections that the same process can be followed, once the state table description has been determined (the creative part!) for any sequential machine.

6.4 Sequential circuits: problem definition

So far we have looked at one particular class of sequential circuit, the counter; let us now consider the problem in general. We begin by examining the methods used to specify and define the initial logical or system requirement—the essential prerequisite to any design problem.

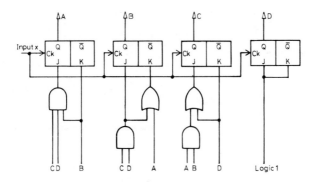

Figure 6.9 JK implementation of excess-3 counter

The classical way of representing a sequential circuit is by either using a *state diagram* or *state table* (also called a state transition table). The state table is perhaps the more important since this is the most convenient means of representing the system requirements prior to realization—most design algorithms start with the state table.

All the bistable circuits we have discussed earlier are in fact examples of simple sequential systems with two internal states. Consequently we can represent the JK bistable in a state table form as shown in Table 6.12(a). Note that we have not assigned a binary value to the internal state (which is the usual procedure at the start of a design) but from earlier work the assignment is obvious as shown in Table 6.12(b).

The entries in the state table are the next states reached by the circuit following any given change of input: similarly the output is specified for all input changes. For example, if the bistable is in present state A and receives the inputs JK = 10 (the set condition), a transition from state A to state B will occur producing an output of 1. The development of the state table, from the original circuit specification, logically formulates the problem in the same way as the truth table for a combinational circuit. It is, in fact, an abstract mathematical representation of a sequential circuit, which lists the outcome of all possible input combinations in terms of the internal states and output states. It has its origins in the function table used in group theory to describe binary operations.

For complicated systems, the table is sometimes difficult to construct because, as no established technique exists, the process is mainly intuitive and relies heavily on past experience. The state diagram is a useful means of expressing problem requirements before constructing the state table; it contains exactly the same information but in a more understandable form. The state diagram is a directed graph, rather like a signal flow diagram,

Table 6.12 State tables for JK-bistable

Present state	Inputs J, K					Output Q			
	Next state					Output Q			
	00	01	11	10		00	01	11	10
A	A	A	B	B		0	0	1	1
B	B	A	A	B		1	0	0	1

(a) Unassigned table

Present state	Inputs J, K					Output Q			
	Next state					Output Q			
	00	01	11	10		00	01	11	10
0	0	0	1	1		0	0	1	1
1	1	0	0	1		1	0	0	1

(b) Assigned table

representing the states of the circuit by circles (nodes) with directed lines
between them showing the transition paths. There are two types of state
diagram, called the *Mealy* model[3] and the *Moore* model.[4] In the former,
each path is labelled with the input which causes the transition and the
resulting output; the circle contains the symbol or code for the internal
state. The Moore model differs in that, although the paths are labelled with
the inputs which cause the transition, the circles contain both the state code
and the output state, i.e. the output state is a function of the internal states
only. When the initial and final states are the same, analogous to a self-
loop in signal flow, we call the transition path a *sling*. This is illustrated in
Fig. 6.10 for the JK bistable. The Mealy state diagram is shown in Fig.
6.10(a) and follows directly from the unassigned state table. Note that in
this case the output and state codes can be made identical (see Table
6.12(b)) and the device is best represented by the Moore model shown in
Fig. 6.10(b).

Both models may be used to represent a sequential machine, the choice
being one of convenience and personal preference. Note, however, that
the Mealy model is more general (and the one used in this text); the Moore
model implies that the output can be made a function of the internal states.
Once the state diagram for a sequential circuit has been produced and
tested, it is an easy matter to convert it to a state table. For simple problems, as we saw in the case of counter circuits, it is convenient to proceed
directly to the state table. The concepts of state tables and state diagrams
apply to sequential systems in general, but slight modifications are necessary when they are used to describe asynchronous logic. Let us now use
these techniques to describe a practical problem. Synchronous sequential
systems are by definition serial in operation and consequently a typical
problem would be to design a machine to recognize a specific pattern in a
serial input sequence. Suppose a circuit is required to recognize a particular three-bit pattern, say 101, and to produce an output whenever it
occurs in the continuous serial input to the circuit. For example, in the
binary sequence 01110111110111 etc., we would want outputs for the two

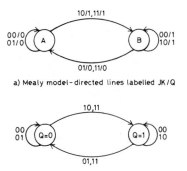

a) Mealy model–directed lines labelled JK/Q

b) Moore model–directed lines labelled JK

Figure 6.10 State diagrams for JK-bistable

occurrences of 101 in the sequence. We shall define the problem first by deriving its state diagram, and second by converting the state diagram to a state table. The state diagram is shown in Fig. 6.11 and the state table in Table 6.13. Since it is a serial input, we need only consider the inputs that can occur during a clock period, i.e. 0 and 1. Starting from an initial waiting state A, if a 0 is received the circuit stays in the same state (indicated by a sling in the state diagram). For a 1, however, there is a transition to state B, indicating the start of the required sequence. If, while in this state, a 0 is received, i.e. sequence (10), the circuit changes to state C. When in state C, if a 1 is received, completing the sequence (101), the circuit changes to state D, giving the required output. An input of 0 returns the circuit to state A to await the start of another sequence. When the circuit is in state D, a 1 returns it to state B, and a 0 returns it to state A. This problem illustrates the difficulty of logic specification. For example, how should the circuit react to an input of the form:

$$1 \quad 0 \quad 1 \quad 0 \quad 1 \quad 0 \quad 1$$

As it stands we would get two outputs for the two separate occurrances of 101. But do we want to recognize the embedded sequence? If so we must redirect the transition from state D (see Fig. 6.11) on input 0 to state C rather than state A.

Clearly, state diagrams and tables are equivalent ways of describing a sequential circuit; there is in fact a one-to-one correspondence between them. However, the state table is a more convenient form for manipulation and it is used expressly for this reason.

Let us consider another example. Suppose we want to develop the state diagram and state table for a circuit that continuously compares two four-bit serial message channels, on a repetitive basis, and gives an output when coincidence is found. That is, we have two message inputs of the form

1101	1111	1000	1110	0001	etc. channel A
0001	1101	0011	1111	0001	etc. channel B
1234	1234	1234	1234	1234	etc. clock timing

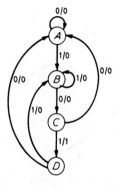

Figure 6.11 State diagram for pattern discriminator

Table 6.13 State tables for pattern discriminator

Present state	Input X Next state 0	1	Output 0	1
A	A	B	0	0
B	C	B	0	0
C	A	D	0	1
D	A	B	0	0
(a)				

Present state	Input X Next state 0	1	Output 0	1
A	A	B	0	0
B	C	B	0	0
C	A	A	0	1
(b)				

Present state	Input X Next state $X = 0$	$X = 1$	Output $X = 0$	$X = 1$
AB	A_+B_+	A_+B_+	A_+B_+	A_+B_+
00	00	01	0	0
01	10	01	0	0
10	00	00	0	1
(c)				

and we want to compare each four-bit word in turn, signalling an output when they are identical, as in the last word above.

The state diagram is shown in Fig. 6.12 and the corresponding state table in Table 6.14. In this case we must consider all possible input combinations available on the two signal channels, i.e. 00, 01, 11, 10, at every stage of the process and for every state. Furthermore, we must have at least four different internal states to allow for the examination of the four-bit message. In deriving state diagrams it is a good idea to follow through the correct sequence of inputs; this, in fact, has been done here, the left-hand half of the diagram being the correct path. Note that, in general, it is assumed that each input combination results in a transition to a new internal state, unless it is obvious that an identical state already exists.

Identical patterns result in a path through states (1), (2, 3), (7, 8), (9, 10,

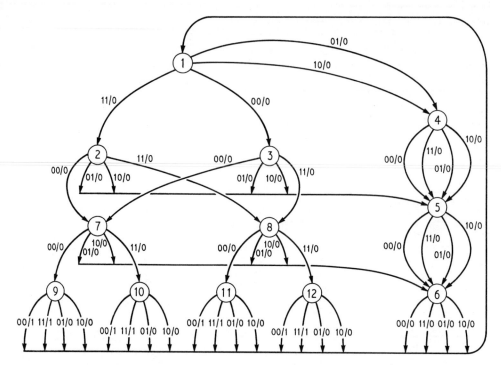

Figure 6.12 State diagram for pattern correlator

11, 12) (note that because of the binary nature of the problem, the state diagram spreads out in the form of a binary, switching, tree). For example, if the pattern 0001 appeared on both input lines, the path through the state diagram, starting from initial state 1, would be via states 3, 7, 9 and back to state 1. However, once we have had different inputs, the messages cannot be identical and we must wait for the next word; this is the reason for the delaying path via states 4, 5, 6—it can be entered at any stage of the four-bit comparison.

It should be obvious by now, particularly if we look at the state table, that this method of problem definition leads to a number of redundant states—in this process we allocate more states than are actually required to specify the logical function. For example, states 9, 10, 11, 12 are clearly identical, as are states (2, 3) and (7, 8). In a large and complicated system it is virtually impossible not to introduce redundant states into the design. Fortunately this does not matter at this stage for, as we shall see later, algorithms exist which can be used to perform the reduction of state tables.

Most of the logical circuits discussed in this chapter have been simple enough to design by intuitive methods alone. The pattern correlator, for instance, could be devised using an exclusive OR circuit to make the comparison, a set–reset bistable to register a non-coincidence, and some means to reset the bistable after every four-bit cycle (a counter circuit).

Table 6.14 State table for pattern correlator

| Present states | Inputs xy | | | | | | | |
| | Next states | | | | Outputs | | | |
	00	01	11	10	00	01	11	10
1	3	4	2	4	0	0	0	0
2	7	5	8	5	0	0	0	0
3	7	5	8	5	0	0	0	0
4	5	5	5	5	0	0	0	0
5	6	6	6	6	0	0	0	0
6	1	1	1	1	0	0	0	0
7	9	6	10	6	0	0	0	0
8	11	6	12	6	0	0	0	0
9	1	1	1	1	1	0	1	0
10	1	1	1	1	1	0	1	0
11	1	1	1	1	1	0	1	0
12	1	1	1	1	1	0	1	0

Nevertheless, it is instructive to follow through the design of simple circuits in order to appreciate fully the basic principles involved, which of course are applicable to larger and more complicated systems.

So far, we have seen that the steps involved in designing synchronous logic circuits are:

(1) Define the problem using a state diagram and/or a state table.
(2) Simplify the state table by eliminating redundant internal states.
(3) Allocate codes to the remaining states.
(4) Determine the input switching equations for the selected storage device.

Steps 2 and 3, unnecessary in the design of counter circuits since the number of states and their coding were implicit in the problem, will be discussed in some detail in the following chapters.

6.4.1 State machines

The state diagram as a tool for specifying sequential logic can become very cumbersome when handling complex systems. In particular the need to fully specify for each state the resultant transitions and outputs for all input combinations can become tedious. What is required is a method which allows a reduced specification and unambiguously reflects the required logic processes. The *state machine* approach enables an algorithmic specification in terms of a flowchart notation (similar to that used in software development) of the required operations of the machine. However, in the final analysis it still remains necessary to consider the operation of the total machine, including the unused states and input conditions. State machines, or to give them their full name algorithmic state machines (ASM), originated in the Hewlett-Packard laboratories and were first described by Clare.[5]

In an ASM flowchart states are represented by rectangular boxes with the symbolic state name enclosed in a small circle adjacent to its left-hand corner (see Fig. 6.13). Outputs may either be associated directly with the state (Moore model) in which case they are written inside the state box or, if generated as a result of some input condition (Mealy model), they are enclosed separately in an oval shape and must follow a conditional symbol. Alternative state transitions as determined by the absence or presence of some input condition, that is conditional branch points, are represented by a diamond shape. Multi-way branches may be depicted by cascading or ganging in parallel the conditionals, in the normal software manner.

These three basic symbols comprise the complete notation for the ASM chart, an example of which is shown in Fig. 6.13(d). The ASM chart shown represents a three-state machine which gives an output OP1 whilst in state A and goes to state C when the input $X = 1$, resetting OP1, and to state B if $X = 0$ giving an output of OP2. When in states B and C the machine will reset back to state A on receiving the next clock pulse. It will be obvious that the machine is assumed to be synchronous with state transitions occurring on the arrival of each clock pulse.

Note that the ASM chart can have the characteristics of both a Moore machine (the output OP1 in state A is independent of input X) and a Mealy machine (the conditional output OP2). Thus it is not possible to represent this machine directly using state diagrams since they require either a Mealy or Moore model to be used exclusively. It is, of course, possible to trans-late state diagrams from one model to another[6] but this does not help in this case. The only recourse in this example would be to restate the design requirements. Note, however, that once the circuit is realized in hardware form it is always possible to produce a state diagram representing its opera-

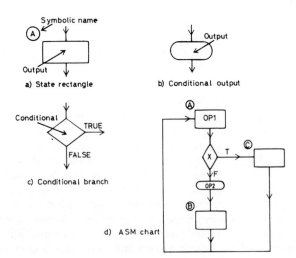

Figure 6.13 State machine notation

tion, but the required algorithm will not necessarily be obvious from such a diagram.

As an example let us consider the design of the control logic for a drink vending machine. The machine dispenses a drink for ten pence and will accept 2p, 5p and 10p coins which sum to the exact amount—otherwise the coins are returned to the customer. This is a very simple design specification, for example additional coins of other denominations could be accepted and when necessary change could be given. However, the specification will suffice for our present purposes. Clearly we require three internal inputs to the logic indicating that a 2p, 5p or 10p piece has been inserted into the machine. These inputs would be derived from other circuits in the vending machine but since the state machine operates in a synchronous mode they must be single pulses in synchronism with the system clock. Note also that the inputs would be mutually exclusive, that is they cannot occur together (there is only one coin slot and coins would be inserted in sequence).

Outputs must be generated to operate the drink dispensing (DISPENSE) and refund (RETURN) mechanisms. In addition indicator lights signalling that the machine is ready for operation (READY) and in operation waiting for the next coin (NEXT COIN) would be required. The ASM chart for the control logic is shown in Fig. 6.14.

The machine has eight states, though in some designs DR and AB could be replaced by conditional outputs. The need for the other states will be apparent since separate states are required to register and count the incoming coins; note also the cascading of the conditional tests which in the machine will all occur simultaneously.

The next step is to realize the machine in hardware form and this may be done using any of the usual methods described earlier (and to be discussed in more detail in Chapter 7). However, since it is usual to implement the ASM in terms of D-type bistables it is more convenient to work from a state transition table. But first the internal states must be encoded by allocating a unique binary code to each state; this is shown in Fig. 6.14 where the values of the state variables A, B and C are inserted at the right-hand corner of the state boxes. The encoded state transition table may now be derived directly by inspection of the ASM chart, as shown in Table 6.15. The large number of don't-cares (cannot happen in this case) arises from the mutual exclusivity of the inputs. However, since the chart only depicts what is required to happen, in practice there are generally a large number of don't-care terms generated.

The application equations for A_+, B_+ and C_+ (that is the input equations for the D-bistables) could be extracted and minimized in the usual way. However, as we have seen it is good design practice to realize combinational logic using an MSI/LSI module such as a PLA, MUX or ROM; in such cases the equations can usually be realized directly without minimization. As we saw earlier in the case of the PLA, providing the module can accommodate all the product terms there is no point in further minimization. Note that each line of the transition table produces a product term (some of which will be common); in our example the equations for

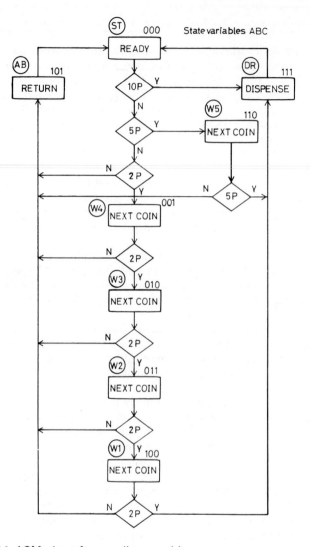

Figure 6.14 ASM chart for vending machine

A_+, B_+ and C_+ can be realized using 15 product terms. An added advantage of deriving the transition table in this form (as shown in Table 6.15) is that the machine as represented can be realized directly using ROMs. This is accomplished by storing the next state and output values as ROM words and using the present state and input values as the address inputs. The outputs of the ROM would be taken directly to a D-type bistable register.

The realization of sequential machines using PLAs to implement the combinational logic for D-type bistables is used extensively in designing LSI/VLSI circuits[7] but in this case reduction techniques can often be

Table 6.15 Transition and output tables for vending machine

Present states A B C	Inputs 2p 5p 10p	Next states A_+ B_+ C_+	Outputs READY . RETURN . DISPENSE . NEXT COIN			
0 0 0	X X 1	1 1 1	1	0	0	0
0 0 0	X 1 X	1 1 0	1	0	0	0
0 0 0	1 X X	0 0 1	1	0	0	0
0 0 1	1 X X	0 1 0	0	0	0	1
0 0 1	0 X X	1 0 1	0	0	0	1
0 1 0	1 X X	0 1 1	0	0	0	1
0 1 0	0 X X	1 0 1	0	0	0	1
0 1 1	1 X X	1 0 0	0	0	0	1
0 1 1	0 X X	1 0 1	0	0	0	1
1 0 0	1 X X	1 1 1	0	0	0	1
1 0 0	0 X X	1 0 1	0	0	0	1
1 0 1	X X X	0 0 0	0	1	0	0
1 1 0	X 1 X	1 1 1	0	0	0	0
1 1 0	X 0 X	1 0 1	0	0	0	0
1 1 1	X X X	0 0 0	0	0	1	0

(a) Encoded state transiton table

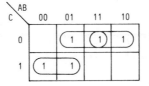

(b) K-map for NEXT COIN output

effective in reducing silicon area. The technique can of course be employed independently of whether state diagrams or ASM charts are used in the original design phase but in most cases the state machine approach will generate the transition tables directly without further manipulation. In the case of VLSI implementation the outputs of the PLA would normally be taken directly to an MOS shift register stage which provides the necessary delay (usually with a two-phase clocking system which gives separate input/output cycles); this will be covered in more detail in Chapter 10.

As we have seen for counters, the timing diagram is an important means of analysing and specifying a sequential system. The timing diagram for the vending machine, showing the correct inputs for dispensing drinks, is given in Fig. 6.15; note how easy it is to relate the waveforms to the original ASM description.

The ASM chart is normally used to describe the *control structure* necessary to perform a required algorithm on a given *data structure*; as we shall see later this is the basis of design at the systems level.

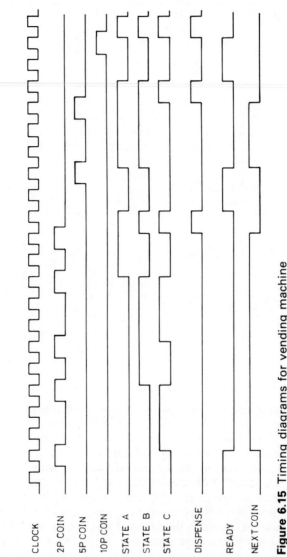

Figure 6.15 Timing diagrams for vending machine

Moreover, in digital systems design there is often a decision to be made as to whether a logic process should be realized in software or hardware. The ASM chart which describes the required algorithm can be used in the conceptial design stages for either form of implementation. Thus, though in essence there is little difference between the state diagram and the state machine approaches, the ASM chart would appear to be a better conceptional design tool and, perhaps more important, establishes a bridge between hardware and software realizations.

6.5 Linear sequential circuits

These are a special class of synchronous sequential machines which consist of two basic elements, modulo-2 adders and a unit delay element (for instance, the D-type bistable) (see Fig. 6.16(a)). They can take the form of either autonomous networks with no input except clock (e.g. counters) or conventional input–output machines. In practice the circuits would normally consist of shift registers with feedback via exclusive OR logic and, for example, can generate long strings of binary digits processing psuedo-random properties. The circuits are used extensively in communication systems, to generate and check error-correcting codes, in control system design as a source of white noise, and in testing logic systems using signature analysis.

The circuits have the properties of linear systems (hence the name) in that they obey the principle of superposition and preserve the scale factor of the inputs. Thus only linear components can be used to realize the machine, that is modulo adders, scaler multipliers and unit delaying elements. Note that the AND/OR functions would be precluded as they are not

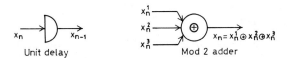

a) Components of linear machine

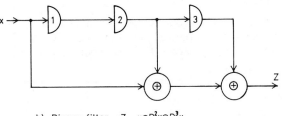

b) Binary filter- $Z = x \oplus D^2 x \oplus D^3 x$

Figure 6.16 Linear sequential machines

linear. Though we shall restrict our discussion to binary machines working in modulo-2, the theory equally applies to machines of any modulus.

The basic delay element (shift-register stage) has one input x and an output which occurs one clock pulse later; binary values are assumed throughout, i.e. logical 0 or 1. Thus we may define the delay element as

Output $x_n = x_{n-1}$ for all n

The modulo-2 adder may be defined as

$$x_n = \left\{ \sum_{j=1}^{k} x_n^j \right\} \text{ mod-2, for k inputs.}$$

Since modulo-2 addition is a linear operation the following algebraic theorems apply. (The symbol $+$ is normally used for modulo-2 addition, but the circle will be omitted later on for convenience; where $+$ occurs modulo-2 addition should always be assumed.)

(1) $A \oplus A \oplus A \oplus \cdots \oplus A = 0$ for an even number of As

$A \oplus A \oplus A \oplus \cdots \oplus A = A$ for an odd number of As

(2) $A \oplus B \oplus C \oplus \cdots = 0$ for an even number of variables value 1

$A \oplus B \oplus C \oplus \cdots = 1$ for an odd number of variables value 1

(3) $A \oplus B = C$ implies

(a) $A \oplus C = B$

(b) $B \oplus C = A$

(c) $A \oplus B \oplus C = 0$

because subtraction and addition modulo-2 are identical.

(4) The operations are associative, commutative and distributive.

6.5.1 Binary filters

The simplest type of linear sequential filter has a single input and output; the output may be expressed as a modulo-2 sum of selected input digits, from the *past* as well as the present. Consider the circuit shown in Fig. 6.16(b); it may be represented as

$$Z = X \oplus D^2 X \oplus D^3 X$$

where the symbol D^n is an algebraic operator the effect of which is to delay by n digits the variable it operates on, and where X and Z can represent either single digits or sequences. The expression for Z may be rewritten

$$Z = X(1 \oplus D^2 \oplus D^3)$$

or as a transfer function

$$Z/X = 1 \oplus D^2 \oplus D^3$$

The '*impulse*' *response* of the filter (i.e. the response to an input

Table 6.16 Impulse response for binary filters

$X \cdots \quad \cdots 0001000000$
$Z \cdots \quad \cdots 0001011000$

Time	Input X	$D1$	$D2$	$D3$	Output Z
1	1	0	0	0	1
2	0	1	0	0	0
3	0	0	1	0	1
4	0	0	0	1	1
5	0	0	0	0	0

(a)

Time	Input X	$D1$	$D2$	$D3$	Output Z
0	0	0	1	1	0
1	1	0	0	1	0
2	0	1	0	0	0
3	1	0	1	0	0
4	1	1	0	1	0
5	1	1	1	0	0
6	0	1	1	1	0
7	0	0	1	1	0

repeats
(b)

sequence containing a single 1) is shown in Table 6.16(a). Note that the length of the 'transient' is 3 bit-times. It is interesting to determine the input sequence that will give all zeros at the outputs, that is, a null sequence X_0 for which

$$X_0(1 \oplus D^2 \oplus D^3) = 0$$

and

$$X_0 = X_0(D^2 \oplus D^3).$$

Thus, to find a digit of the null sequence we must add the second and third digits of the *previous* sequence. There are $2^3 - 1$ non-trivial ways of picking three digits to start the sequence (note that 000 is a trivial case); we shall choose 011. Thus the sequence is

$$\begin{array}{c} 321 \\ X_0 \cdots 0111001 \quad 0111001 \quad 0111001 \cdots \end{array}$$

Note that after seven digit-times, the sequence repeats itself cyclicly; the circuit is analysed in detail in Table 6.16(b).

The filter inverse may also be described by considering Z as the input to the filter and X as the output. Its inverse transfer function may be written

$$\frac{X}{Z} = \frac{1}{1 \oplus D^2 \oplus D^3}$$

Synthesis of this circuit is easily accomplished by changing the direction of information flow in the original circuit. this can only be done for the exclusive OR gates since the flow through the delay elements is unilateral; the inverse filter circuit is shown in Fig. 6.17(a). Note that each feed-forward path in the original circuit now becomes a feedback path. The filters may be cascaded (Fig. 6.17(b)) and if both filters are initially at rest (no stored 1s in the delays) X-out will equal X-in.

Once the output of the filter (impulse response) has been described in terms of a polynomial, the normal theory of rational functions may be applied to simplify the equations, thus yielding a more economic and practical circuit. For example, let us synthesize a circuit which has the impulse response

X $1\,0\,0\,0\,0\,0\,0\,0\,0\,\cdots$
Z $1\,0\,1\,1\,0\,1\,0\,1\,0\,\cdots$

thus

$$Z = (1 + D^2 + D^3 + D^5 + D^7 + D^9 \cdots)X$$

The transfer function is

$$Z = \frac{1 + D^3 + D^4}{1 + D^2}X$$

which can easily be proved by polynomial division.

One simple way of implementation is to rearrange the function as

$$Z(1 + D^2) = X(1 + D^3 + D^4)$$
$$Z = X(1 + D^3 + D^4) + D^2Z$$
$$\therefore Z = X + D^2[Z + D(X + DX)]$$

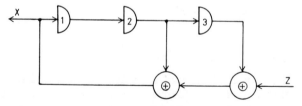

a) Inverse filter circuit

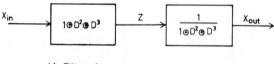

b) Filters in cascade

Figure 6.17 Binary filter circuits

The circuit is shown implemented in Fig. 6.18 and analysed in Table 6.17. An alternative approach is to expand the function as a sum of partial fractions. For example, if we divide the numerator of the transfer function above by $(1 + D^2)$ so that the degree of the numerator of the remaining fraction is less than the denominator we get

$$Z = (D^2 + D + 1) + \frac{D}{D^2 + 1}$$

$$
\begin{array}{r}
D^2 + D + 1 \\
D^2 + 1 \overline{\smash{\big)}\ D^4 + D^3 + 1} \\
\underline{D^4 + D^2} \\
D^3 + D^2 + 1 \\
\underline{D^3 + D} \\
D^2 + D + 1 \\
\underline{D^2 + 1} \\
D
\end{array}
$$

Note:
Modulo-2
subtraction

Resolving the last term into partial fractions (note that $(D + 1)^2 = (D^2 + 1)\text{mod-2}$) we have

$$Z = \left[(D^2 + D + 1) + \frac{1}{(D + 1)} + \frac{1}{(D + 1)^2}\right]X$$

The circuit is implemented in Fig. 6.19 (note the change of information flow for the inverse filters).

It is worth noting that the simple transfer function $Z/X = 1 + D$ (see Fig. 6.20) is equivalent to the trigger bistable circuit. Furthermore, the circuit may be used to convert a serial Gray code input X directly to a binary output Z.

6.5.2 Error-correcting codes

One of the chief uses of binary sequence filters is in the transmission of digital messages, and the error detection and correction of such messages.[8] Consider the arrangement of filters shown in Fig. 6.21. A sequence of seven X digits is fed into a transmitter filter with transfer function T,

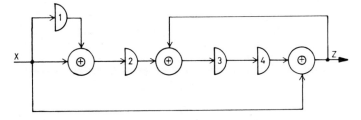

Figure 6.18 $Z = X + D^2[Z + D(X + DX)]$

Table 6.17 Analysis of Fig. 6.18

Time	Input X	D1	D2	D3	D4	Output Z
1	1	0	0	0	0	1
2	0	1	1	1	0	0
3	0	0	1	1	1	1
4	0	0	0	0	1	1
5	0	0	0	1	0	0
6	0	0	0	0	1	1
7	0	0	0	1	0	0

etc.

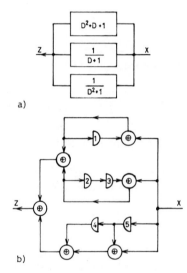

a)

b)

Figure 6.19 Alternative realizations

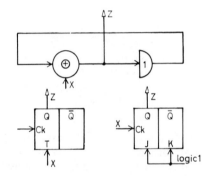

Figure 6.20 Trigger bistables

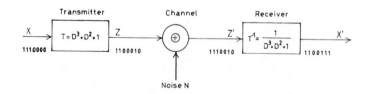

Figure 6.21 Binary sequence filters used for error detection and correction

resulting in a sequence $Z = (T)X$ which is transmitted through the 'noisy' channel. In the channel, a noise sequence, N, is added to Z so that the signal that arrives at the receiver filter is

$Z' = Z + N$

At the receiver, the inverse filter creates from the sequence Z' a sequence

$$\begin{aligned} X' &= (T^{-1})Z' = T^{-1}(Z + N) \\ &= (T^{-1})[(T)X + N] \\ &= X + (T^{-1})N \end{aligned}$$

Thus with no noise in the channel (i.e. $N = 0$) the output $X' = X$. If a *single* noise digit is injected, the sequence X' contains X plus the superimposed impulse response of the receiver filter (Table 6.18(a)).

Table 6.18(b) shows a possible coding and decoding arrangement; note that the first four digits are information digits and may be chosen in $2^4 = 16$ different ways; the remaining three bits are all zeros and are called buffer bits. The seven-bit pattern X is 'scrambled' for transmission in the first filter and 'unscrambled' by the second filter, to reproduce X. If no noise error occurs this would be indicated by all zeros in the last three digit-positions. Thus, if the sequence $X = 1110000$ was transmitted and the noisy channel inserted a 1 in the third position from the left (0010000) the 'un-scrambled' sequence would be 1100111, that is 1110000 + 0010111, the last three bits indicating that an error in transmission has occurred.

Note that the pattern (X') depends on the position of the noise digit and the impulse response of the inverse filter. Thus by observing the three buffer digits, and knowing the impulse response of the filter, we can deter-mine if, and where, an error occurs. Furthermore, by choosing the filter such that its impulse response has a period of seven digits (the length of the sequence), each of the seven possible combinations of three successive digits in the response will be different (Table 6.18(c)). This is governed by the number of buffer bits (three) and the degree of the polynomial. For single-error correction in a block of length n containing b buffer digits and $k = n - b$ information digits, we need a receiver with an impulse response of period n with each b successive digits in the response different from each other. This is possible for the case $n = 2^b - 1$ and the proper polynomial is one of degree b which has a maximal length (null sequence) of $2^b - 1$ digits. (We shall see later that the polynomial must be *primitive* to satisfy this condition.) If we assume that a single noise digit were present, the original sequence (X) can be recreated by adding (modulo-2) the sequence $(T^{-1})N$ to the sequence X'.

6.5.3 Maximum-length sequences

In the running example we have chosen, i.e. the polynomial $1 + z^2 + z^3$, the impulse response repeats itself cyclically with a period $N = 2^n - 1$ where n is the order of the polynomial (this corresponds to the maximum-length null sequence mentioned earlier). Linear sequential machines employing

Table 6.18 Error detecting and decoding network

Time	Input	$D1$	$D2$	$D3$	Z	
1	1	0	0	0	1	⎫
2	0	1	0	0	0	⎪
3	0	0	1	0	1	⎬ period
4	0	1	0	1	1	⎪ 7 bits
5	0	1	1	0	1	⎪
6	0	1	1	1	0	⎪
7	0	0	1	1	0	⎭
8	0	0	0	1	1	
9	0	1	0	0	0	

repeats
(a)

X		$Z = (T)X$
0000	000	0000000
0001	000	0001011
0010	000	0010110
0011	000	0011101
0100	000	0101100
0101	000	0100111
0110	000	0111010
0111	000	0110001
1000	000	1011000
1001	000	1010011
1010	000	1001110
1011	000	1000101
1100	000	1110100
1101	000	1111111
1110	000	1100010
1111	000	1101001

(b)

Impulse response $(T^{-1})N$
Noise occurs in digit position:

1	2	3	4	5	6	7
1	0	1	1	1	0	0
0	1	0	1	1	1	0
0	0	1	0	1	1	1
0	0	0	1	0	1	1
0	0	0	0	1	0	1
0	0	0	0	0	1	0
0	0	0	0	0	0	1

(c)

m-length sequences have very interesting characteristics; for example, consider the circuit of Fig. 6.22, called a *linear feedback shift register* or *chain-code* counter. This is an *autonomous* network with no external input, but providing there is an initial non-zero starting state, the circuit will generate the cyclic sequence $\cdots$ 10111001011100 $\cdots$ the starting point depending on the initial state of the delays (e.g. 001 will generate 10111001 $\cdots$). Furthermore, *all possible combinations of three bits (excluding all zeros)* are generated in the shift register—see Table 6.16(b). M-sequences can be shown to possess *psuedo-random* properties and as such can provide a repeatable source of white noise in the form of a psuedo-random binary sequence (PRBS). Another useful property is that if the outputs of any pair of delays are added modulo-2 the resulting output will be a shifted version of the m-sequence.

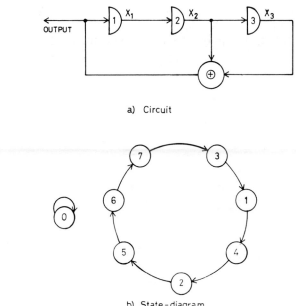

a) Circuit

b) State-diagram

Figure 6.22 Chain code counter

In affect the chain-code counter can be considered as a polynomial divider network[9] performing the general function

$$y(z) = \frac{x(z)}{h(z)}$$

where $x(z) = x_0 + x_1 z + x_2 z^2 + \cdots + x_n z^n = 1$ and $h(z)$ a *primitive* polynomial, that is one that cannot be factorized (irreducible) of the general form

$$h(z) = h_0 + h_1 z + h_2 z^2 + \cdots + h_n z^n$$

where the binary coefficient h represents the absence (0) or presence (1) of a term. The general network for polynomial division is shown in Fig. 6.23;

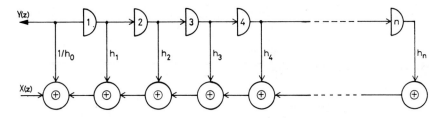

Figure 6.23 General network for polynominal division

note that this reduces to the circuit in Fig. 6.22 for $1/h_0 = h_2 = h_3 = 1$ and $h_1 = 0$. The output $y(z)$ of the network for the input $x(z) = 1$ can be computed by performing the polynomial division (modulo-2):

$$\frac{1}{1 + z^2 + z^3}$$

Thus

$$
\require{enclose}
\begin{array}{r}
1 + z^2 + z^3 + z^4 + z^7 + z^9 + \cdots \\[2pt]
1 + z^2 + z^3 \,\big)\, 1 \phantom{{}+ z^2 + z^3 + z^4 + z^7 + z^9 + \cdots}
\end{array}
$$

$$
\begin{array}{l}
\underline{1 + z^2 + z^3} \\
\quad z^2 + z^3 \\
\quad \underline{z^2 + z^4 + z^5} \\
\qquad z^3 + z^4 + z^5 \\
\qquad \underline{z^3 + z^5 + z^6} \\
\qquad\quad z^4 + z^6 \\
\qquad\quad \underline{z^4 + z^6 + z^7} \\
\qquad\qquad z^7 \\
\qquad\qquad \underline{z^7 + z^9 + z^{10}} \\
\qquad\qquad\quad z^9 + z^{10} \\
\qquad\qquad\quad \underline{z^9 + z^{11} + z^{12}} \\
\qquad\qquad\qquad z^{10} + z^{11} + z^{12} \quad \text{repeats}
\end{array}
$$

or, alternatively, in the binary notation

$$
\begin{array}{l}
\qquad 10111001 \\
1011 \,\big)\, 1 \\
\quad \underline{1011} \\
\quad 001100 \\
\quad\;\; \underline{1011} \\
\quad 01110 \\
\quad\;\; \underline{1011} \\
\quad 01010 \\
\quad\;\; \underline{1011} \\
\quad\;\; 0001000 \\
\quad\;\;\;\;\; \underline{1001} \\
\quad\;\;\;\; 0011 \quad \text{repeats}
\end{array}
$$

Note that as would be expected the sequence is cyclic, i.e.

$$y(z) = 1 + z^2 + z^3 + z^4 + z^7 + z^9 \cdots$$
$$= 1011100101 \cdots$$

Examples of primitive polynomials over a binary field which generate m-sequences are shown in Table 6.19. (References 9, 10 and 11 also give

Table 6.19 Primitive polynomials

Sequence length	Primitive polynomial
3	$1 + D + D^2$
7	$1 + D^2 + D^3$
15	$1 + D^3 + D^4$
31	$1 + D^3 + D^5$
63	$1 + D^5 + D^6$
127	$1 + D^6 + D^7$
255	$1 + D^4 + D^5 + D^6 + D^8$
511	$1 + D^4 + D^9$
1023	$1 + D^3 + D^{10}$
2047	$1 + D^2 + D^{11}$
4095	$1 + D^3 + D^4 + D^7 + D^{12}$
8191	$1 + D + D^3 + D^4 + D^{13}$
16383	$1 + D^4 + D^8 + D^{13} + D^{14}$

lists.) A general treatment of feedback shift registers has been given by Elspas[10] and Peterson and Weldon[9] have described their use in error-correcting codes. In particular since the Hamming codes and any single error-correcting code can be shown to have a basic cyclic property they are

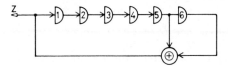

a) FSR for cycle length of 63

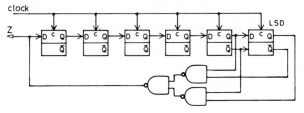

b) Implementation using D-Type bistables

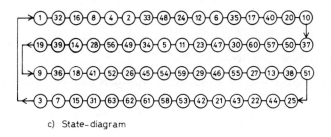

c) State-diagram

Figure 6.24 Chain code generator of length 63

easy to generate and check using linear feedback shift register circuits.

Let us now consider the design of a chain code counter which will generate a pseudo-random binary sequence with a cycle length of 63. Fairly obviously we need a six stage shift register with the feedback being given by the primitive polynomial $1 + D^5 + D^6$ (see Table 6.19). The circuit using D-type bistables is shown in Fig. 6.24 together with the resultant state table; note that the all zeros combination would lead to a locked state condition and that there are $2^n - 1$ unique terms. The Z output which is the serial PRBS is given by

$$1\ 0\ 0\ 0\ 0\ 1\ 1\ 0\ 0\ 0\ 1\ 0\ 1\ 0\ 0\ 1\ 1\ 1\ 1\ 0\ 1\ 0\ 0\ 0\ 1\ 1\ 1\ 0\ 0\ |1\ 0\ 0\ 1\ 0$$
$$1\ 1\ 0\ 1\ 1\ 1\ 0\ 1\ 1\ 0\ 0\ 1\ 1\ 0\ 1\ 0\ 1\ 0\ 1\ 1\ 1\ 1\ 1\ 1\ 0\ 0\ 0\ 0\ 0\ |1\ 0\ 0\ 0\ 0$$

repeats

In a practical situation the shift register must be set to a non-zero starting state (state 1 in the example); this can be easily achieved using the set and clear inputs to the bistables.

References and bibliography

1. Chirlian, P. *Analysis and Design of Integrated Electronic Circuits*. Harper & Row, New York, 1981.
2. Oberman, R. M. *Counting and Counters*. Macmillan, London, 1981.
3. Mealy, G. H. A method for synthesising sequential circuits. *Bell Syst Tech. Journal* **34**, 1045–79 (1955).
4. Moore, E. F. *Gedanken-experiments on Sequential Machines*. Automata Studies. Princeton University Press, 1956, pp. 129–53.
5. Clare, C. R. *Designing Logic Systems Using State Machines*. McGraw Hill, New York, 1973.
6. Friedman, A. D. and Menon, P. R. *Theory and Design of Switching Circuits*. Computer Science Press Inc., 1975.
7. Mead, C. and Conway, L. *Introduction to VLSI Systems*. Addison Wesley, Reading, Mass., 1980.
8. Huffman, D. A. A linear circuit viewpoint on error-correcting codes. *IRE Trans. on Inform. Theory* **IT2**, 20–8 (1956).
9. Peterson, W. W. and Weldon, E. J. *Error-correcting Codes*. MIT Press, Mass., 1972.
10. Elspas, B. The theory of autonomous linear sequential networks. *IRE Trans. on Circuit Theory* **CT6**, 45–60 (1959).
11. MacWilliams, F. J. and Sloane, N. J. A. *The Theory of Error-correcting Codes*. North Holland, Amsterdam, 1978.
12. Muroga, S. *VLSI System Design*. Wiley, New York, 1982.
13. Winkel, D. and Prosser, F. *The Art of Digital Design*. Prentice Hall, Englewood Cliffs, 1980.
14. Gill, A. *An Introduction to the Theory of Finite State Machines*. McGraw Hill, New York, 1962.
15. Dean, K. J. The design of parallel counters using the Map Method. *Radio and Electronic Engineer* **32**, 159–62 (1966).

Tutorial problems

6.1 Design a synchronous five-bit ring counter using the design principles discussed in this chapter. Use both T-FF and SR-FF storage elements.

6.2 Derive the input equations for an SR-FF decimal counter counting and displaying the following number sequence:

	5	4	2	1
0	0	0	0	0
1	0	0	0	1
2	0	0	1	0
3	0	0	1	1
4	0	1	0	0
5	1	0	0	0
6	1	0	0	1
7	1	0	1	0
8	1	0	1	1
9	1	1	0	0

6.3 Design a divide-by-five circuit using JK-bistables and then show how it may be used together with divide-by-two circuits to produce a divide-by-ten counter. Show full waveform diagrams for the circuit.

6.4 A burglar alarm system is to automatically dial the telephone number 01 66 48 95 on detecting an intruder. The number generator comprises a decoder circuit sequentially accessed with the outputs of a clocked 3-bit counter circuit triggered by the intruder detector circuits. The output of this circuit will be the telephone number represented as a sequence of binary coded decimals.

Design a suitable circuit using JK-bistables and NAND gates to effect the control logic.

6.5 Using JK-bistables and discrete gates design a 2-bit synchronous up/down binary counter. A control input X is to be used to select the mode of operation; when $X = 1$ the counter will count down, otherwise it will count up. Repeat the design using the ASM chart method but this time realize the circuit using D-type bistables.

6.6 Construct the state diagram and state table for a logic circuit that gives an output whenever the sum of the 1 digits in a repetitive five-bit serial input sequence is 2.

6.7 Construct the state diagram and ASM chart for the excess-3 BCD counter circuit described in 6.3.1.

6.8 Derive the state diagram and state tables for the clamp-gate or zero-hold circuit used in sampled-data systems. The circuit has two serial inputs x and y and an output z. The characteristics are such that z is made equal to the present value of x if $y = 1$, or to the previous output value if $y = 0$.

Consider the possibility of reducing the number of internal states in this circuit.

6.9 Derive from first principles the state diagram and state tables for a binary serial full-adder circuit.

6.10 Investigate how the output from a simple ON-OFF switch may be used as input to a synchronous digital system.

(HINT: Consider the properties of the SR and D-type bistables.)

Derive the ASM chart for a machine (called a single-shot circuit) that will detect when an input pulse synchronized with clock goes high and as a consequence generates an output signal lasting for one clock pulse only. Note that the duration of the input pulse can be very much greater than the clock period.

6.11 In a certain instrumentation system it is required to know whether balls with a diameter greater than some specified value are moving up or moving down a narrow tube. Two sensors are placed in the tube separated by a distance less than the required minimum diameter of the ball. The sensors S_1 and S_2 indicate logic 1 when the presence of a ball is detected otherwise logic 0. Two outputs are required, $Z_1 = 1$ when a ball of the required diameter is going up and $Z_2 = 1$ when going down. Using the ASM chart method design a logic circuit employing D-type bistables to realize this specification.

6.12 Implement the transition table for the machine obtained in problem 6.11 in terms of ROM.

6.13 Design and realize a linear shift register counter that will generate a psuedo-random binary sequence with a cycle length of 15. Draw the state diagram for the machine and give the PRBS.

6.14 Construct the state table and ASM chart for the sequential circuit shown in Fig. 6.25. Show how the circuit could be redesigned using a D-type bistable and PLA realization.

Give a simple explanation for the operation of the circuit.

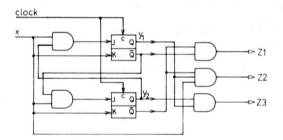

Figure 6.25 Problem 6.14

7 Design of synchronous sequential circuits

7.1 Introduction

We saw in the last chapter that, in general, the formal definition of a sequential problem leads to redundant internal states, that is, the state table contains more states than are actually required for the interpretation of the problem. The number of internal states of a sequential machine is an important parameter, since it determines the amount of hardware required to represent these states in the final circuit. For an s-state machine using binary storage devices, such as bistables, at least n bistables are required to assign these states, where n is the smallest integer greater than or equal to $\log_2 s$, i.e. $s \leq 2^n$. By reducing s, it is possible that n may also be reduced, and thus fewer bistables will be needed in the circuit realization. However, since n is directly dependent on the number of state-variables (r), reduction will only be achieved when s crosses a 2^r threshold. For example, a 14-state machine with 4 state-variables would need to be reduced to 8 states with 3 state-variables before any savings in storage elements would result. As the number of states increases the corresponding difference between s and r also increases; consequently for large machines economies in storage elements seldom result from state reduction.

Moreover reducing the number of states does not automatically reduce the cost of the system, since using fewer storage elements could increase the number of terms in the input and output equations and hence the amount of combinational logic required. For example, in designing a ring-counter (which requires a unique output for every state of the counter), it is often more economical to use a separate bistable for each state, rather than the minimal number of bistables necessary to represent these states, with suitable decoding and input logic. Thus, since a sequential circuit consists of storage plus combinational logic, for an optimum (and hence an economical) design we must also consider the amount of combinational logic required as well as the number of bistables. Nevertheless state reduction can be a worthwhile procedure since the inevitable increase in don't-care values often allows a more economical state-assignment and consequently a reduction in combinational logic. Another advantage is when realizing logic circuits as VLSI circuits; here state reduction can be effective when using PLA structures since the product terms can often be reduced producing overall savings in silicon area.

Another important consideration is the way in which the internal states are assigned an identifying code which determines the amount of combinational logic required. The straightforward approach of allocating each state according to normal ascending binary does not in general lead to an economical solution. It is important to realize that a non-minimal machine will satisfy perfectly all the original design specifications, but the resulting circuit will be costly and, because it contains more components, less reliable.

In this chapter we discuss fundamental techniques which have been evolved to search systematically for redundant (or equivalent) states, and we shall also investigate some of the problems and methods of economical state assignment. For large machines, which are inevitably those encountered in practice, recourse must be made to computer-aided methods of design—these will be discussed further in Chapter 10.

7.2   Reduction of internal states

Let us first consider the problem of state minimization using an intuitive approach. Table 7.1 shows the state table obtained in the last chapter for the pattern correlator problem. Inspection of this table immediately reveals that internal states (9, 10, 11, 12) and (2, 3) are *identical* in all respects, that is, the next-state and output entries in the state table correspond exactly, one with the other, for every input combination. Thus we can replace these identical states by one state in the table (conventionally chosen to be the state with the smallest number), modifying all other entries accordingly. Once this is done it is apparent that states (7, 8) are also identical; replacing these states results in the reduced state table shown in Table 7.2. Using this approach, then, we have evolved a minimal description of the machine which requires only seven states, i.e. (1), (2, 3), (4), (5), (6), (7, 8), (9, 10, 11, 12). Note that in the original table internal states (7, 8) were not identical, i.e. they did not have the same next states. When this condition occurs we generally refer to these states as *equivalent* states.

We shall now attempt a more formal definition of the equivalence property, which is due to Moore.[1] Two states are said to be equivalent if, for all sequences of inputs, the sequential machine produces the same output sequence when it is started in either state. Thus it is impossible to distinguish between the two states by the external, terminal, behaviour of the machine. It follows then that the necessary (but not sufficient) condition for two states to be equivalent is that their output states must be identical. Furthermore, two states may be considered equivalent, even though their next states are not the same, providing it is possible to establish an equivalence between the unlike states. This was demonstrated in the example above (Table 7.1) for states (7, 8).

In order to reduce a state table systematically, we must examine each pair of states having the same output states, and establish if their next states are equivalent. This may be done by direct examination and comparison, e.g. in Table 7.1 we say that $7 \equiv 8$ if $9 \equiv 11$ and $10 \equiv 12$; then by

Table 7.1 State table for pattern correlator

Present states	Inputs x_1x_2							
	Next state				Outputs Z			
	00	01	11	10	00	01	11	10
1	3^2	4	2	4	0	0	0	0
2	7	5	8	5	0	0	0	0
~~3~~	~~7~~	~~5~~	~~8~~	~~5~~	~~0~~	~~0~~	~~0~~	~~0~~
4	5	5	5	5	0	0	0	0
5	6	6	6	6	0	0	0	0
6	1	1	1	1	0	0	0	0
7	9	6	10^9	6	0	0	0	0
8	11^9	6	12^9	6	0	0	0	0
9	1	1	1	1	1	0	1	0
~~10~~	~~1~~	~~1~~	~~1~~	~~1~~	~~1~~	~~0~~	~~1~~	~~0~~
~~11~~	~~1~~	~~1~~	~~1~~	~~1~~	~~1~~	~~0~~	~~1~~	~~0~~
~~12~~	~~1~~	~~1~~	~~1~~	~~1~~	~~1~~	~~0~~	~~1~~	~~0~~

Table 7.2 Reduced state table

Present states	Inputs x_1x_2							
	Next state				Outputs Z			
	00	01	11	10	00	01	11	10
1	2	4	2	4	0	0	0	0
2	7	5	7	5	0	0	0	0
4	5	5	5	5	0	0	0	0
5	6	6	6	6	0	0	0	0
6	1	1	1	1	0	0	0	0
7	9	6	9	6	0	0	0	0
9	1	1	1	1	1	0	1	0

investigating the equivalence of the state pairs (9, 11) and (10, 12) we can establish if, in fact, $7 \equiv 8$. For a completely specified machine, we can make use of the transitive relationship (if $a \equiv b$ and $b \equiv c$, then $a \equiv c$) in order to ascertain equivalence.

An algorithmic technique for finding equivalent states has been described by Paull and Ungar,[2] which holds for both completely and incompletely specified (i.e. containing don't-care conditions) state tables. In this method an *implication chart* is constructed which shows the necessary conditions or *implications* that exist between all possible equivalent state pairs. The implication chart for the correlator problem discussed earlier is shown in Table 7.3(a); note that the chart has as many cells as it has possible equivalent state pairs. To construct the chart, the state table is first inspected for those state pairs which cannot possibly be equivalent because of differing output states, and a cross is entered in the appropriate cell; for example $1 \neq 9$, $1 \neq 10$, etc. These non-equivalent state pairs are called *incompatibles*. Similarly, a tick is inserted in the appropriate cells for identical state pairs. The chart is completed by considering next each

Table 7.3 Implication chart for correlator problem

(a) Initial implication chart

	1	2	3	4	5	6	7	8	9	10	11
2	3,7 / 4,5 / 2,8										
3	✓	3,7 / 4,5 / 2,8									
4	7,5 / 8,5	7,5 / 8,5	3,5 / 4,5 / 2,5								
5	7,6 / 5,6 / 8,6	7,6 / 5,6 / 8,6	5,6	3,6 / 4,6 / 2,6							
6	7,1 / 5,1 / 8,1	7,1 / 5,1 / 8,1	5,1	6,1	3,1 / 4,1 / 2,1						
7	7,9 / 5,6 / 8,10	7,9 / 5,6 / 8,10	5,9 / 5,6 / 5,10	6,9 / 6,10	1,9 / 1,6 / 1,10	3,9 / 4,6 / 2,10					
8	7,11 / 5,6 / 8,12	7,11 / 5,6 / 8,12	5,11 / 5,6 / 5,12	6,11	1,11 / 1,6 / 1,12	9,11 / 10,12	3,11 / 4,6 / 2,12				
9	X	X	X	X	X	X	X	X			
10	X	X	X	X	X	X	X	X	✓		
11	X	X	X	X	X	X	X	X	✓	✓	
12	X	X	X	X	X	X	X	X	✓	✓	✓

(b) After first pass through chart

	1	2	3	4	5	6	7	8	9	10	11
2	3,7 / 4,5 / 2,8										
3	✓	3,7 / 4,5 / 2,8									
4	7,5 / 8,5	7,5 / 8,5	3,5 / 4,5 / 2,5								
5	7,6 / 5,6 / 8,6	7,6 / 5,6 / 8,6	5,6	3,6 / 4,6 / 2,6							
6	7,1 / 5,1 / 8,1	7,1 / 5,1 / 8,1	5,1	6,1	3,1 / 4,1 / 2,1						
7	~~7,9 / 5,6 / 8,10~~	~~7,9 / 5,6 / 8,10~~	~~5,9 / 5,6 / 5,10~~	~~6,9 / 6,10~~	~~1,9 / 1,6 / 1,10~~	~~3,9 / 4,6 / 2,10~~					
8	~~7,11 / 5,6 / 8,12~~	~~7,11 / 5,6 / 8,12~~	~~5,11 / 5,6 / 5,12~~	~~6,11~~	~~1,11 / 1,6 / 1,12~~	9,11 / 10,12	~~3,11 / 4,6 / 2,12~~				
9	⊗	⊗	⊗	⊗	⊗	⊗	⊗	⊗			
10	⊗	⊗	⊗	⊗	⊗	⊗	⊗	⊗	✓		
11	⊗	⊗	⊗	⊗	⊗	⊗	⊗	⊗	✓	✓	
12	⊗	⊗	⊗	⊗	⊗	⊗	⊗	⊗	✓	✓	✓

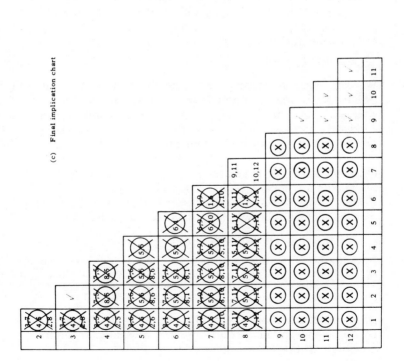

(c) Final implication chart

permissable state pair in turn, say (1, 2), and entering in the chart the necessary equivalent next state pairs, in this case (3, 7), (4, 5) and (2, 8), required to make the initial state pair equivalent.

When the chart is completed in this way, it is examined, starting from the extreme right-hand column, for cells containing a cross. In Table 7.3(a), the first cross occurs in column 8 for the state pair (8, 12). Since these states cannot possibly be equivalent, it follows that any other state pair relying on (8, 12) for equivalence (in practice any cell containing (8, 12)) must also be crossed out, e.g. cells (2, 8) and (3, 8). The chart is systematically processed in this way continuing with (8, 11), (8, 10), (8, 9), (7, 12) etc., noting each time which incompatible has been used, until the final column is reached. The process is then repeated ignoring those incompatibles used in previous searches. A convenient method of noting when an incompatible has been used is to mark the cell by encircling the cross. A number of passes through the chart will be necessary before all the cells have been classified into crossed-out, ticked (identical) or unticked (equivalent) cells. Table 7.3(b) shows the state of the chart after one pass, and the final chart is shown in Table 7.3(c); in practice, of course, only one chart is required.

The final step in the reduction process is to extract the equivalent states, sometimes called *pair-wise compatibles*, from the chart; in this case we have, reading from the right-hand column of Table 7.3(c),

$$11 \equiv 12 \quad 10 \equiv 12 \quad 10 \equiv 11 \quad 9 \equiv 12 \quad 9 \equiv 11 \quad 9 \equiv 10 \quad 7 \equiv 8 \quad 2 \equiv 3$$

These pair-wise compatibles can be grouped into sets of states, each with the same output states and all equivalent to one another, using the transitive relationship, i.e. if $11 \equiv 12$ and $10 \equiv 12$, then $11 \equiv 10$. Thus we may say that the *maximal compatibles* are

(9, 10, 11, 12)(7, 8)(2, 3)

We can define the terms used more explicitly in the following way. A *compatible* is a set of output-consistent internal states, each of which is equivalent to each of the others. A *maximal compatible* is a set of output-consistent states which form a compatible, and whose states do not form a proper subset of any other compatible. For example, (11, 12)(10, 12) (10, 11)(9, 12)(9, 11)(9, 10)(9, 10, 11)(10, 11, 12) are compatibles, but not maximal, since internal states 9, 10, 11 and 12 may be combined to form the maximal compatible (9, 10, 11, 12).

The internal states of a sequential machine can be represented by a set of maximal compatibles which must include *all* states; thus

$$M = (1)(2, 3)(4)(5)(6)(7, 8)(9, 10, 11, 12) \tag{6.1}$$

Note that a compatible can consist of one state equivalent to itself. This set of maximal compatibles forms a partition on the set of internal states:

$$S = (1, 2, 3, 4, 5, 6, 7, 8, 9, 10, 11, 12)$$

since the intersection of any block of the partition M is disjoint. Many partitions may be made on the set S; for example, the following are all possible partitions:

$$P = (1)(2)(3)(4)(5)(6)(7)(8)(9)(10)(11)(12)$$
$$P = (1, 2, 3, 4, 5, 6, 7, 8, 9, 10, 11, 12)$$
$$P = (1, 2, 3, 4)(5, 6, 7, 8, 9, 10, 11, 12)$$
$$P = (1, 2, 3, 4, 5, 6)(7, 8, 9, 10, 11, 12)$$

What distinguishes the particular partition we have arrived at from all the others, and is it in fact unique? An obvious factor, of course, is that the blocks of the partition representing the machine must contain states which are equivalent (called *output-consistent*, since all states in the same block must have identical outputs). Another important characteristic, however, is that the partition must be closed (corresponding to the closure property in group theory), thus satisfying all the *necessary implications*. This means that if the equivalent states in a block depend on the equivalence of some other states, these states must belong to a block in the partition. For the partition M (eqn. (6.1)), reference to the implication charts in Table 7.3 shows that (7, 8) implies the equivalence of (9, 11) and (10, 12) which are both contained in the same block of the partition. For the case of a completely specified machine, the closed partition obtained for M is the unique minimal description of the internal states. An upper bound on the number of internal states in a minimal machine is the number of maximal compatibles. In the case of a fully specified machine, since the maximal compatibles form a partition, i.e. are disjoint, the number of states is equal to the number of maximal compatibles. Moreover, Ginsberg[3] has pointed out that the number of elements in the largest maximal *incompatible* is a lower bound on the number of rows in the reduced state table.

The above result is the same as our intuitive result but we now have a routine tabular method, an algorithm, which may be used for any number of internal states, and which is ideal for automatic calculation on a digital computer.

7.3 Example of state minimization

Let us now consolidate our technique by doing an example. Consider the fully specified sequential machine whose state table is shown in Table 7.4. There are no identical states and an intuitive examination of the state table leads to a long succession of implications, for example $1 \equiv 2$ if $6 \equiv 3$; $6 \equiv 3$ if $8 \equiv 4$ and $1 \equiv 2$; $8 \equiv 4$ if $1 \equiv 5$; $1 \equiv 5$ if $8 \equiv 4$ and $3 \equiv 7$; $3 \equiv 7$ if $8 \equiv 4$ and $1 \equiv 5$; therefore $1 \equiv 2$. The tabular approach is much simpler; the final implication chart is shown in Table 7.5. The table shows, reading from right to left, that

$$6 \equiv 7 \quad 4 \equiv 8 \quad 3 \equiv 7 \quad 3 \equiv 6 \quad 2 \equiv 5 \quad 1 \equiv 5 \quad 1 \equiv 2$$

Thus the sequential machine is represented by

$$M = (6, 7)(4, 8)(3, 7)(3, 6)(2, 5)(1, 5)(1, 2)$$

since all internal states are present. However, this is not minimal since we have not derived the maximal compatibles for the machine. Starting from

Table 7.4

	Input x			
Present state	Next state		Output Z	
	0	1	0	1
1	8	3	0	0
2	8	6	0	0
3	8	1	1	0
4	1	8	1	0
5	4	7	0	0
6	4	2	1	0
7	4	5	1	0
8	5	4	1	0

Table 7.5

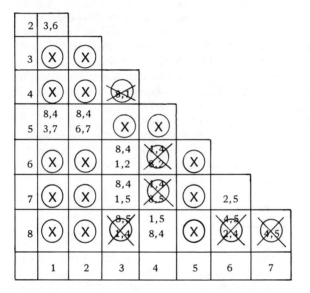

the right-hand side, we can combine equivalent states using the transitive relationship (i.e. $1 \equiv 2$ and $1 \equiv 5$, therefore $2 \equiv 5$) and we may group these together as $(1, 2, 5)$; similarly $3 \equiv 6$ and $3 \equiv 7$, and therefore $6 \equiv 7$, giving $(3, 6, 7)$ and finally $(4, 8)$, resulting in the minimal expression for the machine:

$$M = (4, 8)(3, 6, 7)(1, 2, 5)$$

On checking, we find that the closure property is satisfied since $(4, 8)$ implies $(1, 5)$ and $(8, 4)$; $(3, 6, 7)$ implies $(8, 4)(1, 2)(1, 5)$ and $(2, 5)$; and $(1, 2, 5)$ implies $(3, 6)(8, 4)(3, 7)$ and $(6, 7)$, which are all contained in blocks of the partition. Thus we have reduced the original seven-state machine, requiring three bistables, into a three-state machine which may be represented by two bistables. The reduced state table is shown in Table 7.6; note that the smallest state number in each block has been used to represent the actual states. That is, whenever states 4 and 8 occur in the original state table, they are replaced by 4; likewise 3, 6 and 7 are replaced by 3; and 1, 2 and 5 are replaced by 1.

7.4 Incompletely specified state tables

In a practical system it is very likely that some inputs may never occur, so that under these conditions outputs or next states do not matter or are of no consequence. Thus in these cases the appropriate entries in the state table can be filled in any possible way. This does mean, however, that we must slightly modify our approach for the minimization of state tables. Since it is possible to combine a don't-care condition (either next state or output state) more than once, taking different values each time if necessary, it follows that an internal state can occur in more than one maximal compatible. This means that the final equivalent-states expression is no longer a partition since the blocks (compatibles) can overlap, i.e. the intersection of any two blocks is not necessarily disjoint. In this case the set of maximal compatibles is called a *covering*. Furthermore, the equivalence relationship is not necessarily transitive, i.e. if $a \equiv b$ and $b \equiv c$ it *does not* always follow that $a \equiv c$.

These ideas are best illustrated by means of another example. Consider the incompletely specified state table shown in Table 7.7. From the implication chart (Table 7.8) the pair-wise compatibles are

$$(2, 3)(2, 6)(2, 7)(3, 4)(3, 6)(3, 7)(4, 5)(4, 7)(4, 8)(5, 8)(6, 7)$$

The maximal compatibles are found by combining these state pairs into larger groups with equivalent elements. Now $4 \equiv 7$ and $4 \equiv 8$, but 7 is *not* equivalent to 8, i.e. the transitive relationship does not apply, and $(4, 7, 8)$ is not a maximal compatible. But $4 \equiv 5$, $4 \equiv 8$ and $5 \equiv 8$, and therefore we can combine these to give $(4, 5, 8)$; also $3 \equiv 4$, $3 \equiv 7$ and $4 \equiv 7$, and therefore we have $(3, 4, 7)$; similarly, we can group $(2, 3, 6, 7)$. Thus the machine may be represented by the expression

$$M = (1)(2, 3, 6, 7)(3, 4, 7)(4, 5, 8)$$

Table 7.6

	Input x			
Present state	Next state		Output	
	0	1	0	1
1	4	3	0	0
3	4	1	1	0
4	1	4	1	0

Table 7.7

	Input x			
Present state	Next state		Output Z	
	0	1	0	1
1	2	8	1	1
2	1	7	0	0
3	X	6	0	X
4	4	X	0	1
5	3	4	0	X
6	1	3	X	0
7	X	2	X	X
8	7	5	0	X

Table 7.8

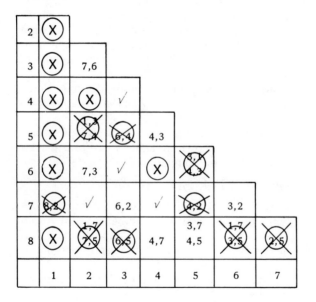

Note that the terms are no longer disjoint, i.e.

$$(2, 3, 6, 7) \cap (3, 4, 7) = (3, 7)$$

and

$$(3, 4, 7) \cap (4, 5, 8) = (4)$$

Thus the expression M forms a covering on the set of original internal states $S = (1, 2, 3, 4, 5, 6, 7, 8)$.

Now to obtain a minimal solution we need a set of maximal compatibles which include all the original states. The higher bound is seen from M, above, to be four; to find the lower bound we must extract from the chart the set of maximal incompatibles; these are

(1, 2, 8)(1, 2, 5)(1, 2, 4)(1, 3, 5)(1, 3, 8)(1, 4, 6)(1, 5, 7)(1, 5, 6)
(1, 8, 6)(1, 7, 8)

giving a lower bound of three.

It would appear at first glance that the covering

$$M = (1)(2, 3, 6, 7)(4, 5, 8)$$

would be a minimal description of the machine since all states are present. But we must also check to see if the covering chosen is closed, thereby satisfying all the necessary implications. Now (2, 3, 6, 7) implies (7, 6) (7, 3)(7, 2)(6, 3)(6, 2) and (3, 2) which is valid; but (4, 5, 8), implying (4, 3)(4, 7)(3, 7) and (4, 5), is invalid since (4, 3) and (4, 7) do not occur in the same blocks. Thus the minimal description is given by the covering:

$$M = (1)(2, 3, 6, 7)(4, 5, 8)(3, 4, 7)$$

corresponding to the higher bound of four. The reduced state table for the machine is shown in Table 7.9. To construct the table, let $A = (1)$, $B = (2, 3, 6, 7)$, $C = (4, 5, 8)$ and $D = (3, 4, 7)$. Now extract from the original state table (Table 7.7) the next-state conditions for each compatible and note in which block they occur, repeating the process for each input condition. For example, the next-state conditions for $A = (1)$ are (2) for input 0, which is contained in B, and (8) for input 1, which is in C. Again, for $B = (2, 3, 6, 7)$, under input 0 we have $(1, X, 1, X) = A$ and for input 1, $(7, 6, 3, 2) = B$; similarly, for $C = (4, 5, 8)$ we have for input 0, $(4, 3, 7) = D$ and for input 1, $(X, 4, 5) = C$. Finally, for $D = (3, 4, 7)$ we obtain for input 0, $(X, 4, X) = C$ or D, and for input 1, $(6, X, 2) = B$. Note that, due to the don't-care condition in the original table, we have a choice of either C or D for the next-state transition of D under input 0.

The minimization of incompletely specified state tables and the derivation of a reduced state table are difficult problems, since there is no direct approach. However, by following the exhaustive search routine, and thereby obtaining all possible maximal compatibles, the best use of the don't-care entries will be ensured.

Table 7.9

Present state	Input x Next state 0	1	Output Z 0	1
A	B	C	1	1
B	A	B	0	0
C	D	C	0	1
D	C or D	B	0	1

In general, the reduction of incompletely specified state tables results in a set of maximal compatibles (a covering of the machine) from which a minimal subset satisfying the state table transitions must be chosen.

7.5 Extraction of maximal compatibles

The procedure for extracting the maximal compatibles can be made more systematic (essential for large-variable problems), and thus more amenable to machine computation. For example, the expansion into maximal compatibles can be performed column by column (see Tables 7.8 and 7.10), starting from the right-hand column (column 7). At each stage, the pair-wise compatibles are recorded and examined for possible groupings; for example, in column 4 we check to see if (4, 5)(4, 7)(4, 8) can be grouped. We do this by examining the other compatibles (in this case we note that $8 \equiv 5$) or by checking through the implication chart. Thus we ascertain that the group (4, 8, 5) may be formed, which consequently replaces the relevant pair-wise compatibles in the table. This procedure is followed through to its natural conclusion on reaching column 1. The final expression is obtained by including those internal states not covered by the maximal compatibles as single-term compatibles.

An alternative approach may be attempted by assuming initially that all the internal states can be accommodated in one grouping, examining the implication chart for contradictions, and splitting the groups where necessary. Referring to Table 7.8, we see that column 1 indicates that state 1 cannot be grouped with states 2, 3, 4, 5, 6, 7 and 8, and thus we separate these terms (see Table 7.11). Column 2 shows that the state 2 cannot be grouped with states 4, 5 and 8, and the group (2, 3, 4, 5, 6, 7, 8) is split into (2, 3, 6, 7) and (3, 4, 5, 6, 7, 8). This process is continued through to the final column; if, as a result of splitting a group, the term obtained can be combined with an existing group, this is done. For example, column 5 splits into groups which include (8, 5) and (6, 7, 8); the former group is already included in an existing group and can therefore be ignored. This technique gives the complete set of maximal compatibles and, furthermore, the systematic procedure is well suited for programming on a digital computer.

Table 7.10

Column	Compatibles
7	
6	(7, 6)
5	(8, 5)(7, 6)
4	(4, 5)(4, 7)(4, 8)(8, 5)(7, 6) = (4, 8, 5)(4, 7)(7, 6)
3	(3, 7)(3, 6)(3, 4)(4, 8, 5)(4, 7)(7, 6) = (4, 8, 5)(3, 4, 7)(3, 7, 6)
2	(2, 7)(2, 6)(2, 3)(4, 8, 5)(3, 4, 7)(3, 7, 6) = (4, 8, 5)(3, 4, 7)(2, 3, 7, 6)
1	(4, 8, 5)(3, 4, 7)(2, 3, 7, 6)
Final	(1)(2, 3, 6, 7)(3, 4, 7)(4, 8, 5)

Another method, an algebraic approach due to Marcus,[4] operates on the incompatible equivalent pairs produced by the implication chart. The incompatible state pairs for each column are expressed as a Boolean OR function, and then all these terms are combined using the AND operation. Thus we have effectively derived a product-of-sums expression describing the incompatible states in the implication chart. From Table 7.8, the incompatible state pairs in column 1 are represented as $(1 + 2)(1 + 3)$ $(1 + 4)(1 + 5)(1 + 6)(1 + 7)(1 + 8)$, and in column 2 as $(2 + 4)(2 + 5)$ $(2 + 8)$, which when continued yields the complete expression

$$MI = (1 + 2)(1 + 3)(1 + 4)(1 + 5)(1 + 6)(1 + 7)(1 + 8)(2 + 4)(2 + 5)$$
$$(2 + 8)(3 + 5)(3 + 8)(4 + 6)(5 + 6)(5 + 7)(6 + 8)(7 + 8)$$

This expression can now be simplified by multiplying out and eliminating redundant terms according to the rules of Boolean algebra. The algebraic work is reduced if the state pairs are grouped by columns:

$$MI = (1 + 2345678)(2 + 458)(3 + 58)(4 + 6)(5 + 67)(6 + 8)(7 + 8)$$

This follows since, if we multiply out the terms $(2 + 4)(2 + 5)(2 + 8)$ for example, we get

$$T = 2.2.2 + 2.2.4 + 2.2.5 + 2.4.5 + 2.2.8$$
$$+ 2.4.8 + 2.5.8 + 4.5.8$$

Now, according to the rules of Boolean algebra, we can say that $2.2.2 = 2$ (from $A . A = A$), and $2 + 2.4 = 2$ [from $A + AB = A(1 + B) = A$]; therefore the expression for T reduces to

$$T = (2 + 458)$$

Thus, multiplying out and ignoring identical terms we have

$$MI = (1 + 2345678)(2 + 458)(3 + 58)(4 + 6)(5 + 67)(6 + 8)(7 + 8)$$
$$= (12 + 2345678 + 1458)(34 + 458 + 36 + 586)$$
$$(567 + 67 + 578 + 678 + 568 + 58)$$
$$= (12 + 2345678 + 1458)(34 + 458 + 36 + 586)(67 + 58)$$
$$= 123458 + 2345678 + 13458 + 123658 + 145836 + 12458$$
$$+ 1458 + 12586 + 14586 + 123467 + 1345867 + 12367$$
$$+ 1245867 + 145867 + 125867$$

Hence

$$MI = 2345678 + 1458 + 12586 + 12367$$

Now for each resultant product we write down the missing internal states, and the resulting set is the set of maximal compatibles:

$$M = (1)(2367)(347)(458)$$

Thus the reduction of a state table consists of generating all possible maximal compatibles and then selecting a minimal cover of the machine which satisfies the closure conditions. This procedure is analogous to the combinational minimization problem and presents similar computational difficulties. The Paull and Ungar method for extracting maximal compatibles

Table 7.11

Column	Compatibles (1, 2, 3, 4, 5, 6, 7, 8)
1	(1)(2, 3, 4, 5, 6, 7, 8)
2	(1)(2, 3, 6, 7)(3, 4, 5, 6, 7, 8)
3	(1)(2, 3, 6, 7)(3, 4, 6, 7)(4, 5, 6, 7, 8)
4	(1)(2, 3, 6, 7)(3, 6, 7)(3, 4, 7)(4, 5, 7, 8)(5, 6, 7, 8)
	(1)(2, 3, 6, 7)(3, 4, 7)(4, 5, 7, 8)(5, 6, 7, 8)
5	(1)(2, 3, 6, 7)(3, 4, 7)(4, 5, 8)(4, 8, 7)(8, 5)(6, 7, 8)
	(1)(2, 3, 6, 7)(3, 4, 7)(4, 5, 8)(4, 8, 7)(6, 7, 8)
6	(1)(2, 3, 6, 7)(3, 4, 7)(4, 5, 8)(4, 8, 7)(6, 7)(7, 8)
7	(1)(2, 3, 6, 7)(3, 4, 7)(4, 5, 8)(4, 8)(4, 7)
Final	(1)(2, 3, 6, 7)(3, 4, 7)(4, 5, 8)

has been successfully programmed for small variable machines; however, the selection of a minimal cover still remains a difficult problem. Various methods have been proposed,[5] in particular the concept of *prime compatibles* first described by Grasselli and Luccio[6] and later extended,[7] but most of the methods described in the literature would be extremely inefficient if programmed for a computer. An exception to this is the work by Bennetts *et al.*[8] who describe a heuristic procedure based on maximal compatibles which obtains a near-minimal solution equivalent to a good manual engineering design.

7.6 State assignment

Once the reduced state table has been obtained, the next step in the design procedure is to allocate a binary code to every internal state, or row, in the table so that input equations for the storage elements (JK or SR bistables etc.) may be derived. Note that a ROM implementation will be independent of assignment codes since it is essentially a table-look-up procedure; the number of states will of course be the determining factor.

Any internal variable assignment which allocates a unique binary combination to each internal state will lead to a legitimate circuit, but the particular assignment chosen will have a considerable effect on the amount of hardware required to implement the circuit equations. To attempt the evaluation of all possible assignments for a particular table quickly leads to an insuperable amount of calculation. For a state table with r rows requiring n state variables, the number of different possible assignments is given by

$$N = \frac{2^n!}{(2^n - r)!}$$

However, many of these assignments are merely rearrangements or permutations of the variables and could be obtained by simply relabelling the variables. In these cases, no advantage can be gained economically by using a different assignment. For example, the four-row assignments $(00-01-11-10)$ and $(11-10-00-01)$ give the same basic circuits, since the latter can be obtained from the former by inverting the variables.

McCluskey[9] has shown that the number of distinct row assignments for a table with r rows using n state variables is

$$N_D = \frac{(2^n - 1)!}{(2^n - r)!n!}$$

This means that for a four-row table requiring two state variables the number of possible distinct assignments is only three. However, the number of assignments rapidly increases to 840 for an eight-row table with three variables; for a nine-row table using four variables the number exceeds 10^7.

In all practical cases (say circuits with more than five states) to attempt to determine an optimal assignment by enumerative methods is obviously impossible and some form of algorithmic technique (programmed for a computer) must be used. Another important consideration is that the type of storage (or bistable) used for the implementation will also determine the optimal coding.[10] There is a considerable literature on the subject of state assignment but very little can be utilized (or is relevant) in designing logic systems. However, although no viable systematic technique exists to date for optimum state assignment there are several useful rules and design methods which can assist in the development of an economical circuit. In the following discussion of these techniques we assume that the best starting point is the reduced state table. Then, by the appropriate coding of the internal states, we attempt to derive near minimal equations for the combinational logic. As an illustration let us continue with the design of the pattern correlator whose reduced state table is shown in Table 7.2. Suppose as a starting point we adopt an arbitrary assignment, say pure binary, for the internal states. The fully assigned table is shown in Table 7.12 where every state number in the reduced table has been replaced by a binary code. Note that we have a seven-row table which requires three variables y_1, y_2 and y_3 in order to allocate a unique code to each state or row. It also follows that three storage devices (bistables) Y_1, Y_2 and Y_3 are

Table 7.12 Pattern correlator—binary assignment

	Present states y_1 y_2 y_3	Input x_1x_2 Next states 00	01	11	10	Output Z 00	01	11	10
1	0 0 0	001	010	001	010	0	0	0	0
2	0 0 1	101	011	101	011	0	0	0	0
4	0 1 0	011	011	011	011	0	0	0	0
5	0 1 1	100	100	100	100	0	0	0	0
6	1 0 0	000	000	000	000	0	0	0	0
7	1 0 1	110	100	110	100	0	0	0	0
9	1 1 0	000	000	000	000	1	0	1	0
Unused	1 1 1								

needed to store these internal states. We shall use set–reset bistables as the internal storage element and derive input equations for the set and reset terminals according to the usual conditions (Table 6.6). The K-maps for the set and reset conditions of bistables Y_1, Y_2 and Y_3 are shown in Table 7.13. These are derived in the manner described in the last chapter, that is the values of y_1 etc. for present and next states are examined in turn for the set and reset conditions, the appropriate values being entered on the K-maps. Considering the set conditions of Y_1, we note from Table 7.12, by comparing the present and next state values of y_1, that a set condition is required when $y_1 y_2 y_3 x_1 x_2$ takes the value of 00100; thus a 1 is entered in the corresponding K-map in this position. The rest of the maps are obtained in the same way by following through this procedure for y_1, y_2 and y_3. There is one unused combination in the assignment (111); this can be incorporated as a don't-care condition in the K-maps, since this state can never occur in practice.

Table 7.13 K-maps pattern correlator—binary assignment

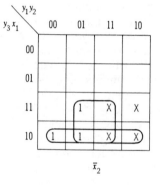

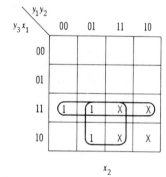

(a) Y_1 Set

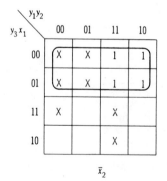

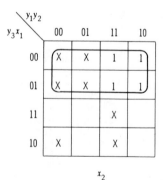

(b) Y_1 Reset

Table 7.13 Continued

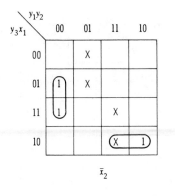
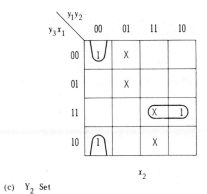

(c) Y_2 Set

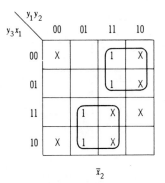
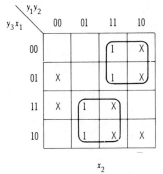

(d) Y_2 Reset

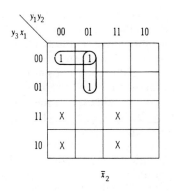
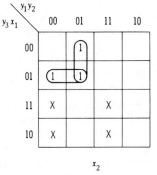

(e) Y_3 Set

Table 7.13 Continued

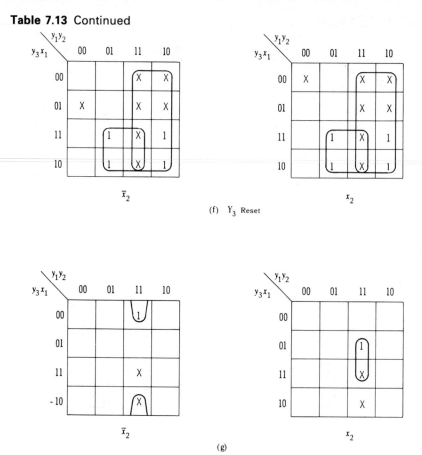

(f) Y_3 Reset

(g)

Implementation requires 19 units

An alternative method of including the don't-care condition is to treat state 9 as being represented by two combinations—110 and 111. This means effectively that, for state 9, y_3 is a don't-care condition throughout the state table. In practice, however, the set and reset inputs to an SR bistable should never occur together (this would appear on the K-maps as a 1 in the same square for both reset and set conditions) since the output under these conditions is indeterminate. Thus, although it does not matter in this case whether y_3 is 0 or 1, since either will correctly represent the state, care must be taken to ensure that the simultaneous presence of set and reset inputs does not adversely affect the bistable circuit. The use of a JK bistable element would obviate this problem.

It should be apparent that the particular storage element chosen for the design will also affect the amount of combinational logic required. If, for example, we had used a JK bistable there would be more don't-care conditions which could effect the minimization. Thus ideally the choice of code assignment for the internal states should take into consideration the type of storage device to be used. In these examples we compare various methods of assignment using the SR bistable as the standard storage element.

If we extracted the input equations for Y_1, Y_2 and Y_3 from the maps and implemented them in hardware, we would have a sequential circuit which would correctly satisfy the design specification. Thus

$$Y_1 \text{ set } = y_2 y_3 + \bar{x}_1 \bar{x}_2 y_3 + x_1 x_2 y_3$$
$$Y_1 \text{ reset } = \bar{y}_3$$
$$Y_2 \text{ set } = x_1 \bar{x}_2 \bar{y}_1 \bar{y}_2 + \bar{x}_1 \bar{x}_2 y_1 y_3 + x_1 x_2 y_1 y_3 + \bar{x}_1 x_2 \bar{y}_1 \bar{y}_2$$
$$Y_2 \text{ reset } = y_1 \bar{y}_3 + y_2 y_3$$
$$Y_3 \text{ set } = \bar{y}_1 y_2 \bar{y}_3 + \bar{x}_1 \bar{x}_2 \bar{y}_1 \bar{y}_3 + x_1 x_2 \bar{y}_1 \bar{y}_3$$
$$Y_3 \text{ reset } = y_1 + y_2 y_3$$
$$Z \text{ output } = \bar{x}_1 \bar{x}_2 y_1 y_2 + x_1 x_2 y_1 y_2$$

However, we have no means of knowing if these equations produce a minimal circuit without trying all possible (840) codes! Is there some simple method we can use to ensure at least some degree of minimization? There is, in fact, a simple rule due to Humphrey[11] which is slightly better than choosing a purely arbitrary assignment. But though in general it leads to simpler input equations, there is no question of it producing a unique best state assignment. Consider Table 7.14(a), which is basically a K-map depicting the present state assignments made above, i.e. pure binary, for all input conditions. Each state appears four times, since there are four possible input conditions. Underneath this map a similar diagram is plotted showing the next states resulting from the prescribed input changes. The positions of the next states in Table 7.14(b) show where the set, reset and don't-care bistable input conditions for a particular state variable will occur when plotted on the K-map. This may easily be confirmed by comparing Table 7.13 with Table 7.14. Thus if similar next states were made adjacent in Table 7.14(b), as for example state 1, the input conditions for transitions to that state would be adjacent and hence easily combined and reduced. For example, if we are in present state 6, coded 100, any input transition takes us to state 1, coded 000, and thus we need to reset Y_1 under these conditions (in general there will, of course, be other conditions as well). Similarly, we need to reset Y_2 for the transition from present state 9 to next state 1. Since all these conditions are adjacent when plotted on the K-map (Table 7.13(b) and (d)) they may be combined into a larger group. Note that the reason these next states are adjacent is that the present states giving rise to them are also adjacent, i.e. present states 9, 6 are adjacent, and hence next state 1 is adjacent. Thus we can formulate the following rules: (a) *if two or more states have the same next states, they should be made adjacent in the assignment*; (b) *it is also advisable to give adjacent coding to states if they are both next states of a present state*. For example, in Table 7.2, states 2 and 4 are the next states of 1, and 7, 5 and 9, 6 are the next states of 2 and 7. Thus these states should be made adjacent; that this should be so is apparent from Table 7.14, since the next states of a state are always adjacent. However, should there be any discrepancy between the two rules, the first should always take precedence. If we examine the state

Table 7.14 Effectiveness of internal state assignment

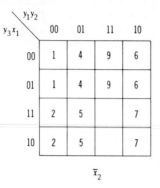

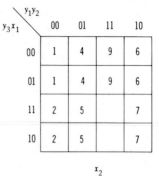

(a)   Present states

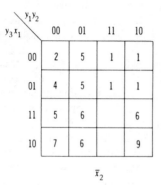

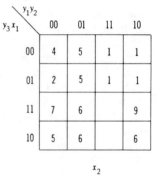

(b)   Next states

Table 7.15 Pattern correlator—
next state assignment

Next states	Present states
1	6, 9
2	1
4	1
5	2, 4
6	5, 7
7	2
9	7
(a)	

Table 7.15 Continued

	Present states y_1 y_2 y_3	Next states 00	01	11	10	Output Z 00	01	11	10
1	0 0 0	010	110	010	110	0	0	0	0
2	0 1 0	101	100	101	100	0	0	0	0
4	1 1 0	100	100	100	100	0	0	0	0
5	1 0 0	111	111	111	111	0	0	0	0
6	1 1 1	000	000	000	000	0	0	0	0
7	1 0 1	011	111	011	111	0	0	0	0
9	0 1 1	000	000	000	000	1	0	1	0
Unused	0 0 1								

(Header: Input $x_1 x_2$)

(b)

Table 7.16 K-maps—next state assignment

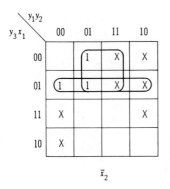

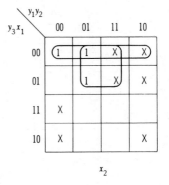

Y_1 Set

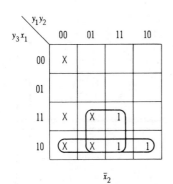

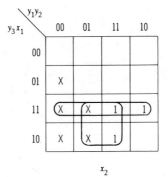

Y_1 Reset

Table 7.16 Continued

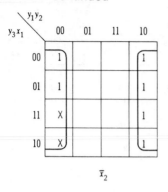

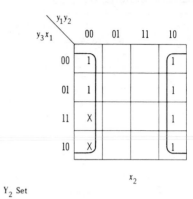

Y_2 Set

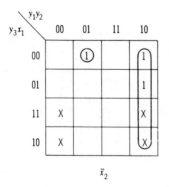

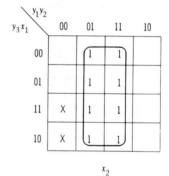

Y_2 Reset

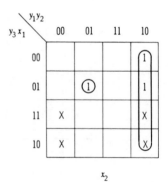

Y_3 Set

Table 7.16 Continued

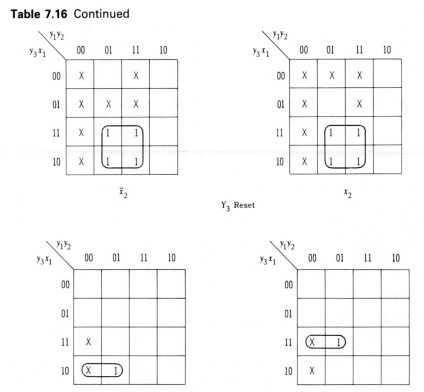

Y₃ Reset

Implementation requires 15 units

table (Table 7.2) to determine the origin of all the next states we arrive at the result shown in Table 7.15(a). This indicates that states (6, 9), (2, 4) and (5, 7) should have adjacent codes, and as this is identical with the result obtained above for the next states of a state, there is no conflict between the two rules. Table 7.15(b) shows the fully assigned table using these two rules as the basis for coding; notice that states (2, 4), (7, 5) and (9, 6) differ by one bit only—they are adjacent. The K-maps for the state variables Y_1, Y_2 and Y_3, using this assignment, are shown in Table 7.16, and lead to the equations

$$Y_1 \text{ set} = y_2\bar{y}_3 + x_1\bar{x}_2\bar{y}_3 + \bar{x}_1x_2\bar{y}_3$$
$$Y_1 \text{ reset} = y_2y_3 + \bar{x}_1\bar{x}_2y_3 + x_1x_2y_3$$
$$Y_2 \text{ set} = \bar{y}_2$$
$$Y_2 \text{ reset} = y_2$$
$$Y_3 \text{ set} = y_1\bar{y}_2 + \bar{x}_1\bar{x}_2\bar{y}_1y_2\bar{y}_3 + x_1x_2\bar{y}_1y_2\bar{y}_3$$
$$Y_3 \text{ reset} = y_2y_3$$
$$Z \text{ output} = \bar{x}_1\bar{x}_2\bar{y}_1y_3 + x_1x_2\bar{y}_1y_3$$

By direct comparison with the maps of Table 7.13, or by examining the equations, it is clear that, using the next state assignment rule, the input functions are much simpler. The method is, however, best suited for D-type bistable implementations (which makes it more useful for LSI logic design) as can be seen from Table 7.17, which yields the equations

$$Y_1 = y_1\bar{y}_3 + y_2\bar{y}_3 + \bar{y}_2x_1\bar{x}_2 + \bar{y}_2\bar{x}_1x_2$$
$$Y_2 = y_2$$
$$Y_3 = y_1\bar{y}_2 + \bar{y}_1y_2\bar{y}_3\bar{x}_1\bar{x}_2 + \bar{y}_1y_2\bar{y}_3x_1x_2$$
$$Z = \bar{y}_1y_3\bar{x}_1\bar{x}_2 + \bar{y}_1y_3x_1x_2$$

7.6.1   State assignment using adjacency conditions

The work of Humphrey has been extended by Armstrong,[12] who has described a programmable algorithmic procedure based on these ideas, as well as additional adjacency conditions. The adjacency conditions may be formalized as follows.

(a) *Type I adjacency.* This occurs when two present states have identical next states for the same given input state. The adjacencies are determined by examining all of the state table columns for the condition

$$N(q_i, I_m) = N(q_j, I_m)$$

where N signifies the next state function and q and I are the internal and input states respectively: q_i and q_j would be assigned adjacent codings to effect a good assignment.

(b) *Type II adjacency.* This is present when two next states have the same present state in the state table and the input codes for the respective columns are adjacent. Thus each row of the state table is evaluated for the condition

$$N(q_i, I_m) = q_j \quad \text{and} \quad N(q_i, I_p) = q_k$$

where I_m and I_p are adjacent input states; in this case q_j and q_k would be given adjacent codes.

(c) *Type III adjacency.* The next output state values are examined, and if the values are the same for two states under the same input conditions they are made adjacent; thus each output column is examined for the condition

$$Z_k(q_i, I_m) = Z_k(q_j, I_m)$$

where Z_k is the output function; q_i and q_j would be given adjacent codes.

(d) *Type IV adjacency.* If two present states occur as their own next states, or as the next states of each other, under the same input conditions, they are made adjacent. Thus the following conditions must be fulfilled for each column:

$$N(q_i, I_m) = q_i \quad \text{and} \quad N(q_j, I_m) = q_j$$

OR

$$N(q_i, I_m) = q_j \quad \text{and} \quad N(q_j, I_m) = q_i$$

Table 7.17 K-maps, D-type bistable

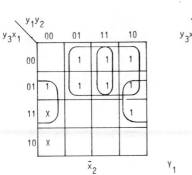
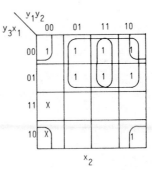

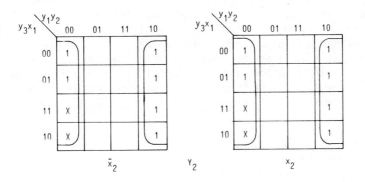

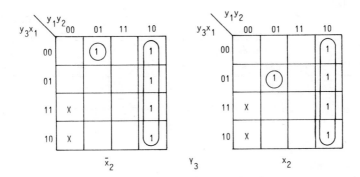

Table 7.17 Continued

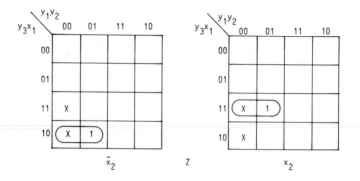

Again, q_i and q_j would be given adjacent codes. Table 7.18 shows the adjacency conditions for the given state table; note that it is possible for the adjacency condition to occur more than once for each state pair. Ideally each internal state pair with an adjacency relationship should be assigned adjacent codes; in most cases, however, this is clearly impossible. Consequently it is necessary to determine those adjacent state pairs which make the greatest contribution to a minimal set of equations—in this context it would appear that Type I is the most important. However, a more effective method is to score each adjacency condition by attributing a

Table 7.18 Adjacency conditions

Present state	Next state Input x_1x_2				Output Z Input x_1x_2			
	00	01	11	10	00	01	11	10
1	3	3	4	1	1	1	0	1
2	1	2 ′	4	4	1	1	0	1
3	3	4	4	2	0	1	1	1
4	3	3	4	1	0	1	1	1

(a) State table

State pairs	Number of occurrences Type 1	Type 2	Type 3	Type 4
1, 2	1	1	4	0
1, 3	2	2	2	0
1, 4	2	3	2	0
2, 3	1	1	2	0
2, 4	1	2	2	0
3, 4	2	3	4	1

(b) Adjacencies

weight $w(q_i, q_j)$ to each state pair. With $w(q_i, q_j)$ initially set equal to zero, the conditions are scored for individual occurrences of each adjacency type according to the following order of merit:

Type I   add $n_0 = |\log_2 s|$ where s = number of states, to $w(q_i, q_j)$
Type II   add 1 to $w(q_i, q_j)$
Type III   add 1 to $w(q_i, q_j)$
Type IV   add $n_0 - 1$ to $w(q_i, q_j)$

The effectiveness of a particular state assignment may be assessed by summing all those weights associated with the state pairs which were given adjacent codings. State assignments with a high total weight will tend to yield an economical realization. A state assignment with a relatively high total weighting may be obtained by allocating adjacent codings to as many as possible of those state pairs with large weights.

Table 7.19 shows the weighting of adjacent state pairs for the state table of Table 7.18(a) which has a value of $n_0 = 2$. A good assignment with a total weighting of 33 would be obtained by giving (1, 4)(4, 3)(2, 3) and (1, 2) adjacent codes as shown in Table 7.20; note that in this particular case the normal binary equivalent would also yield a total weight of 33.

7.7 State assignment using the partition principle

A general and more rigorous approach to the problem of state assignment has been described by Hartmanis.[13] In this method, the state assignment is

Table 7.19 Weighting of state pairs

State pair	Weights Type 1	Weights Type 2	Weights Type 3	Weights Type 4	Total weight
1, 2	2	1	4	0	7
1, 3	4	2	2	0	8
1, 4	4	3	2	0	9
2, 3	2	1	2	0	5
2, 4	2	2	2	0	6
3, 4	4	3	4	1	12

Table 7.20 Assigned state table

Present state		Next state Input $x_1 x_2$ 00	01	11	10	Output Z Input $x_1 x_2$ 00	01	11	10
1	00	11	11	10	00	1	1	0	1
2	01	00	01	10	10	1	1	0	1
3	11	11	10	10	01	0	1	1	1
4	10	11	11	10	00	0	1	1	1

made in such a way that each binary variable describing the next state depends on as few variables of the present states as possible; that is, the next state variables depend on small subsets of the present state variables. It can be shown that these assignments with *reduced dependence* often yield more economical implementations of the circuit equations. We have already seen that the operation of partitioning is the distribution of all the internal states of a sequential machine into blocks, each state belonging to one, and only one, block. We have also shown that many different partitions are possible and, in the case of state reduction, we looked for a partition that was output-consistent and closed under the implication conditions. In applying the partition approach to state assignment, we must search for a partition that possesses the *substitution* property (also called a *stable or SP partition*). *This substitution condition is met if any two internal states in any block of the partition, under the same input combinations, go to next states that are all contained in a single block of the same partition.* Unfortunately the method of obtaining a stable partition is a trial and error process, based on the definition of the substitution property. Furthermore, for any particular sequential machine (as represented by its state table), more than one stable partition can exist. Thus there is no simple way of obtaining a unique stable partition which yields a minimal assignment.

The stable partitions of a machine may be calculated by considering in turn each pair of internal states in the state table, and ascertaining their next states for all possible input conditions. The process is continued by finding the pairs of next states for these pairs, and so on, until a list of state pairs is obtained. The final list is then examined and appropriate pairs combined to form a *least* stable partition, using the fact that the stable condition is transitive. Consider the state table shown in Table 7.2. If we compare the internal state pair (1, 5), we obtain the flow-table shown in Table 7.21(a). State pairs need only be included once in the partition; consequently when a state pair repeats it can be ignored. The final list is

$$P = (1, 5)(2, 6)(4, 6)(7, 1)(2, 9)$$

Note that all the internal states are included in the partition.

The pairs (1, 5) and (7, 1), also the pairs (2, 6), (4, 6) and (2, 9), may be combined using the transitive relationship to give the least stable partition:

$$P_1 = (1, 5, 7)(2, 4, 6, 9)$$

Note that this is a stable partition since, if we consider present states 1 and 7 (Table 7.2), for example, the next states for the input conditions 00, 01, 11 and 10 are (2, 9)(4, 6)(2, 9)(4, 6) which are all contained in a single block of the same partition. For this particular sequential machine, there are only two non-trivial partitions with the substitution property on its set of internal states; they are

$$P_1 = (1, 5, 7)(2, 4, 6, 9)$$

and

$$P_2 = (1)(2, 4)(7, 5)(6, 9)$$

Table 7.21 Extraction of stable partitions

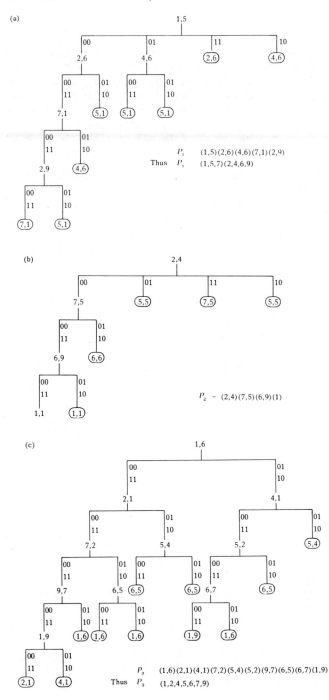

(a)

$$P_1 \quad (1,5)(2,6)(4,6)(7,1)(2,9)$$
Thus $P_1 \quad (1,5,7)(2,4,6,9)$

(b)

$$P_2 = (2,4)(7,5)(6,9)(1)$$

(c)

$$P_3 \quad (1,6)(2,1)(4,1)(7,2)(5,4)(5,2)(9,7)(6,5)(6,7)(1,9)$$
Thus $P_3 \quad (1,2,4,5,6,7,9)$

All the rest are trivial in the sense that all the states are contained in one partition:

$$P_3 = (1, 2, 4, 5, 6, 7, 9)$$

The direct examination of state pairs, as described above, does not necessarily produce the complete set of stable partitions. Fortunately certain algebraic relationships exist between partitions of the same set, and it can be shown that these preserve the substitution property. Thus, once we have found (by examination) all non-trivial partitions, the others may be obtained by algebraic methods without further recourse to the state table.

If P_1 and P_2 are two partitions of the same set, the algebraic relationships between them may be defined as follows:

(a) P_1 is said to be greater than P_2 ($P_1 \geq P_2$) if each block of P_2 is contained in the block of P_1.

(b) The sum of two partitions (least upper bound) ($P_1 + P_2$) is the partition whose blocks are the union of overlapping (i.e. containing common members) blocks of P_1 and P_2.

(c) The product of two partitions (greatest lower bound) ($P_1 . P_2$) is the partition whose blocks are the intersections of blocks of P_1 with blocks of P_2.

As an illustration, consider the two partitions derived above:

$$P_1 = (1, 5, 7)(2, 4, 6, 9)$$

and

$$P_2 = (1)(2, 4)(7, 5)(6, 9)$$

(Note that $P_1 > P_2$.) Now taking each block in turn we have

$$P_1 . P_2 = [(1, 5, 7) \cap (2, 4)][(2, 4, 6, 9) \cap (2, 4)][(1, 5, 7) \cap (7, 5)]$$
$$[(2, 4, 6, 9) \cap (7, 5)][(1, 5, 7) \cap (6, 9)][(2, 4, 6, 9) \cap (6, 9)]$$
$$[(1, 5, 7) \cap (1)][(2, 4, 6, 9) \cap (1)]$$

Hence

$$P_1 . P_2 = (1)(2, 4)(7, 5)(6, 9)$$

Also, for $P_1 + P_2$, the overlapping blocks are

$$[(1, 5, 7) \cup (1) \cup (7, 5)][(2, 4, 6, 9) \cup (2, 4) \cup (6, 9)]$$

Thus

$$P_1 + P_2 = (1, 5, 7)(2, 4, 6, 9)$$

The partition with one block, which is of course the complete set of states, is called the *unit partition* (1), while the partition which contains each member of the set in a separate block is called the *zero partition* (0). A set of partitions which contains the unit and zero partitions, and every sum and product of its members is called a *lattice*.[14,15]

Thus in the example above, since the sum and product of the two partitions do not yield any new partitions, the complete lattice is formed by

$$P_0 = (1)(2)(4)(5)(6)(7)(9) = 0$$
$$P_1 = (1, 5, 7)(2, 4, 6, 9)$$
$$P_2 = (1)(2, 4)(7, 5)(6, 9)$$
$$P_3 = (1, 2, 4, 5, 6, 7, 9) = 1$$

To sum up, the lattice of stable partitions is obtained by deriving the non-trivial partitions from the state table, and then generating all other partitions from this set by combining them using the algebriac operations.

Having now discovered how to obtain stable partitions for a sequential machine, we must next explain how to use them in state assignment. Noting in passing that it is advisable to obtain the complete set of all non-trivial stable partitions and select the ones with the fewest blocks (and/or elements) before commencing the allocation. The first step, then, is to select a suitable partition; suppose in this case we choose

$$P_1 = (1)(2, 4)(5, 7)(6, 9)$$

By inspection, we see that there are four blocks, each block containing a maximum of two elements. Now we also know that, for seven internal states, three bits are required for the state variables; thus two of these can be used to distinguish between blocks, and the remaining bit used to distinguish between elements within a block. There are three distinct ways of making this assignment and they should all be investigated. A possible assignment scheme is shown in Table 7.22(a), and the fully assigned state table is given in Table 7.22(b). Note that, in using this method, we have arranged that the first two bits of the next state $(y_1 y_2)$ can be determined solely from the first two bits $(y_1 y_2)$ of the present state and the inputs $(x_1 x_2)$. Thus we have a self-dependent subset $(y_1 y_2)$ of the state variables set $(y_1 y_2 y_3)$; also Y_1 set, Y_2 set, Y_1 reset and Y_2 reset will be, at most, a function of four variables only. However, Y_3 can still depend on $y_1 y_2 y_3$ and the inputs $(x_1 x_2)$.

The K-maps for this assignment are shown in Table 7.23. This is by far the simplest solution obtained and gives the following input and output equations:

$$
\begin{aligned}
Y_1 \text{ set} &= y_2 \\
Y_1 \text{ reset} &= \bar{y}_2 \\
Y_2 \text{ set} &= \bar{y}_1 \\
Y_2 \text{ reset} &= y_1 \\
Y_3 \text{ set} &= y_1 y_2 + \bar{y}_3 \bar{y}_2 \bar{x}_1 \bar{x}_2 + \bar{y}_3 \bar{y}_2 x_1 x_2 \\
Y_3 \text{ reset} &= \bar{y}_1 x_1 \bar{x}_2 + \bar{y}_1 \bar{x}_1 x_2 + y_3 \bar{y}_2 \bar{x}_1 \bar{x}_2 + y_3 \bar{y}_2 x_1 x_2 \\
Z \text{ output} &= y_1 y_2 y_3 \bar{x}_1 \bar{x}_2 + y_1 y_2 y_3 x_1 x_2
\end{aligned}
$$

In this particular case there is no need to plot a map for Z, since the function can be obtained directly from the assigned state table.

In the discussion above, the partition approach has been developed using the reduced state table as a starting point. However, the processes of state minimization and state assignment may be combined in the one process for a completely specified machine. To do this we must extract from the normal (unreduced) state table all the stable partitions which result

Table 7.22 Pattern correlator
—partition assignment

Block	Allocation	
	y_1	y_2
(1)	1	0
(2, 4)	0	0
(7, 5)	0	1
(6, 9)	1	1

(a)

	Present states			Input x_1x_2					Output Z			
				Next states								
	y_1	y_2	y_3	00	01	11	10		00	01	11	10
1	1	0	1	000	001	000	001		0	0	0	0
2	0	0	0	011	010	011	010		0	0	0	0
4	0	0	1	010	010	010	010		0	0	0	0
5	0	1	0	110	110	110	110		0	0	0	0
6	1	1	0	101	101	101	101		0	0	0	0
7	0	1	1	111	110	111	110		0	0	0	0
9	1	1	1	101	101	101	101		1	0	1	0
Unused	1	0	0									

(b)

Table 7.23 K-maps for partition assignment

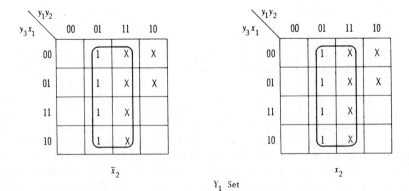

Y_1 Set

Table 7.23 Continued

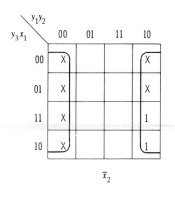

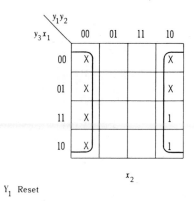

Y₁ Reset

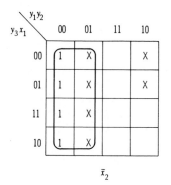

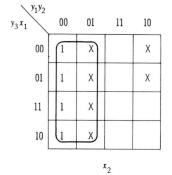

Y₂ Set

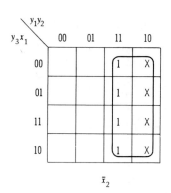

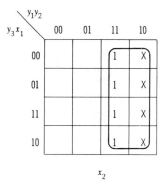

Y₂ Reset

Table 7.23 Continued

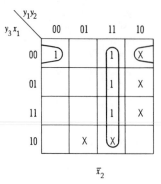
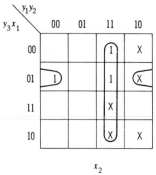

Y₃ Set

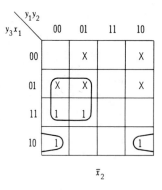
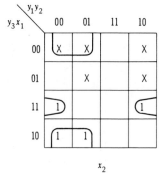

Y₃ Reset

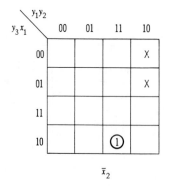
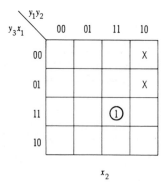

z

Implementation requires 12 units

from the comparison of state pairs, as described earlier. From this set we select those partitions which are output-consistent (output the same for all states in any block of the partition). If there are no output-consistent partitions, the machine cannot be reduced. These output-consistent partitions must also possess the substitution property, since otherwise the resulting machine will not have unique states. The minimal state partition for the machine may be obtained by adding partitions together (using the rule for addition of partitions), and selecting the partition with the fewest blocks.

In some sequential machines, the set of non-trivial stable partitions does not yield a viable assignment (that is, the assignment uses more secondary variables than that required by direct coding). In other cases there are no non-trivial stable partitions for the machine. Stearns and Hartmanis[16] have extended the partition approach to allow assignments, with reduced dependence on the secondary variables, to be obtained from non-stable partitions, using the concepts of *partition pair* algebra.

7.7.1 Partition pair algebra

An ordered pair of partitions (π, π') on the set of states s of a sequential machine is a *partition pair* if, and only if, any two internal states in the same block of π, under any input condition, go to next-states in the same block of π'. Thus we have the substitution property between partitions, and a stable or *SP* partition may be considered as forming a partition pair with itself (also called a *closed* partition) and becomes a special case of the set of partition pairs.

It can be shown[17] that for any partition π on a set of states s there is a smallest partition (greatest number of blocks) π' such that (π, π') is a partition pair; the partition π' is called $m(\pi)$. Thus $m(\pi)$ determines for a given partition π which partition π' can be used to make a partition pair (π, π'). Similarly it can also be shown that there is a largest partition (fewest number of blocks) π such that (π, π') is a partition pair. This partition is called $M(\pi')$ and determines the partition π to match a given π' to form a particular pair.

A partition pair (π, π') is called an *Mm pair* if and only if $\pi = M(\pi')$ and $\pi' = m(\pi)$. In an *Mm* pair π is the largest partition from which we can compute π' and at the same time π' is the smallest partition that can be determined from π. Thus all of the possible partitions of a machine can be found by either taking a refinement of some π from an *Mm* pair or selecting a partition greater than some π'. Thus by generating all possible *Mm* pairs for a given machine we can find any other partition that could be useful in a design. (Note that the set of *Mm* pairs is generally substantially smaller than the set of all partition pairs.) Moreover, the *Mm* lattice so formed completely characterizes the machine and contains all the information concerning its structure.

The computation of $m(\pi)$ can easily be carried out by deriving from the state table of the machine all those sets of next-states generated by the blocks of π under all input conditions, and then constructing the minimial partitions which contain these sets.

For example, consider the state table for a machine M shown in Table 7.24(a); by looking at each state pair in succession and examining the corresponding rows we obtain

$$m(\pi_{1,2}) = (1, 3)(2)(4)(5) = \pi_1'$$
$$m(\pi_{1,3}) = (1, 4)(2)(3)(5) = \pi_2'$$
$$m(\pi_{1,4}) = (1, 2)(3)(4)(5) = \pi_3'$$
$$m(\pi_{1,5}) = (1, 2, 3, 4, 5) = \pi(1)$$
$$m(\pi_{2,3}) = (1, 3, 4)(2)(5) = \pi_4'$$
$$m(\pi_{2,4}) = (1, 2, 3)(4)(5) = \pi_5'$$
$$m(\pi_{2,5}) = (1, 2, 4)(3, 5) = \pi_6'$$
$$m(\pi_{3,4}) = (1, 2, 4)(3)(5) = \pi_7'$$
$$m(\pi_{3,5}) = (1, 2, 3)(4, 5) = \pi_8'$$
$$m(\pi_{4,5}) = (1, 3, 4)(2, 5) = \pi_9'$$

To find the complete set we must form all possible sums (least upper bound) of the $m(\pi)$ partitions, first by forming all pair-wise sums and then the pair-wise sum of the new partitions generated. In this case the only new partition formed is

$$\pi_1' + \pi_7' = \pi_2' + \pi_5' = \pi_3' + \pi_4' = \pi_4' + \pi_7' = \pi_5' + \pi_7' = (1, 2, 3, 4)(5) = \pi_{10}'$$

which with the inclusion of $\pi(0) = (1)(2)(3)(4)(5)$ gives the complete set of $m(\pi)$.

Table 7.24 Machine M

Present state	Next state Input x_1x_2			
	00	01	11	10
1	1	1	4	1
2	3	3	4	1
3	4	1	1	1
4	2	1	4	2
5	5	3	1	2

(a) State table

	Present state			Next state Input x_1x_2			
	y_1	y_2	y_3	00	01	11	10
1	0	0	0	000	000	001	000
2	0	1	0	100	100	001	000
3	1	0	0	001	000	000	000
4	0	0	1	010	000	001	010
5	1	1	1	111	100	000	010
	0	1	1				
	1	0	1	Don't-cares			
	1	1	0				

(b) Assigned table

The partition $M(\pi')$ can also be obtained from the state table by identifying those present states which all go into the same block of π' under all input conditions. That is, we must look up in the state table the states in each block of π', for the same input conditions, and note the present states. However, there is a more convenient method which operates directly on the $m(\pi)$ set by summing all those partitions $\leq \pi'$, for example:

$$M(\pi_1') = \pi_1 = \pi(1,2) = (1,2)(3)(4)(5)$$
$$M(\pi_2') = \pi_2 = \pi(1,3) = (1,3)(2)(4)(5)$$
$$M(\pi_3') = \pi_3 = \pi(1,4) = (1,4)(2)(3)(5)$$
$$M(\pi_4') = \pi_4 = \pi(2,3) + \pi(1,2) + \pi(1,3) \qquad = (1,2,3)(4)(5)$$
$$M(\pi_5') = \pi_5 = \pi(2,4) + \pi(1,2) + \pi(1,4) \qquad = (1,2,4)(3)(5)$$
$$M(\pi_6') = \pi_6 = \pi(2,5) + \pi(1,3) + \pi(1,4) + \pi(3,4) = (1,3,4)(2,5)$$
$$M(\pi_7') = \pi_7 = \pi(3,4) + \pi(1,3) + \pi(1,4) \qquad = (1,3,4)(2)(5)$$
$$M(\pi_8') = \pi_8 = \pi(3,5) + \pi(1,2) + \pi(1,4) + \pi(2,4) = (1,2,4)(3,5)$$
$$M(\pi_9') = \pi_9 = \pi(4,5) + \pi(1,2) + \pi(1,3) + \pi(2,3) = (1,2,3)(4,5)$$
$$M(\pi_{10}') = \pi_{10} = \pi(1,2) + \pi(1,3) + \pi(1,4) + \pi(2,3) + \pi(2,4) + \pi(3,4)$$
$$= (1,2,3,4)(5)$$

Note that if we check back $\pi \to \pi'$ on the state table we obtain the required substitution property. For example, $\pi_6 = (1, 3, 4)(2, 5) \to \pi_6' = (1, 2, 4)$ $(3, 5)$; state pair 1, 3 will generate $(1, 4)(1)(1, 4)(1)$; state pair 1, 4 $(1, 2)(1)$ $(4)(1, 2)$; state pair 3, 4 $(2, 4)(1)(1, 4)(1, 2)$ and state pair 2, 5 $(3, 5)(3)$ $(1, 4)(1, 2)$.

The complete list of Mm pairs for machine M are

$$\pi(1),\ \pi(1)$$
$$(\pi_1, \pi_1') = (1, 2)(3)(4)(5),\ (1, 3)(2)(4)(5)$$
$$(\pi_2, \pi_2') = (1, 3)(2)(4)(5),\ (1, 4)(2)(3)(5)$$
$$(\pi_3, \pi_3') = (1, 4)(2)(3)(5),\ (1, 2)(3)(4)(5)$$
$$(\pi_4, \pi_4') = (1, 2, 3)(4)(5),\ (1, 3, 4)(2)(5)$$
$$(\pi_5, \pi_5') = (1, 2, 4)(3)(5),\ (1, 2, 3)(4)(5)$$
$$(\pi_6, \pi_6') = (1, 3, 4)(2, 5),\ (1, 2, 4)(3, 5)$$
$$(\pi_7, \pi_7') = (1, 3, 4)(2)(5),\ (1, 2, 4)(3)(5)$$
$$(\pi_8, \pi_8') = (1, 2, 4)(3, 5),\ (1, 2, 3)(4, 5)$$
$$(\pi_9, \pi_9') = (1, 2, 3)(4, 5),\ (1, 3, 4)(2, 5)$$
$$(\pi_{10}, \pi_{10}') = (1, 2, 3, 4)(5),\ (1, 2, 3, 4)(5)$$
$$\pi(0),\ \pi(0)$$

Note that $\pi_{10} = \pi_{10}'$ is a non-trivial SP partition. Since if (π, π') is an Mm pair then π is the largest partition from which we can determine π', and at the same time π' is the smallest partition which contains the successor state blocks implied by π, by enlarging π' or refining π we can obtain other partition pairs. Thus there are numerous partitions which can be generated from the Mm partitions, including the SP partitions; in this case, however, only one non-trivial partition exists.

Though in general partition pairs can contain any number of blocks, a maximal reduction in the dependency of the state variables (when used in state assignment) can be achieved if two-block partitions are used. In such a case each state variable would be independent of the remaining state variables. Thus a reduced dependency state assignment may be defined as follows: a set of partitions $(\pi_1, \pi_2, \pi_3, \cdots \pi_r)$ of a set of machine states S is called an assignment (or *r-assignment*) of S if and only if:

(a) each of the partitions contain two blocks
(b) $\pi_1 . \pi_2 . \pi_3 \cdots \pi_r = 0$, the zero partition.

Thus to apply the principles of partition pairs to the state assignment problem, for example to assign machine M shown in Table 7.24(a), we would endeavour to find three partitions (a three-variable code is required) π_1, π_2 and π_3 of two blocks each such that

$$\pi_1 . \pi_2 . \pi_3 = \pi(0)$$

In order to select the partitions we start by looking for two-block partitions in the set of $m(\pi)$ partitions for the machine. An obvious choice in this case would be $\pi'_6 = (1, 2, 4)(3, 5)$ which we call π_1 and assign to variable y_1, the first of the state variables. Moreover, since partition π_6 also consists of two blocks we can assign it to y_2; thus $\pi_2 = \pi_6 = (1, 3, 4)(2, 5)$. Selecting y_3 is straightforward since it must make the product $\pi_1 \pi_2 \pi_3 = \pi(0)$; in this case we can select $\pi_9 = (1, 2, 3)(4, 5)$. In order to determine the structural interdependence of the machine we reconstruct the Mm pairs:

$$M(\pi_1), \pi_1 = (1, 3, 4)(2, 5), (1, 2, 4)(3, 5)$$
$$M(\pi_2), \pi_2 = (1, 2, 3)(4, 5), (1, 3, 4)(2, 5)$$
$$M(\pi_3), \pi_3 = (1, 2, 4)(3, 5), (1, 2, 3)(4, 5)$$

The dependency of any state variable on another may be determined by checking if any other partition used in the assignment is less than or equal to the $M(\pi)$ for that variable or alternatively if the product of any two or more partitions is less than or equal to $M(\pi)$. In the case of our example it will be obvious that Y_1, which is defined by π_1, depends only on y_2 since all the information required is provided by y_2. Similarly Y_2 will depend only on y_3, and Y_3 on y_1. The assigned state table is shown in Table 7.24(b) and the K-maps in Table 7.25; the reduced equations are as follows:

$$Y_1 = y_2 \bar{x}_1$$
$$Y_2 = y_3 \bar{x}_2$$
$$Y_3 = y_1 \bar{x}_1 \bar{x}_2 + \bar{y}_1 x_1 \bar{x}_2$$

To illustrate further the mechanics of establishing dependancy consider a machine assigned in the following way:

$$M(\pi_1), \pi_1 = (1, 2, 4)(3, 5), (1, \cdot 3, 5)(2, 4)$$
$$M(\pi_2), \pi_2 = (1, 4)(2)(3, 5), (1, 2, 4)(3, 5)$$
$$M(\pi_3), \pi_3 = (1, 3)(2)(4, 5), (1, 3, 4)(2, 5)$$

In this case once again it is apparent that Y_1 will depend only on y_2 since the complete information as specified by $M(\pi_1)$ is provided by π_2. However, the dependancies of Y_2 and Y_3 are not so obvious. To ascertain these first check if any partition π_1, π_2 or π_3 is $\leq M(\pi_2)$ or $M(\pi_3)$; it may be seen by observation that this cannot be the case. Next we examine partition products and discover that

Table 7.25 K-maps for machine M

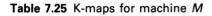

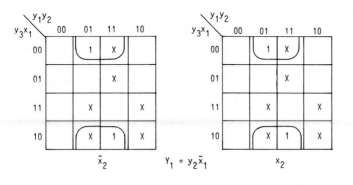

$Y_1 = y_2 \bar{x}_1$

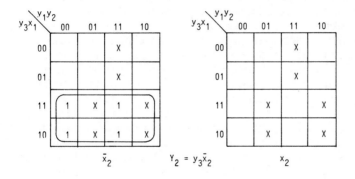

$Y_2 = y_3 \bar{x}_2$

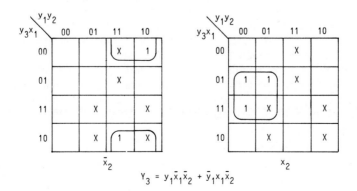

$Y_3 = y_1 \bar{x}_1 \bar{x}_2 + \bar{y}_1 x_1 \bar{x}_2$

$$\pi_2 . \pi_3 = (1, 4)(2)(3)(5) \text{ which is less than } M(\pi_2)$$

and

$$\pi_1 . \pi_3 = (1, 3)(2)(4)(5) \text{ which is less than } M(\pi_3)$$

From this we can state that Y_2 will depend on y_2 and y_3 and Y_3 on y_1 and y_2.

The importance of partition pairs in state assignment is that the corresponding algebra may be used to detect reduced dependencies. As well as defining partition pairs with SP (called an $(S\text{-}S)$ pair) other partition pairs are possible based on input-state $(I\text{-}S)$, state-output $(S\text{-}O)$ and input-output $(I\text{-}O)$ conditions.[17] The $S\text{-}S$ or $S\text{-}O$ partitions imply state variable reduced dependancies, while $I\text{-}S$ or $I\text{-}O$ pairs reduce input dependances. Even more important, the techniques are perfectly general and provide an insight into the structure and information flow of sequential systems.[18–20]

Unfortunately there are many problems in translating this work into a viable design technique, particularly an efficient CAD algorithm. For example, we have only dealt with small, fully specified machines; the computational problems increase enormously with large, incompletely specified systems. Moreover in many cases it is impossible to find enough two-block SP partitions to make a rational choice of assignment. Another difficulty is that the level of redundancy present in a machine will determine the structure and hence the possible partitions. In this sense it is possible that a redundant machine will have a better structure (and hence assignment) than a minimal one. Last, but not least, the ideas of using partition pairs to obtain reduced dependancies is based on a D-type bistable realization, and the results are not necessarily valid for other types of bistable implementations. Its main relevance to designing LSI/VLSI systems (though yet to be exploited) is that it is possible to use the concept of information flow to effect a decomposition of a sequential machine into a number of smaller subsystem modules.

7.8 Decomposition of sequential systems

An important consequence of reduced dependancy is the possibility of decomposing a sequential machine M into two or more smaller component machines suitably interconnected. Only *loop-free* interconnections of the submachine will be considered, that is there must be no external feedback between the components of the decomposition.

The objective of decomposition is to find a subsystem structure that reduces the overall cost and to generate subsystems of such a size that it is possible to produce a good optimized design. The advantages are obvious in developing VLSI systems where modularity and interconnections are a prime consideration.

Two types of decomposition are possible (see Fig. 7.1).

(a) *Serial decompositions*, in which the outputs of any component may be used as inputs to other components. Note that the first sub-

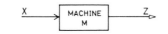

a) Original system

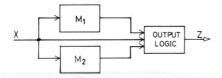

b) Parallel decomposition

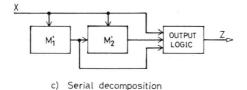

c) Serial decomposition

Figure 7.1 Decomposition of sequential machines

machine (called the *head* machine) in the cascade is independent of
the second, but the second may depend on the first (consequence of
loop-free restriction). The final submachine in a cascade is normally
called the *tail* machine.

(b) *Parallel decompositions*; in this case the submachines are independent of each other, that is the next state depends only on the
external inputs and its own present state.

The system output in both cases will in general depend on the outputs of
the submachines and the external inputs.

In any decomposition each state of the original machine will correspond
to a combination of the states of the submachines, that is the total state of
the system. Note also that all the submachines will effectively operate (that
is change state) simultaneously as a consequence of the external clocking
source (as in a coherent counter). The upper frequency limit of the system,
due to propagation delays, is therefore dictated by the maximum delay
incurred in any individual bistable stage and its associated circuitry. Decomposition depends on the existence of *SP* (closed) partitions for the
machine under consideration. If there are no non-trivial partitions for the
machine decomposition is impossible. However, it can be shown[21] that by
adding extra (redundant) states to the machine, using a technique called
state-splitting, the structure can be changed in such a way as to generate *SP*
partitions.

We are now in a position to formally state the conditions for decomposition. Given a fully specified state table for a sequential machine M there exists

(a) a non-trivial serial decomposition of M if, and only if, there is a set of non-trivial SP partitions $\pi_1, \pi_2, \cdots \pi_n$ on the states of M such that $\pi_1 \geq \pi_2 \geq \cdots \pi_n \geq \pi(0)$;
(b) a non-trivial parallel decomposition of M if, and only if, there exists a set of SP partitions on the states of M such that $\pi_1 . \pi_2 \cdots \pi_n = \pi(0)$.

The term non-trivial is used to indicate a decomposition where each of the component subsystems has fewer states than the original system. Let us now illustrate these concepts with actual design examples.

Consider the machine whose state table is shown in Table 7.26(a); it is in fact the pattern correlator seen originally in Table 7.2. It was previously established that the machine has two non-trivial SP partitions:

Table 7.26 Serial decomposition

Present state	Input x_1x_2 Next state 00	01	11	10	Output Z_0 00	01	11	10
1	2	4	2	4	0	0	0	0
2	7	5	7	5	0	0	0	0
4	5	5	5	5	0	0	0	0
5	6	6	6	6	0	0	0	0
6	1	1	1	1	0	0	0	0
7	9	6	9	6	0	0	0	0
9	1	1	1	1	1	0	1	0

(a) State table machine P

Present state	Input x_1x_2 Next state 00	01	11	10	Output Z_1
$A(1, 5, 7)$	B	B	B	B	0
$B(2, 4, 6, 9)$	A	A	A	A	1

(b) State table machine P_1

Present state y_1	Next state y_{1+}	Output Z_1
0	1	0
1	0	1

(c) Reduced and assigned state table machine P_1

Table 7.26 Continued

(d) State table machine P_2

Present state	Input $Z_1X_1X_2$ Next state							
	000	001	010	110	111	101	100	
$C(1, 2, 4)$	C	C	C	D	D	D	D	
$D(6, 9, 7, 5)$	D	D	D	D	C	C	C	

(e) Reduced and assigned state table machine P_2

Present state y_2	Input Z_1 Next state		Output Z_2
	0	1	
0	0	1	0
1	1	0	1

(f) State table machine P_3

Present states	Inputs $Z_1Z_2x_1x_2$ Next states/output Z_0															
	0000	0001	0011	0010	0110	0111	0101	0100	1100	1101	1111	1110	1010	1011	1001	1000
$E(2, 7, 9)$	X	X	X	X	$F/0$	$E/0$	$E/0$	$E/0$	$F/1$	$F/0$	$F/1$	$F/0$	$F/0$	$E/0$	$F/0$	$E/0$
$F(1, 4, 5, 6)$	$E/0$	$F/0$	$E/0$	$F/0$	$F/0$	$F/0$	$F/0$	$F/0$	$F/0$	$F/0$	$F/0$	$F/0$	$F/0$	$F/0$	$F/0$	$E/0$

(g) Assigned state table machine P_3

Present states y_3	Inputs $Z_1Z_2x_1x_2$ Next states/output Z_0															
	0000	0001	0011	0010	0110	0111	0101	0100	1100	1101	1111	1110	1010	1011	1001	1000
0	X / 0	X / 1	X / 0	X / 1	1 / 1	0 / 1	1 / 1	1 / 1	1 / 1	1 / 1	1 / 1	1 / 1	1 / 0	0 / 1	1 / 1	0 / 0

$$\pi_1 = (1, 5, 7)(2, 4, 6, 9)$$
$$\pi_2 = (1)(2, 4)(7, 5)(6, 9)$$

which together with $\pi(I)$ and $\pi(0)$ complete the lattice. Because $\pi_1 . \pi_2 \neq \pi(0)$ the machine cannot be decomposed into parallel components; however, since $\pi_1 \geq \pi_2 \geq \pi(0)$ a serial decomposition is possible. Moreover the machine (let us call it P) is decomposable into three submachines P_1, P_2 and P_3 connected in series.

The head machine P_1 is derived from the two-block partition π_1; the state table is shown in Table 7.26(b) where state A represents states (1, 5, 7) and state B states (2, 4, 6, 9) of the original machine. The next states are determined by considering the states comprising state A under all input conditions referring to the original state table (in this case the transitions all go to states in B) followed by state B etc. The output Z_1 indicates the individual blocks of the partition which will obviously be the same as the state of the machine. In this particular case machine P_1 can be reduced since it is apparent that the next-state values are independent of the external inputs. The machine requires one state variables, y_1, which we can assign as shown in Table 7.26(c).

The second submachine in the cascade, P_2, is also derived from π_1 such that $\pi_1 . r_1 = \pi_2$; the partition r_1 (of which a number may satisfy the condition) is used to define the states of P_2. For example:

$$(1, 5, 7)(2, 4, 6, 9) . (1, 2, 4)(6, 9, 7, 5) = (1)(2, 4)(5, 7)(6, 9)$$

Using $r_1 = (1, 2, 4)(6, 9, 7, 5)$, the blocks of which we call C and D respectively, we can generate the state table shown in Table 7.26(d). This time the inputs must include the output of P_1, that is Z_1, as well as the external inputs $x_1 x_2$. The state table is generated by intersecting the block of states defined by Z_1 with the block of states corresponding to the present state of P_2 (either C or D) and using the result to access the original state table to determine the next states under the given external input conditions; the block or state of P_2 which contains these states is then entered in the table. For instance, in state C (1, 2, 4) with input $Z_1 \bar{x}_1 \bar{x}_2$ we have (1, 2, 4) $\cap$ (2, 4, 6, 9) = (2, 4); looking up present states (2, 4) for input $\bar{x}_1 \bar{x}_2$ for machine P yields next-state values of (7, 5) which are contained in the block corresponding to state D of machine P_2.

Again in this particular case we can reduce the machine since the next states are independent of the external inputs. Similarly the machine requires one state variable y_2 which we can assign as given in Table 7.26(e).

The final, tail machine, P_3, in the cascade must resolve all the outputs of the preceding machines in such a way as to identify all the individual states of machine P and generate the required system outputs Z_0. To achieve this we much choose a partition r_2 such that

$$r_2 . \pi_2 = \pi(0)$$

In general there are a number of such partitions and where possible it is desirable to choose an *output consistent* partition, that is one in which all the states in a block have identical output values.

In this example we shall choose the partition $r_2 = (2, 7, 9)(1, 4, 5, 6)$ where $(2, 7, 9)(1, 4, 5, 6).(1)(2, 4)(5, 7)(6, 9) = (1)(2)(4)(5)(6)(7)(9)$. Allocating the blocks of r_2 E and F respectively we arrive at the state table shown in Table 7.26(f). Note that the entries are determined exactly as before and also the generation of don't-care (can't-happen in this case) entries—these may be employed in the usual way to effect minimization. Machine P_3 requires one state variable, y_3, which can be assigned as shown in Table 7.26(g).

Realizing the machines in terms of D-type bistables we obtain the following equations:

$$\begin{aligned}
\text{Machine } P_1 \quad & Y_1 = \bar{y}_1, \; Z_1 = y_1 \\
\text{Machine } P_2 \quad & Y_2 = \bar{y}_2 Z_1 + y_2 \bar{Z}_1 \quad Z_2 = y_2 \\
\text{Machine } P_3 \quad & Y_3 = Z_1 Z_2 + Z_2 y_3 + Z_1 y_3 + \bar{x}_1 x_2 + x_1 \bar{x}_2 \\
\text{Output} \quad & Z_0 = Z_1 Z_2 \bar{x}_1 \bar{x}_2 \bar{y}_3 + Z_1 Z_2 x_1 x_2 \bar{y}_3
\end{aligned}$$

The K-maps for machine P_3 are shown in Table 7.27 and a schematic of the total system in Fig. 7.2. The system will of course step through the states of the original machine which are represented by the combined states of the submachines P_1, P_2 and P_3. For example, the starting state of the original machine is state 1, represented by machine P_1 in state A $(1, 5, 7)$, machine P_2 in state C $(1, 2, 4)$ and machine P_3 in state F $(1, 4, 5, 6)$; thus state 1 is represented by $ACF = 001$. An input of $\bar{x}_1 \bar{x}_2$ causes a transition to state 2 which is state E in machine P_3 whilst machine P_1 automatically changes to state B and machine P_2 is unchanged; thus state 2 is represented by $BCE = 100$, etc. Note that the system must be set to the correct starting state, particularly if there are don't-care terms, which in this case will be $Z_1 = 0$, $Z_2 = 0$, $Z_3 = 1$, that is state 1.

Let us now consider a parallel decomposition using the machine as shown in Table 7.28. It can easily be shown that the machine has two non-trivial SP partitions:

$$\begin{aligned}
\pi_1 &= (1, 3, 5)(2, 4, 6) \\
\pi_2 &= (1, 4)(2, 5)(3, 6)
\end{aligned}$$

and also that $\pi_1 . \pi_2 = \pi(0)$.

Table 7.27 K-maps, machine P_3 decomposition

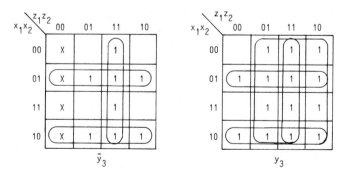

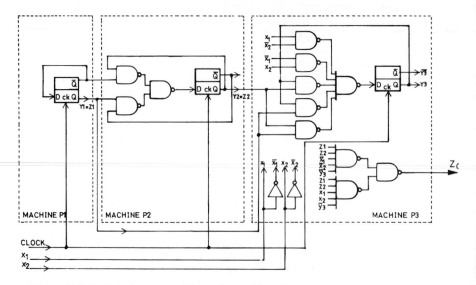

Figure 7.2 Serial decomposition of machine P

Thus we can effect a parallel decomposition, as shown in Fig. 7.1, with two independent submachines S_1 and S_2 based on the partitions π_1 and π_2. The state tables for the two machines are shown in Table 7.29(a) and (b); note that they are derived in exactly the same way as the head machines for serial decomposition. It is constructive to ascertain the total state of the system as indicated by the states of the two submachines. This is shown in Table 7.27(c) where, for example, state 5 of the original machine S is represented by the combination of state A in S_1 and state D in S_2.

Once the state tables have been assigned in the usual way, as shown in Table 7.30(a) and (b), we can design the output logic for the system. Note that the assignment of machine S_2 is not necessarily optimal in this case, but time could usefully be spent, in view of the size of the machine, in obtaining the best assignment (this is of course a general advantage of the decomposition approach).

To obtain the output logic we construct the table shown in Table 7.30(c) using the actual codes for y_1, y_2 and y_3. The input equations for D-type bistable realization (see Table 7.31) are as follows:

$$\text{Machine } S_1 \quad Y_1 = \bar{x}\bar{y}_1 + xy_1$$
$$\text{Machine } S_2 \quad Y_2 = y_3x + y_2y_3\bar{x}$$
$$Y_3 = y_3 + \bar{y}_2\bar{x} + y_2x$$
$$\text{Output} \quad Z = y_2 + y_1x + \bar{y}_1y_3\bar{x}$$

The equations may be implemented in the usual way and the circuit will be seen to have the structure given in Fig. 7.1.

Decomposition could in essence be regarded as simply another method of state assignment (in general a similar number of units will be obtained as in the normal partitioning method of assignment—in the case of our serial

Table 7.28 State table machine S

Present state	Input x Next state 0	1	Output Z 0	1
1	2	3	0	0
2	3	4	0	1
3	4	5	1	1
4	5	6	0	1
5	6	1	1	0
6	1	2	1	1

Table 7.29 State tables for parallel decomposition

Present state	Input x Next states 0	1
$A(1, 3, 5)$	B	A
$B(2, 4, 6)$	A	B

(a) State table machine S_1

Present state	Input x Next states 0	1
$C(1, 4)$	D	E
$D(2, 5)$	E	C
$E(3, 6)$	C	D

(b) State table machine S_2

State machine S	State machine S_1	State machine S_2
1	A	C
2	B	D
3	A	E
4	B	C
5	A	D
6	B	E

(c) Correlation of internal states

Table 7.30 Assigned tables for parallel machines

Present state y_1	Input x Next states 0 y_1^+	1 y_1^+
0	1	0
1	0	1

(a) Machine S_1

Present state y_2	y_3	Input x Next states 0	1
0	0	01	11
0	1	11	00
1	1	00	01
1	0	XX	XX

(b) Machine S_2

Table 7.30 Continued

State	y_1	y_2	y_3	Input x Output Z	
				0	1
1	0	0	0	0	0
2	1	0	1	0	1
3	0	1	1	1	1
4	1	0	0	0	1
5	0	0	1	1	0
6	1	1	1	1	1
	0	1	0	X	X
	1	1	0	X	X

(c) Output table

Table 7.31 K-maps for parallel decomposition

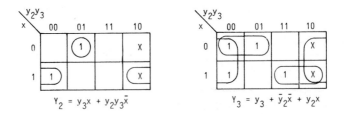

$$Y_2 = y_3 x + y_2 y_3 \bar{x}$$

$$Y_3 = y_3 + \bar{y}_2 \bar{x} + y_2 x$$

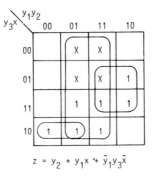

$$z = y_2 + y_1 x + \bar{y}_1 y_3 \bar{x}$$

example the number is the same) and is particularly useful when the *SP* partitions are small, that is containing a large number of blocks. However, there are obvious advantages to be gained from a modular approach to realization, particularly the possibility of decomposing into specified components.

7.9 Analysis of sequential machines

It is often necessary to analyse the operation of an existing sequential machine, perhaps as a prologue to redesign or simply to understand its functions. The inherent feedback property of such a circuit makes this a tedious and error-prone operation unless some systematic technique is employed. Campeau[22] has described an analytical method, which may also be used for synthesis, which uses Boolean matrices and is excellent for such purposes.

Let us first consider how to represent switching equations in a matrix form. Any Boolean switching function, say $f = (a, b, c)$, may be expressed as a polynomial with coefficients of either 0 or 1. For example, consider the general sequential circuit shown in Fig. 7.3; the following equations may be deduced from the terminal conditions:

$$A = b\bar{c} + ac$$
$$B = b\bar{c} + \bar{a}c$$
$$C = \bar{a}b + ab$$

These equations would correspond to the application equations for a sequential circuit, as described in the last chapter. To express these as a set of polynomial equations, we must first expand them into the canonical form, and then denote the presence or absence of actual terms by a 1 or a 0 respectively:

$$A = 0.\bar{a}\bar{b}\bar{c} + 0.\bar{a}\bar{b}c + 1.\bar{a}b\bar{c} + 0.\bar{a}bc + 0.a\bar{b}\bar{c}$$
$$+ 1.a\bar{b}c + 1.ab\bar{c} + 1.abc$$
$$B = 0.\bar{a}\bar{b}\bar{c} + 1.\bar{a}\bar{b}c + 1.\bar{a}b\bar{c} + 1.\bar{a}bc + 0.a\bar{b}\bar{c}$$
$$+ 0.a\bar{b}c + 1.ab\bar{c} + 0.abc$$
$$C = 1.\bar{a}\bar{b}\bar{c} + 1.\bar{a}\bar{b}c + 0.\bar{a}b\bar{c} + 0.\bar{a}bc + 0.a\bar{b}\bar{c}$$
$$+ 0.a\bar{b}c + 1.ab\bar{c} + 1.abc$$

We may now represent these equations in the conventional matrix form:

$$\begin{bmatrix} C \\ B \\ A \end{bmatrix} = \begin{bmatrix} c \\ b \\ a \end{bmatrix} \begin{bmatrix} 1 & 1 & 0 & 0 & 0 & 0 & 1 & 1 \\ 0 & 1 & 1 & 1 & 0 & 0 & 1 & 0 \\ 0 & 0 & 1 & 0 & 0 & 1 & 1 & 1 \end{bmatrix}$$

Note that the least significant digit, in this case c, must be at the top. This may be expressed symbolically as

$$F = BT$$

where B stands for the present state of the sequential sysem, F the next states, and T represents the matrix of 1s and 0s relating present and next states. An alternative decimal form may be used for the vector T:

$$T = [1 \quad 3 \quad 6 \quad 2 \quad 0 \quad 4 \quad 7 \quad 5]$$

Let us now define a vector A_n, of order $n \times 2^n$ such that for $n = 2$,

$$A_2 = \begin{bmatrix} 0 & 1 & 0 & 1 \\ 0 & 0 & 1 & 1 \end{bmatrix}$$

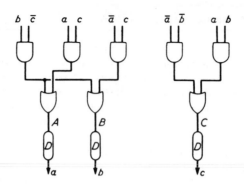

Figure 7.3 Sequential circuit

and for $n = 3$,

$$A_3 = \begin{bmatrix} 0 & 1 & 0 & 1 & 0 & 1 & 0 & 1 \\ 0 & 0 & 1 & 1 & 0 & 0 & 1 & 1 \\ 0 & 0 & 0 & 0 & 1 & 1 & 1 & 1 \end{bmatrix}$$

and so on. In decimal form we have

$$A_3 = [0 \quad 1 \quad 2 \quad 3 \quad 4 \quad 5 \quad 6 \quad 7]$$
$$A_4 = [0 \quad 1 \quad 2 \quad 3 \quad 4 \quad 5 \quad 6 \quad 7 \quad 8 \quad 9 \quad 10 \quad 11 \quad 12 \quad 13 \quad 14 \quad 15]$$

and so on. Thus the vector A may be defined as having components a_s, where $a_s = S - 1$ for $S = 1, 2, 3, 4, \cdots, 2^n$, and n is the order of the matrix (and also represents the number of switching variables).

We are now in a position to define a special Boolean matrix multiplication operation. Suppose we wish to find the product of $XY = Z$, where X and Y are Boolean matrices. The Z columns are obtained one by one by comparing the X columns with the appropriate A matrix, noting which A column is identical to the X column. The Y column corresponding to the number of the identified A column is the resulting Z column. For example, suppose we have

$$\underbrace{\begin{bmatrix} 1 & 1 & 0 & 0 \\ 0 & 1 & 1 & 0 \end{bmatrix}}_{X} \underbrace{\begin{bmatrix} 0 & 0 & 1 & 0 \\ 1 & 1 & 1 & 0 \end{bmatrix}}_{Y} = \underbrace{\begin{bmatrix} 0 & 0 & 1 & 0 \\ 1 & 0 & 1 & 1 \end{bmatrix}}_{Z}$$

and for two variables

$$A_2 = \begin{bmatrix} 0 & 1 & 0 & 1 \\ 0 & 0 & 1 & 1 \end{bmatrix}$$

We take the first column of X,

$$\begin{bmatrix} 1 \\ 0 \end{bmatrix}$$

and compare this with the A_2 matrix, where we find that column 2 contains the identical vector. We now use the vector from column 2 of the Y matrix,

$$\begin{bmatrix} 0 \\ 1 \end{bmatrix}$$

to form the first column of Z. This process is repeated to give the product, shown above, for Z. Note that the operation is not commutative, i.e. $YX \neq Z$. The multiplication can be performed throughout using the decimal notation:

$$\underset{X}{[1 \quad 3 \quad 2 \quad 0]} \underset{Y}{[2 \quad 2 \quad 3 \quad 0]} = \underset{Z}{[2 \quad 0 \quad 3 \quad 2]}$$

where $A_2 = [0\ 1\ 2\ 3]$. In fact, this is a much simpler process, since the required component for Z is the component in column $(X_{component} + 1)$ of Y.

Let us now consider the problem of finding the components of the vector F which satisfy the matrix equation $F = BT$. The appropriate A matrix for the three-variable switching equations is

$$A_3 = \begin{bmatrix} 0 & 1 & 0 & 1 & 0 & 1 & 0 & 1 \\ 0 & 0 & 1 & 1 & 0 & 0 & 1 & 1 \\ 0 & 0 & 0 & 0 & 1 & 1 & 1 & 1 \end{bmatrix}$$

and

$$F = \begin{bmatrix} C \\ B \\ A \end{bmatrix} = \begin{bmatrix} c \\ b \\ a \end{bmatrix} \begin{bmatrix} 1 & 1 & 0 & 0 & 0 & 0 & 1 & 1 \\ 0 & 1 & 1 & 1 & 0 & 0 & 1 & 0 \\ 0 & 0 & 1 & 0 & 0 & 1 & 1 & 1 \end{bmatrix}$$

In this case we must assume starting values, i.e. present state conditions, for the B matrix, say

$$\begin{bmatrix} 0 \\ 0 \\ 0 \end{bmatrix}$$

Then, using this value, we derive the next state condition, F, using the multiplication rule defined above—this is

$$\begin{bmatrix} 1 \\ 0 \\ 0 \end{bmatrix}$$

The process is then repeated, with the next state value becoming the new present state, until the cycle repeats or goes into a loop. For example,

$$F = \begin{bmatrix} 0 & 1 & 1 & 0 & 0 & 1 & 1 & 0 & 0 & 1 \\ 0 & 0 & 1 & 1 & 1 & 1 & 0 & 0 & 0 & 0 \\ 0 & 0 & 0 & 0 & 1 & 1 & 1 & 1 & 0 & 0 \end{bmatrix} \cdots \text{etc.}$$

It will be obvious from inspection that the sequential machine has the characteristics of a cyclic Gray-code counter. It is important to note that the method is applicable only to synchronous (clocked) sequential machines.

If two or more columns of a Boolean matrix are identical, the matrix is said to be *singular*. A singular matrix is said to have a defect of order d, where d is the number of columns of the A_n matrix which are absent. Providing a Boolean matrix is non-singular, it can be inverted:

$$XX^{-1} = A_n$$

The process of inversion is basically the multiplication process in reverse, that is, each column of the A_n matrix is identified in X, and the corresponding X^{-1} term is obtained by subtracting 1 from the column number of the identified term in X. The inverse of the matrix T, for a sequential machine, represents a system for which the output cycle is reversed, assuming identical starting conditions.

A special case of a matrix equation is $B = BT$, and any vector B which satisfies this equation is called a *characteristic vector* of T. In practice this means that the T matrix has an identical column in the same position as the A matrix. Should this condition occur, the sequential machine will automatically lock into a perpetual loop. Thus direct comparison of the A and T matrices will detect any stable loop conditions in the machine. As an example, consider the equation

$$F = \begin{bmatrix} Y_1 \\ Y_2 \end{bmatrix} = \begin{bmatrix} y_1 \\ y_2 \end{bmatrix} \begin{bmatrix} 0 & 1 & 0 & 0 \\ 1 & 1 & 1 & 0 \end{bmatrix}$$

and the relevant A matrix

$$A_2 = \begin{bmatrix} 0 & 1 & 0 & 1 \\ 0 & 0 & 1 & 1 \end{bmatrix}$$

It is apparent that a characteristic vector of T is

$$\begin{bmatrix} 0 \\ 1 \end{bmatrix}$$

and we would expect the machine to lock in this condition. Thus starting from

$$\begin{bmatrix} 0 \\ 0 \end{bmatrix}$$

we have

$$F = \begin{bmatrix} 0 & 0 & 0 & 0 \\ 0 & 1 & 1 & 1 \end{bmatrix} \cdots \text{etc.}$$

So far we have only considered a system where the output is sampled at every clock pulse. It is also possible, however, to determine the output at alternate (or other multiples) of the clock rate by setting up and solving the equation

$$F = BT^n$$

(This assumes, of course, that the circuit continues to change state normally at every clock pulse.) For example, if

$$T = \begin{bmatrix} 1 & 1 & 0 & 0 & 0 & 0 & 1 & 1 \\ 0 & 1 & 1 & 1 & 0 & 0 & 1 & 0 \\ 0 & 0 & 1 & 0 & 0 & 1 & 1 & 1 \end{bmatrix} \quad A_3 = \begin{bmatrix} 0 & 1 & 0 & 1 & 0 & 1 & 0 & 1 \\ 0 & 0 & 1 & 1 & 0 & 0 & 1 & 1 \\ 0 & 0 & 0 & 0 & 1 & 1 & 1 & 1 \end{bmatrix}$$

then

$$T^2 = \begin{bmatrix} 1 & 0 & 1 & 0 & 1 & 0 & 1 & 0 \\ 1 & 1 & 1 & 1 & 0 & 0 & 0 & 0 \\ 0 & 0 & 1 & 1 & 0 & 0 & 1 & 1 \end{bmatrix}$$

and

$$T^3 = \begin{bmatrix} 0 & 0 & 1 & 1 & 1 & 1 & 0 & 0 \\ 1 & 1 & 0 & 1 & 1 & 0 & 0 & 0 \\ 0 & 1 & 1 & 1 & 0 & 0 & 1 & 0 \end{bmatrix}$$

and so on. Then assuming the starting condition

$$B = \begin{bmatrix} 0 \\ 0 \\ 0 \end{bmatrix}$$

we have, for the Gray-code counter described earlier, at every third clock pulse

$$F = \begin{bmatrix} 0 & 0 & 1 & 1 & 0 & & \\ 0 & 1 & 0 & 0 & 1 & \cdots & \text{etc.} \\ 0 & 0 & 1 & 0 & 1 & & \end{bmatrix}$$

Elspas[23] has shown how the matrix method can be applied to the analysis of linear sequential circuits and in particular the determination of the period of a sequence and whether or not it is maximal.

We can represent a linear switching circuit, for example that of Fig. 7.4, as

$$X_1' = 0.X_1 + 1.X_2 + 1.X_3$$
$$X_2' = 1.X_1 + 0.X_2 + 0.X_3$$
$$X_3' = 0.X_1 + 1.X_2 + 0.X_3$$

where X represents present states and X' next states of the circuit. Thus we have the same form of matrix (called by Elspas the T matrix representation) used earlier, i.e.

$$\begin{bmatrix} X_1' \\ X_2' \\ X_3' \end{bmatrix} = \begin{bmatrix} 0 & 1 & 1 \\ 1 & 0 & 0 \\ 0 & 1 & 0 \end{bmatrix} \begin{bmatrix} X_1 \\ X_2 \\ X_3 \end{bmatrix}$$

or

$$X' = TX$$

to represent the operation of linear sequential circuits. However, in this case, since we are dealing only with polynomial equations, we can use normal matrix multiplication methods, but with addition being performed

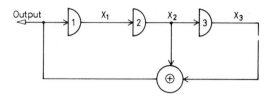

Figure 7.4 Linear switching circuit

modulo-2, to compute the internal states. For example, substituting the arbitrary initial state column vector

$$\begin{bmatrix} 1 \\ 1 \\ 0 \end{bmatrix}$$

into the equation, we have

$$\begin{bmatrix} X_1' \\ X_2' \\ X_3' \end{bmatrix} = \begin{bmatrix} 0 & 1 & 1 \\ 1 & 0 & 0 \\ 0 & 1 & 0 \end{bmatrix} \begin{bmatrix} 1 \\ 1 \\ 0 \end{bmatrix} = \begin{bmatrix} 1 \\ 1 \\ 1 \end{bmatrix}$$

Similarly the complete set of *state vectors* is found to be

$$\begin{bmatrix} 1 \\ 1 \\ 0 \end{bmatrix} \rightarrow \begin{bmatrix} 1 \\ 1 \\ 1 \end{bmatrix} \rightarrow \begin{bmatrix} 0 \\ 1 \\ 1 \end{bmatrix} \rightarrow \begin{bmatrix} 0 \\ 0 \\ 1 \end{bmatrix} \rightarrow \begin{bmatrix} 1 \\ 0 \\ 0 \end{bmatrix} \rightarrow \begin{bmatrix} 0 \\ 1 \\ 0 \end{bmatrix} \rightarrow \begin{bmatrix} 1 \\ 0 \\ 1 \end{bmatrix} \rightarrow \begin{bmatrix} 1 \\ 1 \\ 0 \end{bmatrix}$$

It is possible to determine the period of the sequence, and if it is maximal, by investigating the *characteristic equation* of the matrix. Any matrix T has a characteristic polynomial defined by

$$g(x) = T - xI$$

That is, the determinant of the matrix formed by subtracting an indeterminate x from the diagonal elements of T. In the example above, this is found to be

$$g(x) = \begin{bmatrix} -x & 1 & 1 \\ 1 & -x & 0 \\ 0 & 1 & -x \end{bmatrix}$$

$$\therefore g(x) = -x \begin{vmatrix} -x & 0 \\ 1 & -x \end{vmatrix} - 1 \begin{vmatrix} 1 & 0 \\ 0 & -x \end{vmatrix} + 1 \begin{vmatrix} 1 & -x \\ 0 & 1 \end{vmatrix}$$

$$\therefore g(x) = 1 + x + x^3$$

Note that coefficients must be reduced modulo-2 in this operation and, furthermore, that the *characteristic equation does not necessarily represent the actual hardware circuit*. A basic theorem of matrix algebra states that every square matrix satisfies its own characteristic equation, and thus

$$\varphi(T) = T^3 + T + 1 = 0$$

The matrix period is the smallest integer k for which $T^k = 1$; thus $T^k X = X$ for any initial state X, indicating that all cycle lengths are divisors of k. A general procedure for the determination of the matrix period from the characteristic polynomial depends on polynomial divisibility properties. It is sufficient to find an integer k such that the polynomial $g(x)$ divides $x^k - 1$ without remainder. For, if $x^k - 1 = g(x)h(x)$, then

$$T^k - 1 = g(T)h(T) = 0$$

so that

$$T^k = 1$$

Thus a cyclic code of period k is completely specified by a polynomial $g(x)$ that divides $f(x) = x^k - 1$ without remainder, i.e.

$$f(x) = g(x)h(x)$$

In this particular case we have

$$(x^7 - 1) = (1 + x + x^3)(1 + x + x^2 + x^4)$$

also

$$(x^7 - 1) = (1 + x^2 + x^3)(1 + x + x^3)(1 + x)$$
$$= (1 + x^2 + x^3)(1 + x^2 + x^3 + x^4)$$

Note that both the polynomials $(1 + x + x^3)$ and $(1 + x^2 + x^3)$ are irreducible (prime) factors of $(x^7 - 1)$, so either could be used to generate a cyclic code.

It is apparent that the matrix technique provides a powerful means of analysing clocked sequential systems and, in certain restricted cases, it may also be applied to asynchronous circuits. The use of this technique for synthesis is rather unwieldy and is beyond the scope of this book.

References and bibliography

1. Moore, E. F. Gedanken-experiments on sequential machines. (Automata Studies). *Ann. Math. Stud.* **34**, 129–53 (1955).
2. Paull, M. and Ungar, S. Minimizing the number of states in incompletely specified sequential switching functions. *IRE Trans. electron. comput.* **EC8**, 356–67 (1959).
3. Ginsberg, S. On the reduction of superfluous states in a sequential machine. *J. Ass. comput. Mach.* **6**, 259–82 (1959).
4. Marcus, M. P. Derivation of maximal compatibles using Boolean algebra. *IBM J. Res. Dev.* **8**, 537–8 (1964).
5. Lewin, D. *Computer Aided Design of Digital Systems.* Crane Russak, New York, 1977.
6. Grasselli, A. and Luccio, F. A. A method for minimising the number of internal states in incompletely specified sequential networks. *IEEE Trans. Computers* **EC14**, 350–9 (1965).
7. Luccio, F. Extending the definition of prime compatible classes of states in incomplete sequential machine reduction. *IEEE Trans. Computers* **C18**, 537–40 (1969).

8. Bennetts, R. G., Washington, J. L. and Lewin, D. A computer algorithm for state table reduction. *Radio and Electronic Eng.* **42**, 513–20 (1972).

9. McCluskey, E. J. and Ungar, S. H. A note on the number of internal variable assignments for sequential switching circuits. *IRE Trans. electron. comput.* **EC8**, 439–40 (1959).

10. Harlow, C. and Coates, C. L. On the structure of realisations using flip-flop memory elements. *Inf. and Control* **10**, 159–74 (1967).

11. Humphrey, W. S. *Switching Circuits with Computer Applications.* McGraw Hill, New York, 1958, Chapter 10.

12. Armstrong, D. B. On the efficient assignment of internal codes to sequential machines. *IRE Trans. electron. comput.* **EC11**, 611–22 (1962).

13. Hartmanis, J. On the state assignment problem for sequential machines. I. *IRE Trans. electron. comput.* **EC10**, 157–65 (1961).

14. Levy, L. S. *Discrete Structures of Computer Science.* Wiley, New York, 1980.

15. Farr, E. H. Lattice properties of sequential machines. *J. Ass. comput. Mach.* **10**, 365–85 (1963).

16. Stearns, R. E. and Hartmanis, J. On the state assignment problem for sequential machines. II. *IRE Trans. electron. comput.* **EC10**, 593–603 (1961).

17. Hartmanis, J. and Stearns, R. E. *Algebraic Structure Theory of Sequential Machines.* Prentice-Hall, Englewood Cliffs, N.J., 1966.

18. Friedman, A. and Menon, P. *Theory and Design of Switching Circuits.* Computer Science Press, California, 1975, Chapter 5.

19. Booth, T. L. *Sequential Machines and Automata Theory.* Wiley, New York, 1967.

20. Kohavi, Z. *Switching and Finite Automata Theory.* McGraw Hill, New York, 1978.

21. Hennie, F. C. *Finite State Models for Logical Machines.* Wiley, New York, 1968.

22. Campeau, J. O. Synthesis and analysis of digital systems by Boolean matrices. *IRE Trans. electron. comput.* **EC6**, 230–41 (1957).

23. Elspas, B. The theory of autonomous linear sequential networks. *IRE Trans. on Circuit Theory* **CT6**, 45–60 (1959).

24. Gill, A. *Introduction to the Theory of Finite State Machines.* McGraw Hill, New York, 1962.

25. Miller, R. E. *Switching Theory—Vol. II. Sequential Circuits and Machines.* Wiley, New York, 1965.

26. Givone, D. *Introduction to Switching Circuit Theory.* McGraw Hill, New York, 1970.

27. Harrison, M. A. *Introduction to Switching and Automata Theory.* McGraw Hill, New York, 1955.

Tutorial problems

7.1 Reduce the state table shown in Table 7.32 to a minimal form and investigate possible state assignments. Derive the input and output equations for the circuit using SR bistables.

7.2 Reduce the state table in problem 6 of Chapter 6 (Table S.47) to a minimal form. Carry on and complete the design using JK bistables and derive a logic diagram for the final circuit.

Table 7.32

Present state	Inputs x_1x_2								
	Next states					Output Z			
	00	01	11	10		00	01	11	10
1	4	2	5	1		1	1	0	1
2	2	5	–	3		0	1	1	–
3	1	3	6	5		1	1	0	1
4	2	–	6	–		–	1	–	1
5	–	2	6	1		0	1	1	1
6	2	4	–	4		0	–	1	1

7.3 A synchronous sequential circuit has two inputs x_1, x_2 and an output Z. The output Z is equal to 1 if, and only if, $x_1 = 1$ and the sequence $x_2 = 101$ has occurred immediately after the last time $x_1 = 1$, otherwise the output Z remains equal to 0. Whenever $x_2 = 1$, the output Z is made equal to 0, unless the conditions above are satisfied.

Derive the minimal state diagram for the circuit and then implement the design using JK bistables. Ensure that the input equations are as near optimal as possible.

7.4 Design a synchronous sequential circuit that will compare two serial inputs, x_1 and x_2, and give an output Z whenever any group of five bits in the same clock sequence correspond exactly.

7.5 The sequential machine whose state table is shown in Table 7.33 is required to be realized as a modular structure. Investigate whether a decomposition is possible and if so effect the design using D-type bistables and NAND gates.

7.6 Analyse the circuit shown in Fig. 7.5 using the matrix technique. Investigate the action of the circuit for all starting conditions, and then consider the outputs obtained after alternate clock pulses. Note that y_1 is the least significant input.

Table 7.33

Present state	Next state Input x		Output Z_0	
	0	1	0	1
1	4	3	0	0
2	3	4	0	1
3	5	6	0	0
4	6	6	0	1
5	7	8	0	0
6	8	7	0	1
7	2	1	0	0
8	1	2	0	1

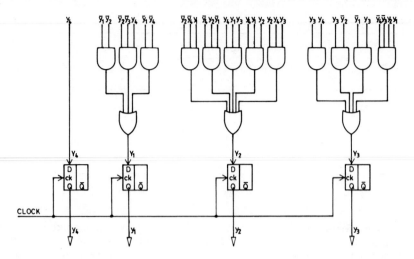

Figure 7.5 Problem 7.6

8 Design of asynchronous sequential circuits

8.1 Introduction

In the last two chapters we have considered synchronous (clocked) sequential circuits; asynchronous circuits are in principle very similar, but special design techniques must be employed to overcome the problems brought about by the absence of any timing pulses. These problems arise mainly as a result of the finite switching time, or propagation delay, of the basic logic modules. In synchronous systems, the clock pulses ensure that the output and input variables are sampled when the circuits have reached a steady state after the delays have settled out. In the absence of any timing pulses, we have to consider two possible conditions for an asynchronous circuit—the *stable* and *unstable* states.

The unstable condition exists when the circuit is changing state in response to an input change, the simplest example is, in fact, the dc set–reset bistable. Consider a bistable with output $Q = 1$ and inputs $S = 0$ and $R = 0$; this is the stable condition. Now, an input change to $S = 0$, $R = 1$ causes the output to change to $Q = 0$, but before the circuit reaches this new stable condition, there is a momentary delay (which varies with each circuit) during which there is an unstable condition of $Q = 1$, with inputs $S = 0$ and $R = 1$. In asynchronous systems we always assume that the circuit will eventually arrive at a stable condition, which implies that the duration of the inputs will always be such as to allow this to occur.

Suppose now we had two such bistables in a circuit, and we were causing the outputs of both of them to go from 1 to 0. Because of the inherent switching delays (which would be different for each bistable circuit) there would be no way of predetermining the output states during the unstable period. Thus the outputs might change $11 \rightarrow 10 \rightarrow 00$; $11 \rightarrow 01 \rightarrow 00$; or, in the ideal case, $11 \rightarrow 00$. Consequently, if these outputs were used as inputs to other circuits erroneous operation would result if we assumed the ideal change of $11 \rightarrow 00$. For this reason, in asynchronous systems all input variable (and also internal state variable) changes are restricted so that only *one* variable can change state at any time. Furthermore, it is also assumed that the internal states have stabilized before another input variable is changed. If these restrictions are ignored circuit 'races' (which may be critical or non-critical) will result; this aspect will be covered in more detail later.

Another problem is that of circuit hazards; as we have seen in Section

4.5 static hazards can exist in combinational logic due primarily to the Boolean expressions $A\bar{A} = 0$ and $A + \bar{A} = 1$ not being true in practice. With clocked systems these hazards can normally be ignored (the circuits are designed to settle out within one clock period) but this is not the case in asynchronous systems which respond directly to every change in the input levels.

It is this characteristic, however, which gives asynchronous logic its main advantage, that of speed of operation, since the circuits work at their own speed and are not constrained to operate within a specified time period dictated by a system clock. This aspect has particular significance for VLSI systems where the problems of distributing and maintaining a synchronous clock increases drastically as the circuits are scaled down to accommodate higher circuit densities. In particular the increased resistivity of the connecting wires means that one of the basic conventions of synchronous logic design—that communication between any two parts of a system can be achieved in a single clock period—is no longer valid unless inordinately long clock periods are used. Seitz[1] has highlighted this difficulty and proposed the use of *self-timed circuits*, which are essentially asynchronous, as a solution to the problem.

As we shall see, many of the design techniques associated with asynchronous logic circuits are concerned with ensuring that critical race and hazard conditions do not materialize in practice. The basic design methods which we describe in the rest of this chapter are due mainly to Huffman,[2] and though originally oriented towards relay circuit design are nevertheless applicable to any switching device, but particularly MOS transfer gates.

8.2   Problem definition: timing diagrams

Because the design technique was originally concerned with relays, much of the terminology used (and retained here since it is still widely used) is different to that evolved for synchronous systems, though in many cases it means the same. Inputs to an asynchronous circuit are generally called *primaries*, originally referring to primary relays directly controlled by the circuit inputs, the states of which are represented by x. The storage characteristics (internal states) of the sequential circuit, represented by secondary relays, are called *secondaries* normally symbolized by Y. The equations for next state Y, in terms of present states (y) and input states (x) are called *excitation equations*, and originally represented the contact circuit necessary to energize relay Y. The output states of the sequential circuit we shall call Z.

As with all logic design problems, the first step is to express the oral or written circuit specification in a formal and unambiguous manner. In earlier chapters, we have used the truth table, state diagram/state table and ASM chart to design combinational and synchronous logic circuits. Similarly, for asynchronous logic we proceed via a state diagram (or timing diagram) to a *flow table*. The flow table fulfils a similar function to the state table in that its construction forces the designer to consider all possible modes of circuit operation.

The ASM chart method can also be used in the initial design stages but as we shall see later the asynchronous state assignment process is quite different to that used for sequential machines and as a result renders the method unwieldly.

Let us illustrate these ideas by considering the design of a divide-by-two counter, i.e. a circuit which changes its output state on alternate input pulses, which will be used as a running example throughout this chapter. The waveform or timing diagram is shown in Fig. 8.1(a). The timing diagram is the usual starting point in the intuitive design of relay circuits,[3] and it is instructive to consider this approach first and then relate the ideas to asynchronous design theory. In Fig. 8.1(a) the vertical timing divisions, not necessarily equal but drawn so for convenience, represent each state of the relay circuit as it operates in a logical sequence. The horizontal lines represent the conditions ON or OFF of the input (x) and the output (Z), which we may also consider, in this particular example, as a secondary (Y_1). Now, in state 1 with input $x = 0$, we require an output $Z = 0$; but with input $x = 0$, in state 3, we also require the output to be 1; this is due to the sequential characteristics of the circuit. Note, moreover, that the actual

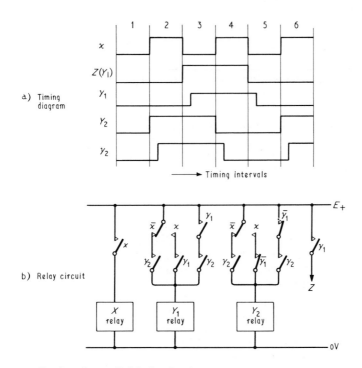

Figure 8.1 Design for a divide-by-2 relay counter

response of the circuit (y_1) to the excitation Y_1 is delayed, because of the time constant of the relay coil. In fact an unstable condition exists, and this must be taken into account in deriving the excitation equations for Y_1 (i.e. the output Z). Thus, from the timing diagram, we have

$$Y_1 = \bar{x}\bar{y}_1 + \bar{x}y_1 + xy_1$$

Unfortunately, if these equations were implemented it would mean that Y_1 could never be de-energized since the condition for this to occur is $\bar{x}y_1$, one of the conditions we used to switch Y_1 on! It is obvious, then, that we require some means of distinguishing between these two conditions. This could be provided by an additional secondary relay Y_2 which, in conjunction with x and Y_1, may be decoded to give the correct output conditions. Thus, when two or more states have the same input conditions, but different output conditions, secondaries (equivalent to internal states) must be used to distinguish between them.

The excitation equations for Y_1, Y_2 from the timing diagram are

$$\begin{aligned} Y_1 &= \bar{x}\bar{y}_1 y_2 + \bar{x}y_1 y_2 + xy_1 y_2 + xy_1 \bar{y}_2 \\ &= \bar{x}y_2 + xy_1 \end{aligned}$$

and

$$\begin{aligned} Y_2 &= x\bar{y}_1 \bar{y}_2 + x\bar{y}_1 y_2 + \bar{x}\bar{y}_1 y_2 + \bar{x}y_1 y_2 \\ &= x\bar{y}_1 + \bar{x}y_2 \end{aligned}$$

Also

$$Z = y_1$$

Inspection of the equations for Y_1 and Y_2 suggests that hazard conditions can arise because if

$$Y_1 = \bar{x}y_2 + xy_1$$

and if

$$y_1 = y_2 = 1$$

then

$$Y_1 = (\bar{x} + x) = 1$$

Now if in the implementation we use break-before-make contacts, there will be a delay between $\bar{x}$ and x which will give rise to a transient drop (0) in the output, causing relay Y_1 to de-energize. We must ensure that this can never happen by including the additional terms $y_1 y_2$ to yield the final excitation equations:

$$\begin{aligned} Y_1 &= \bar{x}y_2 + xy_1 + y_1 y_2 \\ Y_2 &= x\bar{y}_1 + \bar{x}y_2 + \bar{y}_1 y_2 \end{aligned}$$

The full relay circuit is shown in Fig. 8.1(b); in actual relay practice this hazard can be overcome very easily by using make-before-break change-over contacts.

With a simple circuit like this, the design method detailed above is quite successful, but with more complicated circuits involving many variables a more formal approach is desirable, particularly to determine when (and how many) secondaries are required, and to recognize and eliminate hazardous circuit conditions. We now explain how the same circuit may be designed using a more rigorous and versatile procedure.

8.3 Problem definition: state diagrams and flow tables

The first step in the design procedure is to draw up a *primitive flow table* for the counter. In a flow table (Table 8.1(a)) each entry represents either a stable (circled entry) or unstable internal state of the system (or, alternatively, a don't-care condition). A primitive flow table is simply one in which each stable state is allotted a separate row, implying a different secondary state for each stable state. The output Z is recorded for each stable state row at the side of the table. Each stable state is thus uniquely defined by a combination of primaries (input x) and secondaries (present internal states y) of the machine (see Table 8.1(c)). An unstable state results when a particular combination of primaries and secondaries (the excitation Y) requires a transition to a new internal state. This is brought about by a change in the input condition, and for a brief period an unstable condition exists during which the internal states do not correspond to those required by the excitation, as we saw, for example, in the relay design of Section 8.2. Thus the primitive flow table lists all the possible outputs and transitions that can occur when the input variables are changed.

As with synchronous circuits, state diagrams can also be used as a preliminary aid to formalizing the circuit specifications and writing the primitive flow table. However, state diagrams for asynchronous circuits differ from those for synchronous circuits in that each stable state of the circuit must be represented by a *sling*, i.e. a transition path originating and terminating at the same stable state. The reason for this is that, for a synchronous circuit, an unchanging input sequence (say 111 $\cdots$ etc.) will be interpreted as repetitions (one for each clock pulse) of the input, whereas for the asynchronous circuit, concerned only with voltage or current levels, it will be regarded as a single input. Thus, whenever a new input combination causes the circuit to assume a new stable state and remain there (while the input is present) the state diagram must show a sling. The state diagram may be either of the Mealy or the Moore model form, but we shall see later that an attempt should always be made to relate output and internal states (Moore model) to produce more economical output functions. The design procedure is very similar to that employed for synchronous machines; as before when drawing up a state diagram (or flow table) the best plan is to follow through the correct sequence of events to produce the required output. Each time a new input condition occurs, a new internal state is allocated, unless it is obvious that an existing state fulfils the requirements. Again it is of no consequence at this stage if more states are allocated than are actually required to satisfy the circuit specification, as these will be found and eliminated at a later stage of the

Table 8.1 Design tables for
divide-by-2 counter

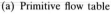

Input		Output	
0	1	Z	
①	2	0	a
3	②	0	b
③	4	1	c
1	④	1	d

(a) Primitive flow table

(b) Transition map

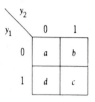

(c) Assigned flow table or Y-map

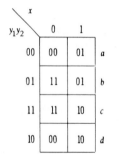

(d) Excitation maps

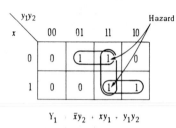

$$Y_1 = \bar{x}y_2 + xy_1 + y_1y_2$$

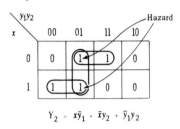

$$Y_2 = x\bar{y}_1 + \bar{x}y_2 + \bar{y}_1y_2$$

(e) Output map

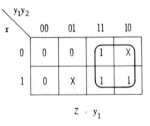

$$Z = y_1$$

design procedure. When the required operation has been met, the state diagram is completed by considering the remaining input transitions. In general, each stable state in the state diagram can have only n directed transitions (where n is the number of input variables) because of the restriction of changing one variable at a time. In the flow table, the restricted input changes would be entered as don't-care or, more realistically, can't-happen conditions. As we have seen when designing a sequential machine it is always necessary to specify some initial starting or resting state; this is conventionally taken as the condition when all the inputs are absent. In practice, provision must be made to reset the machine to this state when switching on.

The Mealy state diagram for the divide-by-2 counter circuit is shown in Fig. 8.2(a). Starting in stable state 1 with no inputs,, i.e. $x = 0$ (note the sling indicating a stable condition), a change of input from 0 to 1 causes a transition to stable state ②, and so on, until in stable state ④ an input change of $1 \rightarrow 0$ returns the counter to the starting condition. It is interesting to observe that the state diagram or flow table can, in this case, be obtained directly from the timing diagram, in which the timing intervals represent the necessary internal states; the unstable states are produced by the finite rise and fall times of the waveforms. This result is generally true for the case of counter and shift register circuits, and any other system that can be conveniently represented by a timing diagram.

The equivalent ASM chart is given in Fig. 8.2(b); note the similarity to the state diagrams and the need for a conditional test and self-feedback loop at each state.

8.4   Internal state reduction and merging

Once the primitive flow table has been established it will, in general, contain more stable states than are actually required, so the next step must be to identify and eliminate these redundant states from the table. This is analogous to the process already described for synchronous systems and the same general philosophy holds good. Thus, for two stable states in a primitive flow table to be identical (or equivalent for the case of incompletely specified tables), the following axioms must be obeyed:

(a) they must have the same output states;
(b) for all possible input changes, their next-state transitions must result in the same (or equivalent) states.

It is important to note that in this case a stable state is specified by both input and secondary conditions; thus for two states to be identical they *must both be in the same column* of the flow table. It appears, then, that this is the identical problem to that encountered with synchronous systems, and consequently it may be solved using the same methods—the implication chart technique.

Let us now consider the primitive flow table shown in Table 8.2(a). This can be reduced by a simple and exhaustive comparison of the stable states in each column; for example, for the input state $\bar{x}_1 \bar{x}_2$ we have ① $\neq$ ⑥ since the output states are different; ① $=$ ⑮ if ② $=$ ⑨, but because their output states are dissimilar ② $\neq$ ⑨, and therefore ① $\neq$ ⑮ , etc. This is a tedious operation (the reader should verify this for himself!) and the best approach is to draw up an implication chart, as shown in Table 8.3. There are many more initial incompatibles in this chart (for the number of internal states involved) than is normal for a synchronous system because of the requirement of column comparison only. The incompatibles should be entered first, followed by the identical states; we then use the procedure adopted earlier for synchronous machines to complete the chart. From the chart the following set of maximal compatibles can be obtained:

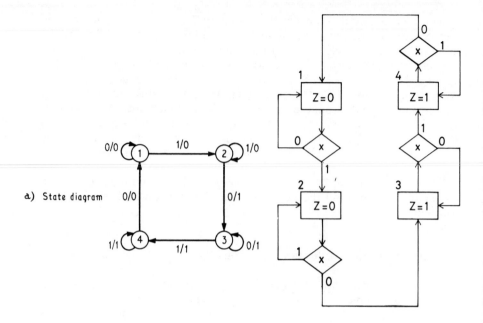

a) State diagram

b) ASM chart

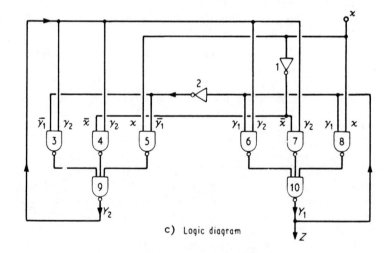

c) Logic diagram

Figure 8.2 Design for a divide-by-2 electronic counter

Table 8.2 Flow table reduction

Inputs x_1x_2				Outputs
00	01	11	10	Z_1Z_2
①	2	–	5	0 0
15	②	3	–	0 0
–	4	③	12	0 0
1	④	11	–	1 0
6	–	13	⑤	1 1
⑥	10	–	7	1 1
16	–	8	⑦	1 1
–	9	⑧	5	0 1
1	⑨	14	–	1 0
1	⑩	11	–	1 0
–	4	⑪	12	0 0
6	–	13	⑫	1 1
–	10	⑬	5	0 0
–	9	⑭	12	0 0
⑮	9	–	5	0 0
⑯	4	–	5	0 0

(a) Primitive flow table

	Inputs x_1x_2				Outputs
	00	01	11	10	Z_1Z_2
a	①	2	–	5	0 0
b	15	②	3	–	0 0
c	–	4	③	5	0 0
d	1	④	3	–	1 0
e	6	–	3	⑤	1 1
f	⑥	4	–	7	1 1
g	15	–	8	⑦	1 1
h	–	4	⑧	5	0 1
i	⑮	4	–	5	0 0

(b) Reduced flow table

Table 8.3 Implication chart

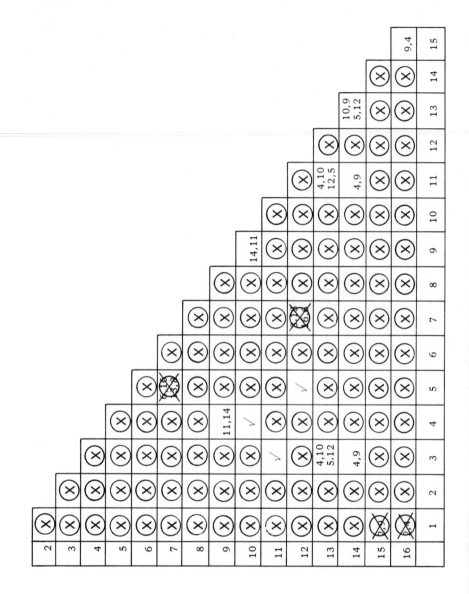

$$M = (1)(2)(6)(7)(8)(3, 11, 13, 14)(5, 12)(4, 9, 10)(15, 16)$$

As the flow table was fully specified (the don't-care conditions due to the input restrictions will always occur in the same places and thus will never be assigned different values) the final result is a partition and will be unique. Incompletely specified flow tables will result in a covering of the machine states and should be treated in the normal way. The elimination of redundant stable states allows us to draw the reduced flow table shown in Table 8.2(b); note that we have simplified the machine to nine stable states, but still expressed in the primitive flow table form of one stable state to a row. If we assigned a code to each row of the table as it stands we would need four secondary variables; indeed, this is the same number required for the original flow table before reduction. Can we reduce the number of rows, and hence secondaries, still further? We can if we remember that a stable state is defined by *both* input and secondary conditions—there is no reason why we should not use the same secondary assignment for different internal states. This means that transitions between stable states in the *same row* will be affected by input changes only. Thus, if we can reduce the number of rows by placing more than one stable state per row, we shall automatically reduce the number of secondary variables required to code the rows; this operation is known as *merging*. Rows may be merged, regardless of output states, if there are no conflicting state numbers (irrespective of stable or unstable states) in any columns, don't-care conditions being used to represent any state. For example, in Table 8.2(b), row c may be merged with row d by combining stable and unstable states 4 and 3 (replaced in the merged row by the relevant stable state), and allowing the don't-care conditions to assume appropriate values. Thus we obtain, as a result of merging rows c and d, the row

1   ④   ③   5

It is interesting to note that during the merging process the circled entries have changed their definition due to our ignoring the output states. In the primitive table they were *internal states* (i.e. state of the feedback loop) concerned only with the input and secondary variables (x, y), whereas in the merged table input, output and secondary variables are represented and the circled entries have now become *total states*. It is possible to indicate the output states associated with each stable state on the merged flow table, but this can become confusing. The best approach is to ignore the output states completely, since they can easily be obtained from the primitive flow table when required.

Generally, there is more than one way of merging the rows of a flow table, and the choice can appreciably affect circuit economy. A unique solution is only possible for a fully specified flow table (one containing no don't-care conditions), but this is an unlikely occurrence in practice. In order to ensure that the best choice is made it is advisable to search for all possible mergers. This may be done in a similar way to the determination of state equivalences by methodically comparing each row with every other row and noting the result.

In Table 8.2(b), by comparing row a with rows (b, c, d, e, f, g, h, i), then row b with rows (c, d, e, f, g, h, i), etc., we can obtain the following pairs of mergeable rows:

$$m = (c, d)(c, e)(c, i)(h, i)$$

The final result must be a partition on the set of all flow table rows, since each row may only be included once. Thus we have

$$M = (a)(b)(c, d)(e)(f)(g)(h, i)$$

Note that the unmergeable rows are included as single element blocks; the fully merged flow table is shown in Table 8.4.

Mergeable rows may be combined into maximal sets (all rows within a block being combinable) but it is important to realize that the relationship is not transitive. For example, consider the reduced flow table shown in Table 8.5; a comparative search yields the following mergeable row pairs:

$$m = (1, 2)(1, 3)(2, 3)(2, 4)(3, 6)(4, 8)(5, 6)(5, 7)(5, 8)(6, 7)$$

These may be combined into maximal sets by examining the row pairs, e.g. rows $(1, 2)$ and $(1, 3)$ can be merged; then if $(2, 3)$ can also be merged (which it can) we may combine to give $(1, 2, 3)$. Applying this technique, we get

$$m = (1, 2, 3)(2, 4)(3, 6)(4, 8)(5, 6, 7)(5, 8)$$

From these sets, we must choose a partition representing all the rows; there are a number of possibilities:

$$M_1 = (1, 2, 3)(4, 8)(5, 6, 7)$$
$$M_2 = (1, 2, 3)(6, 7)(4, 8)(5)$$
$$M_3 = (1)(2, 3)(5, 6, 7)(4, 8), \text{ etc.}$$

The minimal row solution can usually be obtained by including the maximal sets in the partition, though this does not necessarily yield the most economic hardware solution.

An alternative method suggested by Maley and Earle[4] is to merge in

Table 8.4 Merged flow table

Inputs $x_1 x_2$			
00	01	11	10
①	2	–	5
15	②	3	–
1	④	③	5
6	–	3	⑤
⑥	4	–	7
15	–	8	⑦
⑮	4	⑧	5

Table 8.5

Inputs x_1x_2 00	01	11	10	Output Z
①	2	–	3	0
1	②	4	–	1
1	–	–	③	1
–	2	④	5	1
8	–	6	⑤	0
–	7	⑥	–	0
8	⑦	–	–	0
⑧	2	–	5	1

such a way as to minimize or eliminate the output gating. This may be achieved by only merging rows with the same output states, when it may be possible to code the feedback loops (i.e. secondaries) in such a way that the outputs may be obtained directly. This is equivalent to designing a Moore machine with identical output and internal states, but this will become clearer when we discuss the derivation of the output equations. An additional advantage, and perhaps a more important one with high speed logic circuits, is that if the output gating can be eliminated, circuit delays are reduced. Partition M_3 above is output-consistent in this sense, but results in a four-row flow-table.

Another method of establishing the row mergers, which is useful when dealing with a small number of rows, is to draw a *merger diagram*. This is simply a spatial display of all possible mergers (see Fig. 8.3); the rows are represented by the total state numbers and row mergers are indicated by interconnecting lines. In order for two or more rows to be merged, all possible interconnections between the rows must exist. For example, in Fig. 8.3(b), rows 1, 2, 3 can be merged together and all possible interconnecting lines between these rows produce a characteristic triangle pattern in the diagram. For four rows to merge we must look for the pyramid pattern shown in Fig. 8.3(c). In this way, by visual inspection, the best possible row mergers can be chosen. Unfortunately, large merger diagrams are difficult to handle and an algorithmic technique based on the partition principle is a better solution to the problem.

An alternative approach, and one which eliminates the need for merging procedures, is to apply the Paul and Ungar technique to the overall primitive flow table and not just the stable states. By considering each row of the flow table as a total state and comparing both stable and unstable states, including of course the outputs for each row, we can reduce and merge the table in one operation. The method is restricted, however, in that it invariably generates a Mealy machine and requires the output values for the unstable states to be specified prior to the reduction routine. In

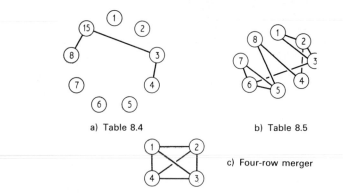

a) Table 8.4   b) Table 8.5

c) Four-row merger

Figure 8.3 Merger diagrams

most cases these limitations are of little consequence compared to the overall advantage. If it is essential to design a Moore machine than the procedures described above with independent stable state reduction and row-merging must be follwed.

The primitive flow table of Table 8.2(a) is repeated in Table 8.6 with its output states fully specified. In general the unstable states are allocated the same output values as that of the next state in the transition: there are, however, other ways of assigning output values to unstable states as we shall see later. Table 8.7(a) and (b) shows the implication charts for the initial and final passes of the reduction procedure. The set of maximal compatibles obtained are

(3, 4, 9, 10, 11, 13, 14)
(4, 9, 10, 11, 13, 14)
(5, 11, 12, 13, 14)
(8, 15, 16)
(11, 13, 14, 15, 16)

which generates the following cover for the machine:

M = (1)(2)(3, 4, 9, 10, 11, 13, 14)(5, 11, 12, 13, 14)(8, 15, 16)(6)(7)

The fully reduced and merged flow table is shown in Table 8.8.

Reverting back to our design for a divide-by-2 counter, we observe from Table 8.1(a) that there are no equivalent states, and therefore the flow table is already in a minimal form. Furthermore, it is also clear that no row mergers are possible.

8.5 Secondary state assignment

This is the process of allocating unique states of a bistable (or relay) to each row of the flow table, or, in other words, assigning a binary code to distinguish between the internal stable states of an asynchronous machine.

Table 8.6 Primitive flow table with output values

	Inputs x_1x_2					Outputs			
	00	01	11	10		00	01	11	10
1	①	2	–	5		00	00	–	11
2	15	②	3	–		00	00	00	–
3	–	4	③	12		–	10	00	11
4	1	④	11	–		00	10	00	–
5	6	–	13	⑤		11	–	00	11
6	⑥	10	–	7		11	10	–	11
7	16	–	8	⑦		00	–	01	11
8	–	9	⑧	5		–	10	01	11
9	1	⑨	14	–		00	10	00	–
10	1	⑩	11	–		00	10	00	–
11	–	4	11	12		–	10	00	11
12	6	–	13	⑫		11	–	00	11
13	–	10	⑬	5		–	10	00	11
14	–	9	⑭	12		–	10	00	11
15	⑮	9	–	5		00	10	–	11
16	⑯	4	–	5		00	10	–	11

Table 8.8 Reduced and merged flow table

		Inputs x_1x_2				Outputs			
		00	01	11	10	00	01	11	10
(1)	A	Ⓐ	B	–	D	00	00	–	11
(2)	B	E	Ⓑ	C	–	00	00	00	–
(3, 4, 9, 10, 11, 13, 14)	C	A	Ⓒ	Ⓒ	D	00	10	00	11
(5, 11, 12, 13, 14)	D	F	C	Ⓓ	Ⓓ	11	10	00	11
(8, 15, 16)	E	Ⓔ	C	Ⓔ	D	01	10	01	11
(6)	F	Ⓕ	C	–	G	11	10	–	11
(7)	G	E	–	E	Ⓖ	01	–	01	11

Table 8.7 Flow table reduction

(a) Initial Pass

	1	2	3	4	5	6	7	8	9	10	11	12	13	14	15
2	1,15														
3	X	X													
4	X	X	3,11												
5	X	X	3,13 5,12	X											
6	X	X	12,7 4,10	X	5,7										
7	5,7 1,16	X	X	X	X	X									
8	X	X	X	X	X	9,10 7,5	7,5								
9	X	X	4,9 3,14	11,14	X	X	X	X							
10	X	X	3,11 4,10	✓	X	X	X	X	11,14						
11	X	X	✓	✓	11,13 5,12	7,12 10,4	X	X	4,9 11,14	4,10					
12	X	X	3,13	X	✓	7,12	X	X	X	X	11,13				
13	X	X	4,10 12,5	11,13 4,10	✓	7,5	X	X	9,10 14,13	11,13	4,10 12,5	12,5			
14	X	X	4,9	11,14 4,9	5,12 13,14	7,12 10,9	X	X	✓	9,10 11,14	4,9	13,14	5,12 10,9		
15	X	X	4,9 12,5	4,9 1,15	X	X	7,5 15,16	✓	1,15	9,10 1,15	12,5 9,4	X	9,10	5,12	
16	X	X	12,5	1,16	X	X	7,5	4,9	1,16 9,4	1,16 10,4	12,5	X	10,4	5,12 9,4	9,4

Table 8.7 Continued

2	X														
3	X	X													
4	X	X	✓												
5	X	X	✓	X											
6	X	X	X	X	X										
7	X	X	X	X	X	X									
8	X	X	X	X	X	X	X								
9	X	X	4,9 3,14	11,14	X	X	X	X							
10	X	X	✓	✓	X	X	X	X	11,14						
11	X	X	✓	✓	✓	X	X	X	4,9 11,14	✓					
12	X	X	✓	X	✓	X	X	X	X	X	✓				
13	X	X	✓	✓	✓	X	X	X	9,10 14,13	✓	✓	✓			
14	X	X	4,9	11,14 4,9	5,12 13,14	X	X	X	✓	9,10 11,14	4,9	13,14	5,12 10,9		
15	X	X	4,9 1,15	X	X	X	X	✓	X	X	12,5 9,4	X	9,10	✓	
16	X	X	✓	X	X	X	X	4,9	X	X	✓	X	✓	5,12 9,4	9,4
	1	2	3	4	5	6	7	8	9	10	11	12	13	14	15

The procedure is once again analogous to that described for synchronous systems, but with one essential difference: with asynchronous machines, the prime objective of assignment is to produce a workable machine free from circuit 'races' and 'hazards', rather than an economic hardware solution. An arbitrary assignment does not necessarily yield a viable design, as is the case with synchronous machines. In practice, of course, one tries to optimize both requirements, but the principal aim must always be to achieve a reliable machine.

A further point is to establish exactly what it is we are assigning. In the case of relay circuits there should be no difficulty, since the secondary assignment variables are, in fact, easily identified with the state of the secondary relays in the circuit. But using electronic logic it is perhaps not so easy to appreciate the situation. We have said above that we are assigning the states of a bistable, which is perfectly true, if a little confusing, since we have not as yet mentioned bistable circuits except in a general sense. What in fact we are doing is evolving equations for the next state (Y) of the circuit in terms of the present states (y) and inputs (x); this is precisely what we did in Chapter 6 when we found the characteristic equations for a bistable from its truth table (Section 6.2). As we saw, the principal characteristic of this type of circuit is that of feedback, for we are taking the outputs of the circuit and feeding them back to the input, a process which requires power gain round the loop. Thus, the excitation equations for the secondary variables (Y) must be implemented using a circuit which incorporates some form of power gain; the NOR/NAND elements are ideal for this purpose. We shall see later that the configuration inevitably produces some form of bistable storage, and hence the analogy to relay circuits is complete.

In the assignment of the secondary variables we must assume, for reasons discussed above, that during a transition from one stable state to another only one secondary variable changes at a time. If more than one secondary variable changed simultaneously, a circuit 'race' could ensue which might cause a transition to the wrong internal stable state. Let us now examine these circuit 'race' conditions in more detail.

Suppose we had arbitrarily allocated the secondary variables $y_1 y_2$ to a flow table as shown in Table 8.9(a). The fully assigned flow table, or

Table 8.9 Circuit 'race' conditions

$y_1 y_2$	Inputs $x_1 x_2$ 00	01	11	10
0 0	(1)	2	(3)	5
0 1	1	2	3	(4)
1 1	1	(2)	6	(5)
1 0	1	(7)	(6)	5

(a) Assigned flow table

$y_1 y_2$	Inputs $x_1 x_2$ 00	01	11	10
0 0	00	11	00	11
0 1	00	11	00	01
1 1	00	11	10	11
1 0	00	10	10	11

(b) Y-map

Y-map, is obtained by replacing the stable entries in the flow table by the appropriate $y_1 y_2$ row values. The map is then completed by giving the unstable state entries, wherever they occur, the same values as their stable states; the complete Y-map is shown in Table 8.9(b). A *critical race* occurs if, when in state ①, the input $(x_1 x_2)$ changes from 00 to 10, and we have the unstable condition in which the present secondary variables show the state 00, but the input variables are dictating a transition to a new stable state (as indicated by the Y-map entry) of 11, i.e. $5 \rightarrow$ ⑤. This necessitates the excitation $(y_1 y_2)$ to change from 00 to 11 which, due to indeterminate delays in the signal path, can take place in three ways: (a) from 00 to 11, if both y_1 and y_2 change together, giving a correct transition to state ⑤; (b) if y_1 changes first, we have $00 \rightarrow 10 \rightarrow 11$ and a *cycle* via $5 \rightarrow 5 \rightarrow$ ⑤; (c) if y_2 changes first, we have $00 \rightarrow 01$ and an incorrect transition to state ④, since the circuit will lock in this stable state. A similar critical race exists if, when in state ①, the inputs change from 00 to 01.

A *non-critical race* is obtained if, when in state ②, the inputs change from 01 to 00, requiring a transition from states 1 to ①, and necessitating the excitation to change from 11 to 00. This time, however, irrespective of which secondary variable changes first, the circuit always cycles to state ①. Since critical races are unpredictable and lead to incorrect or inconsistent operation, they must be avoided at all costs in the design of an asynchronous circuit. However, non-critical races and cycles can sometimes be used to advantage; for example, the inclusion of a non-critical race (or cycle) in a machine can often effect economies in the excitation logic. Also, the fact that a cycle produces a time delay between state transitions can sometimes be utilized.

In order to avoid critical races, any rows of the flow table which have transitions between them must be assigned in such a way that the secondary variables differ in only one variable. Consequently, all possible row-to-row transitions must be examined. A convenient way of doing this is to use a K-map, called in this application a *transition map*, to plot the row adjacencies. In some cases it may be necessary to include redundant rows, i.e. secondary states, in the flow table in order to obtain the required change of one variable; this technique will be discussed later.

Continuing with the counter design, each row of the flow table (Table 8.1(a)) is lettered and then examined for row transitions on the transition map (Table 8.1(b)). It is apparent that the necessary change of one secondary variable can be obtained by using the normal Gray-code assignment. Thus row *a*, state 1 , is arbitrarily assigned $y_1 y_2 = 00$, and the other stable states follow directly from the transition map. Replacing each stable entry with its row assignment code, and each unstable state with the stable state code, we obtain the fully assigned table shown in Table 8.1(c).

8.6 Secondary excitation and output functions

Having produced the fully assigned flow table, the next step is to derive the excitation equations. We can do this directly from the flow table by extracting those combinations of input and present state secondary variables (x, y) that are required to produce the next state secondary variables (Y). For example, in the case of the divide-by-2 counter, examination of the Y-map entries (Table 8.1(c)) reveals that $Y_1 = 1$ for the following condition:

$$Y_1 = \bar{y}_1 y_2 \bar{x} + y_1 y_2 \bar{x} + y_1 y_2 x + y_1 \bar{y}_2 x$$

which is, of course, the excitation equation for Y_1.

For a small number of variables a K-map approach, called an *excitation map*, may be used to plot these functions directly. This is possible since, in most cases, there is a direct correspondence between the Y-map and the excitation map; compare, for example, Table 8.1(c) and (d). However, for a large number of variables, greater than five, say, the excitation equations must be obtained directly from the flow table and minimized using tabular techniques. Furthermore, the excitation equations are, in fact, an example of a multi-terminal circuit and should be treated accordingly.

In obtaining these functions the presence of circuit hazards (due to inherent delaying paths) must always be borne in mind. In fact, the final step in the design procedure is to examine the excitation equations for the possible existence of circuit hazards. For example, the optimized equations for $Y_1 Y_2$ (identical to those obtained for the earlier relay design) are, from Table 8.1(d),

$$Y_1 = \bar{x} y_2 + x y_1$$
$$Y_2 = x \bar{y}_1 + \bar{x} y_2$$

As we showed earlier, a circuit hazard exists for Y_1 when $y_1 = y_2 = 1$ and the input variable changes from $x = 1$ to $\bar{x} = 1$. The problem of designing hazardless relay circuits has been investigated by Huffman[4] and many of the techniques are directly applicable to other forms of switching circuit. The presence of such hazards may be detected (and eliminated) by careful inspection of the K-map. For example, in the map for Y_1 (Table 8.1(d)) there are two distinct groupings of the functions (i.e. loops) corresponding to the optimized equation given above. Hazards can exist if input changes require a transition between two secondary states (represented by the $y_1 y_2$ value) in different groupings on the K-map. This hazard can be eliminated by adding redundancy to the circuit in the form of an additional loop embracing the two hazard loops. This connecting group (shown in Table 8.1(d)) is $y_1 y_2$ and results in the final equation:

$$Y_1 = \bar{x} y_2 + x y_1 + y_1 y_2$$

A similar hazard, which responds to the same treatment, exists in the Y_2 excitation map:

$$Y_2 = x \bar{y}_1 + \bar{x} y_2 + \bar{y}_1 y_2$$

We derive the output map from the primitive and assigned flow tables by plotting the output states as entries in an assigned table. Stable output states are entered first, since the value of unstable output states may be optional, providing they do not cause transient changes in the final output.

The latter can be prevented by noting those stable states which, when involved in a transition, have the same initial and final output states. Any unstable states in the transition path must be assigned the same output states. If the initial and final output states are different, the choice of output state for the unstable states is optional. Bearing in mind, however, that if the output is required as soon as possible, it is necessary for the unstable state to have the same output value as the final stable state. The assigned output table can then be regarded as a K-map and treated accordingly; alternatively, the output functions (including don't-cares) may be extracted directly.

Thus in the counter output map (Table 8.1(e)), since a transition from stable state ③ to ④ involves no change in output states, unstable state 4 must be output 1. A transition from ④ to ① produces a change in output state from 1 to 0, and consequently unstable state 1 can be optional, the choice of output state 1 giving greater simplification. This process is repeated for all stable state transitions until the output map is completed. The output equation obtained from the map is

$$Z = y_1$$

Thus the output may be derived directly from the feedback loop and no additional output logic is required.

The final logic circuit for the counter is shown in Fig. 8.2(c); note that the excitation equations have been implemented directly in terms of NAND elements. From the logic circuit diagram one can easily see how the storage property is obtained—NAND gates (8, 10) and (4, 9), due to the feedback loops, act as bistable stores for the secondaries $Y_1 Y_2$.

This circuit configuration is equivalent to the master–slave bistable circuits briefly mentioned in Chapter 6. In fact it corresponds to the master–slave D-type bistable (see Fig. 8.4(a)) with feedback of the $Q, \bar{Q}$ outputs (from the slave) to the inputs. If we redraw the master–slave circuit (Fig. 8.4(b)) in terms of NAND elements (see Fig. 8.4(c)) and then derive the equations for M and S, we have

$$S = [\overline{(\overline{\bar{x}m})(s)}][\overline{(\overline{\bar{x}m})}] = (\overline{\bar{x}m})s + \bar{x}m$$
$$= s(x + m) + \bar{x}m$$

Hence

$$S = sx + sm + \bar{x}m$$

Now

$$s = y_1 \quad \text{and} \quad m = y_2$$

thus we have

$$Y_1 = y_1 x + y_1 y_2 + y_2 \bar{x}$$

Similarly for M we have

$$M = [\overline{(\overline{\overline{xs}})(m)}][\overline{(\overline{\overline{xs}})}] = (\overline{xs})m + x\bar{s}$$
$$= m(\bar{x} + \bar{s}) + x\bar{s}$$

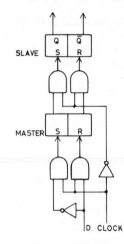

a) D-Type M/S bistable

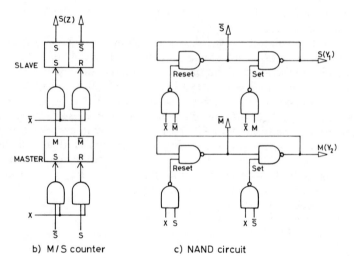

b) M/S counter

c) NAND circuit

Figure 8.4 Master–slave circuits

Thus

$$M = m\bar{x} + m\bar{s} + x\bar{s}$$

and

$$Y_2 = y_2\bar{x} + y_2\bar{y}_1 + x\bar{y}_1$$

In general, the excitation equations (or bistable output equations) can be obtained for any bistable circuit by using the characteristic equation; for an SR bistable: $Q_+ = S + \bar{R}Q$, as derived in Chapter 6. Thus if we substitute $S = \bar{x}y_2$ and $R = \bar{x}\bar{y}_2$ in this equation we obtain the same result for Y_1 as before:

$$Y_1 = \bar{x}y_2 + (\overline{\bar{x}\bar{y}_2})y_1$$

Thus

$$Y_1 = \bar{x}y_2 + xy_1 + x_2y_1$$

We see then that the equations, and hence the circuit functions, are identical; moreover, the master–slave arrangement contains one NAND element less in its implementation.

Consideration of this circuit leads us to a quite different approach for implementing asynchronous logic circuits. Instead of deriving excitation equations as described above, an alternative method is to use dc bistables for the secondary stores and to deduce the set and reset equations for these elements using the techniques developed for synchronous sequential circuits. Starting with a fully assigned flow table, say that for the counter (Table 8.1(c)), we examine the present and next values of Y_1Y_2 (comparing row assignment and table entries) and determine the conditions required to set and reset a dc SR bistable. These conditions are then plotted on a K-map and minimized in the normal way (see Table 8.10), resulting in the equations

$$Y_1 \text{ set } = \bar{x}y_2 \qquad Y_1 \text{ reset } = \bar{x}\bar{y}_2$$
$$Y_2 \text{ set } = \bar{x}y_1 \qquad Y_2 \text{ reset } = xy_1$$

It will be obvious from the equations that the circuit so obtained is identical to that of the master–slave counter. An important point to note with this method is that the hazard conditions are automatically accounted for by the inherent characteristics of the dc SR bistable, e.g. once $Q = 1$ the set inputs have no effect. This can be shown theoretically by the derivation of the excitation equations from the bistable set–reset input conditions. Since this expression always contains the complete set of prime implicant terms, there is no possibility of logic hazards,[5] as can be shown by plotting these terms on a map and noting that all loops overlap. In the case of electronic logic, then, this alternative method of design sometimes leads to a simpler circuit, particularly when NOR/NAND logic is employed. Also, since many integrated circuit logic systems include gated dc bistables (master–slave or dual latch elements) as standard components, this could be a more convenient method of implementation. In addition, it has the considerable advantage of automatically eliminating circuit hazards of the type so far discussed.

Table 8.10 Set–reset maps for divide-by-2 counter

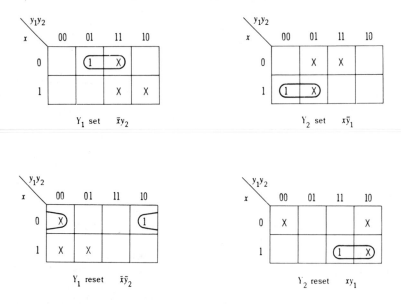

8.7 Design example

The objective of any asynchronous logic design is to evolve optimized hazard-free equations for the secondary excitation and output functions. The process may be divided into several stages:

(a) Construction of a primitive flow table from the oral or written specification—this is generally accomplished with the aid of a state diagram.

(b) If possible the flow table is reduced by eliminating redundant internal states, and the resulting table merged.

(c) A secondary state assignment is made, considering the possibility of circuit 'races'.

(d) Using the derived and primitive flow tables, equations for the secondary excitation and output are obtained; finally these equations are checked for the presence of circuit 'hazards'. Alternatively, the input equations for a dc set–reset bistable may be derived directly from the tables.

Let us consider the following design problem.

A sequential logic circuit is to be controlled by two switches x_1 and x_2. The output Z must go to 1 at the end of the input sequence $00 \rightarrow 01 \rightarrow 11$, which is set up on the switches. This output must be maintained for all input changes until the sequence $11 \rightarrow 10 \rightarrow 00$ occurs, when the output must go to 0.

The circuit has most of the characteristics required of a very simple electronic combination lock. The state diagram is shown in Fig. 8.5(a) together with the primitive flow table in Table 8.11(a).

The design follows very closely the procedures described above; the state diagram is optional—the primitive flow table gives the same information and may be used as the initial starting point. Note that the input sequences which lead to the required outputs are established first, and the table then completed for all other possible inputs. The primitive flow table has no equivalent stable states and therefore cannot be reduced further. Comparing the flow table rows for possible mergers we get

$$m = (1, 2)(1, 5)(2, 4)(3, 4)(3, 6)(5, 7)(5, 9)(6, 8)(6, 10)(7, 9)(8,10)$$
$$= (1, 2)(1, 5)(2, 4)(3, 4)(3, 6)(5, 7, 9)(6, 8, 10)$$

The merger diagram is shown in Fig. 8.5(b). From this, and the set of mergeable pairs above, it is apparent that the best row merge is

$$M = (1, 2)(3, 4)(5, 7, 9)(6, 8, 10)$$

Furthermore, the blocks are output-consistent which will help to reduce the output logic. The merged flow table is given in Table 8.11(b), together with the output states.

The assignment of rows (which requires two secondary variables), so that each transition involves a change of one variable only, is made by examining a transition map (Table 8.11(c)), the fully assigned table being shown in Table 8.11(d). The derivation of the excitation maps is quite standard (Table 8.11(e)) and note that Y_1 cannot contain any hazards since the minimal expression consists of overlapping essential prime implicants:

$$Y_1 = y_1 y_2 + y_1 x_2 + y_1 x_1 + x_1 x_2 \bar{y}_2$$

The minimal expression for Y_2 is given by

$$Y_2 = x_2 \bar{y}_1 y_2 + x_1 \bar{x}_2 \bar{y}_1 + \bar{x}_1 x_2 y_1 + \bar{x}_2 y_1 y_2$$

We now examine the transitions between loops on the excitation map and a convenient way of doing this is to insert the state numbers on the map itself. We find that hazards exist for the transitions ⑦ → ⑤ and ⑥ → ⑧, due to the change in x_2. Thus we must insert two extra terms to ensure hazard-free operation, i.e. $x_1 \bar{y}_1 y_2$ and $\bar{x}_1 y_1 y_2$, which gives

$$Y_2 = x_2 \bar{y}_1 y_2 + x_1 \bar{x}_2 \bar{y}_1 + \bar{x}_1 x_2 y_1 + \bar{x}_2 y_1 y_2 + x_1 \bar{y}_1 y_2 + \bar{x}_1 y_1 y_2$$

These are the only loops that need to be interconnected since there are no ⑨ → ⑥ and ⑤ → ⑩ transitions. The output map is obtained in the normal way and is plotted in Table 8.11(f); due to the merging, the simple expression $Z = y_1$ results. The complete circuit is implemented in NAND elements in Fig. 8.5(c).

Using the alternative method of implementation and starting with the fully assigned flow table in Table 8.11(d), the set and reset conditions for $Y_1 Y_2$ are extracted and plotted directly on K-maps (Table 8.11(g)). This gives the expressions

a) State diagram

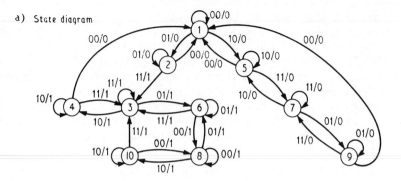

b) Merger diagram

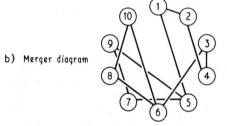

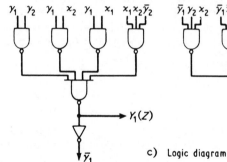

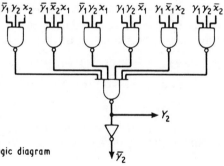

c) Logic diagram

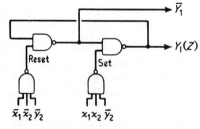

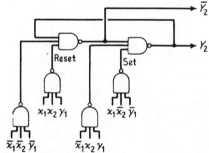

d) Logic diagram using SR bistables

Figure 8.5 Design example

Table 8.11 Design example

Inputs x_1x_2 00	01	11	10	Output Z
①	2	–	5	0
1	②	3	–	0
–	6	③	4	1
1	–	3	④	1
1	–	7	⑤	0
8	⑥	3	–	1
–	9	⑦	5	0
⑧	6	–	10	1
1	⑨	7	–	0
8	–	3	⑩	1

(a) Primitive flow table

	Inputs x_1x_2 00	01	11	10	Output Z
a	①	②	3	5	0
b	1	6	③	④	1
c	1	⑨	⑦	⑤	0
d	⑧	⑥	3	⑩	1

(b) Merged flow table

(c) Row transition map

(d) Fully assigned table

y_1y_2 \ x_1x_2	00	01	11	10
a 00	00	00	10	01
c 01	00	01	01	01
d 11	11	11	10	11
b 10	00	11	10	10

(e) Excitation maps

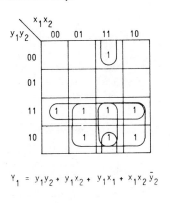

$$Y_1 = y_1y_2 + y_1x_2 + y_1x_1 + x_1x_2\bar{y}_2$$

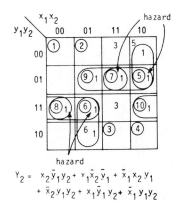

$$Y_2 = x_2\bar{y}_1y_2 + x_1\bar{x}_2\bar{y}_1 + \bar{x}_1x_2y_1 + \bar{x}_2y_1y_2 + x_1\bar{y}_1y_2 + \bar{x}_1y_1y_2$$

Table 8.11 Continued

(f) Output map

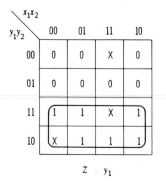

$$z = y_1$$

(g) SR bistable input equations

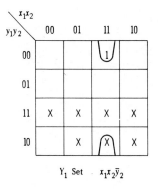

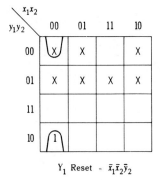

$$Y_1 \text{ Set} = x_1 x_2 \bar{y}_2 \qquad\qquad Y_1 \text{ Reset} = \bar{x}_1 \bar{x}_2 \bar{y}_2$$

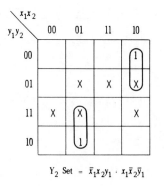

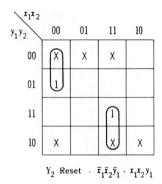

$$Y_2 \text{ Set} = \bar{x}_1 x_2 y_1 \cdot x_1 \bar{x}_2 \bar{y}_1 \qquad\qquad Y_2 \text{ Reset} = \bar{x}_1 \bar{x}_2 \bar{y}_1 \cdot x_1 x_2 y_1$$

$$Y_1 \text{ set} = x_1 x_2 \bar{y}_2 \qquad\qquad Y_1 \text{ reset} = \bar{x}_1 \bar{x}_2 \bar{y}_2$$
$$Y_2 \text{ set} = \bar{x}_1 x_2 y_1 + x_1 \bar{x}_2 \bar{y}_1 \qquad Y_2 \text{ reset} = \bar{x}_1 \bar{x}_2 \bar{y}_1 + x_1 x_2 y_1$$

The implementation of these equations using NAND elements is shown in Fig. 8.5(d); note that only 10 NANDs are required in this circuit as against 14 for the previous one. This reduction of units is assisted by the fact that, with NAND elements forming the bistables, it is possible to have multiple set and reset inputs. Let us analyse this circuit to ensure that it operates correctly and, just as important, that it is hazard-free. From Fig. 8.5(d) we can say that

$$\begin{aligned} Y_1 &= \overline{[\overline{(\bar{x}_1 \bar{x}_2 \bar{y}_2)}(y_1)][\overline{(x_1 x_2 \bar{y}_2)}]} \\ &= [(\bar{x}_1 \bar{x}_2 \bar{y}_2)(y_1)] + x_1 x_2 \bar{y}_2 \\ Y_1 &= x_1 y_1 + x_2 y_1 + y_2 y_1 + x_1 x_2 \bar{y}_2 \end{aligned}$$

which, of course, is the identical excitation equation for Y_1 derived above.

Again, we have

$$\begin{aligned} Y_2 &= \overline{[\overline{(\bar{x}_1 \bar{x}_2 \bar{y}_1)}(x_1 x_2 y_1)}(y_2)][\overline{(\bar{x}_1 x_2 y_1)}][\overline{(x_1 \bar{x}_2 \bar{y}_1)}]} \\ &= (\bar{x}_1 \bar{x}_2 \bar{y}_1)(x_1 x_2 y_1)(y_2) + \bar{x}_1 x_2 y_1 + x_1 \bar{x}_2 \bar{y}_1 \end{aligned}$$

Thus

$$\begin{aligned} Y_2 &= x_1 \bar{x}_2 y_2 + x_1 \bar{y}_1 y_2 + \bar{x}_1 x_2 y_2 + x_2 \bar{y}_1 y_2 + \bar{x}_1 y_1 y_2 \\ &\quad + \bar{x}_2 y_1 y_2 + \bar{x}_1 x_2 y_1 + x_1 \bar{x}_2 \bar{y}_1 \end{aligned}$$

If these terms are plotted on a K-map it becomes obvious that the expression for Y_2 above is the set of all prime implicants and thus no hazards can exist since all loops will interconnect.

8.8 LSI realization using MOS circuits

Implementation of asynchronous circuits using MOS NOR/NAND gates for the excitation equations would not be considered good practice for reasons stated earlier; PLA or ROM structures could, of course, be used. In all cases care must be taken to ensure that there is sufficient gain (and delay) around the feedback loops.

To achieve a more regular structure, though not necessarily an economy in silicon area, MOS set–reset bistables could be used. The schematic for an MOS static memory cell is shown in Fig. 8.6(a) with the corresponding set–reset bistable circuits in Fig. 8.6(b). Note the need for depletion mode transistor loads in the cross-coupled invertors and the use of pass transistor gates for the set and reset inputs.

To set a logic 1 into the memory element (or SR bistable) the $\bar{Q}$ output node must be forced to logic 0 (ground); similarly Q must go to ground to set logic 0. In the SR bistable making either the S or R input go high will effectively put $\bar{Q}$ or Q respectively to ground via the MOS pass transistor.

The master–slave counter circuit designed earlier could be realized directly in MOS using SR bistables as shown in Fig. 8.6(c); note the direct

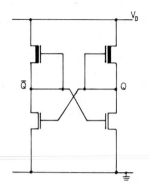

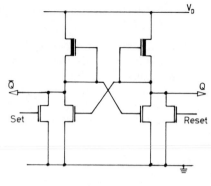

a)   MOS static memory element

b) Set–Reset bistable

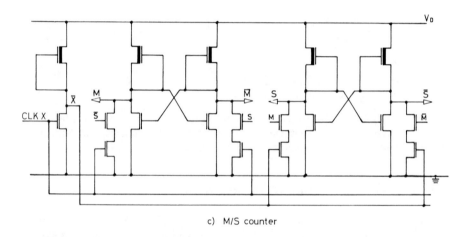

c) M/S counter

Figure 8.6 MOS bistable circuits

correspondence to the circuits shown in Fig. 8.4 and the use of an MOS invertor to generate $\bar{x}$. Note also that it is possible to combine the gated inputs and use them in place of the normal set/reset inputs (analogous to the multiple set and reset inputs in the example in Section 8.7). Thus it is perfectly feasible to implement any asynchronous logic design directly in terms of its equivalent MOS SR bistable circuit.

An alternative method of implementation based on the direct realization of the excitation equations utilizes circuitry of the form shown in Fig. 8.7(a). This is the standard NOR gate configuration but with an AND/OR logic network in place of the single input transistors.

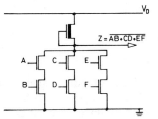

a) Use of AND/OR network in NOR circuit

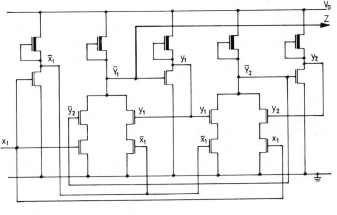

b) MOS network

Figure 8.7 MOS realization of excitation equations

The method is best illustrated by way of an example; consider an asynchronous circuit which has the following excitation and output equations:

$$Y_1 = \bar{x}_1 y_1 + x_1 \bar{y}_2$$
$$Y_1 = x_1 y_2 + \bar{x}_1 y_1$$
$$Z = \bar{y}_1$$

The equations for Y_1 and Y_2 are realized using circuits of the form given in Fig. 8.7(a), the outputs of which give $\bar{Y}_1$ and $\bar{Y}_2$. Thus the y_1 and y_2 variables must be generated by separate MOS invertor circuits and feedback to the AND network. In this way we have obtained the necessary gain and feedback required by an asynchronous system (in essence we have a memory circuit formed by the cross-coupling of the two invertor stages). The final circuit is shown in Fig. 8.7(b); it is interesting to observe the analogy between this circuit and the normal relay implementation where a secondary relay contact is introduced into the excitation circuitry to give the feedback, i.e. memory characteristic.

This form of realization gives a fairly compact circuit but is limited by the usual restriction on the number of transfer gates that can be connected in series (see Section 5.1).

8.9 Further aspects of the state assignment problem

So far in the examples of secondary state assignment that we have chosen, it has always been possible to perform the allocation of row variables in such a way that transitions involve a change of one variable only and critical races are avoided. What can we do, though, if this is not possible? Consider the flow table shown in Table 8.12(a). This is a three-row table and hence requires a minimum of two secondary variables for row coding. If we examine all the possible state transitions, using the transition map (Table 8.12(b)), we find that there are transitions between all three pairs of rows, and thus it is impossible to code the rows so that only one variable changes during a transition. However, secondary assignment can always be obtained for a three-row table using two variables, by making use of the fourth or 'spare' secondary state. Suppose we assign the rows as shown in Table 8.12(c), where we are using the fourth combination $y_1 y_2 = 10$ as a buffer state between transition rows. If the circuit is in stable state ② under the input conditions $x_1 x_2 = 01$, and a change to $x_1 x_2 = 11$ occurs, the machine must go to stable state ③. This is achieved by directing the intermediate secondary changes, using the buffer state, to be $y_1 y_2 = 00$ $\rightarrow 10 \rightarrow 11 \rightarrow$ ⑪. In other words, we have deliberately created a machine cycle by a suitable choice of the $y_1 y_2$ values for the unstable states. Similarly, if in stable state ⑥ under the input $x_1 x_2 = 01$ and x_2 changes to 0, the circuit goes to stable state ① via the directed cycle $y_1 y_2 =$ ⑪ $\rightarrow 10$ $\rightarrow 00 \rightarrow$ ⑩. In some cases it is not necessary to use a buffer state to produce a cycle, for example in Table 8.12(d) the transition ② $\rightarrow$ ③ could have been achieved by coding $y_1 y_2 =$ ⑩ $\rightarrow 01 \rightarrow 11 \rightarrow$ ⑪.

A secondary assignment without critical races is not always possible for a four-row table using two secondary variables, and consequently three secondaries must be used. This will involve selecting four out of the eight possible combinations to represent the rows, and there are $_8C_4 = 8!/4!4! =$ 70 ways of achieving this choice! Fortunately the selection of the four basic combinations is assisted by breaking down the possible combinations into six pattern groups (a full analysis of this problem has been made by Marcus).[6] After this comes the problem of allocating the spare buffer combinations in an appropriate manner to affect the transitions.

The final choice of secondary assignment may be based on circuit economy or on speed of transition. Any assignment which involves directed cycles will entail multiple transmissions through the circuit and thus effectively increases the propagation delay, that is, the time taken to reach a stable state will depend on the number of row transitions. Unfortunately there are as yet no systematic techniques available which give an optimum solution to this assignment problem, and the determination of a minimal circuit must depend on experience and 'trial and error' methods, entailing the comparison of Y-maps.

Table 8.12 State assignment
for three-row table

| | Inputs x_1x_2 | | | |
	00	01	11	10
a	①	②	3	4
b	⑤	2	3	④
c	1	⑥	③	4

(a) Merged flow table

(b) Transition maps

| | | Inputs x_1x_2 | | | |
	$y_1\,y_2$	00	01	11	10
a	0 0	①	②	3	4
b	0 1	⑤	2	3	④
c	1 1	1	⑥	③	4
d'	1 0	1	–	3	–

(c) Assigned flow table

| | Inputs x_1x_2 | | | |
$y_1\,y_2$	00	01	11	10
0 0	00	00	10	01
0 1	01	00	11	01
1 1	10	11	11	01
1 0	00	–	11	–

or

| | Inputs x_1x_2 | | | |
$y_1\,y_2$	00	01	11	10
0 0	00	00	01	01
0 1	01	00	11	01
1 1	10	11	11	01
1 0	00	–	–	–

(d) Y-maps

As an illustration, consider the four-row flow table shown in Table 8.13(a). This is again a 'worst case' condition since there are seven possible transitions between the rows (see Table 8.13(b)i). Table 8.13(b)ii–vii shows six possible assignments using the spare secondary states, one from each of the Marcus pattern groups; many more variations are possible since in each pattern group the basic row combinations may be permutated in $_4P_4 = 4! = 24$ different ways. Table 8.14(a) shows an assigned flow table using the first pattern group (Table 8.13(b)ii). As the unused buffer states are entered as don't-cares, it follows that an economical solution will be given by the assignment that produces the most don't-cares in the table. This can be achieved by ensuring that rows containing a large number of transitions have the smallest cycle or, ideally, are adjacent. In this problem, the majority of transitions occur between rows (a, d) and (c, d), and thus, simply by changing the coding of the basic rows according to these ideas (Table 8.13(b)viii), we can arrive at a much better result. Table 8.14(b) shows the assigned table, and the corresponding Y-map is shown in Table 8.15.

Before using spare secondary states, however, it is as well to check whether critical races can be avoided by the insertion of indigenous cycles in the basic flow table. In this case, due to the existence of a large number of don't-cares, it is perfectly feasible to use indigenous cycles. As can be seen from Table 8.16, by a suitable choice of row coding and directed cycles we can in fact allocate the four-row tables using the minimum of secondary variables, i.e. two. An alternative coding method of avoiding critical races is *multiple secondary state assignment*. This means that a row in the flow table is assigned two secondaries, i.e. combinations; this method is, in fact, another means of utilizing the spare secondary states. Using this approach equivalent stable states are obtained in different rows, so we have introduced 'redundant' rows in the flow table. We must have some means of distinguishing between these states and the usual device is to include subscripts on the stable state figures. For example, the flow table in Table 8.12(a) could be allocated as shown in Table 8.17, where row a has been given two secondaries. The table could equally well have been coded with row c having two assignments; the reason for choosing rows a or c is that a transition is required between these two rows which cannot be obtained (see Table 8.12(b)) using a three-row assignment without changing more than one variable at a time.

In Table 8.17 a transition from stable state ⑥ under the input change $x_1x_2 = 01 \rightarrow 00$ is directed to stable state ①₂ which is in row a_1, involving a change of one variable only. But the transition of stable state ⑤ for the input change $x_1x_2 = 00 \rightarrow 01$ is directed to stable state ②₁ which is in row a_1, thus again ensuring a change of one variable only. Transitions from stable states in both rows a_1 and a_2 must be suitably cycled to the correct stable states, e.g. ②₁ is cycled to ③ via the directed secondary changes $y_1y_2 =$ ⑩⓪ $\rightarrow 10 \rightarrow 11 \rightarrow$ ⑪ .

Though we have only considered three- and four-row tables in the examples above, the same arguments must obviously apply to flow tables containing any number of rows. It is not always possible, for example, to

Table 8.13 State assignment for four-row tables

| | Inputs $x_1x_2x_3$ | | | | | | | |
	000	001	010	011	100	101	110	111
a	(A)	E	G	–	(B)	F	C	–
b	–	–	D	–	K	–	(C)	J
c	A	–	(D)	H	–	–	I	–
d	A	(E)	(G)	(H)	(K)	(F)	(I)	(J)

(a) Merged flow table

(b) Transition maps

(i)
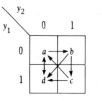

(ii)

y_3 \ y_1y_2	00	01	11	10
0	a	b	c	e'
1	f'	g'	d	

(iii)
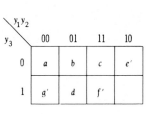

y_3 \ y_1y_2	00	01	11	10
0	a	b	c	e'
1	g'	d	f'	

(iv)

y_3 \ y_1y_2	00	01	11	10
0	a	b	c	e'
1		f'	g'	d

(v)
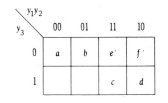

y_3 \ y_1y_2	00	01	11	10
0	a	b	e'	f'
1			c	d

(vi)
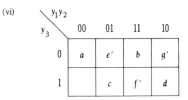

y_3 \ y_1y_2	00	01	11	10
0	a	e'	b	g'
1		c	f'	d

(vii)

y_3 \ y_1y_2	00	01	11	01
0	a	b	h'	g'
1	d	c	e'	f'

(viii)
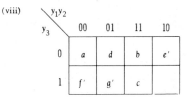

y_3 \ y_1y_2	00	01	11	10
0	a	d	b	e'
1	f'	g'	c	

Table 8.14 Assigned flow tables

	$y_1\,y_2\,y_3$	Inputs $x_1x_2x_3$ 000	001	010	011	100	101	110	111
a	0 0 0	Ⓐ	f'	f'	–	Ⓑ	f'	C	–
b	0 1 0	–	–	D	–	g'	–	Ⓒ	g'
c	1 1 0	e'	–	Ⓓ	H	–	–	I	–
d	1 1 1	g'	Ⓔ	Ⓖ	Ⓗ	Ⓚ	Ⓕ	Ⓘ	Ⓙ
e'	1 0 0	A	–	–	–	–	–	–	–
f'	0 0 1	A	g'	g'	–	–	g'	–	–
g'	0 1 1	f'	E	G	–	K	F	–	J

(a)

	$y_1\,y_2\,y_3$	Input $x_1x_2x_3$ 000	001	010	011	100	101	110	111
a	0 0 0	Ⓐ	E	G	–	Ⓑ	F	e'	–
b	1 1 0	–	–	D	–	K	–	Ⓒ	J
c	1 1 1	g'	–	Ⓓ	g'	–	–	g'	–
d	0 1 0	A	Ⓔ	Ⓖ	Ⓗ	Ⓚ	Ⓕ	Ⓘ	Ⓙ
e'	1 0 0	–	–	–	–	–	–	C	–
f'	0 0 1	A	–	–	–	–	–	–	–
g'	0 1 1	f'	–	–	H	–	–	I	–

(b)

Table 8.15 Y-map

	$y_1\,y_2\,y_3$	Inputs $x_1x_2x_3$ 000	001	010	011	100	101	110	111
a	0 0 0	000	010	010	–	000	010	100	–
b	1 1 0	–	–	111	–	010	–	110	010
c	1 1 1	011	–	111	011	–	–	011	–
d	0 1 0	000	010	010	010	010	010	010	010
e'	1 0 0	–	–	–	–	–	–	110	–
f'	0 0 1	000	–	–	–	–	–	–	–
g'	0 1 1	001	–	–	010	–	–	010	–

Table 8.16 Indigenous cycles

$y_1 y_2$		Inputs $x_1x_2x_3$ 000	001	010	011	100	101	110	111
a	0 0	Ⓐ	E	G	–	Ⓑ	F	C	–
b	0 1	A	–	D	–	K	–	Ⓒ	c
c	1 1	b	–	Ⓓ	H	K	–	I	J
d	1 0	A	Ⓔ	Ⓖ	Ⓗ	Ⓚ	Ⓕ	Ⓘ	Ⓙ

(a) Assigned flow table

$y_1 y_2$		Inputs $x_1x_2x_3$ 000	001	010	011	100	101	110	111
a	0 0	00	10	10	–	00	10	01	–
b	0 1	00	–	11	–	11	–	01	11
c	1 1	01	–	11	10	10	–	10	10
d	1 0	00	10	10	10	10	10	10	10

(b) Y-map

Table 8.17 Multiple secondary state assignment

$y_1 y_2$		Inputs x_1x_2 00	01	11	10
a_1	0 0	①	②	3	4
a_2	1 0	$①_2$	$②_2$	3	4
b	0 1	⑤	2_1	3	④
c	1 1	1_2	⑥	③	4

(a) Assigned flow table

$y_1 y_2$		Inputs x_1x_2 00	01	11	10
a_1	0 0	00	00	10	01
a_2	1 0	10	10	11	00
b	0 1	01	00	11	01
c	1 1	10	11	11	01

(b) Y-map

assign an eight-row table with three variables, and we must resort to four variables. Here the number of possible selections increases considerably ($_{16}C_8 = 16!/8!8! = 12870$) and an exhaustive 'trial and error' procedure is out of the question.

8.9.1 Single transition time assignments

So far we have only considered the case where one state variable is allowed to change at a time for any inter-row transition. It has been shown by Liu[7] that it is possible to have an assignment where multiple changes of state variables can occur without incurring any critical races. Moreover, the transition times with this type of assignment can be the same as those obtained using adjacent coding since the non-critical races which ensue set up a free-running condition.

A *direct* transition between internal states S_i to S_j written (S_i, S_j) may be defined as a transition whereby all internal state variables that are to undergo a change of state are excited simultaneously. A direct transition (S_i, S_j) will race critically with another direct transition (S_m, S_n) if the possibility exists that unequal delays in the transmission paths may cause these two transitions *to share a common internal state* (either stable or unstable). It will be obvious that a direct transition of this type gives a minimum transition time between states. Thus an assignment based on direct transitions called *single transition time* (STT) will approximate in response time to that obtained with a normal adjacently coded flow table.

The STT assignment, of course, must be made in such a way as to ensure that critical races do not occur. This is illustrated in Table 8.18(b) which is an STT assignment of the flow table in Table 8.18(a). A transition from row b to row d simultaneously excites the variables y_2 and y_3 and could involve the state changes

$$001 \rightarrow 011 \rightarrow 010$$
$$001 \rightarrow 010 \rightarrow 010$$
$$001 \rightarrow 000 \rightarrow 010$$

Table 8.18 STT assignment

	Inputs x_1x_2 00	01	11	10
a	①	①	4	2
b	3	②	4	②
c	③	4	5	③
d	1	④	④	3
e	3	2	⑤	⑤

(a) Flow table

	State variables $y_1\ y_2\ y_3$	Inputs x_1x_2 00	01	11	10
a	0 0 0	000	000	010	001
b	0 0 1	111	001	010	001
c	1 1 1	111	010	101	111
d	0 1 0	000	010	010	111
e	1 0 1	111	001	101	101

(b) Assigned flow table

depending on the actual propagation delays in the circuit. However, since there are no stable or unstable states under the input x_1x_2 which have codings corresponding to the transitory state values (except the required one 010) the assignment is free of critical races.

Liu has also shown that a state-variable assignment in which the row assignments correspond to an equidistant error-correcting code will be free of critical races. Note that an equidistant error-correcting code of 2^m message words requires $2^m - 1$ bits, and hence the number of state variables required for a flow table with 2^m rows will be $2^m - 1$! Note also that a Hamming distance of three is the minimum requirement for an error-correcting code.

Based on an extension of Liu's work, Tracey[8] has described an algorithmic method for generating STT assignments which is programmable and does not generate an inordinately large number of state variables, although, of course, the method will always require the use of more state variables than conventional techniques.

Tracey's method is based on determining a set of partitions on the internal states such that stable and unstable states are in separate blocks. For example, in Table 8.18(b) variable y_1 induces the two-block partition

$$r_1 = (a, b, d)(c, e)$$

Similarly, y_2 induces

$$r_2 = (a, b, e)(c, d)$$

and y_3 induces

$$r_3 = (a, d)(b, c, e)$$

The assignment is said to consist of the set of r-partitions r_1, r_2 and r_3 induced by the variables y_1, y_2 and y_3. A row assignment allocating one y-state per row can be used for STT realization of normal flow tables, without incurring critical races, if, and only if, for every transition (S_i, S_j) the following conditions are upheld:

(a) If (S_m, S_n) is another transition in the same input column, then at least one y variable partitions the pairs (S_i, S_j) and (S_m, S_n) into separate blocks.
(b) If S_k is a stable state in the same column, then at least one y variable partitions the pair (S_i, S_j) and the state S_k into separate blocks.
(c) For $i \neq j$, S_i and S_j are in separate blocks of at least one y variable.

It will be seen by examination of Table 8.18(a) and (b) that the above conditions are fulfilled. For example, consider the transitions from row e to b and c to d under input $\bar{x}_1x_2$, that is, $2 \rightarrow ②$ and $4 \rightarrow ④$. Now $(2, ②)$, and $(4, ④)$ are partitioned into separate blocks by y_2; moreover, $(2, ②)$ are in separate blocks of r_1. The pair $(2, ②)$ and the stable state 1 are partitioned into separate blocks by y_3.

The separation of like stable and unstable states into separate blocks ensures that transitions between rows involve only row assignments that contain a non-changing part that is different from the corresponding part (same digit positions) in any other row assignments in the flow table. Moreover, every transition within an input column has a different non-changing part in the row assignments involved. It therefore follows that the set of intermediate unstable states in each column transition must be disjoint. For example, consider the transitions $\textcircled{2} \rightarrow \textcircled{3}$ and $\textcircled{4} \rightarrow \textcircled{1}$ in Table 8.18; we then have

		y_1	y_2	y_3
S_i	$(\textcircled{2})$	0	0	1
S_j	$(\textcircled{3})$	1	1	1
S_m	$(\textcircled{4})$	0	1	0
S_n	$(\textcircled{1})$	0	0	0

Now unequal transmission delays could cause the direct transition (S_i, S_j) momentarily to assume any of the internal states $XX1$ where X can represent either a 0 or 1. Similarly the transition (S_m, S_n) could assume the values $0X0$. The existence of the 0 and 1 values for y_3 ensures that a critical race cannot occur.

Let us now consider how Tracey's STT assignment method will generate assignments such as those shown in Table 8.18(b). The initial step is to generate a *partition list* for the flow table according to the following rule:

If (S_i, S_j) and (S_m, S_n) are transitions in the same input column of a flow table and if S_k is a stable state also in that column then the collection of all two-block partitions $(S_i, S_j)(S_m, S_n)$ and $(S_i, S_j)(S_k)$ comprise the partition list.

The partition list for the flow table in Table 8.18(a) is

$$\pi_1 = (a, d)(b, c) \qquad \pi_6 = (a, d)(c, e)$$
$$\pi_2 = (a, d)(c, e) \qquad \pi_7 = (b, d)(c, e)$$
$$\pi_3 = (a)(b, e) \qquad \pi_8 = (a, b)(c, d)$$
$$\pi_4 = (a)(c, d) \qquad \pi_9 = (a, b)(e)$$
$$\pi_5 = (b, e)(c, d) \qquad \pi_{10} = (c, d)(e)$$

In order to obtain an STT assignment free of critical races it is necessary that each partition on the partition list is $\leq r_i$, where r_i is a member of the set of two-block partitions induced by the internal state variables y_i to give the assignment. Since optimal assignments are those which contain the least number of y variables, it is necessary to derive the minimum number of r-partitions for the flow table.

The r-partitions are generated from the partition list by expressing the list as a Boolean matrix with the π-partitions as rows and the partition elements as columns, and then performing a Boolean reduction process; note that the order is arbitrary. For example, we can express the partition $\pi_1 = (a, d)(b, c)$ as $0110X$ where the elements of block $(a, d) = 0$ and $(b, c) = 1$ and the missing elements are represented by X. The full Boolean matrix is shown in Table 8.19(a). There are various ways of solving this matrix. The simplest technique for small flow tables is to compute the maximal intersectables (analogous to Paull and Ungar's method for maxi-

Table 8.19 Tracey reduction algorithm

Partitions	Elements				
	a	b	c	d	e
π_1	0	1	1	0	X
π_2	0	X	1	0	1
π_3	0	1	X	X	1
π_4	0	X	1	1	X
π_5	X	0	1	1	0
π_6	0	X	1	0	1
π_7	X	0	1	0	1
π_8	0	0	1	1	X
π_9	0	0	X	X	1
π_{10}	X	X	0	0	1

(a) Boolean matrix

	1	2	3	4	5	6	7	8	9	10
A^*	X	X	X			X				
B^*		X				X	X		X	
C^*				X	X			X		X
D			X	X						
E									X	X

(b) Chart method to select covering

mal incompatibles). The method, called by Tracey matrix reduction algorithm 1, starts by compiling a list of pair-wise intersectables; for example

π_1 0 1 1 0 X
π_2 0 X 1 0 1

would intersect since all the declared elements are the same and the X terms can take any value. Note that it is also possible to form an inter-sectable with the inverse of a partition, for example

π_4 0 X 1 1 X
π_{10} X X 0 0 1

since the original designation of terms was arbitrary. The complete list of pair-wise compatibles is

1, 2; 1, 3; 1, 6
2, 3; 2, 6; 2, 7; 2, 9
3, 4; 3, 5; 3, 6; 3, 10
4, 5; 4, 8; 4, 9) 4, 10
5, 8; 5, 10
6, 7; 6, 9
7, 9
8, 9; 8, 10
9, 10

We can now compute the maximal intersectables by combining terms; note that the transitive relationship does not hold due to the X terms; this gives:

$$A \ (1, 2, 3, 6) \qquad D \ (3, 4)$$
$$B \ (2, 7, 9, \overline{6}) \qquad E \ (9, 10)$$
$$C \ (4, 5, 8, \overline{10})$$

The final step is to select a covering of MIs over the partition list matrix. The problem is analogous to the Quine–McCluskey selection of prime implicants and may be solved in the same way using a chart technique as shown in Table 8.19(b). In this case A, B and C are all essential and cover the matrix, that is

$$A \ (1, 2, 3, 6) \ = 0\ 1\ 1\ 0\ 1 = r_3 = (a, d)(b, c, e)$$
$$B \ (2, 6, 7, \overline{9}) \ = 0\ 0\ 1\ 0\ 1 = r_1 = (a, b, d)(c, e)$$
$$C \ (4, 5, 8, \overline{10}) = 0\ 0\ 1\ 1\ 0 = r_2 = (a, b, e)(c, d)$$

Note that, referring back to the partition list,

$$\pi_1, \pi_2, \pi_3 \quad \text{and} \quad \pi_6 \text{ are } \leq r_3$$
$$\pi_2, \pi_6, \pi_7 \quad \text{and} \quad \pi_9 \text{ are } \leq r_1$$
$$\pi_4, \pi_5, \pi_8 \quad \text{and} \quad \pi_{10} \text{ are } \leq r_2$$

which satisfies the conditions mentioned earlier. The table may also be evaluated algebraically using the method described in Section 3.9. The main disadvantage of this algorithm (called *assignment method 1*) is that for large variable problems the amount of computation becomes excessive. Smith[9,10] has described other methods of matrix reduction based on Tracey's work which are more suitable for machine computation.

Two other alternative methods of STT assignment were also proposed by Tracey. *Assignment method 2* is based on the concept of *K-sets*. A K-set exists in a single column of a flow table and consists of all $K - 1$ unstable entries leading to the same stable state, together with the stable state. A direct transition in K-set K_r does not race critically with a direct transition in K-set K_s if an assignment has been made such that at least one y variable partitions the elements of K_r and the elements of K_s into separate blocks. The first step in the assignment procedure is to construct a partition list from the K-sets of a flow table instead of from the transition pairs. From Table 8.18(a) we have

$$\pi_1 = (a, d)(b, c, e) \qquad \pi_5 = (a, b, d)(c, e)$$
$$\pi_2 = (a)(b, e) \qquad\qquad \pi_6 = (a, b)(c, d)$$
$$\pi_3 = (a)(c, d) \qquad\qquad \pi_7 = (e)(a, b)$$
$$\pi_4 = (b, e)(c, d) \qquad\quad \pi_8 = (e)(c, d)$$

Clearly

$$\pi_8 \leq \pi_4; \quad \pi_4, \pi_2 \leq \pi_1; \quad \pi_3 \leq \pi_6; \quad \pi_7 \leq \pi_5$$

Therefore partitions π_2, π_3, π_4, π_7 and π_8 may be removed from the list. If the remaining four π-partitions are converted to Boolean matrix form it is clear that no further reduction is possible and the resulting assignment is

$$y_1 = \pi_1 = (a, d)(b, c, e) = 0\ 1\ 1\ 0\ 1$$
$$y_2 = \pi_6 = (a, b)(c, d)\quad = 0\ 0\ 1\ 1\ X$$
$$y_3 = \pi_5 = (a, b, d)(c, e) = 0\ 0\ 1\ 0\ 1$$
$$y_4 = \pi_4 = (b, c)(c, d)\quad = X\ 0\ 1\ 1\ 0$$

Note that the assignment is incompletely specified, but is nevertheless free of critical races regardless of how the unspecified entries are allocated. In general the method generates less computation but requires more state variables than method 1; it can also be shown that method 2 gives an upper bound on the number of variables for method 1.

Assignment method 3 is very similar to the original Liu method. In this case the state assignment is derived directly from the *column partition* of a flow table and results in even greater savings in computation. A column partition is a partition constructed from a single column of a flow table with each K-set of the column appearing as a separate block. It can be either completely or incompletely specified. A state assignment constructed from the set of column partitions of a flow table contains no critical races, even if all transitions are direct. Before the column partitions are coded to give the assignment they should if possible be reduced in the usual way. From Table 8.18(a) we have

Column 00 $\pi_1 = (a, d)(c, b, e)$
Column 01 $\pi_2 = (a)(b, e)(c, d)$
Column 11 $\pi_3 = (a, b, d)(c, e)$
Column 10 $\pi_4 = (a, b)(c, d)(e)$

By inspection there are no partitions that can be discarded and π_1, π_2, π_3 and π_4 must be used for the assignment. Note that six y variables are required for the assignment, one to distinguish between the blocks of π_1, two for π_2, one for π_3 and two for π_4.

Method 3 can be considered to yield an upper bound on the number of variables required for methods 1 and 2. Note that all the assignment methods work equally well for completely or incompletely specified machines.

8.9.2   One-hot code assignments

A special class of race-free codes which have the advantage of allowing circuit equations to be mapped directly from a state diagram or ASM chart is the *one-hot* code. Moreover, since circuits assigned in this way can be realized using set–reset bistables it produces a regularly structured circuit which is applicable to VLSI design. The one-hot code, which simply allocates one state variable per row of the flow table, was first discussed by Hoffman in terms of 'one-relay-per-row' and later by Ungar[11] and Hollaar.[12]

To assign a flow table using one-hot coding each row is allocated a state variable which goes to 1, with all other variables going to 0, when in that row; this is illustrated in Table 8.20(a) using the flow table for the divide-by-2 counter discussed earlier. The circuit is designed such that each inter-row transition takes place in two steps, with the present row variable and the next row variable both being at 1 during the unstable state. Thus, from Table 8.20(a) for the transition $\text{(1)} \rightarrow 2 \rightarrow \text{(2)}$ we would have $1000 \rightarrow 1100 \rightarrow 0100$. The excitation equations for flow tables coded in this way take the general form

$$Y_i = T_i + \bar{H}_i y_i$$

where T_i is called the *transition term* and $\bar{H}_i y_i$ the *hold term* which keeps state variable y_i true until another state is entered. T_i is determined by examining each row of the flow table and summing all those y variables which initiate a transition from an unstable state to a stable state in row i under the same input conditions. The H_i term is derived in a similar way by considering all the next states reached from row i; the corresponding y values are then summed.

For example, in Table 8.20(a), considering row a there is a transition from row d initiated by y_4 under input $\bar{x}$ thus $T_i = y_4\bar{x}$. Similarly there is a transition from row a to row b giving $H_i = y_2$; the excitation equation for Y_1 is thus $Y_1 = y_4\bar{x} + \bar{y}_2 y_1$. The full set of excitation equations for the divide-by-2 counter is

$$Y_1 = y_4\bar{x} + \bar{y}_2 y_1$$
$$Y_2 = y_1 x + \bar{y}_3 y_2$$
$$Y_3 = y_2\bar{x} + \bar{y}_4 y_3$$
$$Y_4 = y_3 x + \bar{y}_1 y_4$$

Note the similarity of the excitation equation to the characteristic equation of the set–reset bistable, $Q_+ = S + \bar{R}Q$, although of course they have not

Table 8.20 One-hot assignments

	State variables $y_1\ y_2\ y_3\ y_4$	Input x 0	1	Output Z
a	1 0 0 0	①	2	0
b	0 1 0 0	3	②	0
c	0 0 1 0	③	4	1
d	0 0 0 1	1	④	1

(a) One-hot assignment of ÷2 counter

State variables $y_1\ y_2\ y_3\ y_4$	Inputs $x_1 x_2$ 00	01	11	10	Output Z
1 0 0 0	①	②	3	5	0
0 1 0 0	1	6	③	④	1
0 0 1 0	1	⑨	⑦	⑤	0
0 0 0 1	⑧	6	3	⑩	1

(b) One-hot assignment, example 8.7

been derived in the same way. In this case T_i would be the set terms and H_i (because of the y_i term) the reset term. Thus we can rewrite the equations as

$$
\begin{array}{ll}
\text{set } y_1 = y_4\bar{x} & \text{reset } y_1 = y_2 \\
\text{set } y_2 = y_1 x & \text{reset } y_2 = y_3 \\
\text{set } y_3 = y_2\bar{x} & \text{reset } y_3 = y_4 \\
\text{set } y_4 = y_3 x & \text{reset } y_4 = y_1
\end{array}
$$

The circuit is shown implemented using NAND gates in Fig. 8.8; note that the reset terms must be inverted to satisfy the NAND bistable conditions which means that they can be directly obtained from the equations. Note that the next-state variable will always be set to 1 before the current state is reset and in some circuits this could lead to a hazardous operation. Thus, regardless of the fact that we realize the equations in terms of SR bistables, it is still essential that the excitation equations be inspected for possible static hazards and corrected in the usual way.

To illustrate these ideas let us consider implementing the example discussed earlier in Section 8.7 using one-hot coding; the flow table is shown in Table 8.20(b). The excitation equations derived from the flow table are

$$
\begin{aligned}
Y_1 &= y_2\bar{x}_1\bar{x}_2 + y_3\bar{x}_1\bar{x}_2 + y_1\bar{y}_2\bar{y}_3 \\
Y_2 &= y_1 x_1 x_2 + y_4 x_1 x_2 + y_2\bar{y}_4\bar{y}_1 \\
Y_3 &= y_1 x_1\bar{x}_2 + y_3\bar{y}_1 \\
Y_4 &= y_2\bar{x}_1 x_2 + y_4\bar{y}_2
\end{aligned}
$$

(Note that the actual reset terms correspond to the inverse of the terms in the y_i product (excluding y_i); thus for Y_1, $R = \overline{\bar{y}_2\bar{y}_3} = y_2{}' + y_3$.)

If we now examine these equations for static hazards we find, for example for Y_1, that a hazard could occur between $y_2\bar{x}_1\bar{x}_2$ and $\bar{y}_2 y_1\bar{y}_3$ for the condition $\bar{x}_1 = \bar{x}_2 = y_1 = \bar{y}_3 = 1$ and $y_3\bar{x}_1\bar{x}_2$ and $\bar{y}_3 y_1\bar{y}_2$ for $\bar{x}_1 = \bar{x}_2 = y_1 = \bar{y}_2 = 1$. Thus we must add an extra correcting term $y_1\bar{x}_1\bar{x}_2$ to the equation for Y_1 (this can easily be verified by plotting Y_1 on a K-map). Continuing in this way we arrive at the full set of corrected equations:

$$
\begin{aligned}
Y_1 &= y_2\bar{x}_1\bar{x}_2 + y_3\bar{x}_1\bar{x}_2 + y_1(\bar{y}_2\bar{y}_3 + \bar{x}_1\bar{x}_2) \\
Y_2 &= y_1 x_1 x_2 + y_4 x_1 x_2 + y_2(\bar{y}_4\bar{y}_1 + x_1 x_2) \\
Y_3 &= y_1 x_1\bar{x}_2 + y_3(\bar{y}_1 + x_1\bar{x}_2) \\
Y_4 &= y_2\bar{x}_1 x_2 + y_4(\bar{y}_2 + \bar{x}_1 x_2)
\end{aligned}
$$

These are shown implemented using NAND logic in Fig. 8.9 (note that the reset terms can be transferred directly from the excitation equations). It has been shown[12] that the need for hazard correction arises primarily when the state diagram has a cyclic feedback loop due to a state having another state as *both* its predecessor and successor; this structure occurs less frequently in sequencers and counters than it does in, for instance, recognition circuits. The example we have chosen is a complicated one in this respect as can easily be ascertained from the state diagram in Fig. 8.5(a).

The need for hazard correction can be determined by inspection of the state table (looking for cyclic loops) and then corrected by simply ORing the

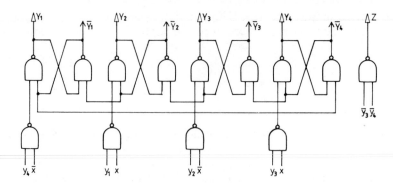

Figure 8.8 One-hot realization of divide-by-2 counter

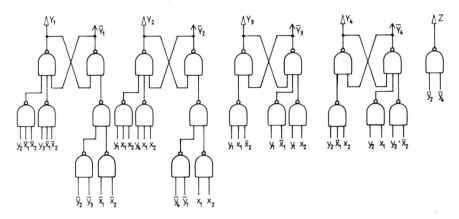

Figure 8.9 One-hot realization of example 8.7

hazardous y_i reset input with the primary input causing the transition. That this is so becomes obvious when we compare the corrected and uncorrected excitation equation; for example

$$Y_1 = y_2\bar{x}_1\bar{x}_2 + y_3\bar{x}_1\bar{x}_2 + y_1\bar{y}_2\bar{y}_3$$
$$= y_2\bar{x}_1\bar{x}_2 + y_3\bar{x}_1\bar{x}_2 + y_1(\bar{y}_2 + \bar{x}_1\bar{x}_2)(\bar{y}_3 + \bar{x}_1\bar{x}_2)$$
$$= y_2\bar{x}_1\bar{x}_2 + y_3\bar{x}_1\bar{x}_2 + y_1(\bar{y}_2\bar{y}_3 + \bar{x}_1\bar{x}_2)$$

Though it is apparent that a circuit realization using one-hot coding will require extra logic the overall cost is not necessarily excessive compared to conventional asynchronous designs. In a practical realization the excitation logic would be best generated using a PLA with separate SR bistables. It is essential with this type of circuit to ensure that the initial starting conditions are set into the machine; this may be done using additional inputs to the set and reset gates of the NAND bistable elements.

One useful feature of the method is that since only one output is active in the stable state the circuit minimizes power drain and hence heat dissipation—a major consideration in VLSI design.

The method is ideal when designing sequencers and counter circuits, when it is possible to work directly from a state diagram or ASM chart, but has a general application in many other areas.

8.10 Circuit hazards

Circuit hazards are a particular problem in asynchronous logic design due to the presence of unclocked feedback loops and the immediate response of the circuit to changes in dc levels. As we have seen, circuit hazards arise predominantly in the combinational logic, that is in the realization of the excitation equations, due to differing delays in the signal paths or the propagation time of the elements (see Section 4.5). Note that, irrespective of the assignment method used, the excitation equations could still contain static hazards. Hazards can be classified into two main groups:

(a) *Single-variable hazards* due to changes in one variable only. These can be corrected logically or by the insertion of appropriate delays in the circuit. There are three main types: *static*, *dynamic* and *essential*.
(b) *Multi-variable hazards* occur because of changes in more than one variable, and can also produce static, dynamic and essential hazards. In general they cannot be completely eliminated by logical means or circuit modification; hence the restriction to single-variable changes in asynchronous circuits.

The type of hazard we met in the design example above was the single-variable static hazard, and occurred in the combinational logic governing the feedback signals, i.e. the excitation equations. We have seen how this type of hazard can be recognized from the K-map (or by algebraic manipulation), and logically corrected by the insertion of additional gates using the technique due to Huffman.[5] However, care must be taken to ensure that the hazard correction terms do in fact cover the variable changes. Consider the Y-map shown in Table 8.21; it could appear that the additional loop x_2y_2 is sufficient to cover the hazards since all loops interconnect, and thus the excitation equations would be

$$Y_1 = \bar{x}_1 x_2 + y_2 x_2 + x_1 y_1 + x_1 y_2$$

But if a transition is required for the conditions $x_1 = x_2 = y_1 = 1$ and $y_2 = 0$, when $x_1 \to 0$ the hazard condition is not covered, and an additional loop is required: $y_1 y_2$. Thus it is essential that all possible transitions are examined on the Y-map and, where necessary, loop terms covering the initial and final states of the transitions should be added. Static hazards can also occur when the output, instead of remaining constant at 0, changes from 0 to 1 to 0 because of a change in the single variable—these are called zero hazards. Huffman has proved, however, that the logical elimination of one type of hazard (either 0 or 1) will automatically correct for the other.

Table 8.21

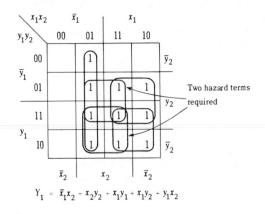

$$Y_1 = \bar{x}_1 x_2 + x_2 y_2 + x_1 y_1 + x_1 y_2 + y_1 x_2$$

Even though static hazards have been eliminated, it is still possible for multi-variable input changes to produce logical hazards. It is worth bearing in mind, though it has in fact been stated earlier, that in order to eliminate all logical hazards in a two-level circuit, whether arising from either single or multi-variable input changes, the complete set of prime implicants must be included in the solution. The dynamic hazard which we have seen occurs in combinational logic due to differing path lengths, usually as a result of a badly factored design, can also cause problems in asynchronous systems and should be avoided.

The *essential hazard* is pertinent only to asynchronous sequential systems, and is basically a critical race between an input signal change and a secondary signal change. The hazard can only exist, for reasons which will be apparent later, in systems with at least two secondaries.

Let us demonstrate this hazard by means of an actual circuit, using the master–slave divide-by-2 counter (Fig. 8.10) as our example. We shall use the same descriptive method to explain this hazard as we used for the dynamic hazard in Chapter 4, i.e. consideration of the basic logic inputs. Figure 8.10(a) shows the counter circuit with the stable state ① conditions $y_1 = y_2 = x = 0$ entered on the logic diagram. In the normal operation of the counter we assume, quite rightly in the majority of cases, that the delay through gates H, I and E, F (the input gates and the Y_2 SR bistable) is very much longer than the delay through the invertor G. This means that when $x \rightarrow 1$, corresponding to the unstable state, the output of gate G responds first and goes to 0, which in turn makes the output of gate C go to 1, gate D output being unchanged. Thus the state of Y_1, since the bistable is already reset, is also unchanged. Meanwhile $x \rightarrow 1$ at gate H has no effect, but at gate I the output goes to 0. This in turn causes the output of gate F to go to 1, thus setting the bistable, and Y_2 goes to 1. The Y_2 output is fed back to gate E whose output then goes to 0, resulting in stable state ② with $y_1 y_2 = 01$. This corresponds to the correct action for the counter as dictated by the flow table shown in Table 8.22.

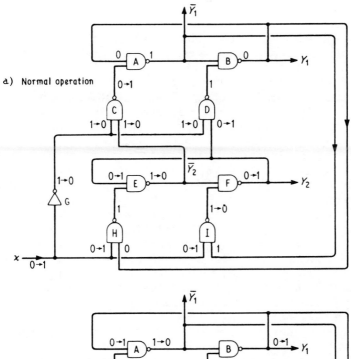

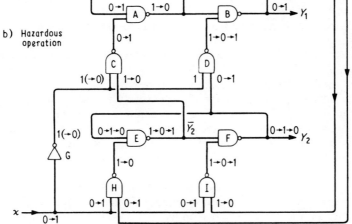

Figure 8.10 Essential hazard in master-slave counter

Now suppose that, due to circuit delays, the response of input gates H, I and the bistable E, F is very much faster than the response of the invertor loop G. (If both circuits have similar responses, we have the condition where a 'critical race' exists between the input signal and the secondary circuit for Y_2). In this case, the output of gate F, and hence Y_2, will have changed before the input change due to $x \rightarrow 1$ has reached gates C and D, and consequently the Y_1 secondary circuit will behave as if it were in the state $y_1 y_2 = 01$ with input $x = 0$. Reference to Table 8.22 shows that this will be unstable state 3 directing $y_1 \rightarrow 1$. The circuit action is then, from Fig. 8.10(b), $\bar{Y}_2 \rightarrow 0$ causes the output of gate C to go to 1, and $Y_2 \rightarrow 1$ will cause the output of gate D to go to 0 since $\bar{x}$ is as yet unchanged and equal to 1. The change in output of gate D will, in turn, cause the output of gate $B(Y_1)$ to go to 1, which is then fed back to gate A producing a 0 output and thus maintaining Y_1. When the input change $x \rightarrow 1$ eventually reaches gates C and D it will have no effect since the bistable has already been set.

Meanwhile, as the outputs of Y_1 are fed back to Y_2 secondary circuits, the circuit will change again, responding as if it were in unstable state 4, i.e. $y_1 y_2 = 11$ and input $x = 1$. Thus the output of gate H goes to 0, which in turn causes gate E output to go to 1. Since the output of gate I has also changed to 1, the output of gate $F (Y_2)$ goes to 0 and a final state is reached with $y_1 y_2 = 10$ and $x = 1$, i.e. stable state ④ in the flow table. This, of course, is incorrect!

The action of the circuit is difficult to understand, and to explain, and the reader is advised to redraw Fig. 8.10 with the initial stable conditions, and then insert the changing values as he reads through the description of the circuit action.

The essential hazard cannot be corrected logically, since it is inherent in the logical structure, as well as depending on the circuit characteristics. The only way of eliminating it is to insert delaying elements (or some form of clock pulse system) in the circuit to ensure that the input signal always wins the 'race'. In the example we considered this would entail a delay in the x signal path to the Y_2 secondary circuit. Ungar[13] has defined the essential hazard in terms of a flow table and has also proved that if a flow table contains an essential hazard, at least one delay element is essential if it is to operate reliably.

The hazard is effectively caused by three changes of input; initially in the counter circuit we had $x = 1$ and secondary Y_2 changed accordingly, giving $y_1 y_2 = 01$; then secondary Y_1 responded with $x = 0$ (due to the input delay) giving $y_1 y_2 = 11$; finally, Y_2 again changed due to $x = 1$ and the new value of Y_1, giving the final condition of $y_1 \dot{y}_2 = 10$. Furthermore, if we examine the flow table for the counter circuit, it is apparent that if the next state of the circuit after the hazard occurred (stable state ④ via unstable state 4 with $y_1 y_2 = 11$ and $x = 1$) had in fact been the same as the starting state (stable state ② via unstable state 2) we would have eventually arrived back at the correct stable state. This structure is shown in Table 8.22(c) and (d). Putting these two facts together, we may now define how an essential hazard may be recognized from the flow table. If, starting from one stable state in the flow table, the state reached after one input change is different

Table 8.22 Essential hazards

	Input x				Input x		
$y_1\,y_2$	0	1		$y_1\,y_2$	0	1	
0 0	①	2		0 0	00 → 01		
0 1	3	②	hazard	0 1	11 ← 01		hazard
1 1	③	4		1 1	11 → 10		
1 0	1	④		1 0	00	10	
(a) Flow table				(b) Y-map			

	Input x			Input x	
$y_1\,y_2$	0	1	$y_1\,y_2$	0	1
0 0	①	2	0 0	00 → 01	
0 1	3	②	0 1	11 ← 01	
1 1	③	2	1 1	11 → 01	
1 0			1 0		
(c) Flow table			(d) Y-map		

to that reached after three changes of the same input, an essential hazard could occur.

The type of flow table structure occurs in counters, shift registers etc. which are extensively used in logic systems, hence the reason for examining the hazard in some detail. Fortunately, though, using medium speed logic circuits with switching times in the order of microseconds, the hazard seldom arises in practice. Moreover essential hazards are less likely to cause problems when the networks are realized on a single microchip, since the device delays in this case are more uniform. Nevertheless, with high speed nanosecond logic systems using integrated circuits, signal delays along a connecting wire may be appreciably longer than the actual switching time of the logic unit, and essential hazards could easily materialize.

With large-variable switching systems it is essential to have some more-systematic method of detecting the presence of hazards in a sequential or combinational circuit. Both McCluskey[14] and Huffman[5] have described methods of detecting and eliminating hazards arising from single input variable changes, and these ideas could be developed into an algorithmic procedure. However, the best approach to date is due to Eichelberger[15] who describes a method which can be used to detect any type of hazard arising from both single- and multi-variable input changes. This uses ternary algebra (i.e. a three-valued Boolean algebra) to describe the transient behaviour of the input switching waveforms.

Though we said earlier that essential hazards cannot be corrected logically it is possible, of course, to use logic gates in a delaying mode to ensure that the input signals always win the race. Armstrong et al.[16] has described such a method based on an alternative technique (to normal invertors) of generating the inverse of the input variable, and thus ensuring that the x variable change is seen before the subsequent change in the y variable. This is achieved by replacing, where necessary, first level AND gates in the SOP form of the equations by NOR/AND pairs. For example, suppose an SOP equation contained the term $x_1\bar{x}_2 y_2 \bar{y}_3$; this would be replaced by

$$x_1(\bar{x}_2 y_2 \bar{y}_3) = x_1(\overline{x_2 + \bar{y}_2 + y_3})$$

which can be realized using a NOR/NAND pair as shown in Fig. 8.11(a). The essence of the approach is to replace each of the AND gates with a logic circuit that realizes the same function but with all the x inputs uncomplemented. In effect what is happening is that those x variables which require to be complemented and the y variables with which they are racing are passed through the same NOR gate. This resolves the race so that the first level gates see the x change before any change in y. If this technique is combined with the SR bistable method of realization it is possible to achieve a hazard-free design. For instance, in our example of Section 8.7 we would transform the equations as follows:

$$Y_1 \text{ set} = x_1 x_2 \bar{y}_2 \qquad Y_1 \text{ reset} = \overline{x_1 + x_2 + y_2}$$
$$Y_2 \text{ set} = x_2(\overline{x_1 + \bar{y}_1}) + x_1(\overline{x_2 + y_1})$$
$$Y_2 \text{ reset} = x_1 x_2 y_1 + \overline{x_1 + x_2 + y_1}$$

The circuit is shown implemented in Fig. 8.11(b).

In our discussion it has been implied that the delays causing the hazards are contained within the logic circuits rather than in the interconnections. It can be shown that the logical correction methods are still valid if the delays are carried by the connecting wires, but with the proviso that the wire delays must not exceed the minimum delay in any path through the logic circuit.

Another method of eliminating static hazards (other than an SR bistable realization) is to implement the circuit using ROM; in this case it is best to use an STT assignment.[18] The procedure is very similar to that described for synchronous circuit realization but in this case care must be taken with the timing. In general the memory access time of the ROM will be greater than the switching times of most logic devices and it is essential to ensure that external inputs do not change until a stable state is reached. This means that the input cycle time of the ROM system must be limited to the maximum transition time that can occur (each transition in the flow table necessitates a read cycle of memory).

Before leaving the subject of asynchronous networks it is worth mentioning another important class of circuit called *speed independent* or *Muller* circuits.[17] In this mode of operation input changes are only allowed when a *ready* signal is indicated. Thus each circuit module must be able to

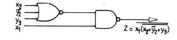

a) NOR–NAND pair

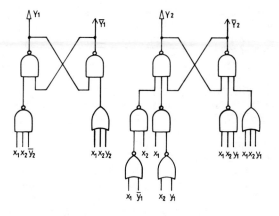

b) Hazard-free realisation of example 8·7

Figure 8.11 Hazard correction

generate indigenously a *completion signal* at the end of its required processing operation. This technique allows circuits, both combinational and sequential, to work at their maximum speeds independent of any system clock.

References and bibliography

1. Seitz, C. *Systems Timing*. Chapter 7, 218–262 Mead, C. and Conway, L. *Introduction to VLSI Systems*. Addison-Wesley, 1980.
2. Huffman, D. A. The synthesis of sequential switching circuits. *J. Franklin Inst.* **257**, 161–90, 257–303 (1954).
3. Keister, W., Ritchie, A. Z. and Washburn, S. H. *The Design of Switching Circuits*. Van Nostrand, New York, 1951.
4. Maley, G. A. and Earle, J. *The Logic Design of Transistor Digital Computers*. Prentice-Hall, Englewood Cliffs, N.J., 1963, Chapter 8 *et seq.*
5. Huffman, D. A. The design and use of hazard-free switching networks. *J. Ass. Comput. Mach.* **4**, 47–62 (1957).
6. Marcus, M. *Switching Circuits for Engineers* 2nd edn. Prentice-Hall, Englewood Cliffs, N.J., 1975, Chapter 16.
7. Liu, C. N. A state-variable assignment method for asynchronous sequential switching circuits. *J. Ass. comput. Mach.* **10**, 209–16 (1963).

8. Tracey, J. H. Internal state assignment for asynchronous sequential machines. *IEEE Trans. electron. comput.* **EC15**, 551–60 (1966).
9. Smith, R. J., Tracey, J. H., Schoeffel, W. L. and Maki, G. K. Automation in the design of asynchronous sequential circuits. *IFIPS SJCC* **32**, 55–60 (1968).
10. Smith, R. J. Generation of internal state assignment for large asynchronous sequential machines. *IEEE Trans. comput.* **C23**, 924–32 (1974).
11. Ungar, S. H. *Asynchronous Sequential Switching Circuits*. Wiley, New York, 1969.
12. Hollaar, L. A. Direct implementation of asynchronous control units. *IEEE Trans. computers* **C31**, 1133–41 (1982).
13. Ungar, S. H. Hazards and delays in asynchronous sequential switching circuits. *IRE Trans. Circuit Theory* **CT6**, 12–25 (1959).
14. McCluskey, E. J. Transients in combinational logic circuits. From *Redundancy techniques for computing systems*, R. H. Wilcox and W. C. Mann (Eds). Spartan Book Co., Washington, D.C., 1962, pp. 9 *et seq.*
15. Eichelberger, E. B. Hazard detection in combinational and sequential switching circuits. *IBM Jl. Res. Dev.* **9**, 90–9 (1965).
16. Armstrong, D. B., Friedman, A. D. and Menon, P. R. Realisation of asynchronous sequential circuits without inserted delay elements. *IEEE Trans. Computers* **C17**, 129–34 (1968).
17. Miller, R. E. *Switching Theory* Vol. II. Wiley, New York, 1965, Chapter 10.
18. Sholl, H. A. and Yang, S. C. Design of asynchronous sequential networks using read only memory. *IEEE Trans. computers* **C24**, 195–206 (1975).
19. Lewin, D. A new approach to the design of asynchronous logic. *Radio Electron. Engn* **36**, 327–34 (1968).
20. Hill, F. J. and Peterson, G. R. *Introduction to Switching Theory and Logical Design*. Wiley, New York, 1981.

Tutorial problems

8.1 Design an asynchronous circuit that has two inputs, $x_1 x_2$, and one output Z. The circuit is required to give an output whenever the input sequence (00), (01) and (11) is received, but only in that order.

8.2 Derive the excitation and output equations for an asynchronous three-bit Gray-code counter which has one input x and three outputs Z_1, Z_2 and Z_3. Implement the design in terms of NAND elements. Redesign the circuit by extracting the set and reset equations for dc SR bistables and then compare and comment on the two circuits.

8.3 Design one stage of an asynchronous shift register, which is a circuit having two inputs x_1 and x_2 and one output Z. Input x_1 is the output of the preceding shift register stage, and x_2 is the shift pulse. When $x_2 = 1$, Z remains unchanged; when $x_2 = 0$, Z takes the previous value of x_1 when $x_2 = 1$.

Derive the excitation equations in NAND logic, and then in terms of input equations for master–slave bistables.

Confirm that the circuit may be connected in cascade to form a multistage shift register and, in so doing, explain its action.

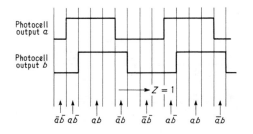

Figure 8.12 Problem 8.4

8.4 In a numerical machine tool control system an optical Moiré fringe device is used to digitize the linear motion of the workpiece. The relative movements of the gratings produce light and dark fringes which are detected photoelectrically using two photocells spaced $\lambda/4$ apart. A waveform diagram is shown in Fig. 8.12; a and b are the squared outputs of the photocells. Note that because of the spacing of the cells the outputs cannot change together.

Design an asynchronous circuit with inputs a and b and output Z, which will detect the direction of motion, left or right, of the workpiece. (*Hint*: note that the sequence is $\bar{a}\bar{b} \to \bar{a}b \to ab \to a\bar{b} \to \bar{a}\bar{b}$ in one direction and is reversed for the other direction.)

8.5 Design an asynchronous circuit that has two inputs x_1 and x_2 and an output Z. Input x_1 is a repetitive square wave or 'clock' pulse signal, and input x_2 originates from a noise-free switch. The action of the circuit is such that when x_2 is pressed at any point in the clock cycle, the output Z must transmit the next complete clock pulse of x_1. This circuit is called a single-shot generator and its function is to produce one clock pulse each time the switch x_2 is pressed, irrespective of the duration of x_2.

Assume that x_1 and x_2 cannot occur together.

8.6 Design an asynchronous version of the clamp-gate circuit described in problem 8 of Chapter 6 and implement the design using master–slave bistables. Compare the relative merits of the synchronous and asynchronous circuits.

8.7 In self-timed and concurrent systems a new process can only be initiated when all the required previous processes have been completed. The circuit used to control this operation, known as the Muller C element, has the characteristics that its output becomes 1 only after all of its inputs are 1, and becomes 0 only after all of its inputs are zero.

Design a 2-input asynchronous version of this element and realize the circuit using NMOS transistor logic. Show how the basic unit may be extended to handle more than two inputs.

8.8 Realize the flow table shown in Table 8.23 using
(a) a one-hot assignment
(b) an STT assignment.
Critically compare the circuits obtained by these methods.

Table 8.23 Problem 8.8

	Input x_1x_2 00	01	11	10	Output Z
a	①	2	3	①	0
b	–	②	3	–	0
c	4	③	③	1	1
d	④	–.	④	5	1
e	1	⑤	4	⑤	0

9 Logic circuit testing and reliable design

9.1 Introduction

With industry's increasing dependence on digital and logic techniques it is imperative that logic systems should work reliably. Moreover, when faults do develop they must be easily detected and located. The design procedures developed in the earlier chapters of this book will lead to a correct realization of the logical specification embodied in the state diagram, ASM chart etc. However, during manufacture or in operation various faults such as devices going open-circuit due to breaks in seals or connections, short-circuits on printed circuit boards, process faults etc. can develop causing the device or systems to malfunction. Hardware faults of this type can generally be detected by devising a test sequence or pattern based on the logically correct circuit. The testing procedure is essentially one of applying specified input signals to the circuit under test and then comparing the output response with some standard reference. We shall describe some formal techniques for generating test pattern sequences in the following sections of this chapter.

There is another important class of 'fault' which is more difficult to detect—these are due to errors in the original logic design process caused, for example, by misunderstanding the original circuit requirements, fundamental errors in constructing an appropriate algorithmic solution or simply not taking into account all the relevant states of the system. Unfortunately in many cases hardware faults and design errors can give rise to the same fault symptoms. This type of 'fault', which is very common in computer software (a set of instructions for a programmable device such as a micro-processor or digital computer) can only be rectified by changes in the original design. Note that software faults in the hardware sense cannot exist; once an algorithm has been correctly programmed it will always operate in the same way providing the hardware is functioning correctly. Design faults are best handled by ensuring that they do not occur in the first place, which means providing good specification tools and a means of evaluating the proposed design before proceeding to a realization.

There are two basic requirements for hardware testing:

(a) testing of the chips, modules and boards after manufacture and prior to assembly or distribution to users;
(b) testing the assembled system in an operational environment, including the repair of faulty boards.

In the first case, particularly with LSI/VLSI chips, it is simply a question of a go–nogo test in that once a fault is detected the devices are discarded. On the systems level, however, it is necessary both to detect and locate a faulty unit so that it can be replaced and/or repaired. One of the major problems in LSI/VLSI manufacture, due to the ever-increasing complexity of the devices, is how to devise reliable test procedures.

Thus, logic circuits and systems are normally tested in one of two ways:

 (a) *Single-flow procedure* in which a complete set of test inputs is applied in sequence to the circuit under test; when a faulty output is detected the testing is terminated. This corresponds to the go–nogo testing referred to above.

 (b) *Multi-flow* testing requires each test input to be applied separately to the circuit with the result of an individual test being used to determine the next test input to be applied.

The multi-flow method is particularly well suited for diagnostic routines and generally tends to be more efficient in operation.

Before attempting to devise test sequences for a logic circuit it is necessary to have a relevant model or description of the network. The description can take two forms—an actual interconnection layout for the implemented circuit in terms of gates and modules or a formal model employing Boolean equations, state diagrams etc. In order to determine the true fault characteristics of the circuit it is essential to use a model that relates directly to the physical circuit. Using a formal model it is possible to check logical correctness but the way the circuit is factored and realized in hardware, for example the inclusion of redundant elements for hazard correction, can affect the checking procedure and possibly mask the presence of faults. However, as we shall see later many of the algorithmic methods for the generation of test sequences require a formal description; in this case a Boolean description of the actual implemented circuit must be generated. Moreover, *functional testing*, using an input–output specification, may well be the only practical method for testing VLSI circuits, since a gate level approach to test generation would not seem feasible. Using this technique the system can be partitioned into various modules such as multiplexers, registers, arithmetic and logic units etc. and an individual test set derived on a functional input/output basis.

To require a design engineer to devise test sequences on an intuitive basis is bad practice since it is almost inevitable that errors or omissions would result. In theory a circuit can always be tested by applying all possible inputs and checking the outputs are as specified—a technique known as *exhaustive testing*. Indeed a standard method of testing is to apply the same inputs to both a good unit and a unit under test and compare the outputs for any disparities; a fast digital computer or special-purpose test engine is normally used for this purpose. However, with large complex circuits containing many variables the exhaustive test approach soon becomes prohibitive in terms of time and cost; the use of randomized input sequences has partially overcome this problem. It would seem essential to determine in some way, preferably using a formal algorithmic procedure, a minimal set of test inputs which when applied to a circuit will produce a defined set of outputs indicating whether or not the circuit is functioning correctly.

Let us now consider in more detail the basic problem of testing; this is illustrated in Fig. 9.1 which shows a typical combinational circuit with primary inputs x_1, x_2, x_3, x_4 and x_5. The objective is to detect a logical fault on any of the interconnection wires by examination of the output z. for example, if input x_2 was permanently HI giving a logic value of 1 it would be necessary to *sensitize* a path from x_2 to Z to enable the fault to be detected. For instance by setting $x_1 = 0$, $x_3 = 1$, $x_4 = 0$ and $x_5 = 0$ the input line x_2 may be tested for being stuck at logic 1 by applying a test input of $x_2 = 0$ since $Z = 0$ under correct conditions and goes to 1 for a fault. In this case x_2 can also be tested for a permanent LO (logic 0) by applying $x_2 = 1$ when Z goes from 1 to 0. The sensitized path is shown blocked in Fig. 9.1; one of the problems in deriving test sequences is how to generate the inputs to set up and maintain a sensitized path.

Almost exclusively the work on testing has assumed a single fault model called the *stuck-at model* though other models such as the bridging fault model[1] have been proposed. The stuck-at model assumes that only single logical faults resulting in an input or output line being permanently stuck at logic 1 or logic 0 can occur in a circuit. Thus no account is taken of multiple or intermittent fault conditions or those arising from physical defects in the circuit. This is the advantage of the bridging fault model which allows for short circuits between connecting paths resulting in a wired-OR or a wired-AND effect. The justification for using such a simple model is that multiple faults are improbable and that testing can be performed frequently enough to detect and rectify a single fault before another can occur. Intermittent faults can be detected by repeated applications of the test inputs. It is also assumed that manufacturing or device faults will emanate as logical faults. In the main these assumptions have been substantiated in practice but it is very unlikely that test sequences based on the stuck-at model can ever achieve anything like a 100% fault cover.

Due to the sheer size of current logic circuits it is inevitable that computer-aided methods must be used for test generation. The literature is full of such techniques[2,3] but a viable procedure has still to evolve. The proper solution to the problem lies in designing systems which are both easily testable and failure-tolerant, including self-diagnosis, rather than attempt to devise tests for unconstrained and often intuitively designed systems.

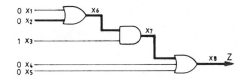

Figure 9.1 Basic test problem

9.2   Generation of test sequences for combinational logic

There are four basic methods of detecting logic fault conditions in combinational networks; these are the *fault matrix, boolean difference, path sensitizing* and *partitioning* techniques. All these methods use the simple stuck-at fault model and consequently apply to single, static, logical faults only. Of all the methods the path sensitizing and partitioning approaches are the predominant techniques used in practice; the other methods, though well founded, are not viable for large circuits.

9.2.1   The fault matrix method

The fault matrix technique was developed by Kautz[4] from work originally described by Chang.[5] In the Kautz method a Boolean matrix, called the *F*-matrix, is used to relate the set of all possible input tests, for a given circuit, to their associated faults. The entries in the *F*-matrix are the output values resulting from applying a given test input under specified fault conditions. For instance, consider the simple exclusive OR circuit shown in Fig. 9.2 and its associated *F*-matrix in Table 9.1(a). There are two inputs X_1 and X_2 and hence four different input tests, termed t_0-t_3. Since the circuit contains 7 connections $C1-C7$ there are 14 possible fault conditions, f_1-f_{14} referred to as $C1/0$, $C1/1$ etc., where $C1/0$ denotes connection $C1$ s-at-0 (stuck-at-0) etc.; the f_0 output entries are those for the fault-free circuit. For example, from Table 9.1(a), under test input $t_2 = 10$ the correct output f_0 would be $Z = 1$, but for a fault f_{10}, that is $C5$ s-at-1, the output is $Z = 0$. For small circuits the table can be manually compiled but for large circuits fault-simulation (see later) is essential.

The *F*-matrix is usually transformed into a G_D-matrix for ease of manipulation; this is done by performing an exclusive OR operation between the correct f_0 column and all the other fault columns. The next step is to select a minimal set of input tests that will cover all possible fault conditions; once again we have a covering problem which is analogous to that of selecting prime implicants. From Table 9.1(b) we can determine the *essential tests*, that is those tests which detect one particular fault only—in practice those columns in the G_D-matrix with a single entry; these are

t_1—only test for f_0f_7 and f_0f_{12} i.e. $C4/0$ and $C6/1$
t_2—only test for f_0f_5 and f_0f_{10} i.e. $C3/0$ and $C5/1$
t_3—only test for f_0f_6 and f_0f_8 i.e. $C3/1$ and $C4/1$

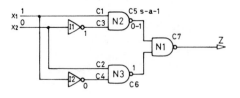

Figure 9.2 Exclusive OR circuit

Table 9.1 Fault matrix method

(a) F-matrix

X_1	X_2	Test	f_0	f_1 C1/0	f_2 C1/1	f_3 C2/0	f_4 C2/1	f_5 C3/0	f_6 C3/1	f_7 C4/0	f_8 C4/1	f_9 C5/0	f_{10} C5/1	f_{11} C6/0	f_{12} C6/1	f_{13} C7/0	f_{14} C7/1
0	0	t_0	0	0	1	0	1	0	0	0	0	1	0	1	0	0	1
0	1	t_1	1	1	0	0	1	1	1	0	1	1	1	1	0	0	1
1	0	t_2	1	0	1	1	0	0	1	1	1	1	0	1	1	0	1
1	1	t_3	0	1	0	1	0	0	1	0	1	1	0	1	0	0	1

(b) G_D-matrix

	f_0f_1	f_0f_2	f_0f_3	f_0f_4	f_0f_5	f_0f_6	f_0f_7	f_0f_8	f_0f_9	f_0f_{10}	f_0f_{11}	f_0f_{12}	f_0f_{13}	f_0f_{14}
t_0		1		1					1		1			1
t_1		1	1				1					1	1	
t_2	1			1	1					1			1	
t_3	1		1			1		1	1		1			1

It can be seen that in this case the essential tests t_1, t_2 and t_3 cover all possible faults, which gives the full test set, expressed in terms of input/output:

01/1;   10/1;   11/0

If the essential tests had not provided a complete fault cover it would, of course, have been necessary to add more tests until a full cover was obtained. To test the circuit the tests would be applied sequentially to the input terminals of the network and the output monitored; any deviation from the specified output sequence would indicate the presence of a fault (this is an example of the single-flow method of testing).

Though the fault matrix method can in principle always be used to generate a minimal test set it is impractical for circuits with a large number of variables since the computational requirements increase drastically. For N connections and n input variables the size of the matrix is $N \times 2^{n+1}$, and thus the computation time will increase exponentially with the number of variables.

9.2.2   The Boolean difference technique

Boolean difference[6,7] is a formal method for analysing the operation of a logic circuit when errors occur in its primary inputs. The method is based on the exclusive OR operation between two Boolean functions, one representing a faulty model of the machine and the other a correct version. If the Boolean difference is equal to one then an error is indicated and the resulting error function can be used as the basis of a test sequence.

For example, consider the function

$$Z = f(x_1, x_2, \cdots x_i, \cdots x_n)$$

where x_n are primary inputs to the circuit. If the input x_i is in error then a new function Z_{x_i} may be defined as

$$Z_{x_i} = g(x_1, x_2, \cdots \bar{x}_i, \cdots x_n)$$

which is formed from Z by replacing x_i by $\bar{x}_i$ and vice versa. The Boolean difference, dZ/dx_i, is defined by

$$\frac{dZ}{dx_i} = Z \oplus Z_{x_i} = h(x_1, x_2, \cdots x_n)$$

where $\oplus$ is the exclusive OR operation. As an example consider the exclusive OR network shown in Fig. 9.2 where the output is given by

$$Z = \bar{C}1C2 + C1\bar{C}2$$

Suppose $C1$ is in error; then

$$Z_{C1} = C1C2 + \bar{C}1\bar{C}2$$

and

$$\frac{dZ}{dZ_{C1}} = (\bar{C}1C2 + C1\bar{C}2) \oplus (C1C2 + \bar{C}1\bar{C}2)$$
$$= \bar{C}1\bar{C}2 + \bar{C}1C2 + C1\bar{C}2 + C1C2$$

The exclusive OR operation can be performed mathematically as above but for a small number of variables it is convenient to use a K-map. The technique is to map the functions Z and Z_{x_i} on separate K-maps and then exclusively OR the two maps together to produce a K-map representing dZ/dx_i. In practice this is performed by comparing corresponding cells on the Z and Z_{x_i} maps and inserting a 1 in the derived dZ/dx_i map if there is a difference in the two values; the method is shown in Table 9.2(a).

In our example Z_{C1} defines the function that is realized by the faulty network when there is a fault either s-at-0 or s-at-1 in the value of $C1$. Under fault conditions the output will differ from the true output for those terms that make $dZ/dC1 = 1$; this can be checked by consulting Table 9.1(a).

Thus $dZ/dC1$ defines the full set of input tests that will cause an observable output if there is a logical fault in the value of $C1$. Since these tests include both types of stuck-at faults, $dZ/dC1$ must be partitioned into separate lists. This is achieved by separating the list of all tests into those containing x_i and those containing $\bar{x}_i$; the former will demand a 1 on x_i and therefore test for x_i s-at-0, and similarly the latter will test for x_i s-at-1. In our example, separating the $dZ/dC1$ terms gives

$$(\bar{C}1\bar{C}2,\ \bar{C}1C2)_{C1\,\text{s-at-1}} \qquad \text{and} \qquad (C1\bar{C}2,\ C1C2)_{C1\,\text{s-at-0}}$$

which corresponds to the tests $(t_0,\ t_1)$ and $(t_2,\ t_3)$ respectively as shown in the G_D-matrix for the circuit in Table 9.1(b).

The technique may also be extended to determine tests for faults on non-primary input lines. For example, let us consider connection C_5 in Fig. 9.2(a); then we have

$$Z = C1\bar{C}2 + \bar{C}1C2 \qquad \text{and} \qquad C5 = C1\bar{C}2$$

Therefore

$$Z = C5 + \bar{C}1C2 \qquad \text{and} \qquad Z_{C5} = \bar{C}5 + \bar{C}1C2$$

which gives

$$dZ/DZ_{C5} = \bar{C}5\bar{C}1\bar{C}2 + \bar{C}5C1C2 + \bar{C}5C1\bar{C}2 + C5\bar{C}1\bar{C}2$$
$$+ C5C1C2 + C5C1\bar{C}2$$

as shown in Table 9.2(b).

Now since $C5 = C1\bar{C}2$ the only time it will be zero will be when $C1 = 1$ and $C2 = 0$. Thus in order to detect $C5$ s-at-1 the input must contain the term $C1C2$; the other combinations will test for $C5$ s-at-0. Thus we have test t_2 for $C5/1$ (the only test and hence essential) and t_0, t_3 for $C5/0$ as confirmed by the G_D-matrix in Table 9.1(b).

The method can also be used to analyse a circuit for specific faults; for example, consider the carry equations for a full adder:

$$C_+ = yC + xC + xy$$

The effect of a fault on input y is given by

$$dC_+/dy = (yC + xC + xy) \oplus (\bar{y}C + xC + x\bar{y})$$
$$= x\bar{C} + xC$$

Table 9.2 Boolean difference using K-maps

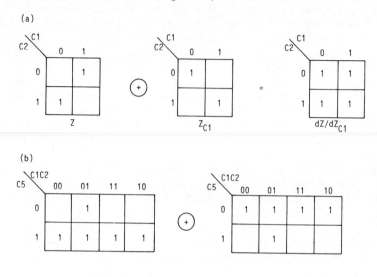

(a)

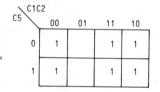

which means that a stuck-at-0 or stuck-at-1 error in y will cause the output to be in error only if $x\bar{C} = 1$ or $\bar{x}C = 1$.

Boolean difference is a useful technique both for fault analysis and the generation of test sequences; unfortunately it is limited to small circuits due to the amount of algebraic computation involved. Its main advantage lies in spotting essential tests since once these are known other methods, such as path sensitizing, can be used to determine all other faults covered by these tests.

9.2.3 Path sensitization methods

The basic one-dimensional path sensitization method[8] has three distinct phases:

(a) the postulation of a specific fault within the circuit structure, for example, $C5$ s-at-1 in Fig. 9.2;
(b) the propagation of the logical effect of this fault from its original site to the output terminals, along a *sensitive path*—this is called the *forward trace*;

(c) a *backward-trace* phase, in which the necessary gate conditions re-quired to propagate the fault along the sensitive path are established.

The inputs to each logic element in the sensitive path are grouped into a *control* input, which is part of the sensitive path and must be allowed to vary in order to detect a fault, and the *static* inputs which are held at a constant value to maintain the sensitive path. Note that the rules for error propagation through individual gate elements are simply derived and summarized in Table 9.3.

To illustrate this let us derive a test input that will detect $C5/1$ in the circuit of Fig. 9.2. The first step is to determine those gates through which the fault must be propagated in order to reach the output Z; in our example this is trivial since only gate $N1$ needs to be considered. However, in order to detect the presence of $C5/1$ on the output of $N1$ the other input to the gate, $C6$, must be held constant at logic 1. Under these conditions with $C5$ specified at 0 and $C6$ held at 1, the output $C7$ would be logic 1; consequently if $C5$ was s-at-1 the output of $N1$ would go to 0, indicating a fault.

Finally it is necessary to establish those primary input conditions which will ensure that $C6$ is held at 1 and $C5$ at 0 for the fault-free circuit, that is the backward trace. The static inputs to gate $N3$ for the output $C6$ to be held at 1 is given by $\bar{C}2 + \bar{C}4$, i.e. $\bar{C}2 + C1$; for $C5$ to be 0 the inputs to gate $N2$ must be $C1C3$, i.e. $C1\bar{C}2$. Thus the input combination 10 (test t_2) is the only test that will detect $C5/1$, as we ascertained earlier using the fault matrix technique. The sensitive path flows through gates $N2$ and $N1$ with gate $N3$ being used to maintain the path.

Once an input test has been established all other faults detected by that test are derived. The process is then repeated using an as yet undetected fault condition until all faults are covered.

Though one-dimensional path sensitization is a viable procedure for test sequence generation it nevertheless has a major drawback which results in some faults not being detectable. This is due to the existence of fan-out paths from the point of failure in the circuit. Should these path reconverge later, and the number of signal inversions that occur along the path be unequal, then the effect of a fault could be masked. For example, the circuit shown in Fig. 9.3 exhibits *reconvergent fan-out* between gates $N1$

Table 9.3 Conditions for error propagation

Type of gate	Value static inputs
AND	All at 1
OR	All at 0
NAND	All at 1
NOR	All at 0
Invertor	NA

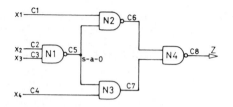

Figure 9.3 Reconvergent fan-out

and N4 which prohibits the setting up of a viable sensitive path to detect an s-at-1 fault occurring on the output of gate N1.

The answer to this problem is simultaneously to sensitize all possible paths from the point of failure to the circuit outputs. The approach, known as *n-dimensional path sensitization*, was first described by Roth.[9]

The basic procedure is as follows:

(i) For each pass through the circuit all possible paths from a chosen fault site to all outputs are generated simultaneously, cancelling any reconvergent fan-out paths that may occur. This operation is called the *D-drive*.

(ii) Using a backward-trace procedure, the primary input conditions required to generate the static inputs for the *D*-drive are derived. This is called the *consistency operation*.

The procedures described above are based on a *calculus of D-cubes* which allows a formal mathematical model of the network, under fault conditions, to be set up. The starting point for the *D*-calculus is the concept of a *singular cover* for a gate or network; the cover can be derived from the truth table and is a compact cubical notation for representing the logical operation of a circuit. Table 9.4 shows the singular covers for two-input AND and OR gates. Note that the gates are represented by set of primitive cubes, where the Xs are don't-care input conditions, that may be expanded in the usual way to yield the vertices. (It will be obvious that if any input to an AND gate is at 0 the output will be 0.) Note that the primitive cubes are directly analogous to prime implicants.

The singular cover for a network can be built up from the singular covers of the individual gates, as shown in Table 9.5(a) for the circuit given in Fig. 9.4. Note that each gate is treated separately, according to its covering table, and that the inputs which do not affect the gate output are assigned don't-cares. For example NAND 1 has inputs N1, N2 and output N4 which correspond to the covering table; input N3 is a don't-care.

In order to establish a sensitive path one gate input must be forced to bear the responsibility for determining the gates output; this is represented in the *D*-calculus by the *propagation D-cube*. The concept is illustrated in Table 9.6 for the basic logic gates; note that a new variable *D* is introduced which may assume either of the Boolean values of 0 and 1, but is constrained to take the *same* value in a particular cube (a kind of constrained

Table 9.4 Singular covers for gates

Input		Output Z		
X_1	X_2	AND	OR	NAND
0	0	0	0	1
0	1	0	1	1
1	0	0	1	1
1	1	1	1	0

(a) Truth tables

X_1	X_2	Z
0	X	0
X	0	0
1	1	1

(b) AND gate

X_1	X_2	Z
1	X	1
X	1	1
0	0	0

(c) OR gate

X_1	X_2	Z
0	X	1
X	0	1
1	1	0

(d) NAND gate

Table 9.5 D-algorithm

	$N1$	$N2$	$N3$	$N4$	$N5$	$N6$
NAND 1	0	X	X	1		
	X	0	X	1		
	1	1	X	0		
NAND 2		0	X	X	1	
		X	0	X	0	
		1	1	X	0	
OR				1	X	1
				X	1	1
				0	0	0

(a) Singular cover for network

	$N1$	$N2$	$N3$	$N4$	$N5$	$N6$
NAND 1	$\bar{D}$	1		D		
	1	$\bar{D}$		D		
NAND 2		$\bar{D}$	1		D	
		1	$\bar{D}$		D	
OR				D	0	D
				0	D	D

(b) Single progagation cubes for network

don't-care term). For example, the cube $\frac{X_1}{D}\frac{X_2}{0}\frac{Z}{D}$ expresses the fact that the output Z is controlled by input X_1 when input X_2 has the value 0 (it is in fact an OR gate).

Whatever value D takes $\bar{D}$ must always be its complement; moreover the dual of propagation D-cubes can be obtained simply by changing all Ds to $\bar{D}$s and vice versa. Note also that D-cubes can always be expanded to give the corresponding vertices e.g. $D\ 0\ D = 0\ 0\ 0,\ 1\ 0\ 1$ and $0\ D\ D = 0\ 0\ 0,$ $0\ 1\ 1$. In addition the D-cubes can also indicate fault-test conditions; for example, in the OR gate the input vector $0\ 1\ 1$ constitutes a test for X_2 s-at-0 and Z s-at-0 and similarly $0\ 0\ 0$ is a test for both lines s-at-1.

Propagation D-cubes can be determined by inspection or derived from the singular cover using an algorithm due to Roth. To apply the algorithm the cubes of the gates singular cover *with differing outputs* are intersected according to the following rules:

$$0 \cap 0 \ = 0 \cap X = X \cap 0 = 0$$
$$1 \cap 1 \ = 1 \cap X = X \cap 1 = 1$$
$$X \cap X = X$$
$$1 \cap D \ = 0; \qquad 0 \cap 1 = \bar{D}$$

For example, from the singular cover of the AND gate as shown in Table 9.4(b) we have

$$(0\ X\ 0) \cap (1\ 1\ 1) = \bar{D}\ 1\ \bar{D} \qquad \text{and} \qquad (X\ 0\ 0) \cap (1\ 1\ 1) = 1\ \bar{D}\ \bar{D}$$

The D-cubes defined above relate to the fault transmission properties of an element rather than to its fault test generation properties. The *primitive D-cube of failure* is used to express fault tests in terms of the input–output vertices of the faulty gate. Suppose, for example, that our two-input NAND gate example had an s-at-1 fault on the output Z; then the corresponding D-cube of failure would be $1\ 1\ \bar{D}$ which states that the correct output is 0 and the faulty output 1 with $X_1 = X_2 = 1$. The primitive D-cubes of failure for the basic gates are shown in Table 9.7.

The primitive D-cubes of failure for a circuit may be deduced from the primitive cubes (prime implicants) of the good and faulty circuits using a cubical algorithm analogous to the Boolean difference approach. Consider the circuit shown in Fig. 9.4; the K-maps for the good and faulty ($N1$ stuck at 1) versions of the circuit are given in Table 9.8(a) and (b). (Note in fact that the circuit is equivalent to a three-input NAND function.)

Table 9.6 Propagation D-cubes

	X_1	X_2	Z	X_1	X_2	Z	X_1	X_2	Z	X_1	X_2	Z
	$\bar{D}$	1	$\bar{D}$	D	0	D	$\bar{D}$	1	D	D	0	$\bar{D}$
	1	$\bar{D}$	$\bar{D}$	0	D	D	1	$\bar{D}$	D	0	D	$\bar{D}$
Duals $\begin{cases} \\ \\ \end{cases}$	D	1	D	$\bar{D}$	0	$\bar{D}$	D	1	$\bar{D}$	$\bar{D}$	0	D
	1	D	D	0	$\bar{D}$	$\bar{D}$	1	D	$\bar{D}$	0	$\bar{D}$	D
	(a) AND gate			(b) OR gate			(c) NAND gate			(d) NOR gate		

Table 9.7 Primitive D-cubes of failures

X_1	X_2	Z	Fault cover		X_1	X_2	Z	Fault cover
0	0	D	$Z/0$		0	0	$\bar{D}$	$X_1/1, X_2/1, Z/1$
0	1	D	$X_1/1, Z/0$		0	1	D	$X_2/0, Z/0$
1	0	D	$X_2/1, Z/0$		1	0	D	$X_1/0, Z/0$
1	1	$\bar{D}$	$X_1/0, X_2/0, Z/1$		1	1	D	$Z/0$
(a) NAND gate					(b) OR gate			

X_1	X_2	Z	Fault cover		X_1	X_2	Z	Fault cover
0	0	$\bar{D}$	$Z/1$		0	0	D	$X_1/1, X_2/1, Z/0$
0	1	$\bar{D}$	$X_1/1, Z/1$		0	1	$\bar{D}$	$X_2/0, Z/1$
1	0	$\bar{D}$	$X_2/1, Z/1$		1	0	$\bar{D}$	$X_1/0, Z/1$
1	1	D	$X_1/0, X_2/0, Z/0$		1	1	$\bar{D}$	$Z/1$
(c) AND gate					(d) NOR gate			

Table 9.8 Primitive D-cubes of failure

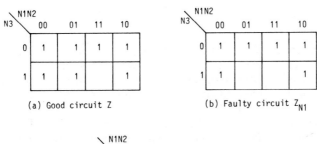

(a) Good circuit Z   (b) Faulty circuit Z_{N1}

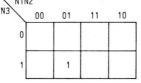

(c) Boolean difference dZ/dZ_{N1}

Now from the K-maps we have

$$Z = \bar{N}1 + \bar{N}2 + \bar{N}3; \qquad \bar{Z} = N1N2N3$$

and

$$Z_{N1} = \bar{N}3 + \bar{N}2 \qquad ; \qquad \bar{Z}_{N1} = N3N2$$

which can be expressed in the cubical notation as

	$N1$	$N2$	$N3$	Z			$N1$	$N2$	$N3$	Z
P_1	0	X	X	1		PF_1	X	0	X	1
P_2	X	0	X	1	and	PF_2	X	X	0	1
P_3	X	X	0	1		$\overline{PF}$	X	1	1	0
$\bar{P}$	1	1	1	0						

We must now intersect the cubes according to the rule that the primitive D-cubes of a fault which result in an output $\bar{D}$ are obtained by intersecting the inputs of each cube in PF with those in $\bar{P}$ and the D-cubes of a fault resulting in D can be obtained by intersecting cubes in $\overline{PF}$ with P. The rules for intersection are

$$0 \cap X = X \cap 0 = 0$$
$$1 \cap X = X \cap 1 = 1$$
$$0 \cap 0 = 0; \ 1 \cap 1 = 1; \ X \cap X = X$$
$$1 \cap 0 = 0 \cap 1 = \varphi \ \text{(that is, there is no intersection)}$$

Now in our example we have

$$
\begin{array}{llll}
PF_1 & X\ 0\ X\ 1 & PF_2 & X\ X\ 0\ 1 \\
\bar{P} & 1\ 1\ 1\ 0 & \bar{P} & 1\ 1\ 1\ 0 \\
\hline
& 1\ \varphi\ 1\ \bar{D} & & 1\ 1\ \varphi\ \bar{D}
\end{array}
$$

$$
\begin{array}{llllll}
\overline{PF} & X\ 1\ 1\ 0 & \overline{PF} & X\ 1\ 1\ 0 & \overline{PF} & X\ 1\ \ 1\ 0 \\
P_1 & 0\ X\ X\ 1 & P_2 & X\ 0\ X\ 1 & P_3 & X\ X\ 0\ 1 \\
\hline
& 0\ 1\ 1\ D & & X\ \varphi\ 1\ D & & X\ 1\ \ \varphi\ D
\end{array}
$$

Thus, discounting those cubes containing φ gives us $0\ 1\ 1\ D$ as the primitive D-cube of failure for the circuit. Note that this result could have been obtained by taking the Boolean difference, as shown in Table 9.8(c), and that in general the method gives D-cubes of failure for both types of logical fault.

Let us now use some of these ideas to show how the D-drive would operate. Assume that the circuit shown in Fig. 9.4 has an s-at-1 fault on the $N1$ input to NAND 1 for which the D-cube of failure is $0\ 1\ D$, that is

$$
\begin{array}{cccccc}
N1 & N2 & N3 & N4 & N5 & N6 \\
0 & 1 & X & D & &
\end{array}
$$

If we now consult Table 9.5(b) or Table 9.6(b) we see that the propagation D-cubes for the OR gate which matches the $N4$ output is $D\ 0\ D$; again

$$
\begin{array}{cccccc}
N1 & N2 & N3 & N4 & N5 & N6 \\
0 & 1 & X & D & & \\
& & & D & 0 & D
\end{array}
$$

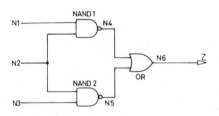

Figure 9.4 Network demonstrating singular cover

Thus the D-drive for the fault $N1/1$ can be defined as $d = 0\ 1\ X\ D\ 0\ D$, that is $N1 = 0$, $N2 = 1$, $N3 = X$, $N4 = D$, $N5 = 0$ and $N6 = D$, which will propagate the fault to the output Z; note that $D = 1$ for correct operation. To consolidate these ideas let us redetermine the sensitized path for the circuit of Fig. 9.1 with the fault $X_2/0$. The primitive D-cube of failure for the stuck-at fault in the OR gate is

X_1	X_2	X_3	X_4	X_5	X_6	X_7	X_8
0	1	X	X	X	D	X	X

Table 9.6(a) gives the propagation D-cubes for the AND gate and in this case to match the X_6 terminal we must choose the dual form $D\ 1\ D$, giving

X_1	X_2	X_3	X_4	X_5	X_6	X_7	x_8
0	1	X	X	X	D	X	X
		1	X	X	D	D	X

Finally we must choose a propagation D-cube for the 3-input OR gate which matches X_7 and does not conflict with X_6; in this case we can choose $D\ 0\ 0$ giving the final drive path

X_1	X_2	X_3	X_4	X_5	X_6	X_7	X_8
0	1	X	X	X	D	X	X
		1	X	X	D	D	X
		0	0			D	D

Thus

$$d = 0\quad 1\quad 1\quad 0\quad 0\quad D\quad D\quad D$$

which can be verified from Fig. 9.1 to be the same as we obtained previously.

In setting up the D-drive it was necessary to search through a list of propagation cubes to establish the required match. As perhaps one would expect, there is an algorithmic method based on the D-calculus and the concept of D-intersection which achieves the same result; the method is once again due to Roth. The intersection rules for combining two propagation cubes can be defined, for our purposes, by the operator table shown in Table 9.9; note that they are very similar to the rules for intersecting primitive D-cubes.

To sum up, the D-algorithm for test pattern generation[10,11] consists

Table 9.9 D-intersection rules

	0	1	X	D	$\bar{D}$
0	0	φ	0	φ	φ
1	φ	1	1	φ	φ
X	0	1	X	D	$\bar{D}$
D	φ	φ	D	D	φ
$\bar{D}$	φ	φ	$\bar{D}$	X	$\bar{D}$

initially of deriving the propagation and failure D-cubes for the network under consideration. Once a fault has been postulated the D-drive takes place which propagates the fault to the primary outputs of the circuit; this is done, using D-intersection, for all possible paths. Finally, a consistency operation is performed which checks out and establishes all secondary and primary input conditions required to support the sensitive paths. This is achieved using the fault-free singular covers and assigning values to the don't-care terms; if the consistency check fails another possible path must be selected and tested.

Though the D-algorithm is a powerful method of test sequence generation it requires a large amount of computation and must inevitably be implemented on a digital computer.[12]

9.2.4   Partitioning method of fault detection

Partitioning is not strictly speaking a method of test generation but it is appropriate to include it in this section. In the partitioning approach a previously generated test set is applied to the faulty network; the set of faults will be partitioned into *equivalence classes* based on the logic values at the primary outputs. Since the members of each equivalent class generate the same output further tests are required to increase the degree of resolution until either f_0 (the correct machine) is identified alone (fault detection) or all faulty versions of the machine are isolated separately (fault diagnosis). The method is obviously best suited to the multi-flow method of testing logic circuits. The value of the method lies in its ability to try different test sets in sequence using the multi-flow procedure, and ascertain which one is best suited for a particular application. This implies the need for a *checkout criteria* to compare the various tests.

As an example consider the circuit of Fig. 9.2 whose fault matrix is shown in Table 9.1. There are fourteen possible faults so that the initial equivalence class is f_0–f_{14}. Suppose it is required to isolate f_0 as quickly as possible, which in essence requires determining a set of tests which when applied in sequence will separate the largest number of faulty circuits from the good circuit at each step. In order to determine which tests should be used we can consult the G_D matrix for the circuit (Table 9.1(b)) and list the number of detectable faults for each test; this is done in Table 9.10 where column $N1$ gives the number of faults detected by each test. From the table, t_3 detects the most faults and is obviously the best choice for the first test in the sequence. After application of t_3 two equivalence classes will be generated determined by their output values (see Fig. 9.5); these are

$$E_1^1 = (f_1, f_3, f_6, f_8, f_9, f_{11}, f_{14})$$
$$E_1^0 = (f_0, f_2, f_4, f_5, f_7, f_{10}, f_{12}, f_{13})$$

The same procedure is repeated on the equivalence class containing f_0 and the corresponding test weightings are shown in column $N2$ of Table 9.10. Note that there are two possible tests, t_1 and t_2, and we shall arbitrarily choose t_1 giving the equivalence classes

Table 9.10 Check-out criteria

Test	No. of faults detected		
	$N1$	$N2$	$N3$
t_0	5	2	1
t_1	5	(4)	–
t_2	5	4	3
t_3	(7)	–	–

$$E_2^1 = (f_0, f_4, f_5, f_{10})$$
$$E_2^0 = (f_2, f_7, f_{12}, f_{13})$$

The procedure is repeated until eventually f_0 is isolated and the full detection set is defined.

Note from Fig. 9.5 that in some cases it is not possible to partition the equivalences classes any further, for example as with E_2^0. This means that the faults contained in this class all respond with the same output to a particular test and hence are *indistinguishable faults*. This characteristic may be utilized to reduce the number of possible detectable faults in a circuit and hence reduce the computational load in test generation. The basic idea of *fault collapsing*[13] is to find sets of faults that cannot be distinguished individually and to replace these sets by a single representative fault. The reduction is performed by analysing the structure of the circuit in terms of gate types and their interconnections looking for specific fault-test relationships.

For example, an n-input gate has $2(n + 1)$ single stuck-at faults (all inputs and the output have s-at-0 and s-at-1 faults). However, it is not possible to distinguish between s-at-0 faults on the inputs and output for an AND gate, and s-at-1 faults for an OR gate. Similarly for a NAND (NOR) gate the set of input faults s-at-0 (s-at-1) and the output s-at-1 (s-at-0) is equivalent. Thus in generating test sets only $(n + 2)$ faults need to be considered for any n-input gate.

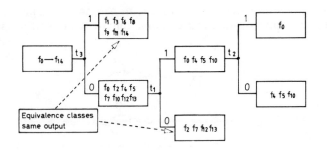

Figure 9.5 Partitioning technique

Again, in an AND gate the output s-at-1 will dominate any input to that gate s-at-1, and likewise for an OR gate, the output s-at-0 will dominate any input s-at-0. Similarly for NAND (NOR) gates the output s-at-1 (s-at-0) fault dominates any input s-at-0 (s-at-1). If both dominance and equivalent fault collapsing is used it is possible to reduce the number of faults for an n-input gate to $(n + 1)$ faults—a considerable reduction.

9.3 Testing sequential circuits

Since logic systems almost invariably consist of both combinational and sequential circuits the need to devise tests for sequential networks is of paramount importance. Unfortunately generating tests for sequential systems is considerably more difficult than for combinational circuits.

In contrast to the combinational case which usually requires a test vector as input, sequential systems by their very nature require a sequence of tests to check out the states of the machine. Moreover, since the initial starting state of the system will determine the response obtained it is essential either to know the state of the machine or to ensure some starting state.

The major problem in testing arises from the feedback loops intrinsic to any sequential system. For example, consider Fig. 9.6; here we have two loops, the feedback in the dc bistable and the $\bar{Z}$ output which is fed back to the preceding and separate combinational logic. In synchronous systems the bistables would normally be clocked so that the internal feedback paths can be ignored (not so of course in asynchronous systems). Now suppose in a synchronous system we attempt to set up a sensitized path through the combinational logic to identify a stuck-at fault at Z; due to the bistable's change of state, which is fed back, the path would consequently be nullified.

However, path sensitization techniques have been applied to sequential circuit testing, particularly the D-algorithm.[14,15] One approach is to represent the circuit in the classical finite-state machine model, that is as combinational logic plus memory. This entails determining the feedback loops (which can be difficult if the circuit is intuitively designed) and 'breaking' them so that they appear as separate inputs (the application or excitation equation description). The circuit can now be considered as a cascaded connection of identical combinational circuits with each circuit

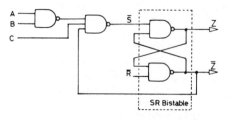

Figure 9.6 Feedback in sequential systems

representing the state of the sequential network at a given instant of time (normally for each clock pulse). In this form combinational test generation procedures can be applied to what is in essence a combinational circuit; note that the same fault must occur for many different time versions of the circuit.

An alternative approach is to separate the combinational and sequential logic either physically in the implementation or architecturally so that they can be independently tested. Sequential circuits such as bistables, shift registers, counters etc. can be tested functionally by checking that the device goes through all of its states and generates the required output responses. Care must be taken to ensure that the devices are correctly initialized either by using appropriate reset inputs or preset input sequences; the method also depends on the availability of external monitor points. One of the advantages of this method is that it imposes a design discipline on the logic designer and forces an early consideration of the testing procedures.

Unconstrained logic design can considerably complicate the testing problem but is a particular pitfall with sequential systems where over-enthusiastic optimizations can introduce races and hazards resulting in timing difficulties.

The testing problem is particularly acute with LSI/VLSI realizations where, due to the extremely large number of gates involved, it is not possible to generate test sequences which will test the total logic system on the basis of any overall input–output response. It is inevitable that the circuits must be designed in such a way that testing of individual sections can take place. Note also that with LSI/VLSI circuits it is not possible or, more strictly speaking, cost-effective to introduce extra monitor connections.

A procedure known as the *scan-path technique* has been evolved which requires partitioning the system in such a way that the memory devices can be tested separately, effectively as one long shift register, while the combinational logic can be checked using normal methods. Another technique is to introduce special logic on the chip itself which facilitates testing; one such method is *signature analysis* which checks out the logic with known test sequences. Both of these methods will be described in later sections. Most of the methods proposed for sequential circuit testing are based on extensions of combinational techniques; what formal theory exists is based on machine identification involving *state-table analysis*. The theory in its present state is too cumbersome to be usefully applied to large systems but nevertheless a knowledge of the techniques gives a useful understanding of the problems involved.

9.3.1 State-table analysis

The problem of testing sequential machines (restricted in our discussion to synchronous finite-state machines) is basically that of determining whether or not the machine is functioning according to specification or, more specifically, can the original state table be reproduced from the input–output behaviour.

In state-table analysis,[16–18] a *fault-detection experiment* must be set up to achieve this purpose. This consists of deriving an input sequence for the machine which takes it through all possible states and state transitions in such a way that faulty operation can be detected by observing the resulting output sequence; note that both the time and order in which the tests are applied are an essential part of the experiment. The fault detection experiment has the limitation that it is a functional test since it is derived from the unassigned state table and thus could apply to many different implementations of the machine. Moreover if the circuit has been designed intuitively it will be necessary to analyse the circuit and construct its state table.

In order to perform a fault-detection experiment it is obviously essential to know what state the machine is in at the start of the experiment. The state of the machine may be determined and then if necessary set to some initial state, by using a *distinguishing* or *homing* sequence. In some cases it is possible to determine a *synchronizing* sequence which will always take the machine to a specified final state regardless of the initial state. It will be obvious that sequences of this type will have a general application in sequential circuit testing since resetting the machine to a known starting state is an essential prerequisite for all forms of testing.

An input sequence I is said to be a distinguishing sequence if when applied to a machine M with n internal states it yields n different output responses depending on the initial states. Thus it follows that the initial state of M at the start of I can be inferred from observing the output response, assuming that the machine is functioning correctly. An input sequence which can uniquely determine the final state of a machine independent of the starting state is said to be a *homing* sequence. Every reduced machine possesses a homing sequence but only a limited number of machines have distinguishing sequences; note also that although every distinguishing sequence is also a homing sequence the converse is not true. Distinguishing and homing sequences can be determined by means of a *successor tree* which shows how the states of a machine can be successively resolved by their output states when an input sequence is applied.

The successor tree for machine M_1 in Table 9.11 is shown in Fig. 9.7. The tree starts with an *initial uncertainty* which is the minimal subset of the set of all machine states which is known to contain the starting state (for machine M_1 this is $ABCD$ since the machine could start in any state). If an input of $x = 0$ is applied we obtain the *uncertainty vector* $(A)(BCC)$ while an input of $x = 1$ gives $(ACD)(B)$. The uncertainty vectors are derived by

Table 9.11 Machine *M*1

Present state	Input x Next state/output	
	$x = 0$	$x = 1$
A	$C/0$	$D/1$
B	$C/0$	$A/1$
C	$A/1$	$B/0$
D	$B/0$	$C/1$

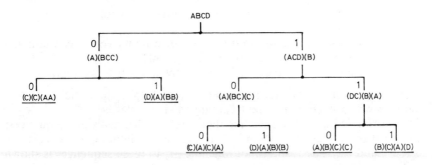

Figure 9.7 Successor tree for machine *M*1

inspecting the state table under the appropriate input, noting the next-state outputs and grouping the states accordingly. The individual uncertainties contained in a vector (bracketed) are called the *components* of the vector. The procedure is repeated using the uncertainty vectors from the previous pass, keeping the states derived from individual components separate and allowing all states to retain their identity (this last condition leads to repeated states in a component). For example, consider the successors of $(ACD)(B)$ in Fig. 9.7. The application of $x = 0$ to machine M_1 when in state B (see Table 9.11) will cause it to go to state C; the successor of component (ACD), however, depends on the outputs. It is (BC) if the output is 0 and (A) if it is 1, and thus the corresponding uncertainity vector is $(A)(BC)(C)$. Similarly the successor for $x = 1$ is $(A)(DC)(B)$; note that *all* next states are included in the vector and the states grouped together as a component all have the same output and are derived from the same component at the preceding level.

An uncertainty vector whose components contain a single state only is said to be a *trivial uncertainty vector*; if a vector contains components with either single states or identical repeated states it is called a *homogeneous uncertainty vector*. Thus in Fig. 9.7 $(C)(C)(AA)$ and $(C)(A)(C)(A)$ are homogeneous and trivial vectors respectively.

The homing and distinguishing sequences for a particular machine can be determined by drawing its successor graph and terminating the procedure according to the following rules:

(a) The *Homing tree* terminates for the following conditions:
   (i) the node is associated with an uncertainty vector whose non-homogeneous components are identical with some node in a preceding level;
   (ii) the node is associated with a trivial or homogeneous vector.
The homing sequence is defined by a path from the initial node to a node terminated by a trivial or homogeneous vector.

(b) The *distinguishing tree* terminates for the following conditions:
   (i) the node is associated with an uncertainty vector whose non-

homogeneous components are identical with some node in a preceding level;

(ii) the node is associated with an uncertainty vector containing a homogeneous non-trivial component;

(iii) the node is associated with a trivial uncertainty vector.

The distinguishing sequence is defined by a path from the initial node to a node in which each component contains a single state.

Referring to Fig. 9.7, we see that there are four distinguishing sequences: 1 0 0, 1 0 1, 1 1 0 and 1 1 1. Similarly there are four homing sequences: 0 0, 0 1, 1 0 and 1 1. The response of Machine M_1 to these sequences is shown in Table 9.12; note that there is a unique output for every component in the vector and that the distinguishing sequence uniquely identifies individual states.

Another more useful but rarer sequence is the *synchronizing sequence* which is an input sequence which always takes the machine to a unique final state irrespective of the starting state or output. An example of a machine which possesses a synchronizing sequence is given in Table 9.13; to determine a synchronizing sequence a simplified form of the successor graph can be used (though the full graph would work just as well) in which the outputs are ignored in the uncertainty vector and repeated states are omitted. Termination occurs when uncertainty vectors are repeated or a vector with a single element is obtained (called a *singleton* uncertainty vector); the synchronizing sequence is the path from the initial node to the node associated with a singleton. Figure 9.8 shows the synchronizing tree for machine M_2; in this case there are a number of sequences but the one with the shortest path is 0 0 0.

Putting the machine into a known initial state is the prelude to all sequential machine testing. In the absence of a synchronizing sequence the homing sequence will enable the state of the machine to be determined and then a *transfer sequence* must be used to put the machine into the required starting state for the fault detection experiment. This is called an *adaptive* procedure since the transfer sequence must obviously be determined by the machine's response to the homing sequence. (Note that it is assumed that any fault in the system will not corrupt this procedure!)

For example, consider machine M_1 in Table 9.11; using the homing sequence 0 1 we could establish from the resulting output that the machine is either in state B (output 0 0), state D (output 1 1) or state A (output 0 1). Assuming that we wished the tests to commence in state A we would need to apply the transfer sequences 1 and 1 1 or 0 1 respectively for the first two cases.

The classical fault detection experiment takes the following form:

(a) the application of a homing sequence which identifies the present state of the machine, followed by a transfer sequence to a specified starting state;

(b) the application of an input sequence based on a distinguishing sequence which causes the machine to visit each of the states and display its response;

Table 9.12 Response of *M*1 to *HS* and *DS* inputs

Initial state	Output			Final state
A	1	0	1	B
B	1	0	0	B
C	0	0	1	A
D	1	1	1	D

(a) Distinguishing sequence 1 0 1

Initial state	Output			Final state
A	1	1	0	B
B	1	1	1	C
C	0	1	1	D
D	1	0	1	A

(b) Distinguishing sequence 1 1 1

Initial state	Output		Final state
A	0	0	B
B	0	0	B
C	1	1	D
D	0	1	A

(c) Homing sequence 0 1

Initial state	Output		Final state
A	1	1	C
B	1	1	D
C	0	1	A
D	1	0	B

(d) Homing sequence 1 1

Table 9.13 Machine *M*2

Present state	Input x Next state/output x = 0	x = 1
A	B/0	E/0
B	E/1	B/0
C	D/1	A/1
D	B/0	C/0
E	E/0	D/1

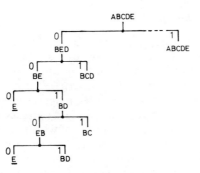

Figure 9.8 Synchronizing tree for machine *M*2

(c) the machine is presented with an input sequence which causes each state transition to be verified for all input conditions.

The start of the preset test begins by assuming that the machine is in some known state, say state A for machine M_1. Now suppose we apply the distinguishing sequence 1 1 1; then we would have

$$
\begin{array}{cccc}
\text{output} & \text{output} & \text{output} & \text{output} \\
\text{State } A \rightarrow \text{State } B & \rightarrow \text{State } C & \rightarrow \text{State } D & \rightarrow \text{State } A \\
1\ 1\ 0 & 1\ 1\ 1 & 0\ 1\ 1 & 1\ 0\ 1
\end{array}
$$

which checks that each state has been visited and does in fact exist. The distinguishing sequence we have chosen is an ideal one in that it covers all states and returns to state A. In a more usual example it would be necessary to apply a transfer sequence to ensure all states are visited.

In the final part of the experiment to verify each state transition the procedure is to apply a single input transfer sequence which causes the desired transition and then to identify it by applying the distinguishing sequence. For example, starting again in state A we could choose to test the 0-transition out of state A to state C. This is done by applying 0 1 1 1 which should put the machine into state D and output 0 0 1 1. Continuing in this way it is possible to derive a complete preset experiment.

A major drawback of the method is that it relies on the machine having a distinguishing sequence; moreover it also assumes that the number of states remain unchanged when a fault occurs. It should be possible, however, to design a machine with a structure such that it possesses a distinguishing sequence. One way of achieving this is to introduce additional output terminals.[19] Perhaps even more important from a practical test point of view is the need to design machines with a synchronizing or homing sequence since this is a major problem in testing sequential systems. This of course presents a strong argument for designing testable logic rather than evolving test methods for unconstrained designs.

Unfortunately there are no algorithms available for generating preset tests and the amount of manual computation involved, particularly with large machines, soon becomes prohibitive. Consequently the method is only of theoretical interest at the present time.

9.3.2 Scan-path techniques

The testing methods we are going to discuss in this and the next section are pragmatic solutions to the problem of testing LSI/VLSI circuits. In both cases they require built-in logic circuits on the chip itself to facilitate the testing procedures. The *scan-path approach*[20,21] is based on establishing a special test mode for the circuit which reconfigures all memory devices into a serial shift register structure. Suitable test patterns can then be shifted through the memory devices, the outputs of which are used to present test inputs to the combinational logic and at the same time gather the output responses.

Figure 9.9 shows the method applied to the testing of a simple 2-bit parallel adder circuit. In normal operation with control signal $TM = 0$ the

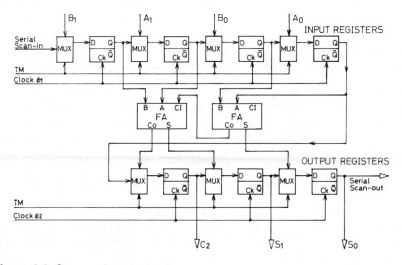

Figure 9.9 Scan-path test circuit

two-to-one multiplexer selects the parallel input data paths A_0A_1, B_0B_1 to load the augend and addend registers (*D*-type bistable) of the adder. Similarly when $TM = 0$ the sum outputs of the adder are loaded into the sum register. Note the use of two phase clocks φ_1 and φ_2 to ensure that the circuit has settled down before the output is registered (input on φ_1, output on φ_2). Note also that the circuit has the structure of a typical finite-state machine.

In the test mode with $TM = 1$ the multiplexers select the serial input; these data will be shifted from input to output of the bistable stages. Since the input and output registers of the adder are connected together one long shift register results with one input and one output to the system. In this way the attendant increased pin count for the chip (external connections) due to testing requirements are minimized. The testing procedure is in two stages:

(a) *Verification that the shift register is functioning correctly*
This normally consists of two tests:
   (i) a 'flush-test' in which all shift clocks are turned on and a signal is flushed through the registers from scan-in to scan-out;
   (ii) a 'shift-test' in which a special test pattern designed to detect all stuck-at faults is shifted through each register (it has been found in practice that the pattern 0 0 1 1 0 0 1 1 0 0 1 1 can be used for this purpose).

(b) *Combinational logic testing*
This presupposes that a test set has been generated using the standard procedures described in earlier sections. In general since individual modules of combinational logic are tested, rather than the overall system, the computational requirements for test generation are considerably reduced. In

the case of the adder with $TM = 1$ the augend and addend registers are loaded serially with the test inputs; this would take four clock pulses in our example. Now at the fifth clock pulse TM is set to 0 and the output of the adder is loaded (in parallel) into the sum register. During the next four clock pulses TM is again set to 1 and the contents of the output register is shifted out and checked for errors. Note that special external test equipment will be required for this purpose.

In practice sequential circuits are mostly implemented in terms of application equations using, for example, PLAs to generate the combinational logic and D-type bistables to provide the memory. Since the scan technique can obviously be used to set any input or internal state value it becomes possible to examine in detail the internal state behaviour of a machine.

Note that another advantage of the method is that by applying the shift input it is possible to initialize every bistable element in the system to any required state. The scan-path technique has the disadvantage of increasing the testing time, because of the serial nature of the test data, and also increasing the hardware overheads (typically in the order of 20% or less).

This structured design approach to testability has been taken a stage further by IBM[22] in their level-sensitive scan design (LSSD) method. Again it is basically a scan-path technique but in order to overcome other problems associated with undisciplined design, such as hazards and races due to the ac characteristics of the logic devices, only special level-sensitive memory devices are employed. In this context level-sensitive refers to constraints on circuit excitation, logic depth (propagation delays) and the handling of clocked circuitry.

A key element in the design is an asynchronous *shift register latch* (SRL) that does not contain any race or hazard conditions and which can support independent scan-in/scan-out paths. The circuit for an SRL is shown in Fig. 9.10(a) and (b); note that it is a master–slave arrangement.

The C and D inputs to the master latch $L1$ form a normal D-type memory mode function in that when $C = 0$ the latch cannot change state but when $C = 1$ the internal state of the latch assumes the value of the data input D. Under normal operation it is arranged that the clock input C is 0 during the time the data signal D is changed, and thus C goes to 1 only when D has settled out and become stable.

The I, A, B inputs and the second latch $L2$ comprise the additional circuitry for the shift register function. The master latch $L1$ has two separate input ports for entry of new data. System data is stored from D when the system clock C goes to 1. Serial data are inputted at I when the A shift clock is energized, taking the data from the previous SRL in the scan path (see Fig. 9.10(c)). The slave latch $L2$ stores the data from the master $L1$ when the B shift clock is energized. Thus when A shift and B shift are both 0 the $L1$ latch operates as a D-type bistable. When the latch is operating as a shift register, data from the preceding SRL are gated into latch $L1$ via input I when A shift $= 1$. After the A shift signal has gone to 0 the B shift signal goes to 1 and gates the data in $L1$ into $L2$; note the master–slave operation. For the shift register to operate correctly shift A and shift B can

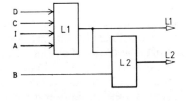

a) Block diagram SRL

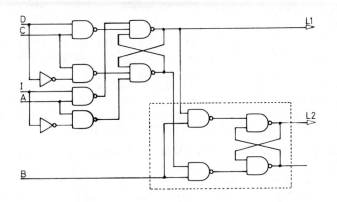

b) NAND implementation of SRL

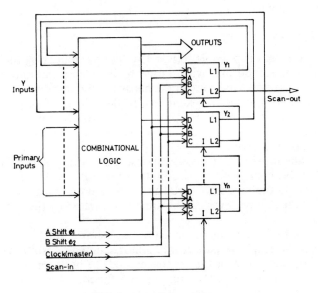

c) FSM structure with two clocks

Figure 9.10 LSSD systems

never both be 1 at the same time, and therefore out-of-phase clocks must be used. The interconnection of the SRLs into a shift register is shown in Fig. 9.10(c); note that the I terminals (input) are connected to the $L2$ terminals (output) and the shift A and shift B clocks are connected in parallel.

It will be apparent that to design an LSSD logic system it is essential to use SRL devices throughout for all internal storage and that the system and shift clocks must be carefully derived and controlled; providing the requisite design rules[22] are followed this does not present any serious difficulties.

The combinational logic may be tested as described previously by shifting the desired test pattern into the SRLs which are applied directly to the primary inputs. After the logic has settled out the system clock is turned on and the outputs stored in the $L1$ master latches. The contents of the $L1$ latches are then shifted out and compared with the expected response.

The LSSD method can be extended to handle very large circuits by additional partitioning of the system; simulation tools are also available to allow checking of the circuits for any violation of the design rules.

On the negative side LSSD has the following disadvantages:

(a) the SRLs are two to three times as complex as simple latches;
(b) up to four additional I/0 points are required at each package level for control of the shift registers;
(c) external asynchronous input signals must not change more than once every clock cycle;
(d) all timing within the module is controlled by externally generated clock signals.

9.3.3 Signature analysis

As we have seen it is possible to test a system by comparing the responses obtained from a good unit with those obtained from the unit under test. Rather than use an actual copy of the hardware the response of a good unit is normally stored, using for example a ROM. The good response may be obtained from a known good copy (called the golden unit!) or by simulation/analysis techniques. Figure 9.11(a) shows the block diagram for a stored response system.

The major drawback of good response generation, particularly if it is required to be incorporated as a built-in test for LSI/VLSI, is the huge amount of data that must be stored and analysed (note also that the speed with which the test patterns can be applied is limited by the access time of the memory).

A solution to this is to use compact testing (shown in Fig. 9.11(b)) which compresses the response data R into a more compact form $f(R)$, and thus only the compact form of the response needs to be stored. Obviously an important requirement of the compression function f is that it should be simple to realize in hardware.

Signature analysis is a compact testing technique developed by Hewlett-Packard for testing LSI systems.[23] In this method each output response is

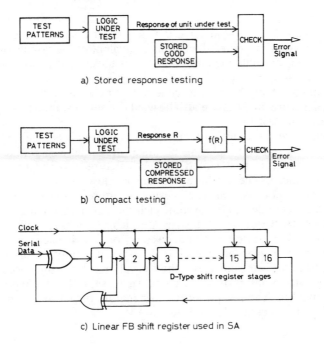

a) Stored response testing

b) Compact testing

c) Linear FB shift register used in SA

Figure 9.11 Compact testing using signature analysis

passed through a linear feedback shift register (see Section 6.5) whose contents $f(R)$ after the test patterns have been applied are called the *signature* (in effect the residue after dividing by a primitive polynomial). Figure 9.11(c) shows a typical linear feedback shift register used for signature analysis; note that in essence we have an inverse filter circuit performing the function $f(R) = R/(1 + D + D^2 + D^{16})$ where the feedback is given by the primitive polynomial $1 + D + D^2 + D^{16}$. Providing the linear feedback shift register has sufficient stages the signatures generated for a good circuit can be regarded as a unique fingerprint and there is little probability of a fault producing a good signature. It can be shown that for any response data stream of length $n > 16$ the probability of missing a faulty response using a 16-bit signature is given by

$$\frac{2^{n-16} - 1}{2^n - 1} \approx 2^{-16}, \qquad \text{for } n \gg 16$$

Hence the possibility of missing an error is in the order of 0.002%.

Though there are 2^{16} possible signatures (in our example) which would require a 65536 (64K) 16-bit word ROM, only the good signatures need to be stored. Moreover the hardware required to realize the compression logic is quite small, and thus signature analysis provides an attractive solution to response evaluation testing.

The technique can be used in a number of ways: as a fault-finding technique for isolating faults that have developed on boards in the field etc., where the method becomes the digital equivalent of the voltage/waveform checking used for analogue equipment; it can also be used as a method of production testing for boards or chips. In both these cases external equipment would normally be used to generate and check the input/output responses. Signature analysis can also be used in a self-test mode using built-in generation and fast logic.

As with all testable design techniques, however, it is essential that the decision to use signature analysis occurs at the initial design phase. Moreover the method is best suited to bus-structured architectures such as memory or microprocessor based designs because of the ease of applying and analysing the tests. The key to using signature analysis for testing LSI/VLSI circuits is to design the system such that it can stimulate itself; this is where microprocessor-based designs are ideal. Otherwise self-stimulation would normally require additional circuitry or ROM space. For example, to check out a microprocessor chip the basic functions of the microprocessor must be working correctly (the Kernal) which are then used to check out in sequence other parts of the circuit such as the RAM, I/0 ports, interrupt facilities etc. The test patterns and signature responses can be resident in ROM or if necessary loaded into RAM.

9.4   Fault-tolerant and testable logic design

Fault-tolerance in logic systems is normally obtained by incorporating protective redundancy in the system usually in the form of additional hardware. The underlying principle of *fault-tolerant design* is that machines must be able to tolerate a fault (called *fault-masking*) in such a way that the required system behaviour is seemingly unaffected. This is quite different to *fail-safe design* where a system is simply required to operate under fault conditions without *adverse* effects on the overall system.

Testable logic design on the other hand has the prime objectives of producing a logically irredundant circuit which has a small and easy-to-generate test set able to locate faults to a desired degree.

These topics are an extremely important aspect of logic design since they represent a means of overcoming the reliability and testing problems brought about by the complexity of current LSI/VLSI systems. Considerable work has taken place in this area[24,25] and much more is still in progress. We shall endeavour in this section to present some of the basic ideas and useful concepts as an introduction to the subject.

9.4.1   Testable logic design

The main approach that has been adopted in the work on designing testable combinational logic is to look for circuit structures that can be easily tested and then to translate any arbitrary logic function into this form. For example, it is easier to devise tests for an n-input parity check circuit if it is realized in terms of a cascaded array of exclusive OR gates than as a two-level AND/OR network (as we have seen only three tests are required for the 2-input EXOR gate). Reddy[26,27] has proposed a design technique based on the Reed–Muller (RM) expansion[28] which will realize any arbitrary n-variable function using a cascaded connection of AND and exclusive OR gates. Any arbitrary logic function can be expressed by a generalized Reed–Muller canonical expansion of the form

$$f(x_1, x_2, \cdots, x_n) = C_0 \oplus C_1 x_n \oplus C_2 x_{n-1} \oplus C_3 x_{n-1} x_n \oplus \cdots$$
$$\oplus C_4 x_{n-2} \oplus \cdots \oplus C_{2^n-1} x_1 x_2 \cdots x_n$$

where x_i are the input variables in the true or uncomplemented form, C_i is a binary coefficient having the value 0 or 1 and $\oplus$ is the modulo-2 sum. Thus for a three-variable function the corresponding RM expansion is

$$f(A, B, C) = C_0 \oplus C_1 C \oplus C_2 B \oplus C_3 BC \oplus C_4 A \oplus C_5 AC$$
$$\oplus C_6 AB \oplus C_7 ABC$$

which can be realized using the general circuit configuration shown in Fig. 9.12(a). Note that each AND gate corresponds to a product term in the expansion for which $C_i = 1$. As we shall see later the binary coefficients C_i are derived from the minterms (ON terms) for the actual function to be implemented.

The RM expansion has even greater generality; for example, rather than express the function with the input variables in true form we could have chosen to use the complemented form (but one or the other, and not both together). Moreover a similar expression could be developed for each of

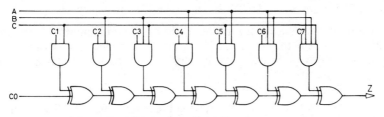

a) Generalised RM circuit for 3 variables

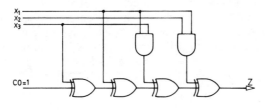

b) $Z = \bar{x}_1 \bar{x}_3 + x_1 x_2$

Figure 9.12 Reed–Muller circuits

the 2^n possible sets of true and complemented variables; the coefficients, of course, will differ for each set.[29]

The binary coefficients C_i in the RM expansion can be derived from the minterms f_i in the truth table for the function. For the 3-variable function $f(A, B, C)$ note that

C_1 is the coefficient of C,	$f_1 = 0\ 0\ 1$
C_2 is the coefficient of B,	$f_2 = 0\ 1\ 0$
C_3 is the coefficient of BC,	$f_3 = 0\ 1\ 1$
C_4 is the coefficient of A,	$f_4 = 1\ 0\ 0$ etc.

In order to determine the coefficients the *sub-numbers* of each binary equivalent of f_i must be formed; this can be done by replacing all 1s and 0s in all possible ways; for example

$$f_3 = 0\ 1\ 1 = 3 \qquad \text{and} \qquad f_5 = 1\ 0\ 1 = 5$$
$$\ 0\ 0\ 1 = 1 \qquad\qquad\qquad\ 1\ 0\ 0 = 4$$
$$\ 0\ 1\ 0 = 2 \qquad\qquad\qquad\ 0\ 0\ 1 = 1$$
$$\ 0\ 0\ 0 = 0 \qquad\qquad\qquad\ 0\ 0\ 0 = 0$$

note that $1 \subseteq 3$, $2 \subseteq 3$, $0 \subseteq 3$ etc.

Following this procedure leads to the following rules for a 3-variable function:

$$C_0 = f_0 \qquad ; \quad C_1 = f_0 \oplus f_1$$
$$C_2 = f_0 \oplus f_2; \quad C_3 = f_0 \oplus f_1 \oplus f_2 \oplus f_3$$
$$C_4 = f_0 \oplus f_4; \quad C_5 = f_0 \oplus f_1 \oplus f_4 \oplus f_5$$
$$C_6 = f_0 \oplus f_2 \oplus f_4 \oplus f_6;$$
$$C_7 = f_0 \oplus f_1 \oplus f_2 \oplus f_3 \oplus f_4 \oplus f_5 \oplus f_6 \oplus f_7$$

The values of the coefficients are obtained by summing modulo-2 the value of the f_i terms for the actual function under consideration. Consider the Boolean function $Z = \bar{x}_1\bar{x}_3 + x_1x_2$; this has the Reed–Muller expansion

$$Z = 1 \oplus x_3 \oplus x_1 \oplus x_1x_3 \oplus x_1x_2$$

which may easily be checked by drawing the truth table. The RM circuit for this function is shown in Fig. 9.12(b).

The starting point for the Reddy method of designing testable logic is to implement the function in the Reed–Muller form using the AND/exclusive OR array. Now Kautz[30] has shown that to detect a single faulty gate with a s-at-0 or a s-at-1 fault in a cascade of exclusive OR gates it is sufficient to apply a set of tests which will exercise all possible input conditions to each gate; the test matrix for a 3-variable RM circuit is given by

$$T_1 = \begin{array}{c} \begin{array}{cccc} C_0 & x_1 & x_2 & x_3 \end{array} \\ \begin{bmatrix} 0 & 0 & 0 & 0 \\ 0 & 1 & 1 & 1 \\ 1 & 0 & 0 & 0 \\ 1 & 1 & 1 & 1 \end{bmatrix} \end{array}$$

The test assumes that the primary inputs are fault-free and that faults can only occur on the gate inputs. Note that the structure of the test set and the number of test vectors is always the same and that it is independent of the function being realized. For instance a four-variable test set would have the form

$$T_1' = \begin{array}{c} \\ \end{array} \begin{array}{ccccc} C_0 & x_1 & x_2 & x_3 & x_4 \\ \left[\begin{array}{ccccc} 0 & 0 & 0 & 0 & 0 \\ 0 & 1 & 1 & 1 & 1 \\ 1 & 0 & 0 & 0 & 0 \\ 1 & 1 & 1 & 1 & 1 \end{array}\right] \end{array}$$

Reddy has also shown that an s-at-0 fault on the input or output of any AND gate in the RM circuit can be detected by applying either one of the test inputs 0 1 1 1, 1 1 1 1. Similarly an s-at-1 fault on the output of an AND would be detected by the test inputs 0000 and 1000. However, a s-at-1 fault at any of the inputs to the AND gates can only be detected separately using the test-set

$$T_2 = \begin{array}{c} \\ \end{array} \begin{array}{cccc} C_0 & x_1 & x_2 & x_3 \\ \left[\begin{array}{cccc} X & 0 & 1 & 1 \\ X & 1 & 0 & 1 \\ X & 1 & 1 & 0 \end{array}\right] \end{array}$$

where the X is a don't-care input. The justification for this test set is that, for any input vector of T_2, one input is held at 0 while all the other inputs are set to 1. Thus for a particular test an s-at-1 fault on any input will be propagated through the network to the output terminal. For an n-variable function the full test-set will now consist of $T_1 + T_2$ and will contain $(n + 4)$ tests.

To detect the presence of primary input faults it is necessary to sensitize an odd number of paths from the faulty input to the output (with exclusive OR gates an even number would cancel out). The $(n + 4)$ tests above will also detect input faults which appear in an odd number of product terms in the original RM expansion. For instance inputs x_1 and x_2 in our example appear an odd number of times and hence s-at-0 faults on these lines will be detected by either one of the test inputs 0 1 1 1 or 1 1 1 1; an s-at-1 fault on x_1 will be detected by either 1 0 1 1 or 0 0 1 1 and an s-at-1 fault on B by 1 1 0 1 or 0 1 0 1. To detect faults in input variables appearing an even number of times requires additional tests which increases the total number of tests by $2n_c$, where n_c is the number of input variables appearing an even number of times, in our example, since x_3 occurs twice, $n_c = 1$. However, it is possible to dispense with this test by including extra logic and output terminals. The inputs which appear an even number of times are simply ANDed together to give an additional output; in our example this means providing an additional output terminal for x_3.

The technique is an interesting one and because of its regular structure and the possibility of self-testing (the test vectors could be stored in a ROM) it could be of application to LSI design. However, it has the usual drawback of cascaded circuits in that it considerably increases the propagation delay and also requires rather excessive amounts of hardware.

The method has, however, been extended to two-dimensional arrays[31] based on AND/exclusive OR networks again using RM canonical forms for realization. It has been shown that a stuck-at fault in a single cell can be detected using $2n + 5$ tests.

Numerous other techniques have been described; again, many of these are based on a particularly easily tested configuration. The approach of deriving a universal test-set[32] for a particular circuit structure, say the two-level AND/OR realization, is particularly attractive since it considerably reduces the test generation problem and would seem applicable to large systems.

The size of the universal test set is, however, dependent on the degree of unateness of the function (a *unate function* is one in which the variables are either all in true or complemented form but not both together—the RM expansion used above is a positive unate function). If the variables must appear in a mixed form (due to the characteristics of the functions) very little reduction in the size of the test set can be made but if unate functions, or functions with a large degree of unateness, can be used considerable reductions result.

In practice switching functions are not generally unate, but if a double-rail logic system is employed (both true and complemented logic signals are distributed round the system) then any function could be converted to, say, a positive unate function by considering complemented variables to be independent. For example, the function $Z = A\bar{C} + AB$ is not unate but by considering $\bar{A}$ and $\bar{C}$ as independent variables a and c, the function can be expressed as $Z = ac + AB$ where $a = \bar{A}$ and $c = \bar{C}$.

Many of these methods also depend on extending the circuits' *observability* (the ability to observe the effect of a fault at primary outputs) and *controllability* (the extent to which internal conditions can be controlled by the primary outputs).

One technique due to Hayes[33] which illustrates this is based on designing conventional two-input NAND and invertor gate circuits and then adding control inputs by inserting two-input exclusive OR gates into the input line of all NAND gates and replacing invertors by exclusive OR gates. This is shown in Fig. 9.13(a) and (b) for the function $Z = \bar{A}\bar{C} + AB$ used in our earlier example. Since only one input of the exclusive OR gate is required to propagate the input variable the other input can be brought out as a primary input and used as a control terminal. Note that since $x \oplus 0 = x$ and $x \oplus 1 = \bar{x}$ for normal operation the control inputs K_i for NAND inputs and invertors would be 1 and 0 respectively. Now the basic exclusive OR/NAND configuration, shown in Fig. 9.13(c), has the universal test set shown in Table 9.14. In order to test the module it is necessary to apply all four input combinations to each exclusive OR gate and ensure that their outputs (I_1 and I_2) produce the four different input combinations to the following NAND gate; this may be done with the first four of the tests shown in Table 9.14. If I_0, the output of the module, becomes an input to a following exclusive OR gate, as will normally be the case, it is necessary to generate an additional zero on I_0 to satisfy the following exclusive OR's

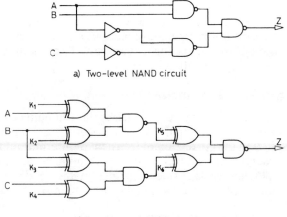

a) Two-level NAND circuit

b) Transformed EXOR circuit

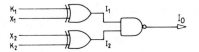

c) Basic logic circuit

Figure 9.13 Haye's technique

Table 9.14 Test vectors for EXOR/NAND

x_1	x_2	K_1	K_2	I_0	I_1	I_2
0	0	0	0	1	0	0
0	0	1	1	0	1	1
1	1	1	0	1	0	1
1	.1	0	1	1	1	0
1	1	0	0	0	1	1

input requirement of two 1s and two 0s; the fifth test in Table 9.14 will satisfy this requirement. Thus each module requires a minimum of five tests to ensure full fault detection.

Hayes also described a method of specifying five-bit sequences on all primary and control lines such that module outputs and successor module inputs are compatible in the sense that the testing requirements are met, thus enabling the testing of large networks.

Though this method is a good illustration of the techniques involved in designing testable logic it is not a practical method for LSI/VLSI circuits because of the large number of extra input terminals required. Again the inclusion of extra control circuitry will also increase the propagation delays in the system.

The design of testable sequential circuits has mainly concentrated on pragmatic solutions, as we saw in the last section. Some work has been done, however, on improving state-table methods by adding additional outputs to produce distinguishing sequences[34,35] or additional control inputs to increase the controllability.[36]

9.4.2   Failure tolerant logic design

Error-detecting and correcting codes, such as the Hamming code etc., play a large part in the design of fault tolerant systems[37] since they are particularly well suited to the correction of transient or intermittant faults. They are used, for example, to test ROM and RAM modules under operational conditions employing error-correcting codes in the read/write circuitry. Also, as we shall see later they can be used to produce self-checking counter designs.

This section will deal with the design of machines possessing integral redundancy using either error-correcting codes for state assignment or independent self-checking circuitry. It is essential in these designs that the tolerated faults can easily be detected in an operational environment. Without fault detection capability it becomes impossible to determine whether or not a unit is completely fault-free or to maintain the reliability of the system by rectifying the tolerated faults as soon as it is conveniently possible. It is also important that any additional logic added to the design to effect fault-tolerance be kept to a minimum.

Using minimum-distance-3 codes for state assignments and considering error-states when deriving the bistable input equations it is possible to design synchronous counters that can tolerate single faults and thus automatically correct for one-bit errors.[38] The state assignment must be such that the set of required (correct) states is disjoint with the set of possible error states; thus when an error occurs it will cause a transition to an error state which can then be detected and corrected. Since all the possible error states for a given transition can be determined, the counter can be designed to correct automatically for the error by ensuring that a transition back to the correct state is effected on the next clock pulse. Note that the output of the counter when in an error state must be assigned a safe value. It follows from the use of a minimum-distance-3 code that only one bistable can be in error at any one time (which implies that the input circuitry controls only one bistable) and thus will only correct for single errors.

Table 9.15 Error-tolerant counter

	Present state					Next state				
	Y_1	Y_2	Y_3	Y_4	Y_5	Y_{1+}	Y_{2+}	Y_{3+}	Y_{4+}	Y_{5+}
A	0	0	0	0	0	0	0	1	1	1
B	0	0	1	1	1	1	1	0	1	1
C	1	1	0	1	1	1	1	1	0	0
D	1	1	1	0	0	0	0	0	0	0

To illustrate the design technique consider the transition table for a 4-state counter as shown in Table 9.16. The states have been assigned using a distance-3 code such that each state transition requires a change of three variables (separated by distance-3).

Consider the transition from state A to state B: the occurrence of a single s-at-0 or s-at-1 fault would result in the following error states:

$$\text{State } A \quad 0\ 0\ 0\ 0\ 0 \rightarrow \text{State } B \quad 0\ 0\ 1\ 1\ 1$$

$$\text{Error states} \begin{cases} 1\ 0\ 1\ 1\ 1 \\ 0\ 1\ 1\ 1\ 1 \\ 0\ 0\ 0\ 1\ 1 \\ 0\ 0\ 1\ 0\ 1 \\ 0\ 0\ 1\ 1\ 0 \end{cases}$$

Note that the error states are all distance-1 from the correct state and at least distance-2 from each other. For fault-tolerant operation it is necessary to ensure that if an error has occurred then the correct next state will nevertheless be reached. This is achieved by arranging that the next-state equations include the error states as well as the correct state. Table 9.16 shows the full transition table including all possible error states, while

Table 9.16 Transition table showing error states

	Present state							Next states					
	Y_1	Y_2	Y_3	Y_4	Y_5			Y_{1+}	Y_{2+}	Y_{3+}	Y_{4+}	Y_{5+}	
A	0	0	0	0	0			0	0	1	1	1	
	1	0	0	0	0			1	0	1	1	1	
	0	1	0	0	0			0	1	1	1	1	
	0	0	1	0	0			0	0	0	1	1	Error states
	0	0	0	1	0			0	0	1	0	1	
	0	0	0	0	1			0	0	1	1	0	
B	0	0	1	1	1			1	1	0	1	1	
	1	0	1	1	1			0	1	0	1	1	
	0	1	1	1	1			1	0	0	1	1	
	0	0	0	1	1			1	1	1	1	1	Error states
	0	0	1	0	1			1	1	0	0	1	
	0	0	1	1	0			1	1	0	1	0	
C	1	1	0	1	1			1	1	1	0	0	
	0	1	0	1	1			0	1	1	0	0	
	1	0	0	1	1			1	0	1	0	0	
	1	1	1	1	1			1	1	0	0	0	Error states
	1	1	0	0	1			1	1	1	1	0	
	1	1	0	1	0			1	1	1	0	1	
D	1	1	1	0	0			0	0	0	0	0	
	0	1	1	0	0			1	0	0	0	0	
	1	0	1	0	0			0	1	0	0	0	
	1	1	0	0	0			0	0	1	0	0	Error states
	1	1	1	1	0			0	0	0	1	0	
	1	1	1	0	1			0	0	0	0	1	

Table 9.17 shows the K-maps for a D-type bistable realization; note that the entries E are the error terms and X the don't-cares, i.e. unused combinations. The application equations derived from the K-map are

$$Y_1 = Y_2 = Y_4Y_5 + Y_1\bar{Y}_3Y_5 + \bar{Y}_1Y_3Y_5 + Y_1\bar{Y}_3Y_4 + \bar{Y}_1Y_3Y_4$$
$$Y_3 = \bar{Y}_3\bar{Y}_4Y_5 + \bar{Y}_1\bar{Y}_3\bar{Y}_5 + \bar{Y}_2\bar{Y}_3\bar{Y}_5 + \bar{Y}_1\bar{Y}_2\bar{Y}_4\bar{Y}_5$$
$$+ \bar{Y}_3Y_4\bar{Y}_5 + Y_2\bar{Y}_3Y_5 + Y_1Y_2Y_4Y_5 + Y_1\bar{Y}_3Y_5$$
$$Y_4 = Y_5 = \bar{Y}_1\bar{Y}_2 + \bar{Y}_1\bar{Y}_3\bar{Y}_5 + \bar{Y}_2\bar{Y}_3\bar{Y}_5 + \bar{Y}_1Y_3Y_5 + \bar{Y}_2Y_3Y_5$$

which are shown implemented in Fig. 9.14. Though the method is conceptually very simple it suffers from the obvious disadvantage of very high implementation costs (it has been suggested some 5–6 times the cost of a conventional counter). Moreover the assignment procedure would become very complex for a large system.

A similar procedure for designing fail-safe machines uses k-out-of-n codes ($_nC_k$ codes where k is the weight and n the number of bits) for the state assignment.[39] Fail-safe operation is such that if a fault occurs the machine goes into an error state cycle which generates a safe output. Again it is essential to ensure that the set of required states and the set of all error states are disjoint since the procedure depends on a fault causing a transition to an error state. The advantage of using $_nC_k$ codes over, for example, a Hamming code (which would satisfy the basic requirements) is that

Table 9.17 K-maps for failure-tolerant counters

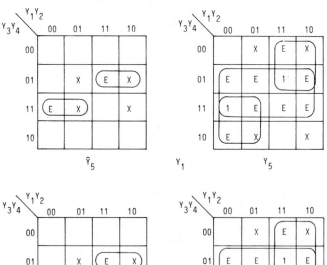
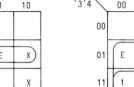

Table 9.16 Continued

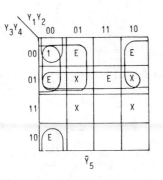

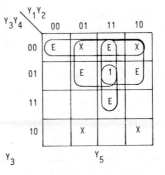

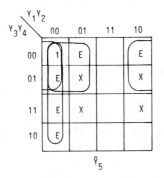

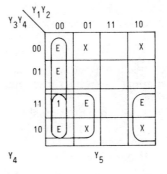

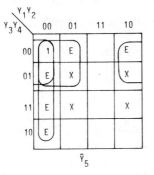

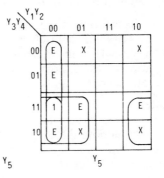

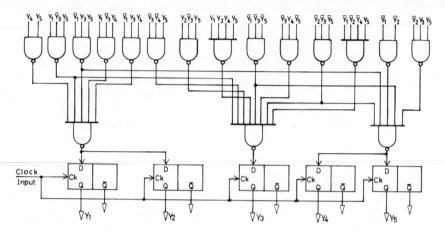

Figure 9.14 Failure-tolerant counter

a unate function (true variables only) can be derived for the application equations thereby simplifying the problem of fault detection and
realization.

Using unate equations enables the circuit behaviour to be predicted
under single fault conditions. For example a fault affecting a single state
variable and causing a false 1 can only force other state variables to
incorrect 1s but never to an incorrect zero. Similarly a zero fault can force
other state variables to incorrect 0s but never to an incorrect 1.

The design procedure to be described assumes that

(a) faults can only occur in the input logic to the memory element and
the bistables themselves;
(b) only single faults are possible emanating as s-at-1 or s-at-0 on the
output of a memory element;
(c) both primary inputs and clock lines are fault-free.

Unateness is obtained by arranging the assignment such that any state
variable y_i which has a value $y_i = 0$ has its inverse $\bar{y}_i$ included in a don't-care
term. Thus suppose the code 1 0 0 1 was used for assignment; then 1 1 0 1
and 1 0 1 1 could not be used for assignment and hence would be included
in the don't-care set. In this way the complemented variable y_i can be
eliminated from the reduced form of the equations; note that this procedure would follow automatically as a result of using an $_nC_k$ code.

Consider for example the transition table shown in Table 9.18(a) which
has been assigned using a 2-out-of-4 code. The excitation equations
derived from the K-maps in Table 9.18(b) are

$$Y_{1+} = Y_1 Y_4 + Y_3 Y_4$$
$$Y_{2+} = Y_1 Y_2 + Y_1 Y_4$$
$$Y_{3+} = Y_1 Y_2 + Y_2 Y_3$$
$$Y_{4+} = Y_2 Y_3 + Y_3 Y_4$$

Table 9.18 $_nC_k$ assignment

	Present states				Next states			
	Y_1	Y_2	Y_3	Y_4	Y_{1+}	Y_{2+}	Y_{3+}	Y_{4+}
A	1	1	0	0	0	1	1	0
B	0	1	1	0	0	0	1	1
C	0	0	1	1	1	0	0	1
D	1	0	0	1	1	1	0	0

(a) Transition table

(b) K-maps

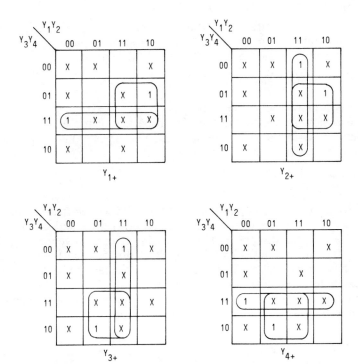

Note that the don't-care terms exclude members of the $_nC_k$ class. The excitation equation (the ON-set realization) may be obtained by direct inspection of the transition table; for example Y_{1+} has the terms $\bar{Y}_1\bar{Y}_2Y_3Y_4$ and $Y_1\bar{Y}_2\bar{Y}_3Y_4$, and thus $Y_{1+} = Y_3Y_4 + Y_1Y_4$.

The derivation of unate equations for $_nC_k$ coded tables can be generalized to include state tables with inputs x_j. The *ON-set realization* can be defined as

$$Y_i = F_i(x, y_1, y_2, \cdots y_n)$$
$$= \sum_j x_j(M_{j1} + M_{j2} + \cdots + M_{jp})$$

where

$$M_{jh} = y_{h1}y_{h2}y_{h3} \cdots y_{hk} \text{ and } h = 1 \sim p$$

$y_{h1}, y_{h2}, \cdots h_{hk}$ are k state variable having values 1 in a present state q_h that transfers to a state with $Y_i = 1$ under the input x_j (note in any $_nC_k$ code there will be k 1s and $n - k$ 0s).

The dual *OFF-set realization* can be similarly defined:

$$\bar{Y}_i = \bar{G}_i(x, y_1, y_2, \cdots y_n)$$
$$= \sum_j x_j(N_{j1} + N_{j2} + \cdots + N_{jr})$$

where

$$N_{jr} = \bar{y}_{h1}\bar{y}_{h2} \cdots \bar{y}_{h(n-k)} \text{ and } h = 1 \sim r$$

$\bar{y}_{h1}, \bar{y}_{h2} \cdots \bar{y}_{h(n-k)}$ are $n - k$ state variables having value 0 in a present state that transfers to a state with $Y_i = 0$ under input x_j.

By complementing the equation for $\bar{Y}_i$ we obtain

$$Y_i = \prod_j (\bar{x}_j + \bar{N}_{j1}\bar{N}_{j2} \cdots \bar{N}_{jr})$$

which is a more useful function.

Let us now return to the failsafe properties of the machine shown in Table 9.18. Using an $_nC_k$ code there are two types of fault that can result from a single error; these have the effect of

(a) decreasing k to $k - 1$
(b) increasing k to $k + 1$.

Now from the state diagram shown in Fig. 9.15 we see that any increasing error will cause the machine to go into an error cycle and stay there even after the error has disappeared; a decreasing error will eventually go to and stay in the all zero state. This follows from the fact that the set of correct states and error states are disjoint and allows the fault to be detected using, for example, a k-out-of-n checker circuit.

Unfortunately this is not always the case: consider the assignment shown in Table 9.19 which yields the ON-set equations

$$Y_1 = y_3y_4$$
$$Y_2 = y_1y_4 + y_2y_4$$
$$Y_3 = y_2y_4 + y_2y_3$$
$$Y_4 = y_1y_4 + y_2y_3 + y_3y_4$$

This time if an error causes the machine to go to state 1 1 1 0 the next $x = 1$ input will cause a transition to state 0 0 1 1 which is in the correct state set.

To overcome this defect it is necessary either to place restrictions on the state assignment[40] or to use both ON and OFF realizations in deriving the excitation equations. This is achieved by partitioning the set of state variables into two blocks $B1$ and $B2$ such that $B1$ has k members of state variables and $B2$ $n - k$ members (without losing generality it can be assumed that $B1$ includes $y_1, y_2; \cdots y_k$ and $B2$ $y_{k+1}, y_{k+2}, \cdots y_n$). The OFF-set realization is used for $B1$ and the ON-set for $B2$. If assignment codes

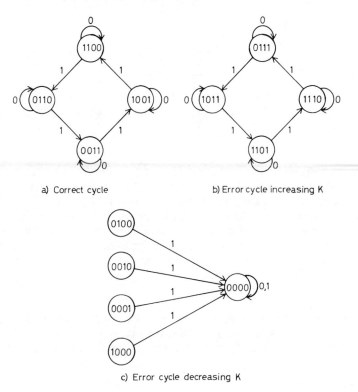

a) Correct cycle

b) Error cycle increasing K

c) Error cycle decreasing K

Figure 9.15 State cycles failsafe counter

of the form $y_1 = y_2 = \ldots y_k = 1$ and $y_{k+1} = y_{k+2} = \ldots y_n = 0$ are used, that is for example 1 1 0 0 in the case of a 2-out-of-4 code (called a $K1$ assignment), they must be treated as special cases. However, in the majority of cases it is possible to omit codes of this type from the assignment. Now let us reconsider the transition table shown in Table 9.19, and let $B1$ consist of the set (Y_1, Y_2) and $B2$ the set (Y_3, Y_4). Note that the $K1$ assignment is 1 1 0 0, which is not used. Extracting the OFF-sets for Y_1 and Y_2 and the ON-sets for Y_3 and Y_4 we have

$$Y_1 = (y_2 + y_3)(y_1 + y_3)(y_1 + y_4)$$
$$Y_2 = (y_1 + y_4)(y_1 + y_2)$$
$$Y_3 = y_2 y_4 + y_2 y_3$$
$$Y_4 = y_1 y_4 + y_2 y_3 + y_3 y_4$$

Referring back to our example, the error state 1 1 1 0 will now go to state 1 1 1 1 and stay there irrespective of input.

If this procedure for extracting the excitation equations is followed (with the exception of the $K1$ constraint), errors will always force a transition to a closed error state cycle of all 1s for an increasing fault and all 0s for a decreasing fault.

Table 9.19 $_cC_k$ coding

Present state				Next state			
Y_1	Y_2	Y_3	Y_4	Y_{1+}	Y_{2+}	Y_{3+}	Y_{4+}
1	0	0	1	0	1	0	1
0	1	0	1	0	1	1	0
0	1	1	0	0	0	1	1
0	0	1	1	1	0	0	1

A more systematic method for designing failsafe synchronous circuits[41] and one which gives considerable savings in hardware over other methods uses the concepts of partition theory similar to that previously described for asynchronous logic. As with other methods, only single logic faults in the excitation circuitry are considered, with both primary and clock inputs being uncorrupted. The first step in the design procedure is to derive the *p-sets* (predecessor sets) for each input column of the state table. The *p*-set of a state S_K for input I_j consists of all

$$S_i \in S | \delta(I_j, S_i) = S_K$$

where S is the set of all internal states and δ the state transition function.

For example, consider Table 9.20(a) for machine $M3$; $p_1 = (AD)$ is the set of predecessors of state B in column $\bar{x} = 0$; the full set of *p*-sets is

$$\text{column } \bar{x} \quad p_1 = (AD) \qquad \text{column } x \quad p_4 = (AB)$$
$$p_2 = (B) \qquad\qquad\qquad\quad p_5 = (D)$$
$$p_3 = (C)$$

Table 9.20 Machine $M3$

Present state	Next states $x = 0$	$x = 1$
A	B	C
B	D	C
C	A	B
D	B	A

(a) State table

Present states					Next states $x = 0$					$x = 1$				
	y_1	y_2	y_3	y_4	y_5	Y_1	Y_2	Y_3	Y_4	Y_5	Y_1	Y_2	Y_3	Y_4	Y_5
A	1	0	0	1	0	0	1	0	1	0	0	0	1	0	0
B	0	1	0	1	0	1	0	0	0	1	0	0	1	0	0
C	0	0	1	0	0	1	0	0	1	0	0	1	0	1	0
D	1	0	0	0	1	0	1	0	1	0	1	0	0	1	0

(b) Assigned table

The p-set list can in some cases be reduced (which will reduce the number of state variables required) by examining for the condition that if p_2 contains a subset of states contained by p_1 then p_2 may be deleted if the remaining states of p_1 are unspecified for the column of the state table from which p_2 was taken. This does not apply in the case of our example.

A state assignment can now be generated by associating an r_i-partition with each unique p_i in the p-set list such that r_i is coded 1 for each state in p_i and 0 for all states not in p_i. The full set of r-partitions is

$$r_1 = (AD)(BC) \qquad r_4 = (AB)(CD)$$
$$r_2 = (B)(ACD) \qquad r_5 = (D)(ABC)$$
$$r_3 = (C)(ABD)$$

The fully assigned table for machine $M3$ is shown in Table 9.20(b).

The next step is to extract the *next-state partitions* n_i^p which partition all the states to be specified 1 in the assignment for a next-state variable Y_i under input I_p from the states to be specified 0. This is done by examining the assigned state-table entries column by column:

$$\text{column } \bar{x} \quad n_1^1 = (BC)(AD) \qquad \text{column } x \quad n_2^1 = (D)(ABC)$$
$$n_1^2 = (AD)(BC) \qquad\qquad n_2^2 = (C)(ABD)$$
$$n_1^3 = (\varphi)(ABCD) \qquad\qquad n_2^3 = (AB)(CD)$$
$$n_1^4 = (ACD)(B) \qquad\qquad n_2^4 = (CD)(AB)$$
$$n_1^5 = (B)(ACD) \qquad\qquad n_2^5 = (\emptyset)(ABCD)$$

The reason for extracting the next-state partitions is to allow the next-state equations, the excitation equations, to be derived in a unate (monotonic) form.

The next-state equations for a sequential machine can be expressed in the following manner:

$$Y_1 = f_1^1(y_i)I_1 + f_1^2(y_i)I_2 + \cdots + f_1^m(y_i)I_m$$
$$Y_2 = f_2^1(y_i)I_1 + f_2^2(y_i)I_2 + \cdots + f_2^m(y_i)I_m$$
$$\vdots$$
$$Y_n = f_n^1(y_i)I_1 + f_n^2(y_i)I_2 + \cdots + f_n^m(y_i)I_m$$

where y_i are the present state variables, $Y_1 \cdots Y_n$ are the next-state variables, $I_1 \cdots I_m$ the input states and $f_1^1 \cdots f_1^m$ are functions of the internal state variables only.

The design equation coefficient f_i^p is related to the next-state partition n_i^p by the following theorems:[41,42]

(i)   If the next-state partitions $n_i^p \leq r_i$ then $f_i^p = y_i$.
(ii) If all the elements of n_i^p are in the same block of the partition then $f_i^p = 0$ or 1.
(iii) Let r_i partition p-sets p_i of input I_p (where $i = 1, 2, \cdots x$); then for any $n_i^p = (p_1 p_2 p_3 \cdots p_m)(p_{m+1} \cdots p) \, m \leq x$, then $f_i^p = \Sigma y_i, y_i \in y_j$ $j = 1, 2, \cdots, m$.

For example, from the list of p-sets and r-partitions for machine $M3$ we have

$$n_1^1 = (BC)(AD) = (p_2p_3)(p_1)$$

now

$$r_2 = (B)(ACD) = (p_2)(p_1p_3)$$

and

$$r_3 = (C)(ABD) = (p_3)(p_1p_2)$$

Thus r_2/r_3 partition the p-sets of n_1^1 and $f_1^1 = y_2 + y_3$ again

$$n_2^1 = (D)(ABC) = r_5; \text{ thus } f_2^1 = y_5$$

and

$$n_3^1 = (\emptyset)(ABCD) \text{ giving } f_3^1 = 0$$

It follows from these theorems that if y_i is a partitioning variable under input I_p then $f_i^p = y_i$. This means that partitioning variables cannot share common logic circuits and fault conditions must therefore be contained within individual circuits. Moreover it will be seen that each non-partitioning variable is a function solely of the partitioning variables (in the case of partitioning variables $n_i^p = r_i$; theorems (ii) and (iii) apply to non-partitioning variables). Furthermore it can be shown that the next-state design coefficient f_i^p of any partition set of I_p are realizable using the partition variable of I_p once only. The significance of this is that each partitioning variable is only required to feedback once (or not at all) to realize the f_i^p term of every partition set. It thus follows that the next-state equations will be unate and consequently no invertors are required in the feedback paths.

The full set of excitation equations for machine $M3$ are

$$Y_1 = y_2\bar{x} + y_3\bar{x} + y_5x$$
$$Y_2 = y_1\bar{x} + y_3x$$
$$Y_3 = y_4x$$
$$Y_4 = y_1\bar{x} + y_3\bar{x} + y_3x + y_5x$$
$$Y_5 = y_2\bar{x}$$

which is shown realized in Fig. 9.16. Note that the fault detection circuits function by examining the partition set of the circuit's present input state. If the parity is even a fault has occurred; in this case the parity of $(y_1y_2y_3)$ and $(y_3y_4y_5)$ must be checked for $\bar{x}$ and x respectively.

To illustrate the operation of the circuit under fault conditions consider Fig. 9.17. Now for an increasing fault, i.e. Y_1 s-at-1, the machine will go into an error cycle for $x = 1$ of $1\,1\,0\,1\,0 \leftrightarrow 1\,0\,1\,0\,0$ and for $x = 0$ will go to the error state $1\,1\,0\,1\,1$ and remain there independent of input. For a decreasing fault, i.e. Y_2 s-at-0, the machine will go to an all 0s state and remain there. For $x = 1$ the machine will go in fact to the correct state, and similarly for subsequent inputs of $x = 0$ and $x = 1$; this is due to the particular stuck-at-0 fault. However, for all other inputs a fault state will

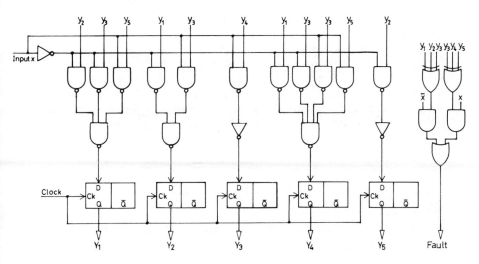

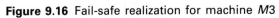

Figure 9.16 Fail-safe realization for machine $M3$

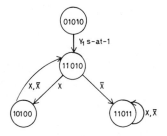

a) Increasing fault

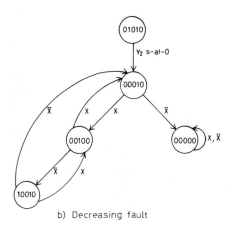

b) Decreasing fault

Figure 9.17 Error-cycles partition method

result with the machine eventually reaching the permanent all 0s state. Note that the machine indicates a fault immediately an error state (which has even parity) is entered.

The method may be applied directly to asynchronous machines[42,43] operating in the fundamental mode with the usual restrictions on single input variable changes. In this case the p-sets are called *destination sets* but they are derived in exactly the same way. Consider Table 9.21(a) which shows the flow table for a divide-by-2 counter circuit; the destination sets are

$$\text{column } \bar{x} \qquad\qquad\qquad \text{column } x$$
$$\begin{array}{ll} d_1 = (AD) & \qquad d_3 = (AB) \\ d_2 = (BC) & \qquad d_4 = (CD) \end{array}$$

which yield the r-partitions

$$\begin{array}{ll} r_1 = (AD)(BC) & r_3 = (AB)(CD) \\ r_2 = (BC)(AD) & r_4 = (CD)(AB) \end{array}$$

and the assignment shown in Table 9.21(b).

As before, to derive the excitation equations we must first extract the next-state partitions; these are

$$\begin{array}{llll} \text{column } \bar{x} & n_1^1 = (AD)(BC) & \text{column } x & n_1^2 = (CD)(AB) \\ & n_2^1 = (BC)(AD) & & n_2^2 = (AB)(CD) \\ & n_3^1 = (AD)(BC) & & n_3^2 = (AB)(CD) \\ & n_4^1 = (BC)(AD) & & n_4^2 = (CD)(AB) \end{array}$$

Table 9.21 Fail-safe asynchronous counter

	Inputs x		Output
	0	1	Z
A	(A)	B	0
B	C	(B)	0
C	(C)	D	1
D	A	(D)	1

(a) Flow table

					Inputs $x = 0$				$x = 1$			
	y_1	y_2	y_3	y_4	y_1	y_2	y_3	y_4	y_1	y_2	y_3	y_4
A	1	0	1	0	1	0	1	0	0	1	1	0
B	0	1	1	0	0	1	0	1	0	1	1	0
C	0	1	0	1	0	1	0	1	1	0	0	1
D	1	0	0	1	1	0	1	0	1	0	0	1

(b) Assigned flow table

Applying the rules for computing f_i^p we obtain the excitation equations

$$Y_1 = y_1\bar{x} + y_4x \qquad Y_3 = y_1\bar{x} + y_3x$$
$$Y_2 = y_2\bar{x} + y_3x \qquad Y_4 = y_2\bar{x} + y_3x$$

As before faults can easily be detected by examining the partition set of the circuit's present input state for even parity; in this case we have for $\bar{x}$, variables y_1 and y_2 and for x, y_3 and y_4; the circuit is shown implemented using NAND gates in Fig. 9.18.

The method can in general also be applied to the detection of faults in the output logic. This is done by generating the output equations in a similar way to the excitation equations, treating the output variables as state variables.

Using this technique it is also possible to design single fault-tolerant systems by replicating the machines and operating them in parallel, as shown in Fig. 9.19. When the currently operating machine experiences a fault it switches itself out and allows the other machine to provide the system output. Note that the amount of common logic required for fault detection and switching (called the *hardcore*) is minimal. Moreover the bistable switch could itself if necessary be designed as a failsafe circuit.

As with all fault-tolerant and failsafe circuits it is essential that appropriate precautions are taken to ensure that the initial start conditions are correct. If the circuit powers-up and sets in a failure cycle it will of course stay there. Other work has been done and much more is still in progress on the design of fault-tolerant networks. Notable in this field is the work of Pradham and Reddy[44] and Mukai and Tohma.[45] The techniques have also been applied to ROM designs[46] and could easily be extended to PLAs.

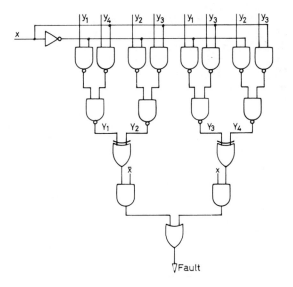

Figure 9.18 Fault-tolerant counter

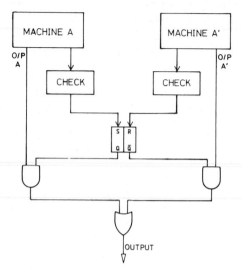

Figure 9.19 Single fault-tolerant circuit

References and bibliography

1. Mei, K. C. Y. Bridging and stuck-at faults. *IEEE Trans. Computers* **C23**, 720–6 (1974).
2. Van Cleemput, V. M. *Computer Aided Design of Digital Systems: a Bibliography*. Vol. 1 1960–74, Vol. 2 1975–76, Vol. 3 1976–77, Vol. 4 1977–79. Computer Science Press, Maryland.
3. Lewin, D. *Computer Aided Design of Digital Systems*, Crane Russak, New York, 1977.
4. Kautz, W. H. Fault testing and diagnosis in combinational digital circuits. *IEEE Trans. Computers* **C17**, 352–66 (1968).
5. Chang, H. Y. An algorithm for selecting an optimum set of diagnostic tests. *IEEE Trans. Electron. Computers* **EC14**, 706–11 (1965).
6. Sellers, F. F., Hsiao, M. Y. and Bearnson, L. W. Analysing errors with the Boolean difference. *IEEE Trans. Computers* **C17**, 678–83 (1968).
7. Amar, V. and Condulmari, V. Diagnosis of large combinational networks. *IEEE Trans Electron Computers* **EC16**, 675–80 (1967).
8. Armstrong, D. B. On finding a nearly minimal set of fault detection tests for combinational logic nets. *IEEE Trans. Electron Computers* **EC15**, 66–73 (1966).
9. Roth, J. P. Diagnosis of automata failures—a calculus and method. *IBM J. Res. Dev.* **10**, 278–91 (1966).
10. Friedman, A. D. and Menon, P. R. *Fault Detection in Digital Circuits*. Prentice Hall, Englewood Cliffs, N.J., 1971.
11. Bennetts, R. G. *Introduction to Digital Board Testing*. Crane Russak, New York, 1982.
12. Roth, J. P., Bouricius, W. G. and Schneider, P. G. Programmed algorithms to compute tests to detect and distinguish between failures in logic circuits. *IEEE Trans. Electron Computers* **EC16**, 567–80 (1967).

13. McCluskey, E. J. and Clegg, F. W. Fault equivalence in combinational logic networks. *IEEE Trans. Computers* **C20**, 1286–93 (1971).
14. Kriz, T. A. A path sensitising algorithm for diagnois of binary sequential logic. *Proc. IEEE Computer Conf.* Vol. 70-C-23-C, pp. 250–9 (1970).
15. Putzolu, G. R. and Roth, J. P. A heuristic algorithm for the testing of asynchronous circuits. *IEEE Trans. Computers* **C20**, 631–47 (1971).
16. Gill, A. *An Introduction to the Theory of Finite State Machines.* McGraw Hill, New York, 1962.
17. Hennie, F. C. *Finite State Models for Logical Machines.* Wiley, New York, 1968.
18. Kohavi, Z. *Switching and Finite Automata Theory.* McGraw Hill, New York, 1978.
19. Kohavi, Z. and Lavellee, P. Design of sequential machines with fault detection capabilities. *IEEE Trans. Computers* **C16**, 473–84 (1967).
20. Williams, M. J. Y. and Augell, J. B. Enhancing testability of large scale integrated circuits via test points and additional logic. *IEEE Trans. Computers* **C22**, 46–60 (1973).
21. Eichelberger, E. B. and Williams, T. W. A logic design structure for LSI testability. Proc. 14th Design Automation Conf., New Orleans, 1977, pp. 462–8.
22. Jones, H. E. and Schauer, R. F. *An Approach to a Testing System for LSI.* Computer Aided Design of Digital Electronic Circuits and Systems. North Holland, Amsterdam, 1979.
23. Signature analysis. *Hewlett Packard Journal* **28**(9) (May 1977).
24. Bennetts, R. G. and Scott, R. V. Recent developments in the theory and practice of testable logic design. *IEEE Computer* **9**(6), 47–63 (1976).
25. Special Issue Failure Tolerant Computing. *IEEE Computer* **13**(3) (1980).
26. Reddy, S. M. Easily tested realisations for logic functions. *IEEE Trans. Computers* **C21**, 1183–8 (1972).
27. Reddy, S. M. A design procedure for fault locatable switching circuits. *IEEE Trans. Computers* **C21**, 1421–6 (1972).
28. Muller, D. E. Application of Boolean algebra to switching circuit design and to error detection. *IRE Trans. Electron Computers* **EC3**, 6–12 (1954).
29. Fisher, L. T., Unateness properties of AND–exclusive OR logic circuits. *IEEE Trans. Computers* **C23**, 166–72 (1974).
30. Kautz, W. H. Testing faults in combinational cellular logic arrays. Proc. 8th Ann. Symp. Switching and Automata Theory, Oct. 1971, pp. 161–74.
31. Saluja, K. K. and Reddy, S. M. Easily testable two-dimensional cellular logic arrays. *IEEE Trans Computers* **C23**, 1024–7 (1974).
32. Akers, S. B. Universal test sets for logic networks. *IEEE Trans. Computers* **C22**, 835–9 (1973).
33. Hayes, J. P. On modifying logic networks to improve their diagnosability. *IEEE Trans. Computers* **C23**, 56–62 (1974).
34. Kohavi, Z., Rivierre, A. and Kohavi, I. Machine distinguishing experiments. *BCS Computer Journal* **16**, 141–7 (1973).
35. Fujiwara, H. and Kinoshita, K. Design of diagnosable sequential machines utilizing extra outputs. *IEEE Trans. Computers* **C23** 138–45 (1974).
36. Murakami, S., Kinoshita, K. and Ozaki, H. Sequential machines capable of fault diagnosis. *IEEE Trans. Computers* **C19**, 1079–85 (1970).
37. Pradhan, D. K. and Stiffler, J. J. Error-correcting codes and self-checking circuits. *IEEE Computer* **13**(3), 27–37 (1980).
38. Russo, R. L. Synthesis of error-tolerant counters using minimum distance three state assignments. *IEEE Trans. Electronic Computers* **EC14**, 359–66 (1965).

39. Tohma, Y., Ohyama, Y. and Sakai, R. Realisation of fail-safe sequential machines by using a k-out-of-n code. *IEEE Trans. Computers* **C20**, 1270–5 (1971).
40. Diaz, M., Geoffroy, J. C. and Courvoisier, M. On-set realisation of fail-safe sequential machines. *IEEE Trans. Computers* **C23** 133–8 (1974).
41. Sawin, D. H. III. Design of reliable synchronous sequential circuits. *IEEE Trans. Computers* **C24**, 567–9 (1975).
42. Sawin, D. H. III and Maki, G. K. Asynchronous sequential machines designed for fault detection. *IEEE Trans. Computers* **C23**, 239–49 (1974).
43. Maki, G. K. and Sawin, D. H. III. Fault-tolerant asynchronous sequential machines. *IEEE Trans. Computers* **C23**, 651–7 (1974).
44. Pradhan, D. K. and Reddy, S. M. Fault-tolerant asynchronous networks. *IEEE Trans. Computers* **C23**, 662–8 (1973).
45. Mukai, Y. and Tohma, Y. A method for the realisation of fail-safe asynchronous sequential circuits. *IEEE Trans. Computers* **C23**, 736–8 (1974).
46. Pradhan, D. K. Fault-tolerant asynchronous networks using read-only memories. *IEEE Trans. Computers* **C27**, 674–9 (1978).
47. Sellers, F. F., Hsiao, M. and Bearnson, L. W. *Error Detecting Logic for Digital Computers*, McGraw Hill, New York, 1968.
48. Breuer, M. and Friedman, A. *Diagnosis and Reliable Design of Digital Systems*. Computer Science Press, Woodland Hills, California, 1976.
49. Chang, H., Manning, E. and Metze, G. *Fault Diagnosis of Digital Systems*, Wiley, New York, 1970.
50. Roth, P. *Computer Logic, Testing and Verification*. Computer Science Press, Woodland Hills, California, 1980.
51. *Design Automation of Digital Systems*. Vol. 1. *Theory and Techniques*. M. Brewer (Ed.). Prentice Hall, Englewood Cliffs, N.J., 1972.
52. Williams, T. W. and Parker, K. P. Testing logic networks and designing for testability. *IEEE Computers* **12**(10), 9–21 (1979).
53. Abadir, M. S. and Reghbati, H. K. LSI testing techniques. *IEEE Micro* **3**(1) 34–51 (1983).
54. Raymond, T. C. LSI/VLSI design automation. *IEEE Computers* **14**(7), 89–101 (1981).
55. Bennetts, R. G. and Scott, R. V. Recent developments in the theory and practice of testable logic design. *IEEE Computers* **9**(6), 47–63 (1976).
56. Bennetts, R. G. *Design of Testable Logic Circuits*. Addison Wesley, London, 1984.
57. Lala, P. K. *Fault Tolerant and Fault Testable Hardware Design*. Prentice Hall, Englewood Cliffs, N.J., 1985.

Tutorial problems

9.1 For the circuit shown in Fig. 9.20 derive the full fault matrix and determine the test-set for the circuit. Check your result by using Boolean difference methods and confirm the essential tests for the circuit.

9.2 Using the cubical technique determine the primitive D-cubes of failure for the circuit shown in Fig. 9.21 assuming that input B is in error. Check your result using Boolean difference methods.

9.3 Derive the Reed–Muller expansion for the function

$$Z = x_2 x_4 + x_1 \bar{x}_2 \bar{x}_3 + \bar{x}_2 x_3 \bar{x}_4$$

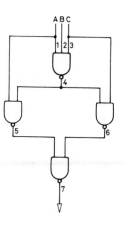

Figure 9.20 Problem 9.1

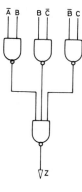

Figure 9.21 Problem 9.2

and show how the resulting expression may be realized using cascaded AND/exclusive OR modules.

9.4 The transition table for a clocked divide-by-2 counter is shown in Table 9.22; note that the states have been assigned using a minimum-distance-3 code. Show how the table can be realized as a failure-tolerant counter using JK bistables and draw the resulting state diagrams.

9.5 Realize the synchronous machine whose state table is shown in Table 9.23 as a failsafe circuit using the partition approach and show that the resulting circuit is failsafe in operation.

9.6 Repeat problem 9.5 using a 2-out-of-4 assignment to obtain a failsafe circuit and compare the results.

9.7 For the asynchronous machine whose flow table is shown in Table 9.24, design a failsafe circuit using NAND gates. Show how the error condition would be detected.

Table 9.22 Problem 9.4

Present state			Next state			Output
A	B	C	A_+	B_+	C_+	Z
0	0	0	1	1	1	0
1	1	1	0	0	0	1

Table 9.23 Problem 9.5

Present state	Next state Input x/output 0	1
A	$C/0$	$B/0$
B	$C/0$	$A/0$
C	$D/0$	$B/0$
D	$B/0$	$D/1$

Table 9.24 Problem 9.6

	Input $x_1 x_2$ 00	01	11	10
A	Ⓐ	B	Ⓐ	Ⓐ
B	A	Ⓑ	C	Ⓑ
C	Ⓒ	Ⓒ	Ⓒ	B

10 Design methods and tools

10.1 Introduction

The traditional methods of designing logic systems using part intuition part theoretical knowledge, realizing the network by connecting together (breadboarding) the appropriate gates or bistables (in the form of MSI or LSI packages), and then finally testing on the bench, are no longer appropriate to today's technology. The major difference is the complexity of the systems which we are now capable of designing and in particular the need to realize logic systems directly in terms of integrated circuit technology— *designing in silicon.*

These new objectives necessitate major changes in design methods and a requirement for computer-aided design (CAD) techniques. Software tools have now become essential to the design process, enabling the engineer to model and evaluate his designs using simulators and hardware description languages, to determine and specify test requirements and to assist in the physical design process of layout and interconnection. In addition current research in the development of silicon compilers, software tools which enable mask sets for LSI/VLSI circuits to be generated directly from a functional specification, will radically change the design process.

Another important design issue is the type of technology chosen to realize the logic system and the design requirements induced by this decision. Table 10.1 shows currently available implementation methods; the information in the table is greatly simplified and the final choice must be decided by the actual system to be developed, taking into account performance factors, volume of production, available resources etc. Moreover, the development time costs will reflect the design technologies employed and the availability of appropriate software aids. For example, in a microprocessor system program development time is the prime consideration whilst for custom and semi-custom LSI design it is the layout and testing of the chip which is the major factor. In both cases specialist software tools would be required as well as the general need for tools to assist in the logic design and fault test generation stages.

Though each technology imposes its own particular design characteristics there are a number of basic concepts which are fundamental to all logic systems design. First and foremost there is the requirement to specify as precisely as possible the system to be designed. In many cases this will necessitate frequent consultations with the users or 'customers' to deter-

Table 10.1 Comparison of alternative realizations

Characteristic	Technology			
	Custom LSI	Semi-custom	Discrete SSI/MSI	Micro-Processor
Speed	Fastest	Fast	Medium	Slow
Size	Smallest	Small	Large	Medium
Flexibility	Lowest	Low	Medium	High
Reliability	Highest	High	Low	Medium
Unit Cost:				
High Vol	Lowest	Low	Medium	Highest
Low Vol	Highest	High	Medium	Lowest
Dev Time	Longest	Long	Medium	Medium

mine just exactly what is required. At this stage the actual means of realization is far less important than establishing the correct behavioural or algorithmic description of the logic processes.

The next stage is to attempt a functional decomposition of the system requirements, normally in some structured hierachial form, based on well defined input–output specifications. Note that the manner in which the system functions are partitioned will determine both the overall performance and the reliability (including testability) of the system.

This top-down approach to design clearly has many advantages over the alternative bottom-up method which concentrates on the actual realization and the exploitation of available components. In practice the design process would iterate between the top-down and bottom-up approaches (and even middle-out!) but it is always good practice to start at the top.

Thus it will be seen that the design process is essentially hierarchal in practice, consisting of a number of levels, commencing with the *systems design* stage, as shown in Table 10.2. This is followed by the *logic design* phase where the actual hardware/software components are defined and interconnected. Finally, in the case of LSI/VLSI systems, there is the *physical design* stage or circuit level where the structural representation is translated into a physical network.

Table 10.2 Hierarchial design levels

Design stages	Design procedures
System design	Behavioural—system described in terms of algorithmic processes specifying required behaviour: note description independent of implementation.
Logic design	Functional—system partitioned into subsystem components with defined logic functions. Structural—description of actual hardware/software implementation.
Physical design	Physical—translation of structural representation into physical circuits.

A complication at the physical design stage is that the use of conventional design techniques produces logic networks in terms of gates, bistables and/or standard LSI modules which must then be translated into the appropriate technology, say MOS circuits, before they can be realized in silicon. With very complex circuits, and certainly with VLSI, this transformation process can and does lead to errors and the inefficient use of silicon area. It is essential in this case to employ a design method that relates the logic design process to the actual technology to be used for the realization of the system. Such techniques should take into account the layout geometries and physical design rules such as the line widths and separation of the diffused regions. Moreover, the design must be structured in such a way as to minimize interconnections and area by generating a regular, modular architecture. As we shall see the problem has been solved to some extent by Mead and Conway[1] who have proposed a procedure based on a structured design and the use of stick diagrams, which relate logic function to layout, and a simplified set of design rules.

Another basic design approach is to partition the system into a *control structure* and a *data structure*; this is the essence of the structured design technique mentioned above. The data structure, which would normally consist of registers, memory, processing elements and data paths, realizes the required logic processes as dictated by the algorithm embodied in the control structure. The control algorithm may be specified using a state machine (ASM chart) or a state table. Note that by partitioning the system into functional modules, each with its own data/control structure architecture, the need for very large state machines or state tables is obviated.

In this chapter we shall attempt to establish the basic design concepts and techniques involved in realizing logic systems in particular technologies and the computer-aided methods available and required for this purpose.

10.2   System design using IC modules

We have already seen how combinational and sequential logic networks can be realized using ROMs, PLAs, MUXs etc. However, there are many more modules available to the designer, in particular, arithmetic and logic units, RAMs, controllers, counters, decoder and encoder circuits, interface and communication units etc. Many of these special LSI devices have been developed specifically for use with microprocessor chips to enable microcomputer architectures to be easily designed.

In contrast to most of the design methods covered in earlier chapters, which have been primarily concerned with the development of specific networks, the availability of complex LSI components enables design at a higher, systems, level to be performed (see Table 10.2). Designing a system using these devices necessitates the setting up of a suitable data structure and then specifying and implementing the required control structure. The data structure will consist primarily of registers (n-bit memories) with combinational and sequential circuits used to perform computations on the contents of the registers. The sequencing and control of these operations are effected by a separate control structure which, as we shall see, in many cases uses programmable techniques. A schematic diagram of the general data/control structure architecture for a clocked system is shown in Fig. 10.1. Note that the controller will in general require status information, such as completion or overflow signals, to be sent back from the data structure.

As a simple illustration of these techniques let us consider the design of a 4-bit parallel multiplier as specified by the following algorithm:

(a) The least significant digit (LSD) of the multiplier is examined. If it is a 1 the multiplicand is added to the product register; no action is taken if the LSD of the multiplier is zero.
(b) The contents of the product and multiplier registers are shifted one place right.
(c) A check is made for the end of the operation. If it is complete, the process stops, otherwise step (a) is repeated.

Figure 10.2 shows a block diagram of a possible data structure which could be implemented from standard MSI modules. Note that the multiplication of two 4-bit numbers generates an 8-bit product, and consequently an 8-bit shift register is required for the product (PR). The multiplicand (MD) and multiplier (MR) are held in 4-bit registers; since the multiplier register is shifted right at each operation it may be used as the least significant four-bits of the product register. A parallel 4-bit arithmetic unit is used to add the multiplicand to the product, the input and output to the device being controlled by standard gates. Note the need to include an extra stage for the carry overflow from the adder unit. Alternatively, if a two-phase clocked system was used, the gating could be performed directly at the inputs to the bistable stages (the D input in the case of a latch) with phase 1 clock being used for Add 1 and phase 2 for Add 2. It is also necessary to incorporate some form of loop counter; in this case a simple 2-bit counter is used, since there are four distinct cycles of operation to be performed on the 4-bit numbers (in general an n-bit count is required where n is the number of bits in the word).

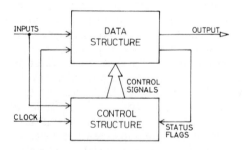

Figure 10.1 Data-control structures

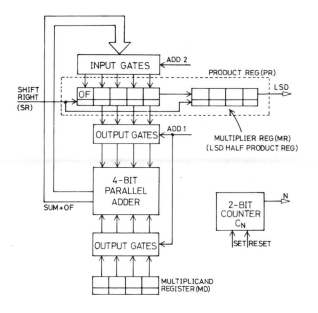

Figure 10.2 Data structure for parallel multiplier

The control of the multiplier circuit is effected by arranging that the Set CN, Shift Right (SR), Add 1, Add 2 (necessary to allow for the propagation time through the adder) are generated in the correct sequence; this is of course the function of the control structure. The control algorithm may be defined in terms of a state machine (as shown in Fig. 10.3) and realized in the usual way. Note that the algorithm assumes that the required data are resident in the registers prior to commencing the multiplication. The outputs of the control unit are the signals Set CN, Reset CN, ADD 1, ADD 2 and SR while the inputs are LSD and N fed back from the multiplier; the input MULT is an external start signal. It will be clear that in a real design the control signals required to be generated would be determined by the actual LSI units employed.

An alternative method of implementing the control structure is to store the required signals in a ROM; these signals are then accessed in sequence, timed by a system clock. Note that whatever method is employed each control signal requires a corresponding physical connection to the data structure. Thus, if a ROM realization is employed each output would need a separate bit in the ROM word.

If we now consider the data structure for the parallel multiplier it will be apparent that it could, by defining an appropriate control algorithm, be used to perform a number of different logic processes. In fact the combination of registers, transfer gates and highways coupled with the arithmetic and logic unit forms the basis of the central processing unit (CPU) in a general purpose digital computer.[2] Moreover, if a ROM implementation

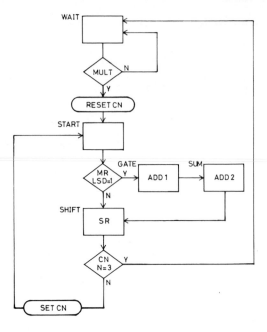

Figure 10.3 ASM chart for the multiplier control

is used the processing function can easily be changed by suitable repro-
gramming—thus we have the concept *of programmable electronics* or
microprogramming[3]

10.2.1   The microprogram concept

The concept of microprogramming was first proposed by Wilkes[4] for the
design of the control unit in a digital computer. In this case the machine's
internal control signals are called *micro-orders* and a collection of such
orders to perform a specific machine-code function such as ADD (the
contents of two registers), LOAD (a register from memory), STORE
(from a register to memory) being called a *microprogram*. It will be
obvious that the task of designing a microprogramming unit for a digital
computer is considerably more complex than that of the multiplier, multiplication being only one of the many such functions required; however, the
basic approach is identical.

   In designing general purpose microprogrammed logic systems,[5] whether
for special applications or as part of a computer architecture, it is necessary
to optimize the set of micro-orders. That is, the smallest set of micro-
orders which is basic and general enough to provide for the required
number of different machine-code orders must be chosen. Thus micro-
orders are typically signals which open or close gates, to put data into
registers or onto highways, read/write initiation signals for memory units

etc., control inputs for arithmetic and logic units, shift signals, set and clear inputs for registers and flags, cycle counting etc.

In the microprogram approach the specific register transfer operations and inter-register processing, which go to make up a control algorithm, are broken down into individual microinstructions. For example, a transfer of data between register A and register B at time step Tk would require control waveforms to open the output gates of register A and the input gates of register B to be generated at timeslot Tk. The microinstructions then must indicate the appropriate input/output gating controls, and any other controls required at time Tk, plus the address of the next micro-instruction in the ROM. In the microinstruction word the micro-orders are encoded using either a unary code, with each order being allocated a specific bit in the word, or with mutually exclusive groups of micro-orders being represented in binary.

In operation the ROM words are read down in sequence and the output bits, either directly or after decoding, are transmitted to the data structure. This sequencing continues until a transfer to a new location, out of the current sequence, is effected as the result of an input (called the *conditions*) from the data structure. The microprogram would consist of a number of microinstruction steps of this type which would be initiated by accessing the ROM at a particular starting address determined by the algorithm required to be performed (in the case of a microprocessor this would be the machine-code instruction).

The block diagram for a typical static (ROM based) microprogram unit is shown in Fig. 10.4. The process is initiated by the instruction or order specifying the algorithm to be performed being sent to the ROM controller which then generates the starting address of the microprogram in the ROM. The microinstruction is then read down and decoded and the out-

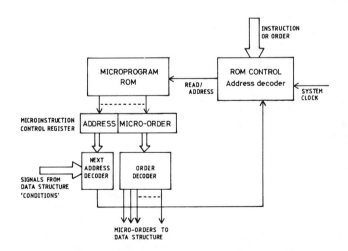

Figure 10.4 Static microprogram unit

puts distributed to the data structure. The encoded next address, suitably modified if necessary, goes to the ROM control unit and the process repeated. The Next Address encoder and the ROM Control Unit would normally be realized using PLAs.

The full microprogramming approach would only be viable if there were a large number of distinct operations to be performed on the same data structure—which is of course exactly the case in the CPU of a micro-computer. However, a simplified form of the technique can be used to design special-purpose logic systems with the added advantage that the function can easily be changed by using different ROMs. The realization of a logic process in this way using microprogramming techniques is normally called *firmware engineering*.

In this simple account of microprogramming many questions have been left unanswered, particularly the question of conditional loops and addres-sing and the problems of encoding and optimizing the microinstruction words; the reader is referred to more specialised texts on the subject for a fuller treatment of these topics.[6-8]

10.2.2 Design methodology

The microprogramming approach, using the ASM chart to define the algorithm, is the ideal method for designing at the systems level using LSI modules. In using the ASM chart a state is usually allocated for each micro-instruction; the micro-orders appear either as unconditional state outputs (for example, SR in Fig. 10.3) or conditional outputs (SET ON). The 'conditions' inputs, used to determine the sequencing, appear as input variables in the conditional diamonds. Each path through the ASM chart from one state to another, via the conditionals, gives rise to a particular microinstruction. The next address is obtained in the usual way by logically combining state and conditional inputs. Table 10.3 shows the assigned transition table for the multiplier control circuit discussed earlier, while Table 10.4 gives the corresponding ROM layout. Note that a 64×8-bit word ROM would be required which due to the large number of don't-cares would be very sparsely occupied. The equations for a PLA realiza-tion are given below and will be seen to be much more economical.

$$Y_1 = \bar{y}_1 y_2 y_3$$
$$Y_2 = \bar{y}_1 \bar{y}_2 y_3 + y_1 \bar{y}_2 \bar{y}_3$$
$$\text{RSCN} = \bar{y}_1 \bar{y}_2 \bar{y}_3 \text{ MULT}$$
$$\text{SETCN} = \bar{y}_1 y_2 \bar{y}_3 \ (N = 3)$$
$$\text{ADD1} = \bar{y}_1 y_2 y_3$$
$$\text{ADD2} = y_1 \bar{y}_2 \bar{y}_3$$
$$\text{SR} = \bar{y}_1 y_2 \bar{y}_3$$

Moreover, the PLA realization would in general be faster than a ROM microprogram unit, the fastest of course being a direct implementation using random logic. In a small system the direct application of the ASM chart method using PLAs or ROMs for the realization would be sufficient to generate the control structure.

Table 10.3 Multiplier control transition table

State name	Present state y_1 y_2 y_3	Inputs (conditions) MULT LSD $N=3$	Next state Y_1 Y_2 Y_3	Outputs RSCN SETCN ADD1 ADD2 SR
WAIT	0 0 0	0 X X	0 0 0	0 0 0 0 0
	0 0 0	1 X X	0 0 1	1 0 0 0 0
START	0 0 1	X 0 X	0 1 0	0 0 0 0 0
	0 0 1	X 1 X	0 1 1	0 0 0 0 0
SHIFT	0 1 0	X X 1	0 0 0	0 0 0 0 1
	0 1 0	X X 0	0 0 1	0 1 0 0 1
GATE	0 1 1	X X X	1 0 0	0 0 1 0 0
SUM	1 0 0	X X X	0 1 0	0 0 0 1 0

Table 10.4 ROM word layout

ROM address Y_1 Y_2 Y_3	(conditions) MULT LSD $N=3$	ROM output word Next Add. Y_1 Y_2 Y_3	Micro-orders RSCN SETCN ADD1 ADD2 SR
0 0 0	0 X X	0 0 0	0 0 0 0 0
0 0 0	1 X X	0 0 1	1 0 0 0 0
0 0 1	X 0 X	0 1 0	0 0 0 0 0
0 0 1	X 1 X	0 1 1	0 0 0 0 0
0 1 0	X X 1	0 0 0	0 0 0 0 1
0 1 0	X X 0	0 0 1	0 1 0 0 1
0 1 1	X X X	1 0 0	0 0 1 0 0
1 0 0	X X X	0 1 0	0 0 0 1 0

It will be fairly obvious that the microprogramming function is analogous to the normal programming, at machine-code level, of a digital computer. Consequently it will be no surprise that software methods and tools are used for the design of more complex microprogrammed systems. For instance, one essential tool is a micro-assembly language which enables micro-order bit patterns to be generated from a basic memonic description of the microprogram. It is also essential that microprograms should be checked out before implementation; this can be performed using a software simulation of the data structure and its controller. As we shall see in a later section, Register Transfer languages are used for the specification and evaluation of a register structured architecture.

10.3 VLSI design methods

Realizing a logic system in terms of a custom built integrated circuit requires a different design stategy and necessitates a strong reliance on computer-aided design tools. As an example of the method and the problems involved let us consider the overall design of a typical VLSI chip such as a microprocessor;[9] a schematic of the design process is shown in Fig. 10.5.

The first stage is to define the operational requirements specification for the device and then attempt an architectural description in terms of standard modules and structures. In contrast to other design methods it is essential at this stage to relate the architecture to chip area and the overall layout, called the *floor plan*; Fig. 10.6 shows a typical layout used at the system design stage.

After defining the architecture the next step is to specify the required logical operations of the various component modules, usually in the form of ASM charts, state tables, logic equations etc. In the case of a micro-processor the micro-code for the machine-code instruction set must be specified and provision made for their sequencing and control, and similarly for the input–output bus control etc.

The final step in the design process is the translation of the logic specification into appropriate circuit structures and their physical layout in terms of the design rules for the technology. Note that the final performance of the chip, particularly in terms of speed, will be dependent on these physical dimensions. The main consideration in converting logic equations to an MOS transistor network is to keep the circuit diagram as close as possible to the final chip layout, which considerably eases the problem of estimating internal delay times. The translation process is one of the problem areas in LSI design; a particular difficulty arises from the need to check back to ensure that the MOS circuits reproduce the required logic equations.

It is essential that the design process should be top-down but iterative

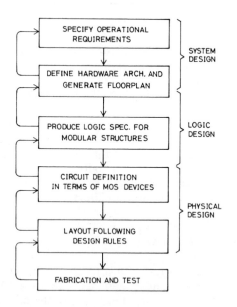

Figure 10.5 Top-down design of LSI chip

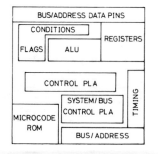

Figure 10.6 Typical floor-plan

feedback between all levels will naturally occur during the design. It is also essential, throughout the design process, to keep in mind the structure and layout of the chip and the consequences of making particular design decisions—for example, the effect of choosing random logic as against microprogram control for the execution of an instruction set. CAD tools are required at all levels of design, logic simulators to check out the functional descriptions and generate test procedures, circuit analysers and simulators to check delays etc. and tools for layout and design rule checking; these aspects will be covered in a later section.

With VLSI the objective of providing greater processing power on smaller surface areas invalidates to a large extent the use of classical switching theory. The area occupied on a silicon chip is more a function of the topology of circuit interconnections than the number of gates—thus 'wires' constitute the main problem in VLSI design. Physical characteristics such as surface area, power consumption and delay times are the new parameters of optimization, with the objective of minimizing circuit connections. Consequently, a major consideration in designing VLSI circuits is the achievement of a regular structured architecture, that is logic circuits which are built up from repeated patterns of identical devices and interconnections. Unfortunately there is very little formal theory to assist in this objective and current methods depend heavily on the concept of controlled data paths, where the controller is usually a finite-state machine realized using a PLA. In MOS technology the data path would normally consist of combinational logic blocks separated by input–output shift registers. In essence we have, once again, the concept of register-to-register transfers with inter-register processing. However, in this case since the registers plus combinational logic blocks are connected in series it is possible to perform simultaneous operations on different sets of data words as they progress through the data-path—in other words *pipe-line processing*.[10] A block schematic of a data path is shown in Fig. 10.7(a); note that a two-phase clocking system is used to avoid problems due to propagation delays with data being entered to the combinational logic in phase $\varphi1$ and the output extracted in phase $\varphi2$; the required clocks are shown in Fig. 10.7(b). Note that the clocks are non-overlapping and that $\varphi1 . \varphi2 = 0$ for all time. The

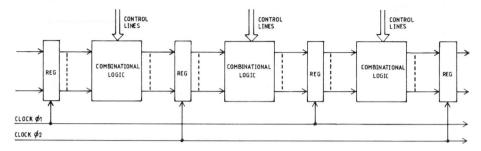

a) Register/combinational logic elements

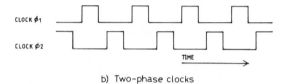

b) Two-phase clocks

Figure 10.7 Data-path structures

combinational logic is, where possible, implemented using regular networks of MOS pass transistors or, in the more general case, PLA structures.

Synchronous machines can easily be implemented using this technique by simply including additional feedback paths as shown in Fig. 10.8(a). Since the combinational logic required by a finite-state machine is normally very irregular, PLAs would almost invariably be used for the realization; Fig. 10.8(b) gives the typical structure for an FSM using PLAs.

Since the PLA is a basic structure in VLSI design, being used extensively to implement random logic and in many applications constituting a large proportion of the total silicon area, it is essential to minimize its size. The factors which affect the area of an embedded PLA are as follows:

(a) the number of product terms,
(b) the number of inputs,
(c) the number of outputs,
(d) the number of invertor circuits,
(e) propagation delays,
(f) the technology design rules.

However, the PLA area is mainly determined by the following:

Length = (number of inputs to the AND matrix)
         + (number of outputs)
Width = the number of product term lines.

The number of inputs to the AND matrix is at most twice the number of input variables due to inversion.

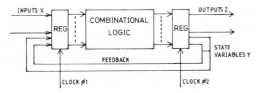

a) FSM as part of register transfer path

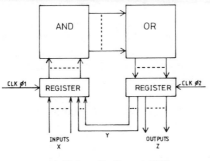

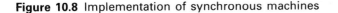

b) PLA realisation of FSM

Figure 10.8 Implementation of synchronous machines

The number of inputs and outputs would normally be determined at design time as a system requirement, hence dictating the length of the PLA. However, the width of the array is dependent on the number of multiple output product terms and when used as an FSM the number of states; to a certain extent their assignment will also effect the input and product terms. A useful technique is to extract monotonic next-state equations (that is the uncomplemented form) which eliminates the need for invertors and hence reduces the length of the array. Note that this would also increase the number of state variables required.

In many instances standard logic minimization programs can be employed, with suitably modified criteria, to effect an initial reduction of the product terms; a typical program used for this purpose, based on heuristic principles using the cubical notation and algebra, is MINI developed at IBM.[11]

Though the techniques described above will reduce the overall area of the PLA the actual percentage of used MOS device in the array is still very small—something like 10% for the AND array and 4% for the OR. Moreover, in the conventional PLA if the number of inputs and outputs with independent functions are increased by a factor of N the array area increases by N^2. Considerable savings in silicon area can be achieved by using a technique known as *folding*.[12] The folded PLA structure is shown in Fig. 10.9(b); note that in essence there are two PLAs back-to-back with multiple folds which necessitate making cuts in the metalization to allow interconnections to be brought out and routed.

Using folding techniques considerable reduction can be made in the area of the PLA (at least 25% has been suggested). In Fig. 10.9 it will be obvious that the area has been drastically reduced since only seven columns are required in the folded array compared to twelve for the normal case. The determination of suitable folding geometries for a particular logic personality is not an easy task to undertake manually; however, CAD methods have been described[13] which could easily be incorporated into a PLA generator.

The design of an FSM using PLAs follows fairly conventional lines as described earlier. The controller would be specified using either an ASM chart or state diagram and the product terms extracted from the resulting assigned state-transition table. In many cases the product terms are used directly without further reduction though a restricted form of multiple-output minimization could usefully be employed. Alternatively, an attempt could be made to reduce the number of states thereby reducing the number of state variables required. Unfortunately for large machines considerable

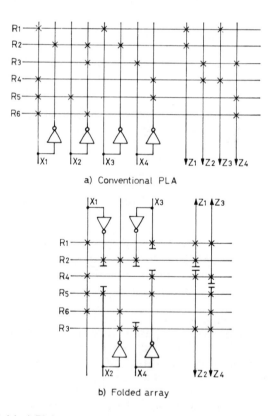

a) Conventional PLA

b) Folded array

Figure 10.9 Folded PLAs

reduction would be required before any real economies could be effected. However, it is possible that the generation of more don't-care terms could effect a reduction in the product terms required. It will be obvious that the problems of synthesizing optimal PLA structures are far from being solved or for that matter even well specified.

An alternative design aid for PLAs is the *PLA generator*.[14] This is a software tool (also called a *silicon compiler*) which accepts a specification of the logic requirements for the PLA in terms of formal logic equations (using a high-level language) or an ASM chart. The specification is then translated into an actual circuit description and layout. In some cases the circuit is also optimized using the design rules for the technology.[15] PLAs can also be used in a hierarchial mode (similar to the decomposition of sequential machines) to overcome the problems inherent with very large structures.

10.3.1 The physical design stage

This is a very comprehensive subject[1,16] and in this text little more than an overview of the basic principles relevant to logic design can be given.

An integrated circuit system using MOS technology consists of three levels of conducting material, separated by layers of insulating material, deposited on the surface of a silicon wafer. The levels are termed, from top to bottom, *metal*, *polysilicon* and *diffusion*. Points in the various levels may be connected together by making contact holes (cuts) through the insulating material. The actual paths of the conducting material, on and between the levels, is determined during fabrication by using *masks*. A mask is similar to a photographic negative in that it has opaque and clear areas representing the path patterns required; it is used during fabrication as a template to allow the insulating materials to be etched away to form a path. In the absence of any inter-layer connections (contact cuts), paths on the metal level may cross over paths on either the diffusion or polysilicon levels with no significant effects. When, however, a polysilicon path crosses over a path on the diffusion level a transistor switch is produced such that a voltage on the polysilicon level will control the current in the diffusion path.

At the physical design level the objective is to generate the actual physical geometry of the device, that is the location of paths in the various layers, in order to satisfy the logic specification. Thus it will be evident that the logic and physical design stages are intimately connected and ideally should be one and the same process. The designer is constrained in this task by the *design rules*, appertaining to a particularly technology and the processing methods employed, which specify the allowable geometric shapes and their dimensions. Design rules then are a set of regulations which define the acceptable dimensions and the electrical parameters for a particular process without violating the device physics.

The geometric rules specify the minimum allowable values, usually expressed in microns, for certain *widths*, *separations*, *extensions* and *overlaps* for patterns in the various levels of the chip (elemental components and their dimensions are usually referred to as *features*). Though these rules vary considerably from process to process (even for the same technology) it is possible to produce a normalized set of rules. Mead and Conway[1] have used a common metric λ to express the design rules, where λ represents the possible misalignment of any geometric feature between levels. The metric λ is based on typical mask tolerances and other process deviations and for a standard MOS process would be of the order of 2.5 microns.

Using the generalized λ-based design rules ensures that LSI circuits can be reliably and simply designed, albeit with a very conservative safety factor which leads to increased silicon areas and possibly lower performance. Providing the overall system requirements can be satisfied this is of no consequence; however, for maximum performance it is necessary to optimize the layout using more specific design rules.

To illustrate these design rules let us consider the layout for an MOS NAND gate; Fig. 10.10(a) shows the circuit schematic and Fig. 10.10(b) a simplified version of the actual geometric shape which reproduces the function in MOS technology. Note the input enhancement mode transistors formed by the polysilicon paths crossing the diffusion paths and similarly for the pull-up transistor operating in the depletion mode. The separation and widths etc. of the various components forming the NAND gate geometry are determined by the design rules, an example of which is shown in Fig. 10.11.

It will be obvious that the manual translation from a conventional logic diagram to an MOS circuit layout in terms of constrained geometries is not an easy task. Moreover, the normal logic diagram symbolism bears no relationship to the actual structures to be laid down on silicon. To overcome this problem it is necessary to use a different representational system coupled with CAD methods. One approach, though now almost

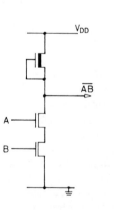

a) MOS NAND gate

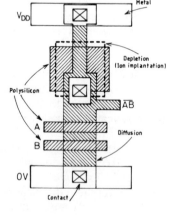

b) NAND gate geometry

Figure 10.10 MOS NAND circuits

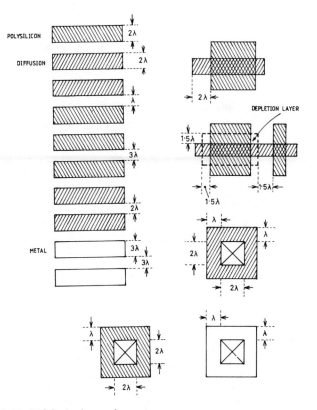

POLYSILICON 2λ

DIFFUSION 2λ

λ

3λ

2λ

METAL 3λ 3λ

2λ

DEPLETION LAYER

$1{\cdot}5\lambda$ $1{\cdot}5\lambda$

$1{\cdot}5\lambda$

λ λ

2λ 2λ

λ λ

λ

2λ 2λ

2λ

Figure 10.11 MOS design rules

standard practice, is to use a *sticks diagram*[17] as the symbolic input to a computer-aided layout system. The sticks diagram, shown in Fig. 10.12, allows logic circuits to be described in a format which closely relates to the topology of the physical layouts to be employed. For example, Fig. 10.12(b) shows a sticks diagram for the NAND gate which if compared with Fig. 10.10(b) shows distinct structural similarities.

It is customary to represent each different layer with its own individual colour—red for polysilicon, green for diffusion, blue for metal and yellow for an implant area; however, in this text we shall use the conventions shown in Fig. 10.12(a).

Path-closing switches are easily represented in the sticks notation as shown in Fig. 10.12(c); note that the MOS gate input is at the cross-over point between the polysilicon and the diffusion paths. The direct application of this is shown in Fig. 10.12(d) for a one-out-of-four multiplexer circuit. Figure 10.12(e) shows a stick diagram for a 3-to-8 decoder circuit, that is decoding a binary input ABC to individual outputs N_0–N_7. The circuit effectively comprises eight separate MOS NOR gates but arranged in an orthogonal pattern (note that $N_0 = \overline{A + B + C} = \bar{A}\bar{B}\bar{C}$ etc.). An important

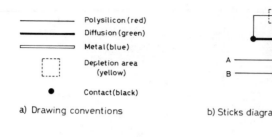

Polysilicon (red)
Diffusion (green)
Metal (blue)
Depletion area (yellow)
Contact (black)

a) Drawing conventions

AB

A
B

b) Sticks diagram for NAND gate

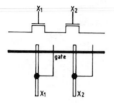

c) Path-closing circuits

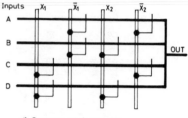

d) One out-of-four MUX

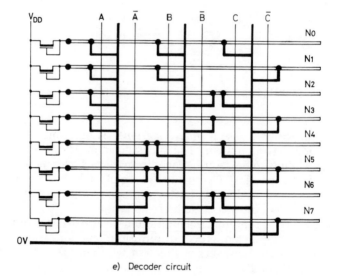

e) Decoder circuit

Figure 10.12 Logic design using stick diagrams

point to note with all these circuits is their regular orthogonal shape; in fact most MOS circuits can be represented by a set of polygons.

Stick diagrams can be used by the logic designer to obtain a realistic toplogical circuit diagram and also as the input (graphical in many cases) to a CAD system which then ensures that the design rules are appropriately administered.

An important development in this area is the silicon compiler, the main function of which is to remove the time consuming and error prone task of producing a chip design. The silicon compiler is analogous to the software compiler in that they both accept a high-level functional description and convert it to a specific low level working realization, in the case of the silicon compiler converting the formal description into an appropriate assembly of cells and interconnections to form the actual chip design.

10.4 Semi-custom IC design

An alternative[18] to full custom IC design is to use chips with pre-processed components or cells, such as an array of MOS gates, which require only the metal interconnection layers to be added to produce the final LSI circuit. Thus the designer needs only to produce masks for connections and contacts; in many cases only one mask is required, since the majority of the processing steps have already been completed—for this reason the technique is often called *mask programmable design*. The main disadvantage of this approach is the loss of performance and the decreased packing density resulting from the inflexibility of the laid down structure.

The ROM and PLA modules discussed in earlier chapters also fall into this category of mask programmable devices. However, in general PLAs are limited in the number of gates available, usually a few hundred, and ROMs, though widely used, are not considered secure enough for many applications since they can be easily copied; both devices are of course slow in operation.

The *uncommitted logic array* or *ULA* is probably the most widely used type of semi-custom LSI. The gate array consists of rows of basic logic cells in a matrix format with space for interconnection highways between the rows. Surrounding the logic cells, at the edge of the chip, are basic peripheral devices such as buffer amplifiers which provide the input/output interface. Logic networks are realized by connecting these gates together via the interconnection highways; the pattern of interconnecting paths so produced is used to generate the metalization mask which then completes the fabrication process. Note that the chip design process is considerably simplified since only one or two masks are required to be produced as against a full mask set, at least six in number, for a custom LSI chip.

Figure 10.13(a) shows the layout of a typical gate array and Fig. 10.13(d) the connections required to construct a basic CMOS NAND gate. In practice a cell library is normally provided which gives standard constructions for NAND/NOR gates, bistables, arithmetic units etc. which can be used as basic elements when designing logic circuits.

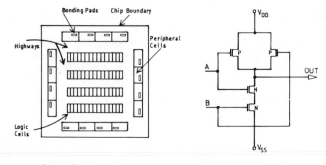

a) ULA Chip layout   b) CMOS NAND gate

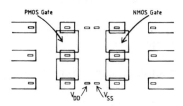

c) Section of basic cell

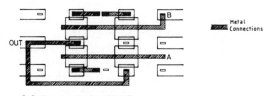

d) Cell connected as NAND gate

Figure 10.13 CMOS gate array

The layout and interconnections can be performed manually using a special layout grid showing the underlayers of the chip with the location of the 'resistors' and transistors; decals are provided for the standard cell constructions. After positioning the necessary gates and devices the designer then interconnects them using techniques similar to those used for the layout of a printed circuit board; the final layout is then used to produce the masks. Manual methods are not viable with systems containing more than 500 gates when it becomes essential to use CAD methods.

Using an interactive CAD system the designer either inputs his network directly from the logic diagram (using manual digitizing techniques) or simply provides a textual description of the required interconnections (the wire list). The design system enables appropriate gates or devices to be selected from the library and makes available a placing and routeing routine—most systems use interactive graphics so that the designer can manually insert connection edits etc. as required.

Gate arrays are obtainable in most technologies but the final choice will obviously depend on application. CMOS arrays with up to 10,000 gates and 2–6 ns delay times have the advantage of low power comsumption. ECL gives the highest performance with switching times of <1 ns and arrays of up to 2000 gates but has a high power dissipation. TTL and I^2L logic have also been used; these give medium power consumption and packing density and delay times in the order of 2–6 ns.

Gate arrays have the advantage that only one cell or module needs to be carefully designed and laid out, since the logic network will consist of repetitions of these basic units. The designer is concerned primarily with the problems of logic design, such as maximum fan-out from the gates, with the routeing between cells being performed semi-automatically using CAD tools. The routeing problem is easier in the case of semi-custom design since the gates occupy fixed positions on the chip. This gives rise, however, to a major difficulty—that of keeping the delay paths uniform, since unequal paths can generate spurious outputs.

Another approach to semi-custom design is the *standard cell* technique. In this method component layouts are pre-produced by expert designers for basic gates and circuits (such as shift-registers, decoders etc.) and stored in a *cell library*. In the case when every cell has the same height, though its width may vary, the technique is called *polycell* design. The logic design technique is analogous to the conventional approach using MSI/LSI modules laid out on a printed circuit board. For a given system design appropriate cells are selected from the cell library and placed in rows across the chip, leaving sufficient space for the interconnections. The process is highly automated and relies heavily on CAD tools, with the initial layout and placement being performed using a graphics display terminal and the interconnections and final mask layouts generated by the software. In essence this is a psuedo-custom design method since a full production mask set is required (a minimum of some six masks) and consequently the capital investment required for the CAD tools approaches that for full custom design.

10.5 Computer-aided design

It will be obvious by now that computer-aided design tools are essential for the design of complex logic systems and absolutely vital when these systems are to be realized as LSI/VLSI circuits. In fact the tremendous advances in microelectronics are due as much to the developments in software tools as they are to solid state physics and chemistry. The software tools required range over the whole hierarchial design process, as shown in Fig. 10.14.

As would be expected from a hierarchial approach, the degree of detail required to describe the logic system increases enormously as the design proceeds from top to bottom. At the systems level the specification is independent of the implementation, and thus it can be abstract and represented using formal mathematical methods such as *directed graphs*.[19] As

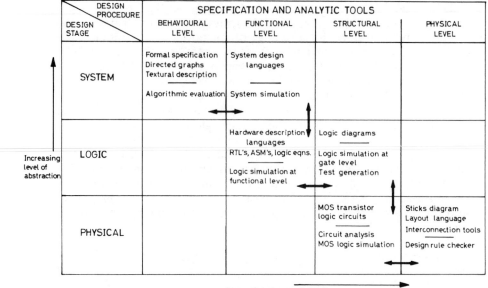

DESIGN STAGE \ DESIGN PROCEDURE	SPECIFICATION AND ANALYTIC TOOLS			
	BEHAVIOURAL LEVEL	FUNCTIONAL LEVEL	STRUCTURAL LEVEL	PHYSICAL LEVEL
SYSTEM	Formal specification Directed graphs Textural description ——— Algorithmic evaluation	System design languages ——— System simulation		
LOGIC		Hardware description languages RTL's, ASM's, logic eqns. ——— Logic simulation at functional level	Logic diagrams ——— Logic simulation at gate level Test generation	
PHYSICAL			MOS transistor logic circuits ——— Circuit analysis MOS logic simulation	Sticks diagram Layout language Interconnection tools Design rule checker

Increasing level of abstraction

Increasing degree of detail

Figure 10.14 Design tools for logic systems

the design proceeds so this specification is expanded, first into a functional representation using logic equations, microprograms ASM charts etc. The functional requirements can be simulated using, for example, a hardware description language[20] the most common example of which is the *register transfer language (RTL)*. Register transfer languages enable a register structure, together with the required inter-register transfers and processing, to be specified and evaluated. The detailed design at the logic gate level, the actual logic diagram, can also be modelled and evaluated using *logic level simulation*. It is at this stage that test generation, if required, would need to be implemented.

At the physical design stage the physical properties of the MOS circuits must be modelled to ascertain their electrical performance in terms of drive capabilities, propagation delays, loading effects of the inputs etc. CAD tools for chip layout and the routeing of module interconnections, including mask generation, must also be available as must some means of checking that the final layouts obey the technology design rules.

An important requirement of the overall CAD tool kit, shown in Fig. 10.14, is the need for two-way iteration between each level and stage in the design process and the requirement for the evaluation procedures to extend across the design boundaries. This is essential in order to ensure that the algorithmic content of the original behavioural description is preserved after the necessary transformations between levels. For example, it is essential to be able to back-track from the MOS circuit realization to check that the logic requirements are still satisfied and to detect any errors that may have occured during the translation process. It is also necessary to correlate the functional performance at the logic gate level with the actual performance of the physical circuit in terms of circuit delays etc.

The actual CAD requirements will depend on the particular technology to be used for the realization of the logic system. For example, using MSI/LSI units mounted on printed circuit boards the major need is for functional/logic level simulators and fault test generation tools; PC board layout and interconnections would normally be performed by hand but could if necessary be done semi-automatically.

Semi-custom design would have similar CAD requirements for logic design and testing but the placement, routeing and interconnection tools now become essential and more sophisticated. Full-custom design requires all these tools and in addition means of analysis at the circuit and process levels and for design rule checking. As the degree of integration increases so the requirements for semi-custom and custom LSI design become almost identical. Many of the tools now being developed for custom VLSI are directly applicable to semi-custom design, for example the silicon compiler, which is now used to define and construct the logic functions to be realized as a gate array.

The need for specification and evaluation exists at all levels in the design hierarchy but it is at the behavioural level where software tools are most urgently needed. The importance of the behavioural level is that it is here that the system requirements are translated into a formal specification that determines the overall design and the final form of the system. If errors or misunderstandings occur at this stage the repercussions will pervade the entire design process. Moreover, it is also necessary to demonstrate to the customer that the proposed interpretation of his requirements does in fact meet his needs; hence the vital importance of evaluation at this stage.

An ideal specification system at the behavioural level should have the following characteristics:

(a) able to represent logical processes independent of eventual realization;
(b) able to evaluate formally the control and data flow in large variable systems;
(c) capable of handling concurrent processes and alternative partitionings of the system;
(d) can be mapped into the functional level without loss of algorithmic content;
(e) acts as a mean of communication between customers, designers and implementers.

Most of these properties will also apply to design languages used at lower levels but in these cases the languages can be more specific and directly related to hardware/software realizations.

Many methods have been proposed for the description and design of logic systems but three main techniques predominate: these are *directed graphs*, *hardware description languages* (in particular RTLs) and *simulation* techniques. We shall now look at each of these methods in more detail.

10.5.1   Directed graph methods

Graph theory is finding increasing application in the design and analysis of logic systems due primarily to its implementation-free notation and the use of formal methods of evaluation. A directed graph is a mathematical model of a system showing the relationships that exists between members of its constituent set. The elements of the set are normally called *vertices* or *nodes* with the relationship between them being indicated by *arcs* or *edges*. An example of a directed graph is shown in Fig. 10.15(a) where the set of nodes is given by

$$N = \{n_1, n_2, n_3, n_4, n_5\}$$

and the set of edges by

$$E = \{e_1, e_2, e_3, e_4, e_5\}$$

Graphs may be classified into various types according to their properties, for example, a *net* shown in Fig. 10.15(b) is a directed graph consisting of a finite non-empty set of nodes and a finite set of edges. Note that a net may have parallel edges, that is two nodes connected by two different edges but both acting in the same direction. Again, a net which does not contain parallel edges but with assigned values to its edges is called a *network* as shown in Fig. 10.15(c).

A directed graph approach which has found considerable application in the description and analysis of digital systems is the *Petri net*.[21,22] The Petri net is an abstract formal graph model of information flow in a system consisting of two types of node, *places* drawn as circles and *transitions* drawn as bars, connected by directed arcs. Each arc connects a place to a transition or vice versa; in the former case the place is called an *input place* and in the latter an *output place* of the transition. The places correspond to system conditions which must be satisfied in order for a transition to occur. A typical Petri net is shown in Fig. 10.16; note that the net is able to depict concurrent operations.

In addition to representing the static conditions of a system the dynamic behaviour may be visualized by moving markers (called *tokens*) from place to place round the net. It is usual to represent the presence of tokens by a black dot inside the place circle; a Petri net with tokens is called a *marked* net. A Petri net *marking* is a particular assignment of tokens to places in the net and defines a state of the system, for example, in Fig. 10.16(a) the marking of places B and C defines the state where the conditions B and C hold and no others.

Progress through the net from one marking to another, corresponding to state changes, is determined by the *firing* of transitions according to the following rules:

(a) a transition is enabled if all of its input places hold a token;
(b) any enabled transition may be fired;
(c) a transition is fired by transferring tokens from input places to output places; thus firing means that instantaneously the transition inputs are emptied and *all* of its outputs filled.

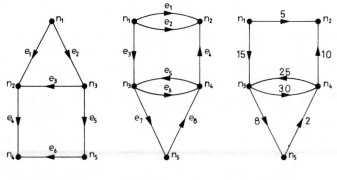

| a) Directed graph | b) Net | c) Network |

Figure 10.15 Directed graphs

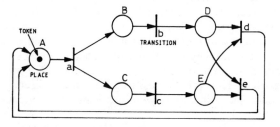

a) Marked net

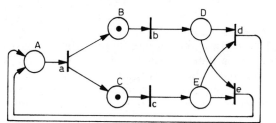

b) Net after firing

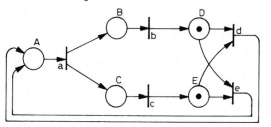

c) Conflict situation

Figure 10.16 Petri nets

Note that the transitions cannot fire simultaneously, and thus only one transition can occur at a time. This is illustrated in Fig. 10.16, where (a) shows the original marked net and (b) the state of the net after firing transition *a*. After two further firings the net would assume the marking shown in Fig. 10.16(c); here the net is said to be in *conflict* since firing either of the transitions *d* or *e* would cause the other transition to be disabled. In general a conflict will arise when two transitions share at least one input place; Petri net models are normally constrained to be *conflict free*.

Another limitation imposed on the model is that a place must not contain more than one token at the same time; this condition leads to a *safe* net. This restriction is essential when the Petri net is used to represent a set of interrelated events and conditions in a practical environment. In this case conditions would normally be represented by places and a particular condition considered to hold if and only if the place contains a token; thus to have more than one token would be irrelevant.

A *live* Petri net is defined as one in which it is possible to fire any transition of the net by some firing sequence, irrespective of the marking that has been reached: note that a live net would still remain live after firing. Liveness is an important property of the Petri net if transitions are to be interpreted as representing logical processes. The liveness of a transition means that there is no way in which a sequence of process executions can cause the system to get into a state from which the given process can never again be executed. Thus the liveness of a Petri net is directly related to the concept of deadlock or the 'deadly embrace' situation encountered in logic systems.

Other sub-clases of the Petri net may be defined but of particular interest is the *marked graph* shown in Fig. 10.17. A Petri net is called a marked graph if, and only if, each place has exactly one output transition. In this case the graph can be simplified by absorbing each place into an edge and letting the place marking be represented by a marking on the edge. If a further restriction is imposed such that each transition has exactly one input and output place the net is reduced to a finite-state machine. This may be seen by simplifying the graph such that each transition is represented by a directed edge from its input place to its output place when the net assumes the structure of a state diagram, as shown in Fig. 10.17(b).

If the initial marking of the net is such that only one place holds a single token than state transitions will correspond to transition firings. Note that while every finite-state machine has an equivalent Petri net the reverse does not hold, except of course for the restricted model.

The Petri net is considerably more powerful than the FSM model in that it can represent concurrent operations and provide a natural representation of a system with distributed control and state information. The use of FSMs to represent systems of this type would result in unmanageable large single states, being the set of all distributed states.

One of the major uses of Petri nets is to model and evaluate the control structure of a logical system. When used to specify hardware systems transitions in the Petri net would normally relate to the processing elements in the data structure and places to the control links between the control and

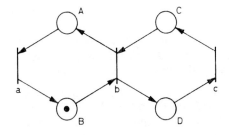

a) Marked graph

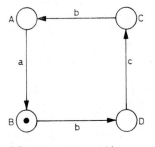

b) Equivalent state machine

Figure 10.17 Marked graphs

data structures. Note that in order to handle concurrent processes, asynchronous operation must be assumed, which necessitates passing control signals between the control and data structures. Thus a ready signal must be sent from the control structure, that is from a place, to initiate operations in the data structure (such as addition, multiplication etc.). When the required operation has been completed the processing unit must respond by transmitting an acknowledge signal back to the control structure. Note that this corresponds to the normal 'handshake' procedure used in asynchronous interfaces. The same procedure would be used for a conditional test performed on a register in the data structure but in this case the return signal would indicate whether the test had been true or false.

Let us consider how this technique may be used to process the function $(x + y)(x^2 + y)/z^2$. Figure 10.18 shows the control Petri net and the associated data structure. Note that the data structure contains registers $(x, y, z, A, B, C, D, E, F)$ and processing units for the arithmetic operations. The control procedure requires the firing rules to be modified as follows:

(a) Remove tokens from input places.
(b) Send a ready signal to the processing unit.
(c) Wait for acknowledge signal.
(d) Put tokens in output places.

Initiation of the process shown in Fig. 10.18 is accomplished by putting a token in place Q which corresponds to sending a ready signal to the data structure to indicate that the register x, y and z should now be loaded. When the registers are loaded the data structure responds with an acknowledge signal; applying the firing rules to the Petri net we have:

(a) A token is put in place Q which sets up transition a ready for firing.
(b) Remove token from Q and send ready signal to register circuits.
(c) Wait for acknowledge signal; when it is received place tokens in places R, S and T.

Application of the firing rules to transitions b, c and d initiate the next stages of the process; this procedure is repeated until the computation is complete, as indicated by a token in place z. Note that each processing unit contains its own registers and that parallel processing can be easily handled. For example, transition f cannot fire until places X and Y are filled, that is the independent operations of addition and multiplication and addition are completed.

Petri nets have been used to model and evaluate the control structures of logical systems in both hardware and software design. In addition it has been shown[23] that it is possible to replace the individual elements of a Petri net by hardware equivalents, thus providing a direct realization of the control circuits. Petri nets can also be used to model hierarchal systems, since an entire net can be replaced by a single place or transition at a higher level. Thus Petri nets can be used at all levels of design, from behavioural through to structural, and are able to span the full range of specification and evaluation requirements.

One of the major advantages of the directed graph approach is that its formal structure makes it amenable to mathematical analysis and many authors[24,25] have described algorithmic methods for their analysis. In the main these techniques apply to the control graph only, known as *uninterpreted* analysis, and no allowance is made for operations performed in conjunction with the data structure.

An important property of a control graph is whether or not the implied algorithm is *determinate*. Determinate systems are such that for any two inputs which have the same values the system always gives identical outputs. For example, in a parallel process multiple accesses to a common resource, for a given initial control state and data values, must always result in the same set of final values. In practice this can be achieved by satisfying the following conditions:

(a) No two data operations can simultaneously write into the same data location.
(b) Data operations must not be allowed to read and write simultaneously into the same location.

The obvious advantage of using Petri nets in logic system design is that once the control graph is generated and evaluated it can readily be converted to a hardware/software realization. Unfortunately Petri nets are difficult to analyse and though systems with a small number of variables

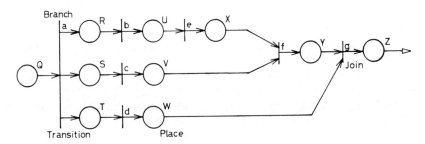

a) Petri net control structure

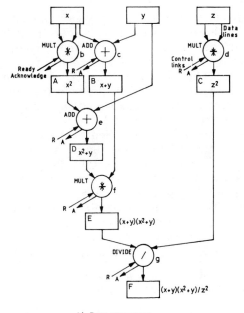

b) Data structure

Figure 10.18 Petri net structure

can be effectively evaluated the computation in the general case rapidly becomes unmanageable.

10.5.2   Hardware description languages

As we have seen, design at the systems level using MSI/LSI modules (and particularly computer design) is normally centred around a predefined register and highway configuration. The execution of a required system function, such as a machine-code instruction in a computer, is then interpreted in terms of a sequence of micro-orders, the microprogram, which controls the necessary transfers and the logic and arithmetic operations between registers. The same basic technique is also employed in the desgin of VLSI systems where an FSM is normally used to specify the micro-orders.

Register transfer languages are based on this heuristic design procedure and are similar to high level programming languages in that both carry out register assignments and support the specification of algorithms. Using an RTL a register structure can be declared (the data structure) and the required data and logical operations between the registers specified using program statements. Thus the declarative section of the program gives a formal description of the block diagram of the machine while the RTL program statements specify the control algorithms for executing specific instructions.

In a programmable electronics system, such as a digital computer, the required operation is specified in an *instruction word* held in memory (RAM) which must be read down, decoded and obeyed; this sequence of operations is usually called the *load/execute* cycle. The address of the instruction in memory is held in the *program counter* which must be incremented after each load/execute cycle to obtain the required sequential execution of instructions which are stored in contiguous locations in memory. This concept of storing machine-code instructions in specific areas of memory (a set of which is called a *program*) and then obeying them in sequence is fundamental to computer operation. The computer will carry on obeying instructions in this way until told otherwise, usually by transferring control to another part of the memory.

The RTL program shown in Table 10.5, based on CDL (computer design language),[26] illustrates how a machine-code instruction is obtained from memory and obeyed. The particular instruction calls for the contents of a location in memory to be added to an arithmetic register in the main machine. The instruction, suitably encoded, takes the form:

Add the contents of the location whose address is specified in the instruction; to the contents of register X

The program commences with the declaration of the register structure including the main memory, register fields and flags (single bits). For example, after declaring instruction register I to be 16-bits in length it is also possible to specify an address field I(ADR) comprising bits 6–15 of register I, and similarly for I(OP) which specifies the actual operation to be performed. Note that in declaring main memory the total number of words and the wordlength must be specified.

The timing of the operations in a set of RTL statements is represented by the order in which the operations are evaluated without regard to the real time delays involved in the register transfers. The execution sequence is determined by whether the RTL is a *procedural* or a *non-procedural* language. In a procedural language an explicit ordering is imposed on the statements, with a statement or operation being evaluated on the completion of the preceeding one. Non-procedural languages (of which CDL is an example) have no meaning associated with the textural ordering of the

Table 10.5 CDL program for Add instruction

Statement		Comment
		DATA STRUCTURE
Registers,	M(0–15)	Buffer register, Memory MN
	A(0–11)	Address register, Memory MN
	X(0–15)	Arithmetic register
	D(0–5)	Operation decode register
	P(0–11)	Program counter
	I(0–15)	Instruction register
	K(0–7)	State control register
Sub-regs,	I(ADR) = I(6–15)	Address digits register I
	I(OP) = I(0–5)	Operation code digits register I
	G	Stop/Start control bit
	RF	Read cycle finished
Memory,	MN(A) = MN(0–4095, 0–15)	Main memory
Terminals,	READ	Initiate memory read cycle
	WRITE	Initiate memory write cycle
Clocks,	Ck	System clock
		CONTROL STRUCTURE
	(K0.Ck) A: = P, READ, K: = 1	Load cycle
	(K1.Ck) P = P + 1, K: = 2	Increment P
	(K2.Ck) IF RF = 0 THEN I: = M, K = 3	Wait for Read to end
	(K3.Ck) D: = I(OP), K = 4	Decode Instruction
	(K4.Ck) A: = I(ADR), READ, K = 5	Obey cycle
	(K5.Ck) IF RF = 0 THEN X: = X + M, K = 6	
	(K6.Ck) IF G = 1 THEN K: = 0	Wait for start

(a) Non-procedural

Statement	Comment
	CONTROL STRUCTURE
T1 A: = P, READ	
P: = P + 1	
T2 IF RF = 1 THEN GOTO T2	Wait for READ to end
D: = I(OP)	
A: = I(ADR), READ	
T3 IF RF = 1 THEN GOTO T3	
X: = X + M	
T4 IF G = 0 THEN GOTO T4	Wait for start
GOTO T1	

(b) Procedural

statements and the sequence of operations is determined by specifying a conditional, state value or clock period as a prefix to each statement; this is illustrated in Table 10.5.

For the non-procedural case, in state $K0$ (determined by the state control counter K) and when clock Ck is present the contents of the Program counter P is transferred to the Memory address register and a READ cycle initiated. Note the manner in which this is specified: $A:=P$ signifies that the contents of Register P is transferred to Register A; the state control counter is set to 1 by $K:=1$. During the READ cycle, in state $K1$, the contents of the Program counter are incremented by 1. The end of the READ cycle is determined by examining the flag RF using an IF THEN instruction; note that if $RF = 1$ the state control counter is not updated and the operation will repeat until $RF = 0$. When $RF = 0$ the contents of the Memory register M are transferred to the Instruction register. In the next time-slot $K3$ the instruction is decoded by transferring the order bits of the Instruction register, $I(OP)$, to the decoder unit. This sequence of operations completes the Load cycle of the computer. In the next time-slot, $K4$, the address digits of the Instruction word, $I(ADR)$, are used to address the memory; again the IF THEN statement is used to wait for the end of the READ operation. When the READ is completed the contents of registers X and M are added together and control goes to time-slot $K6$; if $G = 1$ the process repeats taking the next instruction in sequence from the memory.

The same sequence of events can be programmed using a procedural RTL; this is shown in Table 10.5(b). In this case the state control register is not required and labels $T1 - T4$ are used to identify particular statements. Note that GOTO statements are used to effect a READ wait state (by jumping back to itself) and to repeat the procedure if $G = 1$.

Register transfer languages are principly used for functional simulation (in particular the evaluation of microcode) though *Boolean translators* have been described[24] for generating logic equations; note that the language is at too high a level to generate microcode directly. Unfortunately, from the viewpoint of LSI circuits these equations are generally derived using NAND/NOR logic; the register operations can, however, be regarded as a primitive set of circuit functions. Moreover, the logic realizations produced are generally very redundant due to the fact that each algorithm is specified separately in the RTL description, and the identification and merging of similar functions is not as yet possible.

Though considerable effort has been expended on the development of register transfer and hardware description languages[28-30] very few have been adopted for use in a real engineering design, as distinct from simulation, environment. Even when used for evaluation, hardware description languages suffer from the inherent disadvantage that they have no formal mathematical structure. Since a system is normally modelled in terms of functional components and their interconnections the behaviour must be interpreted indirectly from program performance whilst operating on specified data types. Again, in order to evaluate logic systems modelled in this way it is necessary to perform a step-by-step examination of all relevant input–output conditions; this can be time consuming as well as requiring large amounts of storage. Moreover, since most design languages are based on machine independent languages like ALGOL, APL etc., there are problems with hardware features such as timing, operating speed, parallelism and asynchronism. These parameters are often very difficult to describe and hence the correspondence between the model and the actual hardware is often ambiguous.

Nevertheless, hardware description languages provide the basis for useful design tools, particularly in functional level simulation. The techniques have also been applied to overall system design, embracing behavioural, functional and structural levels. In particular the multi-level design system LALSD[31] which consists of three separate languages—a graphical one based on Petri nets, a register transfer language and a logic level language, SABLE[32]—a system developed at Stanford University and the CMU-DA system[33] at Carnegie-Melon are worth mentioning.

10.5.3   Simulation techniques

Simulation is a process whereby it is possible to model, either mathematically or functionally, the behaviour of a real system; experiments can then be conducted on the model and related back to the actual system. It is important to realize that the simulated system is just a model and not necessarily an exact representation of the real physical world owing to the constraints and necessary approximations inherent in constructing the model.

In the simulation of microcomputer systems and logic systems generally the structures and processes to be modelled are so complex that formal mathematical modelling is normally impractical. The exception to this is the use of Petri nets and other directed graph methods to model the behaviour of a system. Consequently functional simulation is employed. For example, though it is possible to describe the logical operation of a circuit using Boolean equations and thereby generate the input/output characteristics in the form of a truth table no information about the performance of the implemented circuit, in terms of propagation delays, fault performance etc. can be obtained. Moreover the manipulation of Boolean equations with a large number of variables (the usual situation in practice) presents a very difficult computational problem.

In functional modelling there is usually some direct correspondence between the real system and the simulated system. For example a logic circuit would be represented by a topological model with specific words in the computer memory being allocated to represent each gate and its interconnections. To evaluate a functional model it must be exercised, that is required to emulate the actions of the real system, starting from specified initial conditions and a known system environment. The information derived from observations on such a model depends on the way the experiments are run and the interpretation placed upon the results. Mathematical models on the other hand *compute* a result depending on initial conditions and the system parameters. Most functional models are stochastic in that the system variables cannot be previously defined and probabilistic methods must be employed; a system which has completely predictable variables, that is deterministic, is usually amenable to mathematical modelling.

The major use of simulation in logic systems design is to verify and check designs, both at the functional and logic gate levels, prior to actual manufacture. Another important use is to generate test schedules for manufacturing and operational testing procedures. Owing to the limitations mentioned earlier, and the methods used in exercising the model, simulation does not entirely eliminate the need for debugging the actual equipment, but it does considerably reduce the problems involved.

Digital systems may be simulated at basically three different levels; these are as follows:

(i) *Systems level.* This consists at the moment of using high-level general-purpose simulation languages, such as GPSS[34] and SIM-SCRIPT[35] and deals mainly with a timing analysis of the system modelled in terms of subsystem components such as arithmetic and logic units, memory modules, peripherals, etc.

(ii) *Functional level.* In this case data flow at the register level is modelled, thereby enabling microprograms etc. to be evaluated. This form of simulation using RTLs was dealt with in the last section.

(iii) *Gate or logic level.* Here the actual logic gates, or modules, and their interconnections are functionally modelled in the computer. Each signal line is restricted to binary values, and time is usually quantized to gate-propagation delays.

The major benefits to be accrued using simulation techniques are savings in time and cost caused by eliminating fundamental design errors. Some of the more specific advantages of using simulation in digital systems design are as follows:

(a) Operational specifications can be validated.
(b) Corrections and modifications can be made during the early design stages.
(c) Alternative procedures and designs may be evaluated.
(d) The simulation description can serve as design documentation.

Simulation at the logic level is currently the most frequently employed technique, and has become a valuable tool for logic designers. In particular the technique is used to generate and verify fault-test procedures and to discover the presence of circuit races and hazards. It is essential, however, that the simulator should accurately model the timing of the circuit devices, to allow prediction of hazards and races, in order for it to be effective in design verification.

The major disadvantages of simulation are that it can be very time consuming, particularly if detailed results about large systems are required, and that it is also capable of misinterpretation.

In *gate level simulation*[36,37] the logic diagram is described in terms of logic modules, such as NAND/NOR gates etc., and their input/output connections using some form of problem orientated language. This topological description of the circuit is then translated and stored in the computer using a suitable data structure. To perform a simulation the model must be exercised and the results examined. This requires the user to specify the initial state of the network, input values and sequences, gate delays, monitor points etc.

The simulation process is performed by computing the logic states (as determined by the inputs and module type) at each output node of a circuit at descrete time intervals or states of the system (note that this refers to machine-time not real-time, though the time periods will be related to propagation delay values). The simulation continues in this way, printing the logic value of specified output nodes for each state of the system, until the required number of iterations have been completed. Thus by gate level simulation is meant the process of generating a logic state-time map (truth or transition table) for the logic network given the initial starting values and specified input sequences.

To illustrate this technique let us consider the simulation of the logic circuit shown in Fig. 10.19; this circuit contains a dynamic hazard and was fully discussed in Section 4.5 (the circuit is in fact the same as shown in Fig. 4.19). The simulation program is shown in Table 10.6(a): note the simple nodel specification, for example $N1$ NAND 1 $X1$ $NY2$ states that the output $N1$ of a NAND gate with unit delay 1 has inputs $X1$ and $NY2$. Having described the network the simulation controls must now be specified: the initial conditions are set (in this example) by the FLOW and FHIGH commands, and thus the inputs $X1$, $X2$ and $Y1$ are all set to logic 1 and $Y2$ to logic 0; all other output nodes are set to logic 0 using the DFLT (Default) command. The input $X1$ is further defined as a clock which starts at logic 1 and changes to logic 0 at time (state) 4 of the simulation. The total simulation time required is defined as TIME 16, i.e. 16 total states of the network are required to be investigated. Finally the output nodes to be examined must be specified: this is done using PLOT or PRINT statements.

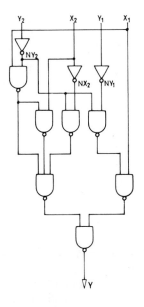

Figure 10.19 Circuit with dynamic hazard

Table 10.6 Simulation of Fig. 10.19

1	$DYNAMIC HAZARD
2	NX2 INV 1 X2
3	NY2 INV 1 Y2
4	NY1 INV 1 Y1
5	N1 NAND 1 X1 NY2
6	N2 NAND 1 X2 N1
7	N3 NAND 1 NX2 NY2
8	N4 NAND 1 NY1 NY2
9	N5 NAND 1 N1 N2 N3
10	N6 NAND 1 X1 N4
11	Y NAND 1 N5 N6
12	$CONTROLS
13	.FLOW .DFLT
14	.FHIGH X1 X2 Y1
15	.FLOW Y2
16	.PLOT X1 X2 Y1 Y2 Y
17	.TIME 16
18	$CLOCKS
19	X1 CLK1 4
20	.END

(a) Program

$DYNAMIC HAZARD

TIME	X_1	X_2	Y_1	Y_2	Y
0	1	1	1	0	1
1	1	1	1	0	1
2	1	1	1	0	1
3	1	1	1	0	1
4	0	1	1	0	1
5	0	1	1	0	1
6	0	1	1	0	0
7	0	1	1	0	1
8	0	1	1	0	0
9	0	1	1	0	0
10	0	1	1	0	0
11	0	1	1	0	0
12	0	1	1	0	0
13	0	1	1	0	0
14	0	1	1	0	0
15	0	1	1	0	0
16	0	1	1	0	0

(b) Output

The result of the simulation run is shown in Table 10.6(b). Note that the computed logic values of the specified monitor points are plotted for each time period; to assist the interpretation of the results the 1 and 0 values are displaced to give the effect of a timing diagram. It is of course possible to obtain actual waveform diagrams using graphics or digital plotters. The results obtained are identical with those deduced using the 'walkthrough' method in Section 4.5, showing that for the given initial conditions a dynamic hazard exists on output Y with the output changing $1 \to 0 \to 1 \to 0$ when $X1$ changes $1 \to 0$.

The simulator described above is very basic and to provide a realistic design tool other facilities are essential. For example, additional output information is often required such as loading (fan-out) statistics and the maximum number of logic levels a signal must pass through for each particular clock time. In simulating a large circuit to require a print out on each state of the system (as specified in the TIME command) could be very time wasting. Consequently facilities must be included to print-out only when particular outputs change value; for example, by using a PRINT-ON-CHANGE statement with either all or just particular output nodes specified. Again the possibility of 'spikes' or circuit hazards occurring, such as when two inputs are requested to change simultaneously, should also be signalled to the output.

With logic level simulation it is essential to have a *macro* facility which allows a previously defined circuit to be treated as a module in its own right. For example, in simulating a counter circuit using JK master–slave bistables each stage would need to be expressed in terms of basic NAND gates (as shown in Table 10.7). Using the macro facility, once the basic circuit has been defined (or better still already available in a predefined library of standard modules such as SN74 series logic) the counter can be specified by calling up the macro and using it as a basic element; this is shown in Table 10.8 for a simple two-bit asynchronous counter circuit. As a further example, this time using a standard library of SN74 series logic units, Table 10.9 shows the simulation program for the chain-code counter given in Fig. 10.20.

The most common method of incorporating propagation delays into a simulator is called *unit delay* in which each circuit element is assumed to take one unit (or some multiple) of machine-time to switch. In the case of the simple simulator described above the propagation delay is usually taken as the nominal or worst case delay for the device. The propagation delay is specified separately for each element in terms of machine-time steps, which can be arbitrarily related to real-time. For example, for devices with 100–200 ns delay times and assuming the network is required to be examined over a 10 ms period then 1000 machine steps would be required. For some applications simple delay models of this type are inadequate, for instance when checking out race conditions a more comprehensive set of delay parameters is required. These would include the propagation delays for both a $1 \to 0$ and a $0 \to 1$ change of state, the minimum time allowable between the edges of an input pulse to operate a device and, in the case of bistables, the minimum time an input pulse must maintain its state both before and after a clock pulse. The very minimum delay parameters that are required to fully check out circuit hazards would be the rise and fall time delays for $0 \to 1$ and $1 \to 0$ changes.

The presence of hazards and race conditions in logic networks may be detected using *three valued simulation*[38] based on the concepts of ternary

Table 10.7 Simulation of JK-bistable

		$JK-MS BISTABLE		
1	$JK-MS BISTABLE		C	N	Q
2	$CIRCUIT	L	3	
3	Q NAND 1 N1 NQ	K		
4	NQ NAND 0 N2 Q			
5	N1 NAND 0 N7 N3	TIME		
6	N2 NAND 0 N7 N4	0	0	0	0
7	N3 NAND 1 N5 N4	1	0	0	0
8	N4 NAND 0 N3 N6	2	1	0	0
9	N5 NAND 0 NQ J CLK	3	1	1	0
10	N6 NAND 0 Q K CLK	4	0	1	0
11	N7 INV 0 CLK	5	0	1	1
12	$CLOCKS	6	1	1	1
13	CLK CLK0 2 4 R 2	7	1	0	1
14	J CLK1	8	0	0	1
15	K CLK1	9	0	0	0
16	$CONTROLS	10	1	0	0
17	.TIME 16	11	1	1	0
18	.FLOW .DFLT	12	0	1	0
19	.PLOT CLK N3 Q	13	0	1	1
20	.END	14	1	1	1
		15	1	0	1
(a) Program		16	0	0	1

(b) Output

Table 10.8 Two-bit counter using macros

1	$TWO-BIT COUNTER
2	.MACRO JKFF J K CLK Q NQ
3	$NETWORK
4	Q NAND 1 N1 NQ
5	NQ NAND 0 N2 Q
6	N1 NAND 0 N7 N3
7	N2 NAND 0 N7 N4
8	N3 NAND 1 N5 N4
9	N4 NAND 0 N3 N6
10	N5 NAND 0 NQ J CLK
11	N6 NAND 0 Q K CLK
12	N7 INV 0 CLK
13	.EOM
14	*M1 JKFF L1 L1 CLKIN A NA
15	*M2 JKFF L1 L1 A B NB
16	$CLOCKS
17	L1 CLK1
18	CLKIN CLK0 2 4 R 2
19	$CONTROLS
20	.TIME 64
21	.FLOW .DFLT
22	.PLOT CLKIN A B
23	.END

Table 10.9 Program for chain code counter

```
 1   $CHAIN CODE COUNTER
 2   $CLOCKS
 3   VCC CLK1
 4   VSS CLK0
 5   CLKIN CLK0 2 4 R 2
 6   SET CLK0 1
 7   $NETWORK
 8   *MACR0 LS74 VCC N3 CLKIN SET P5A P6A VSS P8A P9A VCC CLKIN P5A
     VCC VCC
 9   *MACR1 LS74 VCC P9A CLKIN VCC P5B P6B VSS P8B P9B VCC CLKIN
     P5B VCC VCC
10   N1 NAND 0 P5B P8B
11   N2 NAND 0 P6B P9B
12   N3 NAND 0 N1 N2
13   $CONTROLS
14   .FLOW .DFLT
15   .PRINT CLKIN SET N3 P5A P9A P5B P9B
16   .TIME 64
17   .END
```

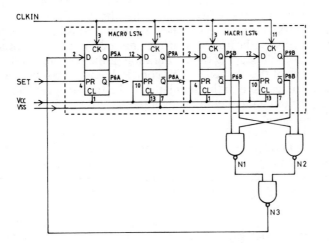

Figure 10.20 Chain-code counter

algebra. In this method a third value X, which can assume either logic 0 or logic 1, is used to represent unspecified initial logic states and hazardous conditions. Basic logic gates, such as NAND/OR etc., may be redefined in terms of its ternary function using the values 0, 1 and X; Table 10.10 gives the truth table for the functions AND, OR and NAND; note that with three variables there are $3^2 = 9$ possible input combinations. The ternary function of a logic gate can easily be determined by putting the X inputs to the gate alternately at 0 and 1 and noting whether the gate output (Z) changes ($Z = X$) or remains at 0 or 1.

Table 10.10 Ternary logic functions

| Inputs | | Outputs | | |
A	B	$Z = A.B$	$Z = A + B$	$Z = \overline{A.B}$
0	0	0	0	1
0	1	0	1	1
1	0	0	1	1
1	1	1	1	0
X	0	0	X	1
0	X	0	X	1
X	1	X	1	X
1	X	X	1	X
X	X	X	X	X

To detect hazards using three-valued simulation the X value must be included in the primary inputs to a circuit between every $0 \rightarrow 1$ and $1 \rightarrow 0$ transition. The simulation then proceeds in the normal way, except that the ternary function of the logic element is used to compute the output values. If a gate output takes the value X then this is an indication that a *possible hazard* could exist in the network; note that the X value will in general be propagated through the circuit to the final output. It is important to realize that the hazard so indicated may or may not occur in the actual hardware implementation, depending on the duration of the transient and the actual response times of the elements. This comes about as a consequence of the actual delay times used in the simulator; if the three-valued simulator employs unit or nominal delay times then the hazard predictions will tend to be very pessimistic.

Most current simulators employ at least a three-valued model and in many cases a multi-valued system is employed which allows precise delay times. For example, if minimum or maximum delay times are used it is possible to have a six-valued simulation with the values 0, 1, X, U (rising signal), D (falling signal) and E (potential hazard).

As well as hazard detection multistate simulators are also used for the modelling of wired-OR connections and in particular path-closing networks such as MOS transfer gate circuits. The logic output from a wired-OR connection is dependent both on the logic levels and signal strengths of all the elements connected to the node. Logic levels would be modelled in the usual way using logic 0 and logic 1 values and with X representing a don't-care value; an indeterminate logic level (between 0 and 1 values) must also be represented using, for example, the $*$ symbol. The output strength of the element is usually defined in terms of low impedance (LZ) or high impedance (HZ) and whether the signal originates from a forcing (F) source such as ground or a supply voltage. Thus we have the states 0, 1, X, $*$, LZ, HZ, F and XS (any signal strength).

In performing a simulation both logic levels and signal strengths must be taken into account according to a predefined truth table; Table 10.11 shows such a table for the wired-OR connection. Note that a forcing level would

Table 10.11 Truth table wired-OR circuit

Input 1 Signal strength	Logic level	Input 2 Signal strength	Logic level	Output node Signal strength	Logic level
F	*	LZ	X	F	*
F	1	LZ	X	F	1
F	0	LZ	X	F	0
F	*	HZ	X	F	*
F	1	HZ	X	F	1
F	0	HZ	X	F	0
LZ	*	HZ	X	LZ	*
LZ	1	HZ	X	LZ	1
LZ	0	HZ	X	LZ	0
F	*	F	*	F	*
F	*	F	1	F	*
F	1	F	*	F	1
F	1	F	1	F	1
F	1	F	0	F	*
F	0	F	1	F	*
F	0	F	0	F	0
LZ	*	LZ	*	LZ	*
LZ	*	LZ	1	LZ	*
LZ	1	LZ	*	LZ	*
LZ	1	LZ	0	LZ	*
LZ	0	LZ	1	LZ	*
LZ	0	LZ	0	LZ	0
HZ	*	HZ	*	HZ	P
HZ	*	HZ	1	HZ	P
HZ	1	HZ	*	HZ	P
HZ	1	HZ	0	HZ	P
HZ	0	HZ	1	HZ	P
HZ	0	HZ	0	HZ	P

over-ride any low or high impedence level, and any weak levels over-ride high impedence levels; with signals of equal strengths an indeterminate logic level will arise if the signals have different logic levels. Note also that in Table 10.11 the signal P indicates that the previous logic level will persist.

The consideration of signal strengths and circuit delays in a logic simulation can of course only be done on a qualitative basis. To produce a realistic model of the circuit the actual electrical performance of the network and its elements must be investigated using an AC/DC circuit analysis program such as SPICE.[39] Better still, hybrid circuit analysis and logic simulation programs should be employed; these allow accurate circuit and timing analysis to be performed for critical areas of the network (such as long chains of pass transistors or high fan-out nodes) in conjunction with logic stimulation.[40]

Two basic techniques are used to implement a simulator program: these are *compiled-code* and *table-driven*. In a compiled-code simulator each logic function, such as NAND/NOR etc., would have a corresponding program segment or macro, which performs the required logic operations. The logic network to be simulated would be translated by a software compiler into a series of interconnected macros which are executed in the order they are specified in the network description. Using a compiled-code simulator combinational circuits can be evaluated in one pass through the network; sequential circuits would normally require several iterations, performing a number of passes through the circuit in one time period until a stable condition has been reached. Timing is under the control of a machine generated clock which is usually synchronous in the sense that nodal outputs (which form the input to the elements) are only sampled at definite time periods. The major disadvantage of the compiled-code technique is that all logic elements are evaluated in sequence, even if a signal path is unactivated (i.e. no change in the input conditions) and consequently the time required to simulate a large circuit can be excessive. Note also that any changes in the circuit configuration requires the network description to be recompiled.

In a table-driven simulator a *data structure* is set up in the computer memory which stores, in a tabular form, the particular parameters of each element in the network. Each table entry in the element list would comprise data on logic function, output value, propagation delays and pointers (memory addresses) to separate Input and Output tables (see Fig. 10.21). In this way the source language description of the circuit is translated into a data-structure which reflects the topology of the network.

The simulation proceeds by operating on the data structure following the linkages (as defined by the pointers) from one element to another, interpreting the information on the element lists in accordance with the simulator command statements. Note that in this technique, for specified input changes, only activated paths through the network are computed (by considering those elements which actually change state) rather than the entire circuit. This process of simulating only *significant events* in a network gives a considerable increase in computational efficiency and would be essential when simulating very large networks.

For example from Fig. 10.21, starting with a particular gate on the Element list the inputs, determined by their pin numbers, would be obtained from the Input list; note that this necessitates referring back to the Element list to determine the current output values. The new output of the gate would then be computed using the appropriate macro for the gate type and the result written back into the Element list. The next step is to determine from the Input list those elements which accept this output as their input (the destination elements). This information is obtained from the Output list which points back to the element on the Element table and the process repeats. Note that the simulation is driven by the data structure and that the circuit structure can easily be modified by simply changing pointers in the list structure.

Logic level simulation is also used extensively to generate test sets for logic networks, based on the stuck-at single fault model. Fault simulation can also be used to validate fault detection or diagnostic tests, create fault-

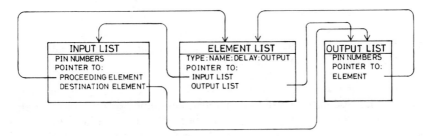

Figure 10.21 Table-driven simulator

dictionaries, and to assist in the design of diagnosable and fault-tolerant logic. The obvious approach to test pattern generation is first to simulate the fault-free circuit for all input conditions and then to repeat the simulation for all possible faulty versions of the circuit containing single logic faults. Comparison of the outputs obtained with the faulty and fault-free models yields the set of identifiable faults. That this is a time consuming process will be obvious and is in practice very seldom used due to the excessive computation involved for very large circuits, as for example encountered in LSI design. As a consequence, full fault cover is very rarely attempted and resort is made to obtaining a good or effective cover employing, for instance, statistical testing using random input sequences.

To evaluate the fault-mode characteristics of a circuit it is necessary to be able to introduce the effect of a gate failure into the circuit—known as *fault injection*.[41] Using a table-driven simulator the pointers in the data structure can be modified to allow a conditional choice between input and output values; alternatively a special table can be used to store the fault conditions. Special commands are required to set and reset the fault conditions which are simply treated as different cases of the same model. As we have seen checking out test sequences by performing separate simulation runs can be an expensive and time-consuming process. Single fault simulation has been replaced in practice by *parallel fault* and *deductive fault* simulation, both techniques giving considerable reductions in computation time.

Parallel fault simulation[42] is the simultaneous simulation in parallel of a number of copies of the circuit under test. One network is designated the correct circuit while each of the other copies represents a single fault version of the network. This is achieved by arranging that each bit in the computer word represents a fault condition and performing the appropriate element logic functions between words. Note that this would normally require using a compiled-code form of simulator and that the number of circuits that can be simulated would be restricted by the computer word length. This approach was adopted for the TEGAS simulator[43] which is still used extensively in practice. Deductive fault simulation[44] considers the fault-free circuit and all faults in one pass of the simulator using a *fault-list*. A fault-list is a list of all those faults which will change the output value of a gate at the present instant of simulated time. The list contains one entry for each fault which is detectable on the output of the gate and is updated as the simulation proceeds; the fault-list for a gate output is computed from the fault lists associated with its inputs. All faults are considered in one simulation pass with the fault lists being propagated from the site of the failures to the circuit outputs.

It has been suggested that the parallel technique is more cost effective for small highly sequential networks or for circuits with a small number of faults. The deductive method, on the other hand, would appear better suited for circuits having a large number of faults or where sequential circuits form a small part of the total system. On average the parallel fault simulator uses less storage but more processing time than a deductive simulator. Thus if storage requirements are an important factor the parallel fault simulation approach is better. In general the deductive approach would, however, be more efficient.

General purpose simulation languages, such as GPSS and SIMSCRIPT, can be used to simulate any system which changes in a descrete manner and consequently they can be used for modelling logic systems at the systems or behavioural level.[45] For example, it is possible to express the structure of a system, its behaviour and the required input conditions, with timing controls, output formats and diagnostics being built into the software. In particular GP simulators are used to model the overall performance characteristics of computer based systems such as

(a) workload studies on certain classes of computer;
(b) comparison of computer systems;
(c) processing capability for various system configurations;
(d) investigation of different partitionings of the system;
(e) 'tuning' the system by determining and eliminating critical areas etc.

The most commonly used languages are SIMSCRIPT which is a FORTRAN based event driven simulator, GPSS which is an interpretive (compiled code) block structured language and GASP[46] which uses an event scheduling approach.

Simulation has become an essential software tool for the design and testing of logic systems at both the systems and logic gate levels. However, the tools still need to be perfected, particularly if called upon to handle LSI/VLSI circuits which have a capability of at least $\frac{1}{2}$ million gates per chip; most of the available simulators are woefully inadequate to handle circuits of anything like this complexity. In this case it is essential to develop powerful multi-level simulators for both functional and logic levels, including timing and circuit analysis. Fault-test generation where applicable must be considered as part of the overall design process and fully integrated in an interactive manner with functional and logic level simulation—the HILO system[47] goes a long way to achieving these objectives.

Finally, simulation at all levels can involve very large investments in both time and manpower resources—to justify the adoption of a simulator package as a design tool it must be shown to be accurate (for the purpose required) and cost effective.

10.5.4 Tools for physical design of LSI circuits

CAD tools for the physical design and fabrication of custom LSI/VLSI circuits fall into three main categories:

(a) layout and interconnection aids;
(b) layout verification and design rule checking;
(c) functional and parametric testing.

Integrated circuit layout is the process of translating a description or specification of the integrated circuit into the photolithographic masks used to fabricate the circuit. The procedure usually commences by partitioning the logic system into functional blocks or cells, such as decoders, ALUs etc., and allocating them, with due attention to connectivity paths and buses, to areas on the chip surface—the floor-plan. As the design proceeds a detailed physical specification for the function blocks is evolved together with the actual interconnection paths between blocks.

Mask geometries are comprised almost entirely of rectangular shapes and wires (defined by their centre line and width). Consequently, and in order to handle the extremely large numbers of rectangles involved (some 50–100,000 for a chip containing a few thousand gates) it is essential to have some computer-assisted method of specifying their size, shape and relative positions. This can be done in a number of ways:

(a) Digitizing hand produced drawings—this involves manually tracing over a scaled diagram and automatically entering the digital coordinates into a computer data-file (the procedure is rather akin to using a digital plotter in reverse).
(b) Using a high-level geometric programming language, called a layout language to produce a textual description of scaled drawings.
(c) Graphical input via an interactive visual display unit which allows shapes to be entered, positioned and edited using an on-screen cursor and command menu.

Most CAD workstations would provide a combination of these techniques, for example a layout language would be available for the initial input and an interactive graphics editor to display and modify the mask designs. In addition some means of producing a hard copy print-out, such as a colour plotter, would also be required for visual checking of the layout.

All layout languages would provide a basic means of specifying shape coordinates but more sophisticated facilities are required to make it a useful design tool. A typical layout statement would have the format

Layer	Shape	Lower left coordinates XY	Width X	Length Y
POLY	RECT	6, 10	2, 5	

where the XY coordinates would be referred to a λ-based grid—typically $\lambda = 3$ microns with a grid size of 1000. In our example the statement would generate a rectangle 2 microns by 5 microns in the polysilicon layer, as shown in Fig. 10.22. It should also be possible to *transform* symbols, that is

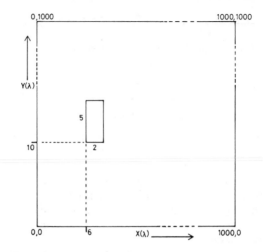

Figure 10.22 LSI layout languages

perform such operations as *mirroring*, *translation* and *rotation*.[48] The minimum requirement for a layout language would be rotation by fixed multiples of 90°, mirroring (or reflection) about the *XY* axis and translation (moving the symbol through a distance *X* or *Y* from its origin).

The order to facilitate the design process other primitives which allow replication, deletion, use of a standard cell library etc., would also be required. An effective way of implementing a layout language is to embed it within an existing programming language as a set of macros or procedure cells, thus making available the full power of the host language.

The layout language description would normally be assembled into a *design file*, such as the CALTECH Intermediate form CIF,[1] and used as the input to output devices such as plotters, VDUs and mask generation machines.

The full interactive graphics approach allows the layout design to be created and modified, using a *graphics editor*, directly on a CRT visual display screen. There are two ways of using a symbolic graphics system— either in a *static* or *dynamic* mode.

In a static symbolic system the graphics are used to draw and inter-actively edit a layout description normally derived from layout drawings. In this case the designer has complete responsibility for generating and placing the symbols and as a consequence must ensure that the relevant design rules are not violated.

The dynamic symbolic system, on the other hand, allows the designer to produce a rough sketch, normally in stick diagram form, of the circuit which then forms the input to the graphics system. Once the overall topology of the cell or circuit has been specified the system will then com-plete the design in detail working to a pre-defined set of parametized design rules. Commencing with a floor-plan in the usual way the design would proceed with individual cells being developed and added to the overall layout—facilities would normally be available for *stretching* the cell dimensions to ensure allignment with the input/output connections to other cells. The main disadvantage of dynamic systems of this type is that the component density would normally be less than that achievable by manual or static symbolic methods. However, a technique known as *compaction* can be used to reduce the overall area of the cells: this is a software routine which 'squeezes' the geometry in orthogonal directions until a minimal design rule constraint is reached.

The next logical development in the layout process is *silicon compilation*, such as Bristle Blocks,[49] which accepts a high-level description of specific architectures and produces a layout, stick diagrams and diagrams for functional blocks. Bristle Blocks is primarily a system for constructing the constituent parts of a microprocessor such as the arithmetic and logic unit, register structures, bus shifters (to effect register transfers) etc. presented in terms of a standard floor-plan. Each element is considered in terms of 1-bit slices, for instance a single-bit register stage, which are stacked vertically to form a functional unit. The functional units are arranged horizontally across the chip to connect with bus interfaces situated on the edges. The control portion of the microprocessor, i.e. the micro-order decoder and sequencer, must be laid out separately using other means. PLAs are almost exclusively used for this purpose often using a special PLA generator which is also a form of silicon compiler.

Bristle Block cells, unlike those in static systems which are stored with fixed geometries, are in effect subroutines which can reproduce and stretch the dimensions of the specified cell, compute its power requirements etc.

The input to Bristle Blocks has three main components:

(a) the number of bits in each micro-code word that is assigned to, for instance, the register select field, the ALU field etc.;
(b) the data-word width and list of buses;
(c) a list of the core elements, exclusive of pads (input/output connections), such as instruction decoders, the main upper and lower buses, and the micro-code.

The system performs a layout in three distinct stages—a core pass to lay out the core elements, a control pass to add the instruction decoders and a pad pass to place the pads on the perimeter of the chips and to route wires to points in the core and decoders.

The wiring together or interconnection of randomly sized circuit blocks in custom LSI design is usually performed as an integral part of the overall layout procedure. The routeing problem was one of the first areas of logic systems manufacture to benefit from CAD techniques originating in the need to layout and interconnect MSI packages on printed circuit boards; consequently the basic techniques are well established.[50] The current major application area is the auto-routeing of uncommitted logic arrays where the gates are normally situated on a defined grid and connected via channels of a fixed width.[51,52]

One of the most important aspects of LSI design, and certainly one which absorbs a major part of the total development time, is that of testing the layout design prior to fabrication. The ideal, of course, would be to employ synthesis tools which would eliminate the need for error-checking —unfortunately the technology has not yet reached this stage.

Mask designs must be checked in two ways: first to ensure that the technology design rules have been properly observed and, second, to verify that the actual physical circuit (as described by the layout) will fulfil the required logic functions. Software tools are available for both these purposes but sight-testing is still a valid and well used procedure. This requires the manual inspection of a print-out of the mask patterns including subsets of the individual layers such as metal and contacts. Normally a high contrast enlargement or colour plot would be used for this purpose (clearly complex circuits would require a lot of paper!). Even when automatic design rule checkers are used they do not necessary show up all fatal faults; for example, transistors could obey the design rules but be wired incorrectly, breaks in bus lines could go undetected—in many cases visual inspection can easily detect these errors.

A design rule checker embodies the rules for a particular process, as described in Section 10.3, and operates on the formal description of the pattern (usually in the layout language form). Each feature is examined for the correct dimensions and spacing, both in the same and different layers, overlap tolerances etc., reporting each violation together with its coordinates (which can also be indicated on a VDU). Design rule checking would appear to be primarily a geometric problem but many design rules are difficult to specify and compute and a satisfactory solution to the checking problem is still being sought.[53,54]

In order to verify that the mask design faithfully reproduces the required logic specification circuit extraction programs are needed which automatically interpret from the mask geometry data the circuit to be realized. The program is required to extract both a logic diagram (which can then be used either directly or indirectly as the input to a simulator) and circuitry data such as resistance and capacitance values to analyse the electrical performance.

All the software tools described above have the common objective of reducing design errors; however, faults can also occur as the result of bad processing. Consequently it is necessary to arrange for both functional and parametric testing of the processed wafer. Functional testing of logic systems has already been considered in depth in Chapter 9 and of course the same principles apply to IC design. Process and circuit (device) parameters, for example invertor delays, can be tested by including a small test pattern, consisting of a simple structure like a transistor or ring oscillator, on the wafer. Provision must also be made for additional bonding pads to allow access, normally via special probing equipment, to internal nodes for debugging and test purposes.

References and bibliography

1. Mead, C. and Conway, L. *Introduction to VLSI Systems*. Addison Wesley, Reading, Mass., 1980.
2. Lewin, D. *Theory and Design of Digital Computer Systems*. Van Nostrand Reinhold (UK), Wokingham, 1980.
3. Rosen, R. Contemporary concepts of microprogramming and emulation. *ACM Computing Surveys* **1**, 197–212 (1969).
4. Wilkes, M. and Stringer, C. Microprogramming and the design of control circuits in an electronic digital computer. *Proc. Camb. Phil. Soc.* **49**, 230–8 (1953).
5. Vandling, G. C. and Waldecker, D. E. The microprogram control technique for digital logic design. *Computer Design* **8**, 44–51 (1969).
6. Husson, S. *Microprogramming: Principles and Practice*. Prentice Hall, Englewood Cliffs, N.J., 1970.
7. Agrawala, A. K. and Ranscher, T. G. *Microprogramming: Concepts and Implementations*. Academic Press, New York, 1974.
8. Andrews, M. *Principles of Firmware Engineering in Microprogram Control*. Computer Press, Potomac, Maryland, 1980.
9. Shima, M. Demystifying microprocessor design. *IEEE Spectrum* **16**(7), 22–30 (1979).
10. Ramamoorthy, C. V. and Li, H. F. Pipeline architecture. *ACM Computing Surveys* **9**, 61–102 (1977).
11. Hong, S. J., Cain, R. G. and Ostapko, D. L. 'MINI': A Heuristic Approach for Logic Minimisation. *IBM J. Res. Dev.* **18**, 443–58 (1974).
12. Wood, R. A. A high density programmable logic array. *IEEE Trans. Computers* **C28**(9), 602–8 (1979).
13. DeMicheli, G. and Sangiovanni-Vincentelli, A. Multiple constrained folding of programmable logic arrays: theory and applications. *IEEE Trans. Computer Aided Design of ICAS* **CAD2**(3), 151–67 (1983).
14. Ayres, R. Silicon compilation—A hierarchial use of PLAs. *Proc. Cal. Tech. Conf. VLSI* 311–26 (1979).
15. Anceau, F. CAPRI: A design methodology and a silicon compiler for VLSI circuits specified by algorithms. *Third Cal. Tech. Conf. VLSI*, R. Bryant. (Ed.). Computer Science Press, Potomac, Maryland, 1983, pp. 15–31.
16. Mavor, J., Jack, M. A. and Denyer, P. B. *Introduction to MOS LSI Design*. Addison Wesley, London, 1983.
17. Williams, J. D. Sticks—A new approach to LSI Design. MSEE Thesis Dept. EE and CS, MIT, 1977.
18. Tobias, J. R. LSI/VLSI building blocks. *IEEE Computer* **14**(8), 83–101 (1981).
19. Stigall, P. O. and Tasar, O. A review of directed graphs as applied to computers. *IEEE Computer* **7**(10), 39–47 (1974).
20. Hardware Description Languages (special issue). *IEEE Computer* **7**(12) (1974).
21. Peterson, J. L. Petri nets. *ACM Computing Surveys* **9**, 223–52 (1977).
22. Agerwala, T. Putting Petri nets to work. *IEEE Computer* **12**, 85–94 (1979).
23. Patil, S. S. On structured digital systems. *Proc. Int. Sym. on CHDLs and Their Applications*. New York, 1975, pp. 1–6.
24. Karp, R. M. and Miller, R. E. Properties of a model for parallel computation: Determinacy, termination, queueing. *J. Appl. Math.* **14**. 1300–1411 (1966).
25. Peterson, J. *Petri Net Theory and the Modelling of Systems*. Prentice Hall, Englewood Cliffs, N.J., 1980.

26. Chu, Y. An ALGOL-like computer language. *Comm. ACM* **8**, 605–15 (1965).
27. Mesztenyi, C. K. Computer design language, simulation and Boolean translation. Tech. Rep. 68–72, Computer Science Centre, University of Maryland, 1968.
28. Barbacci, M. R. A comparison of register transfer languages for describing computers and digital systems. *IEEE Trans. Computers* **C24**, 137–50 (1975).
29. Duley, J. R. and Dietmeyer, D. L. A digital system design language (DDL). *IEEE Trans. Computers* **C17**, 850–61 (1968).
30. Bell, C. G. and Newell, A. The PMS and ISP descriptive system for computer structures. *AFIPS, SJCC*, **36**, 351–74 (1970).
31. Baray, M. B. and Su, S. Y. H. LALSD—A language for automated logic and system design. *Proc. Int. Sym. CHDL* 30–31 (1975).
32. Hill, D. and Van Cleemput, W. M. SABLE: A tool for generating structural multilevel simulations. *Proc. 16th Design Automation Conf.*, San Diego, 1979, pp. 172–9.
33. Director, S. W., Parker, A. C., Siewiorek, D. P. and Thomas, D. E. A design methodology and computer aids for digital VLSI systems. *IEEE Trans. Circuits, Systs.* **CAS-28** 634–5 (1981).
34. Efron, R. and Gordon, G. General purpose digital simulation and examples of its application. *IBM Systems J.* **3**, 22–34 (1964).
35. Wyman, F. P. *Simulation Modelling: A guide to using SIMSCRIPT*. Wiley, New York, 1970.
36. Hays, G. G. Computer aided design—Simulation of digital design logic. *IEEE Trans. Computers* **C18**, 1–10 (1969).
37. Lake, D. W. Logic simulation in digital systems. *Computer Design* **9**, 77–83 (1970).
38. Duke, K. A., Schnurman, H. D. and Wilson, T. I. System validation by three-level modelling synthesis. *IBM, J. Res. Dev.* **15**, 166–74 (1971).
39. Nagel, L. SPICE 2: A computer program to simulate semiconductor circuits. Electronics Research Lab., Rep. No. ERL-M520, University of California, Berkeley, May 1975.
40. Richard Newton, A. The simulation of large scale integrated circuits. Electronics Research Lab. Mem. No. UCB/ERL M78/52, University of California, Berkeley, July 1978.
41. Szygenda, S. A. and Thompson, E. W. Fault injection techniques and models for logical simulation. *AFIPS FJCC* **41**, 875–84 (1972).
42. Szygenda, S. A. and Thompson, E. W. Digital logic simulation in a time-based table driven environment. Pt 2—Parallel fault simulation. *IEEE Computer* **8**(3), 38–49 (1975).
43. Szygenda, S. A. TEGAS2—Anatomy of a general purpose test generator and simulation system for digital logic. *Proc. 9th ACM IEEE Design Automation Workshop*, June, 1972.
44. Chang, H. Y. and Chappell, S. G. Deductive techniques for simulating logic circuits. *IEEE Computer* **8**(3), 52–9 (1975).
45. Jayakumar, M. S. and McCalla, T. M. Simulation of microprocessor emulation using GASP-PL1. *IEEE Computer* **10**(4), 20–6 (1977).
46. Alan, A., Pritsker, B. and Young, R. *Simulation with GASP-PL1*. Wiley, New York, 1975.
47. Flake, P. L., Musgrave, G. and White, I. J. A digital system simulator—HILO. *Digital Processes* **1**, 39–53 (1975).
48. Newman, W. M. and Sproull, R. F. *Principles of Interactive Graphics*. McGraw Hill, New York, 1979.

49. Johannsen, D. Bristle Blocks: A silicon compiler. *Caltech. Conf. on VLSI*, Jan. 1979, pp. 303–310.
50. Breuer, M. (Ed.) *Design Automation of Digital Systems*. Vol. 1: *Theory and Techniques*. Prentice Hall, Englewood Cliffs, N.J., 1972.
51. Yoshimura, T. and Kuh, E. S. Efficient algorithms for channel routeing. *IEEE Trans. Computer Aided Design of ICAS*. **CAD1**, 25–35 (1982).
52. Pinter, R. Y. River routeing: Methodology and analysis. *Third Caltech Conf. on VLSI*, R. Bryant (Ed.). Computer Science Press, Rockville, Maryland, 1983, pp. 141–64.
53. Losleben, P. and Thompson, K. Topological analysis for VLSI circuits. *16th Design Automation Conf*. June 1979, pp. 461–73.
54. Mitsuhashi, T., Chiba, T., Takashiwa, M. and Yoshida, K. An integrated mask artwork analysis system. *17th Design Automation Conf*. June 1980, 277–84.
55. Trimberger, S. Automating chip layout. *IEEE Spectrum* June 1982, 38–45.
56. Raymond, T. C. LSI/VLSI Design Automation. *IEEE Computer* **14**(7), 89–101 (1981).
57. Hicks, P. J. (Ed.) Semi-custom IC design and VLSI. Peregrinus, London, 1983.
58. Bergmann, N. A case study of the F.I.R.S.T. Silicon Compiler. Third Caltech Conf. on VSLI. R. Bryant (Ed.). Computer Science Press, Rockville, Maryland, 413–30, 1983.

Tutorial problems

10.1 Figure 10.23 shows the ASM chart for the controller unit of a synchronous accumulating adder system which is to be implemented as an LSI circuit; CLR, ADD and CAR are inputs and RASC, RSOF, SUM and COF the required control outputs. Design a PLA circuit in the normal way and then determine the effect of the overall size of the array by assigning such that monotonic equations may be derived for the next-state equations.

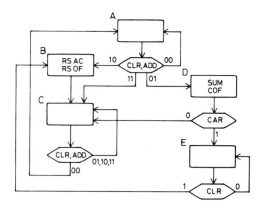

Figure 10.23 Problem 10.1

10.2 Translate the parity check circuit shown in Fig. 5.5 into a stick diagram and then convert it into the symmetric network $S_{(0,1,2,3)}$ ($X1$, $X2$, $X3$).

10.3 For the simulation program shown in Table 10.7 draw the corresponding logic diagram and ascertain that the JK-bistable is functioning correctly.

What other tests would be required to check out the circuit completely?

Explain why the NANDs in line 3 and 7 of the program have 1-unit delays. (Note: C CLK0 2 4 R 2 means that C starts at logic 0 changes to logic 1 at time 2 and back to logic 0 at time 4 then repeats from time 2).

Answers with worked solutions

Chapter 1

1.1 **(a)** It is best to convert the integral and fractional parts separately. To find the integer equivalent we successively divide by 2, noting the remainder each time; the binary equivalent appears least significant digit first.

```
2/2397
   1198  1   l.s.d.
    599  0
    299  1
    149  1
     74  1
     37  0
     18  1
      9  0
      4  1
      2  0
      1  0
         1
```

Thus, binary number is 100101011101.

The fraction is converted by successively multiplying by 2, and noting the integer value each time.

```
              0.55
m.s.d.   1    1.1
         0    0.2
         0    0.4
         0    0.8
         1    1.6
         1    1.2
         0    0.4
```

Thus, binary number is 0.1000110011.

Final binary number 100101011101.1000110011 recurring.
(b) Proceeding as **(a)** for binary fractions gives

```
              0.79
m.s.d.  1   1.58
        1   1.16
        0   0.32
        0   0.64
        1   1.28
        0   0.56
        1   1.12
        0   0.24
        0   0.48
        0   0.96
        1   1.92
```

Binary number is 0.11001010001, etc.

(c) Negative numbers are usually represented in either 2s or 1s complement form. First convert 90 to binary equivalent as in **(a)**.

```
Divide by 2 | 90
             45  0
             22  1
             11  0
              5  1
              2  1
              1  0
              1
```

Thus, binary number is 1011010.

We must allow one digit position to act as the sign digit. Thus, using eight-bit numbers, we have 01011010. The 1s complement form is the inverse, 10100101, and the 2s complement form is obtained by adding 1 to this at the least significant end, i.e. 10100110.

2 (a) As before, convert the binary fractions and integers separately. The decimal integer may be obtained by successively dividing by binary ten, and mentally converting the remainder at each stage to decimal equivalent.

```
              1001
     1010)1011011
          1010
          ─────
           1011
           1010
           ─────
              1
```

remainder 1 (l.s.d.)

```
1010)1001
remainder 9
```

Decimal integer is 91.

The fraction is converted by successive multiplication by binary ten, and mentally converting the integer at each stage to decimal

$$
\begin{array}{r}
0.101 \\
1010. \\
\hline
1.01 \\
101.00 \\
\hline
110.01 \quad \text{integer 6 (m.s.d.)} \\
\hline
0.01 \\
1010. \\
\hline
0.1 \\
10.0 \\
\hline
10.1 \quad \text{integer 2} \\
\hline
0.1 \\
1010. \\
\hline
1.0 \\
100.0 \\
\hline
101.0 \quad \text{integer 5} \\
\hline
\end{array}
$$

Decimal fraction is 0.625
Thus, decimal number is 91.625

(b) Proceeding as (a) for binary integer conversion we have

```
            10010101                    1110
   1010)10111010111       1010)10010101
        1010                    1010
        ────                    ────
         1101                   10001
         1010                    1010
         ────                    ────
          1101                    1110
          1010                    1010
          ────                    ────
           1111                    1001
           1010                    ────
           ────
            101
            ───
```

remainder 5 (l.s.d.)   remainder 9

```
         1
   1010)1110
        1010
        ────
         100              1010)1
         ───
```

remainder 4   remainder 1

Thus, decimal number is 1495

(c) Converting the binary fraction as in **(a)** we have

0.111011	0.00111
1010.	1010.

$$\frac{\begin{array}{l}1.11011\\111.011\end{array}}{1001.00111}$$

integer 9 (m.s.d.)

$$\frac{\begin{array}{l}0.0111\\1.11\end{array}}{10.0011}$$

integer 2

0.0011	0.111
1010.	1010.

$$\frac{\begin{array}{l}0.011\\1.1\end{array}}{1.111}$$

integer 1

$$\frac{\begin{array}{l}1.11\\111.00\end{array}}{1000.11}$$

integer 8

0.11	0.1
1010.	1010.0

$$\frac{\begin{array}{l}1.1\\110.\end{array}}{111.1}$$

integer 7

$$\frac{\begin{array}{l}1.0\\100.0\end{array}}{101.0}$$

integer 5

Decimal fraction is 0.921875

1.3 *Hexadecimal multiplication table*

	1	2	3	4	5	6	7	8	9	A	B	C	D	E	F
1	1	2	3	4	5	6	7	8	9	A	B	C	D	E	F
2	2	4	6	8	A	C	E	10	12	14	16	18	1A	1C	1E
3	3	6	9	C	F	12	15	18	1B	1E	21	24	26	2A	2D
4	4	8	C	10	14	18	1C	20	24	28	2C	30	34	38	3C
5	5	A	F	14	19	1E	23	28	2D	32	37	3C	41	46	4B
6	6	C	12	18	1E	24	2A	30	36	3C	42	48	4E	54	5A
7	7	E	15	1C	23	2A	31	38	3F	46	4D	54	5B	62	69
8	8	10	18	20	28	30	38	40	48	50	58	60	68	70	78
9	9	12	1B	24	2D	36	3F	48	51	5A	63	6C	75	7E	87
A	A	14	1E	28	32	3C	46	50	5A	64	6E	78	82	8C	96
B	B	16	21	2C	37	42	4D	58	63	6E	79	84	8F	9A	A5
C	C	18	24	30	3C	48	54	60	6C	78	84	90	9C	A8	B4
D	D	1A	26	34	41	4E	5B	68	75	82	8F	9C	A9	B6	C3
E	E	1C	2A	38	46	54	62	70	7E	8C	9A	A8	B6	C4	D2
F	F	1E	2D	3C	4B	5A	69	78	87	96	A5	B4	C3	D2	E1

(a)   B7
      53
      ──
      15
      21
      23
      37
      ──
      3B55

(b)   9C
      CD
      ──
      9C
      75
      90
      6C
      ──
      7CEC

(c)   FF
      FF
      ──
      E1
      E1
      E1
      E1
      ──
      FE01

Repeating **(b)** with binary equivalents we have

$$
\begin{array}{ll}
1001 & 1100 \\
1100 & 1101 \\
\hline
1001 & 1100 \\
0111\quad 0101 & \\
1001\quad 0000 & \\
0110\quad 1100 & \\
\hline
0111\quad 1100 & 1110\quad 1100 \\
\end{array}
$$

The full binary multiplication is

```
          1 0 0 1 1 1 0 0
          1 1 0 0 1 1 0 1
          ───────────────
          1 0 0 1 1 1 0 0
      1 0 0 1 1 1 0 0
    1 0 0 1 1 1 0 0
1 0 0 1 1 1 0 0
1 0 0 1 1 1 0 0
─────────────────────────
1 1 1 1 1 0 0 1 1 1 0 1 1 1 0 0
```

The decimal equivalent of the number is $31{,}980 = 156 \times 205$.

1.4 (a) The number 149 would be represented as

   00001   00100   11001

(b) 149 in ASCII code:

   31   34   39

(c) To represent 149 in Hamming code requires 15 bits, thus:

1	2	3	4	5	6	7	8	9	10	11	12	13	14	15
C	C	M	C	M	M	M	C	M	M	M	M	M	M	M

Decimal 149 has the binary equivalent 10010101. The parity checks are

(1) 1, 3, 5, 7, 9, 11, 13, 15;
(2) 2, 3; 6, 7; 10, 11; 14, 15;
(3) 4, 5, 6, 7; 12, 13, 14, 15;
(4) 8, 9, 10, 11, 12, 13, 14, 15.

Inserting the even-parity bits results in

   010100110010101

(d) Diamond $3n + 2$ code representation is

    00101   01110   11101   as binary coded groups

or 111000001, i.e. $3 \times 149 + 2 = 449$

(e) 10001   10100   01010

1.5 The multiplication and addition tables for mod-3 arithmetic are as follows:

+	0 00	1 01	2 10
0 00	00	01	10
1 01	01	10	00(1)
2 10	10	00(1)	01(1)

×	0 00	1 01	2 10
0 00	00	00	00
1 01	00	01	10
2 10	00	10	01

Note the carry outputs given in parentheses.

Similarly for mod-5 we have

+	0 000	1 001	2 010	3 011	4 100
0 000	000	001	010	011	100
1 001	001	010	011	100	000(1)
2 010	010	011	100	000(1)	001(1)
3 011	011	100	000(1)	001(1)	010(1)
4 100	100	000(1)	001(1)	010(1)	011(1)

×	0 000	1 001	2 010	3 011	4 100
0 000	000	000	000	000	000
1 001	000	001	010	011	100
2 010	000	010	100	001	011
3 011	000	011	001	100	010
4 100	000	100	011	010	001

1.6 (a) Error in third group.

   (b) Performing parity checks we have

     (1) digits 1, 3, 5, 7.   incorrect 1
     (2) digits 2, 3; 6, 7.   correct 0
     (3) digits 4, 5, 6, 7.   incorrect 1

Message in error and digit 5 should be a 0.

(c) Decimal equivalent is 106, subtract 2 gives 104, divide by 3 leaves remainder of 2, therefore error occurred.

1.7 (a) $F_0 = A\bar{B} + \bar{A}B$ (b) $F_1 = AB\bar{C} + ABC = \bar{A}BC$

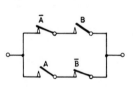

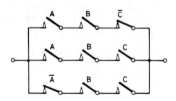

Circuit (b) can be simplified to

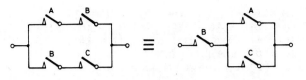

$$F_2 = AB + BC = B(A + C)$$

We shall discover how to do this formally in the next chapter; you could check the circuit is correct by deducing the output for all possible input values and so form a truth table.

Chapter 2

2.1 (a) $F = (A + B + C)(\bar{A} + B + C)(\bar{A} + B + \bar{C})$
$= (A\bar{A} + AB + AC + \bar{A}B + BB + BC + \bar{A}C + BC + CC)(\bar{A} + B + \bar{C})$
$= [C(A + \bar{A} + B + 1) + B(A + 1) + \bar{A}B][\bar{A} + B + \bar{C}]$
$= [C + B(1 + A)][\bar{A} + B + \bar{C}]$
$= (C + B)(\bar{A} + B + \bar{C})$ (from $A\bar{A} = 0$, $A + 1 = 1$, $AA = A$,
$\qquad\qquad\qquad\qquad\qquad\qquad\qquad A + \bar{A} = 1$, $A.1 = A$)

$F = \bar{A}C + BC + C\bar{C} + \bar{A}B + BB + B\bar{C}$
$= \bar{A}C + B(C + \bar{C} + 1) + \bar{A}B$

Thus

$\qquad F = \bar{A}C + B$

The truth table is shown in Table S.1.

(b) $F = \bar{A}\bar{C}D + A\bar{C}D + A\bar{B}\bar{C} + A\bar{B}C + \bar{A}CD$
$= \bar{C}D(\bar{A} + A) + A\bar{B}(\bar{C} + C) + \bar{A}CD$
$= \bar{C}D + A\bar{B} + \bar{A}D(C + \bar{C})$ (add redundant $\bar{A}\bar{C}D$ term)
$F = \bar{C}D + A\bar{B} + \bar{A}D$

The truth table is shown in Table S.2.

(c) $F_1 = D(\bar{A} + B + C + \bar{D})(A + B + \bar{C} + \bar{D})$
$= D(\bar{A}A + \bar{A}B + \bar{A}\bar{C} + \bar{A}\bar{D} + AB + BB + B\bar{C} + B\bar{D} + AC$
$+ BC + C\bar{C} + C\bar{D} + A\bar{D} + B\bar{D} + \bar{C}\bar{D} + \bar{D}\bar{D})$
$= D(B + \bar{A}\bar{C} + AC + \bar{D})$ (from $A\bar{A} = 0$, $A + AB = A$ and $AA = A$)

Thus,

$\qquad F_1 = DB + \bar{A}\bar{C}D + ACD$

Again,

$\qquad F_2 = (D + A\bar{C} + \bar{A}C)(\bar{A}\bar{C} + BD + AC)$
$\qquad\quad = \bar{A}\bar{C}D + BDD + ACD + A\bar{C}\bar{A}\bar{C} + A\bar{C}BD + A\bar{C}AC$
$\qquad\qquad + \bar{A}C\bar{A}\bar{C} + \bar{A}CBD + \bar{A}CAC$

Table S.1 Problem 2.1(a)

A	B	C	$\bar{A}C + B$	F
0	0	0	0	0
0	0	1	1	1
0	1	0	1	1
0	1	1	1	1
1	0	0	0	0
1	0	1	0	0
1	1	0	1	1
1	1	1	1	1

Table S.2 Problem 2.1(b)

A	B	C	D	$\bar{C}D + A\bar{B} + \bar{A}D$	F
0	0	0	0	0	0
0	0	0	1	1	1
0	0	1	0	0	0
0	0	1	1	1	1
0	1	0	0	0	0
0	1	0	1	1	1
0	1	1	0	0	0
0	1	1	1	1	1
1	0	0	0	1	1
1	0	0	1	1	1
1	0	1	0	1	1
1	0	1	1	1	1
1	1	0	0	0	0
1	1	0	1	1	1
1	1	1	0	0	0
1	1	1	1	0	0

$$= \bar{A}\bar{C}D + BD + ACD + AB\bar{C}D + \bar{A}BCD$$
$$= \bar{A}\bar{C}D + BD(1 + A\bar{C} + \bar{A}C) + ACD$$

Thus,

$$F_2 = \bar{A}\bar{C}D + BD + ACD$$

The truth table for these functions is shown in Table S.3.

2.2 (a) Using De Morgan's theorem we have

$$T = \overline{[\overline{ab}.a].[\overline{ab}.b]} = (\overline{\overline{ab}.a}) + (\overline{\overline{ab}.b})$$
$$= (\bar{a} + \bar{b})a + (\bar{a} + \bar{b})b$$
$$= a\bar{a} + a\bar{b} + \bar{a}b + b\bar{b}$$

Thus,

$$T = a\bar{b} + \bar{a}b$$

It is interesting to note that the original equation represents the NAND con-figuration for the exclusive-OR circuit (see Fig. S.1).

(b) $T = \overline{(a + b + \bar{c})(\overline{ab + cd})} + (\overline{bcd})$
$$= [(a + b + \bar{c}) + (ab + cd)](bcd)$$
$$= [(a + b + \bar{c}) + (\bar{a} + \bar{b} + \bar{c} + \bar{d})](bcd)$$
$$= (a + \bar{a} + b + \bar{b} + \bar{c} + \bar{d})(bcd)$$

Thus,

$$T = bcd$$

(c) $T = \overline{(abc + b\bar{c}d) + \overline{(acd + \bar{b}\bar{c}d + bc\bar{d})}}$
$$= (abc + b\bar{c}d)(\overline{acd + \bar{b}\bar{c}d + bc\bar{d}})$$
$$= (abc + b\bar{c}d)(\bar{a} + \bar{c} + \bar{d} + \bar{b}\bar{c}d + bc\bar{d})$$
$$= (abc + b\bar{c}d)(\bar{a} + \bar{c} + \bar{d})$$
$$= abc\bar{a} + abc\bar{c} + abc\bar{d} + \bar{a}b\bar{c}d + b\bar{c}\bar{c}d + b\bar{c}d\bar{d}$$
$$= abc\bar{d} + \bar{a}b\bar{c}d + b\bar{c}d$$
$$T = abc\bar{d} + b\bar{c}d$$

Table S.3 Problem 2.1(c)

A	B	C	D	F_1	F_2
0	0	0	0	0	0
0	0	0	1	1	1
0	0	1	0	0	0
0	0	1	1	0	0
0	1	0	0	0	0
0	1	0	1	1	1
0	1	1	0	0	0
0	1	1	1	1	1
1	0	0	0	0	0
1	0	0	1	0	0
1	0	1	0	0	0
1	0	1	1	1	1
1	1	0	0	0	0
1	1	1	1	1	1
1	1	0	0	0	0
1	1	1	1	1	1

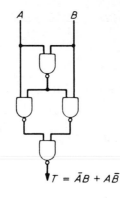

$$T = \bar{A}B + A\bar{B}$$

Figure S.1 Problem 2.2

2.3 (a) The equation for the circuit, including all nets, is

$$T = \bar{A}BD + \bar{A}\bar{B}CD + ABCD + A\bar{B}D + ABD$$

Expanding into canonical sum-of-products we have

$$T = \bar{A}BD(C + \bar{C}) + A\bar{B}D(C + \bar{C}) + ABD(C + \bar{C}) + \bar{A}\bar{B}CD + ABCD$$

Thus

$$T = \bar{A}BCD + \bar{A}B\bar{C}D + A\bar{B}CD + A\bar{B}\bar{C}D + ABCD + AB\bar{C}D + \bar{A}\bar{B}CD$$

Alternatively, the following formula may be used:

$$
\begin{aligned}
T =\ &(\bar{A}\bar{B}\bar{C}\bar{D}.0) + (\bar{A}\bar{B}\bar{C}D.0) + (\bar{A}\bar{B}C\bar{D}.0) + (\bar{A}\bar{B}CD.1)\\
&+ (\bar{A}B\bar{C}\bar{D}.0) + (\bar{A}B\bar{C}D.1) + (\bar{A}BC\bar{D}.0) + (\bar{A}BCD.1)\\
&+ (A\bar{B}\bar{C}\bar{D}.0) + (A\bar{B}\bar{C}D.1) + (A\bar{B}C\bar{D}.0) + (A\bar{B}CD.1)\\
&+ (AB\bar{C}\bar{D}.0) + (AB\bar{C}D.1) + (ABC\bar{D}.0) + (ABCD.1)
\end{aligned}
$$

Hence,

$$T = \bar{A}\bar{B}CD + \bar{A}B\bar{C}D + \bar{A}BCD + A\bar{B}\bar{C}D + A\bar{B}CD + AB\bar{C}D + ABCD$$

To obtain the canonical product-of-sums we again apply the formula

$$
\begin{aligned}
T =\ &(\bar{A} + \bar{B} + \bar{C} + \bar{D} + 1)(\bar{A} + \bar{B} + \bar{C} + D + 0)(\bar{A} + \bar{B} + C + \bar{D} + 1)\\
&(\bar{A} + \bar{B} + C + D + 0)(\bar{A} + B + \bar{C} + \bar{D} + 1)(\bar{A} + B + \bar{C} + D + 0)\\
&(\bar{A} + B + C + \bar{D} + 1)(\bar{A} + B + C + D + 0)(A + \bar{B} + \bar{C} + \bar{D} + 1)\\
&(A + \bar{B} + \bar{C} + D + 0)(A + \bar{B} + C + \bar{D} + 1)(A + \bar{B} + C + D + 0)\\
&(A + B + \bar{C} + \bar{D} + 1)(A + B + \bar{C} + D + 0)(A + B + C + \bar{D} + 0)\\
&(A + B + C + D + 0)
\end{aligned}
$$

And

$$
\begin{aligned}
T =\ &(\bar{A} + \bar{B} + \bar{C} + D)(\bar{A} + \bar{B} + C + D)(\bar{A} + B + \bar{C} + D)\\
&(\bar{A} + B + C + D)(A + \bar{B} + \bar{C} + D)(A + \bar{B} + C + D)\\
&(A + B + \bar{C} + D)(A + B + C + \bar{D})(A + B + C + D)
\end{aligned}
$$

The equation may be simplified to

$$
\begin{aligned}
T =\ &BCD(A + \bar{A}) + B\bar{C}D(A + \bar{A})\\
&+ \bar{B}CD(A + \bar{A}) + A\bar{C}D(B + \bar{B}) \quad \text{(by adding term } AB\bar{C}D)\\
=\ &BD(C + \bar{C}) + CD(B + \bar{B}) + A\bar{C}D \quad \text{(by adding term } BCD)\\
=\ &BD + D(C + \bar{C}A)
\end{aligned}
$$

Hence,

$$T = BD + CD + AD = D(A + B + C)$$

The resulting circuit diagram is shown in Fig. S.2(a).

(b) The circuit is represented by the switching equation

$$T = B\bar{C}\bar{D} + B\bar{D}C + A\bar{B}D + \bar{A}\bar{B}D + B\bar{C}D$$

Expanding into canonical sum-of-products, we have

$$T = B\bar{C}\bar{D}(A + \bar{A}) + B\bar{D}C(A + \bar{A}) + A\bar{B}D(C + \bar{C})$$
$$+ \bar{A}\bar{B}D(C + \bar{C}) + B\bar{C}D(A + \bar{A})$$
$$T = AB\bar{C}\bar{D} + \bar{A}B\bar{C}\bar{D} + ABC\bar{D} + \bar{A}BC\bar{D} + A\bar{B}CD + A\bar{B}\bar{C}D$$
$$+ \bar{A}\bar{B}\bar{C}D + \bar{A}\bar{B}CD + AB\bar{C}D + \bar{A}B\bar{C}D$$

Having found this canonical form, it is easier perhaps to obtain the product-of-sums using De Morgan's theorem; writing down the excluded terms and finding the inverse we have

$$T = \overline{\bar{A}\bar{B}\bar{C}\bar{D} + \bar{A}\bar{B}C\bar{D} + \bar{A}BCD + AB\bar{C}D + ABC\bar{D} + ABCD}$$

Thus,

$$T = (A + B + C + D)(A + B + \bar{C} + D)(A + \bar{B} + \bar{C} + \bar{D})$$
$$(\bar{A} + B + C + D)(\bar{A} + B + \bar{C} + D)(\bar{A} + \bar{B} + \bar{C} + \bar{D})$$

The expression may be simplified thus:

$$T = B\bar{C}\bar{D} + B\bar{D}C + A\bar{B}D + \bar{A}\bar{B}D + B\bar{C}D$$
$$= B\bar{C}(\bar{D} + D) + \bar{B}D(\bar{A} + A) + B\bar{D}C$$
$$= \bar{B}D + B(\bar{C} + C\bar{D})$$

and

$$T = \bar{B}D + B\bar{C} + B\bar{D}$$

or

$$T = B\bar{D}(C + \bar{C}) + \bar{B}D(A + \bar{A}) + B\bar{C}D$$
$$= B\bar{D} + D(\bar{B} + B\bar{C})$$

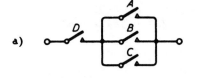

a)

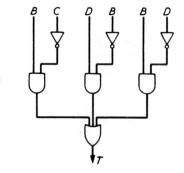

b)

Figure S.2 Problem 2.3

Hence

$$T = B\bar{D} + \bar{B}D + \bar{C}D$$

Both are equally valid minimal solutions; the circuit is shown in Fig. S.2(b).

2.4 The subsets of the set $A = \{x_1, x_2, x_3, x_4\}$ are

$$(\bar{x}_1\bar{x}_2\bar{x}_3\bar{x}_4), (\bar{x}_1\bar{x}_2\bar{x}_3x_4), (\bar{x}_1\bar{x}_2x_3\bar{x}_4), (\bar{x}_1\bar{x}_2x_3x_4), (\bar{x}_1x_2\bar{x}_3\bar{x}_4), (\bar{x}_1x_2\bar{x}_3x_4),$$
$$(\bar{x}_1x_2x_3\bar{x}_4), (\bar{x}_1x_2x_3x_4), (x_1\bar{x}_2\bar{x}_3\bar{x}_4), (x_1\bar{x}_2\bar{x}_3x_4), (x_1\bar{x}_2x_3\bar{x}_4), (x_1\bar{x}_2x_3x_4),$$
$$(x_1x_2\bar{x}_3\bar{x}_4), (x_1x_2\bar{x}_3x_4), (x_1x_2x_3\bar{x}_4), (x_1x_2x_3x_4)$$

The Venn diagram is shown in Fig. S.3. The diagram may be used to simplify switching equations as adjacent terms differ in one variable, and hence it can be clearly recognized that $x_2x_1 + x_2\bar{x}_1 = x_2$ applies. See Chapter 3 for further description.

2.5 The problem is first stated in truth table form; let the three on/off switches be A, B and C and the 'light-on' condition represented by $Z = 1$. The truth table is shown in Table S.4.

In this way we have considered all possible input conditions for the three on/off switches A, B and C. The next step is to read down the switching function from the table; this is for the terms $Z = 1$:

$$Z = \bar{A}\bar{B}C + \bar{A}B\bar{C} + A\bar{B}\bar{C} + ABC$$

The equation will not simplify, but will factorize as

$$Z = \bar{C}(\bar{A}B + A\bar{B}) + C(\bar{A}\bar{B} + AB)$$

Let $\bar{A}B + A\bar{B} = X$; then $Z = \bar{C}X + C\bar{X}$ since $\bar{A}\bar{B} + AB = \bar{A}B + A\bar{B}$. The circuit may be implemented using AND/OR, exclusive-OR or changeover contact circuits (see Fig. S.4).

2.6 The truth table is shown in Table S.5 where A, B, C and D represent the panel members' on/off buttons. Extracting from the truth table those input terms that cause a 'hit' we have

$$H = \bar{A}BCD + A\bar{B}CD + AB\bar{C}D + ABC\bar{D} + ABCD$$

Similarly

$$M = \bar{A}\bar{B}\bar{C}\bar{D} + \bar{A}\bar{B}\bar{C}D + \bar{A}\bar{B}C\bar{D} + \bar{A}B\bar{C}\bar{D} + A\bar{B}\bar{C}\bar{D}$$
$$T = \bar{A}\bar{B}CD + \bar{A}B\bar{C}D + \bar{A}BC\bar{D} + A\bar{B}\bar{C}D + A\bar{B}C\bar{D} + AB\bar{C}\bar{D}$$

Table S.4 Problem 2.5

A	B	C	Z
0	0	0	0
0	0	1	1
0	1	0	1
0	1	1	0
1	0	0	1
1	0	1	0
1	1	0	0
1	1	1	1

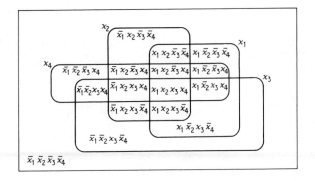

Figure S.3 Problem 2.4

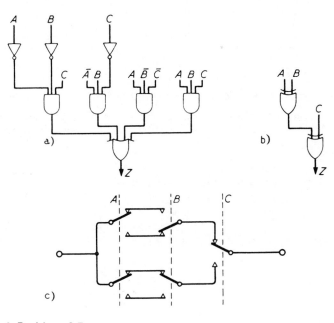

Figure S.4 Problem 2.5

Simplifying the function, we have

$$H = ABD(C + \bar{C}) + ABC(D + \bar{D}) + ACD(B + \bar{B})$$
$$+ BCD(\bar{A} + A) \qquad \text{(adding redundant terms } ABCD)$$

Hence

$$H = ABD + ABC + ACD + BCD$$

Also

$$M = \bar{A}\bar{B}\bar{C}(D + \bar{D}) + \bar{A}\bar{B}\bar{D}(C + \bar{C}) + \bar{A}\bar{C}\bar{D}(B + \bar{B})$$
$$+ \bar{B}\bar{C}\bar{D}(A + \bar{A}) \qquad \text{(adding redundant terms } \bar{A}\bar{B}\bar{C}\bar{D})$$

Table S.5 Problem 2.6

A	B	C	D	Hit	Miss	Tie
0	0	0	0	0	1	0
0	0	0	1	0	1	0
0	0	1	0	0	1	0
0	0	1	1	0	0	1
0	1	0	0	0	1	0
0	1	0	1	0	0	1
0	1	1	0	0	0	1
0	1	1	1	1	0	0
1	0	0	0	0	1	0
1	0	0	1	0	0	1
1	0	1	0	0	0	1
1	0	1	1	1	0	0
1	1	0	0	0	0	1
1	1	0	1	1	0	0
1	1	1	0	1	0	0
1	1	1	1	1	0	0

Table S.6 Problem 2.7

A	B	C	D	F
0	0	0	0	1
0	0	0	1	0
0	0	1	0	0
0	0	1	1	0
0	1	0	0	1
0	1	0	1	0
0	1	1	0	0
0	1	1	1	0
1	0	0	0	1
1	0	0	1	1
1	0	1	0	0
1	0	1	1	0
1	1	0	0	1
1	1	0	1	0
1	1	1	0	0
1	1	1	1	0

Thus

$$M = \bar{A}C\bar{D} + \bar{A}\bar{B}\bar{C} + \bar{B}C\bar{D} + \bar{A}\bar{B}\bar{D}$$

The equation for the 'tie' condition cannot be simplified; however, the 'tie' condition occurs when there is not a 'hit' ($\bar{H}$) and not a 'miss' ($\bar{M}$), i.e.

$$T = \bar{H}\bar{M}$$

The logic diagram is shown in Fig. S.5. If only two-input gates were available, it would be better to factorize the equations as

$$M = \bar{A}\bar{C}(\bar{B} + \bar{D}) + \bar{B}\bar{D}(\bar{A} + \bar{C})$$

and

$$H = AB(C + D) + CD(A + B)$$

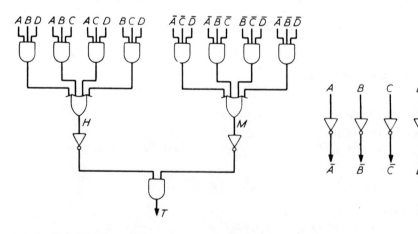

Figure S.5 Problem 2.6

2.7 The truth table for this problem is shown in Table S.6, where A, B, C and D represent the absence or presence of Hubert, Joe, Sid and Don in the bar. The function F represents the state of no drinking, i.e. $F = 1$. Reading down the terms from the truth table that occur for $F = 1$ gives

$$
\begin{aligned}
F &= \bar{A}\bar{B}\bar{C}\bar{D} + \bar{A}B\bar{C}\bar{D} + A\bar{B}\bar{C}\bar{D} + AB\bar{C}\bar{D} + AB\bar{C}D \\
&= \bar{A}\bar{C}\bar{D}(B + \bar{B}) + A\bar{C}\bar{D}(B + \bar{B}) \\
&\quad + AB\bar{C}(D + \bar{D}) \qquad \text{(adding redundant term } A\bar{B}\bar{C}\bar{D}) \\
&= \bar{C}\bar{D}(A + \bar{A}) + AB\bar{C}
\end{aligned}
$$

Thus

$$
F = \bar{C}\bar{D} + AB\bar{C} = \bar{C}(\bar{D} + AB)
$$

Thus we may say that no drinking occurs if Sid is absent from the bar, *and* either Don is absent *or* Hubert is present *and* Joe is absent.

2.8 The hypercube is shown in Fig. S.6.

$$
\begin{aligned}
f(a, b, c, d) &= \Sigma(1, 5, 9, 11, 12, 15) \\
&= \bar{a}\bar{b}\bar{c}d + \bar{a}b\bar{c}d + a\bar{b}\bar{c}d + a\bar{b}cd + ab\bar{c}\bar{d} + abcd \\
&= 0001 + 0101 + 1001 + 1011 + 1100 + 1111
\end{aligned}
$$

$$
\text{ON array} = \left\{ \begin{matrix} 0001 \\ 0101 \\ 1001 \\ 1011 \\ 1100 \\ 1111 \end{matrix} \right\}
\qquad
\text{OFF array} = \left\{ \begin{matrix} 0000 \\ 0010 \\ 0011 \\ 0100 \\ 0110 \\ 0111 \\ 1000 \\ 1010 \\ 1101 \\ 1110 \end{matrix} \right\}
$$

Note the ease of identifying adjacent terms from the hypercube.

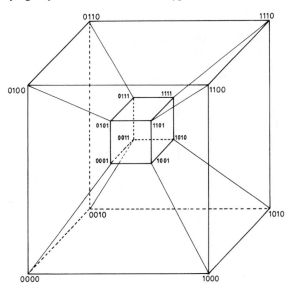

Figure S.6 Problem 2.8

Chapter 3

3.1 The truth table for this problem is shown in Table S.7, where M represents the add/subtract control signal, x and y the digits to be added (or subtracted), b/c the borrow (or carry), and S/D the sum (or difference). Note that the sum and difference outputs are identical, i.e. independent of M. Thus

$$S/D = \bar{M}\bar{x}\bar{y}b/c + \bar{M}\bar{x}y\overline{b/c} + \bar{M}x\bar{y}\overline{b/c} + M\bar{x}\bar{y}b/c + M\bar{x}\bar{y}b/c$$
$$+ M\bar{x}y\overline{b/c} + Mx\bar{y}\overline{b/c} + Mxyb/c$$
$$= (\bar{x}\bar{y}b/c + \bar{x}y\overline{b/c} + x\bar{y}\overline{b/c} + xyb/c)(M + \bar{M})$$

Thus

$$S/D = \bar{x}\bar{y}b/c + \bar{x}y\overline{b/c} + x\bar{y}\overline{b/c} + xyb/c$$

The borrow/carry output is given by

$$b_+/c_+ = \bar{M}\bar{x}\bar{y}b/c + \bar{M}\bar{x}y\overline{b/c} + \bar{M}\bar{x}yb/c + \bar{M}xyb/c$$
$$+ M\bar{x}yb/c + Mx\bar{y}b/c + Mxy\overline{b/c} + Mxyb/c$$

Table S.8 shows the K-map for the problem; thus the minimal expression for the borrow/carry is

$$b_+/c_+ = \bar{M}\bar{x}b/c + Mxb/c + \bar{M}\bar{x}y + Mxy + yb/c$$

The equations may be realized in the usual way.

3.2 From the circuit diagram (Fig. 3.7) we have six variables A, B, C, D, E and F. However, A is common to all terms and may be ignored in the minimization. Ideal contacts are assumed throughout, and all possible paths are included in the transmission function F. Thus

$$F = AB\bar{C} + AB\bar{E} + ABD + AE\bar{D} + AF$$
$$= A(B\bar{C} + B\bar{E} + BD + E\bar{D} + F)$$

Table S.7 Problem 3.1

M	x	y	b/c	S/D	b_+/c_+
0	0	0	0	0	0
0	0	0	1	1	1
0	0	1	0	1	1
0	0	1	1	0	1
0	1	0	0	1	0
0	1	0	1	0	0
0	1	1	0	0	0
0	1	1	1	1	1
1	0	0	0	0	0
1	0	0	1	1	0
1	0	1	0	1	0
1	0	1	1	0	1
1	1	0	0	1	0
1	1	0	1	0	1
1	1	1	0	0	1
1	1	1	1	1	1

Table S.8 Problem 3.1

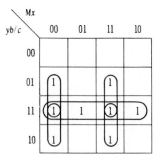

The function may be plotted on a five-variable K-map shown in Table S.9. This gives the minimal expression

$$F = A(F + B + \bar{D}E)$$

There are many ways of implementing logic in terms of NAND/NOR elements, and these techniques are discussed in Chapter 4. We may, however, deduce a circuit from first principles bearing in mind that the NAND unit is logically equivalent to a cascaded AND/NOT arrangement.

Figure S.7(a) shows the NAND equivalent of AND/OR gates; Fig. S.7(b) shows the switching function implemented directly, and Fig. S.7(c) the final circuit after eliminating redundant elements. This configuration is a sequential switching circuit; note the feedback of F, and the bistable element formed by NANDS II and III.

3.3 (a) The K-map is shown in Table S.10. From this the minimal expression is given by

$$T = D + AB + \bar{B}\bar{C}$$

All these terms are essential prime implicants.
(b) Table S.11 shows the K-map for the switching function. The complete prime implicant set is $(BE, \bar{D}\bar{E}, \bar{C}\bar{E}, \bar{A}\bar{B}\bar{D}, A\bar{C}, A\bar{B}D\bar{E}, \bar{C}\bar{D}, \bar{B}\bar{C})$. The minimal form of the function is

$$T = A\bar{B}D\bar{E} + \bar{A}\bar{B}\bar{D} + \bar{C}\bar{E} + \bar{D}E + A\bar{C} + BE$$

These are all prime implicants. It is illustrative to repeat this example using the McCluskey tabular method; it is more time-consuming but ensures the best use of the unspecified conditions. Furthermore the minimal prime implicant set may be more easily arrived at by using the chart technique.
(c) Product-of-sums equations may be plotted directly on a K-map in exactly the same way as sums-of-products. Table S.12 shows the function plotted on a map. The terms are read from the map in the normal way, but expressed as a product-of-sums. Thus from Table S.12 we have

$$T = (\bar{A} + D + \bar{E})(\bar{A} + B + \bar{C} + D)(C + \bar{D} + E)(B + \bar{D} + \bar{E})$$
$$(A + \bar{B} + D + E)(A + B + \bar{D})(A + \bar{B} + C)$$

Table S.9 Problem 3.2

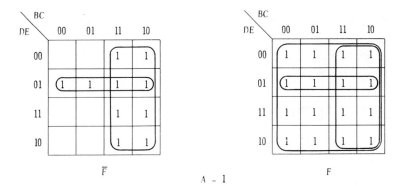

$A = 1$

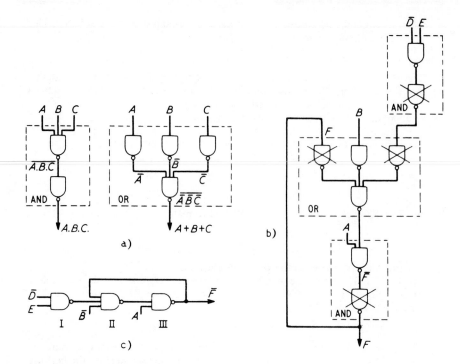

Figure S.7 Problem 3.2

3.4 Table S.13 shows the tabular arrangement; note that the don't-care conditions have been included in the table. Many identical terms are evolved in the computation process; these are simply ignored in succeeding comparisons. From the table, the complete prime implicant set is given by

$$PI = (\bar{B}\bar{C}\bar{D}E\bar{F}, AB\bar{C}\bar{D}F, AB\bar{C}\bar{E}F, ABC\bar{D}\bar{F}, AB\bar{D}EF, ABC\bar{D}E, ABCEF,$$
$$\bar{A}\bar{B}CD, BC\bar{E}\bar{F}, \bar{A}CDF, \bar{A}B\bar{D}F, BCDF, D\bar{E})$$

Table S.10 Problem 3.3(a)

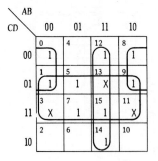

Table S.11 Problem 3.3(b)

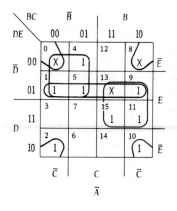
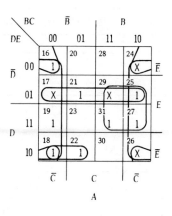

Table S.12 Problem 3.3(c)

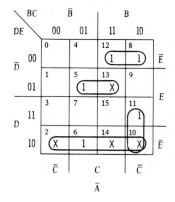
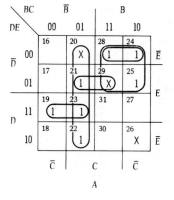

Table S.13 Problem 3.4

	A	B	C	D	E	F			A	B	C	D	E	F	
2	0	0	0	0	1	0	✓	37	1	0	0	1	0	1	✓
4	0	0	0	1	0	0	✓	44	1	0	1	1	0	0	✓
5	0	0	0	1	0	1	✓	49	1	1	0	0	0	1	✓
12	0	0	1	1	0	0	✓	52	1	1	0	1	0	0	✓
20	0	1	0	1	0	0	✓	56	1	1	1	0	0	0	✓
24	0	1	1	0	0	0	✓	15	0	0	1	1	1	1	✓
34	1	0	0	0	1	0	✓	23	0	1	0	1	1	1	✓
36	1	0	0	1	0	0	✓	29	0	1	1	1	0	1	✓
13	0	0	1	1	0	1	✓	45	1	0	1	1	0	1	✓
14	0	0	1	1	1	0	✓	51	1	1	0	0	1	1	✓
21	0	1	0	1	0	1	✓	53	1	1	0	1	0	1	✓
28	0	1	1	1	0	0	✓	58	1	1	1	0	1	0	✓

Table S.13 Continued

	A	B	C	D	E	F
60	1	1	1	1	0	0 √
31	0	1	1	1	1	1 √
59	1	1	1	0	1	1 √
61	1	1	1	1	0	1 √
63	1	1	1	1	1	1 √

List 1

	A	B	C	D	E	F
15,31	0	–	1	1	1	1 √
23,31	0	1	–	1	1	1 √
29,31	0	1	1	1	–	1 √
29,61	–	1	1	1	0	1 √
45,61	1	–	1	1	0	1 √
51,59	1	1	–	0	1	1 E
53,61	1	1	–	1	0	1 √
58,59	1	1	1	0	1	– F
60,61	1	1	1	1	0	– √
31,63	–	1	1	1	1	1 √
59,63	1	1	1	–	1	1 G
61,63	1	1	1	1	–	1 √

List 2

	A	B	C	D	E	F
2,34	–	0	0	0	1	0 A
4,5	0	0	0	1	0	– √
4,12	0	0	–	1	0	0 √
4,20	0	–	0	1	0	0 √
4,36	–	0	0	1	0	0 √
5,13	0	0	–	1	0	1 √
5,21	0	–	0	1	0	1 √
5,37	–	0	0	1	0	1 √
12,13	0	0	1	1	0	– √
12,14	0	0	1	1	–	0 √
12,28	0	–	1	1	0	0 √
12,44	–	0	1	1	0	0 √
20,21	0	1	0	1	0	– √
20,28	0	1	–	1	0	0 √
20,52	–	1	0	1	0	0 √
24,28	0	1	1	–	0	0 √
24,56	–	1	1	0	0	0 √
36,37	1	0	0	1	0	– √
36,44	1	0	–	1	0	0 √
36,52	1	–	0	1	0	0 √
13,15	0	0	1	1	–	1 √
13,29	0	–	1	1	0	1 √
13,45	–	0	1	1	0	1 √
14,15	0	0	1	1	1	– √
21,23	0	1	0	1	–	1 √
21,29	0	1	–	1	0	1 √
21,53	–	1	0	1	0	1 √
28,29	0	1	1	1	0	– √
28,60	–	1	1	1	0	0 √
37,45	1	0	–	1	0	1 √
37,53	1	–	0	1	0	1 √
44,45	1	0	1	1	0	– √
44,60	1	–	1	1	0	0 √
49,51	1	1	0	0	–	1 B
49,53	1	1	0	–	0	1 C
52,53	1	1	0	1	0	– √
52,60	1	1	–	1	0	0 √
56,58	1	1	1	0	–	0 D
56,60	1	1	1	–	0	0 √

	A	B	C	D	E	F
4,5/12,13	0	0	–	1	0	– √
4,5/20,21	0	–	0	1	0	– √
4,5/36,37	–	0	0	1	0	– √
4,12/5,13	0	0	–	1	0	–
4,12/20,28	0	–	–	1	0	0 √
4,12/36,44	–	0	–	1	0	0 √
4,20/5,21	0	–	0	1	0	–
4,20/12,28	0	–	–	1	0	0
4,20/36,52	–	–	0	1	0	0 √
4,36/5,37	–	0	0	1	0	–
4,36/12,44	–	0	–	1	0	0
4,36/20,52	–	–	0	1	0	0
5,13/21,29	0	–	–	1	0	1 √
5,13/37,45	–	0	–	1	0	1 √
5,21/31,29	0	–	–	1	0	1
5,21/37,53	–	–	0	1	0	1 √
5,37/13,45	–	0	–	1	0	1
5,37/21,53	–	–	0	1	0	1
12,13/14,15	0	0	1	1	–	– H
12,13/28,29	0	–	1	1	0	– √
12,13/44,45	–	0	1	1	0	– √
12,14/13,15	0	0	1	1	–	–
12,28/13,29	–	–	1	1	0	–
12,28/44,60	–	–	1	1	0	0 √
12,44/13,45	–	0	1	1	0	–
12,44/28,60	–	–	1	1	0	0
20,21/28,29	0	1	–	1	0	– √
20,21/52,53	–	1	0	1	0	– √
20,28/21,29	0	1	–	1	0	–
20,28/52,60	–	1	–	1	0	0 √
20,52/21,53	–	1	0	1	0	–
20,52/28,60	–	1	–	1	0	0
24,28/56,60	–	1	1	–	0	0 I

Table S.13 Continued

	A	B	C	D	E	F
24,56/28,60	–	1	1	–	0	0
36,37/44,45	1	0	–	1	0	– √
36,37/52,53	1	–	0	1	0	– √
36,44/37,45	1	0	–	1	0	–
36,44/52,60	1	–	–	1	0	0 √
36,52/37,53	1	–	0	1	0	–
36,52/44,60	1	–	–	1	0	0
13,15/29,31	0	–	1	1	–	1 J
13,29/15,31	0	–	1	1	–	1
13,29/45,61	–	–	1	1	0	1 √
13,45/29,61	–	–	1	1	0	1
21,23/29,31	0	1	–	1	–	1 K
21,29/23,31	0	1	–	1	–	1
21,29/53,61	–	1	–	1	0	1 √
21,53/29,61	–	1	–	1	0	1
28,29/60,61	–	1	1	1	0	1 √
28,60/29,61	–	1	1	1	0	–
37,45/53,61	1	–	–	1	0	1 √
37,53/45,61	1	–	–	1	0	1
44,45/60,61	1	–	1	1	0	– √
44,60/45,61	1	–	1	1	0	–
52,53/60,61	1	1	–	1	0	– √
52,60/53,61	1	1	–	1	0	–
29,61/31,63	–	1	1	1	–	1 L
29,31/61,63	–	1	1	1	–	1
29,61/31,63	–	1	1	1	–	1

List 3

	A	B	C	D	E	F
4,5/12,13/20,21/28,29	0	–	–	1	0	– √
4,5/12,13/36,37/44,45	–	0	–	1	0	– √
4,5/20,21/12,13/28,29	0	–	–	1	0	–
4,5/20,21/36,37/52,53	–	–	0	1	0	– √
4,5/36,37/12,13/44,45	–	0	–	1	0	–
4,5/36,37/20,21/52,53	–	–	0	1	0	–
4,12/20,28/5,13/21,29	0	–	–	1	0	–
4,12/20,28/36,44/52,60	–	–	–	1	0	0 √
4,12/36,44/5,13/37,45	–	0	–	1	0	–
4,12/36,44/20,28/52,60	–	–	–	1	0	0
4,20/36,52/5,21/37,53	–	–	0	1	0	–
4,20/36,52/12,28/44,60	–	–	–	1	0	0
5,13/21,29/37,45/53,61	–	–	–	1	0	1 √
5,13/37,45/21,29/53,61	–	–	–	1	0	1
5,21/37,53/13,29/45,61	–	–	–	1	0	1
12,13/28,29/44,45/60,61	–	–	1	1	0	– √
12,13/44,45/28,29/60,61	–	–	1	1	0	–
12,28/44,60/13,29/45,61	–	–	1	1	0	–
20,21/28,29/52,53/60,61	–	1	–	1	0	– √

Table S.13 Continued

	A	B	C	D	E	F
20,21/52,53/28,29/60,61	–	1	–	1	0	–
20,28/52,60/21,29/53,61	–	1	–	1	0	–
36,37/44,45/52,53/60,61	1	–	–	1	0	– √
36,37/52,53/44,45/60,61	1	–	–	1	0	–
36,44/52,60/37,45/53,61	1	–	–	1	0	–

List 4

	A	B	C	D	E	F
4,5/12,13/20,21/28,29 36,37/44,45/52,53/60,61	–	–	–	1	0	– M
4,5/12,13/36,37/44,45/ 20,21/28,29/52,53/60,61	–	–	–	1	0	–
4,5/20,21/36,37/52,53/ 12,13/38,39/44,45/60,61	–	–	–	1	0	–
4,12/20,28/36,44/52,60/ 5,13/21,29/37,45/53,61	–	–	–	1	0	–

List 5

Drawing the prime implicant table, ignoring the don't-cares, we find that the essential prime implicants are

$$\text{PIE} = (BC\bar{E}\bar{F}, \ \bar{A}BDF, \ D\bar{E})$$

This leaves terms 15, 51, 58, 59 and 63 still to be covered, with the resulting PI chart having a cyclic form. The algebraic approach will give all possible solutions:

$$\begin{aligned}
\text{PI} = &(M)(M + H)(M + J + H)(J + H)(M + K)(K)(I)\\
&(M + L + K + J)(L + K + J)(M)(M)(M)(M)(E + B)(M)\\
&(M + C)(I + D)(F + D)(G + F + E)(M + I)(G + L)
\end{aligned}$$

Simplifying we have

$$\begin{aligned}
\text{PI} = &(M)(J + H)(K)(I)(E + B)(F + D)(G + F + E)(G + L)\\
= &(MKI)(JE + JB + HE + HB)(F + D)(G + EL + FL)\\
= &(MKI)(JE + JB + HE + HB)(FG + FL + DG + DEL)\\
= &(MKI)(JEFG + JEFL + JEDG + JEDL + JBFG + JBFL\\
&+ JBDG + HEFG + HEFL + HEDG + HEDL + HBFG\\
&+ HBFL + HBDG)
\end{aligned}$$

All these are possible solutions; however, the prime implicant sets (*JEFL, HEFL, JBFL, HBFL, JEDL, HEDL*) contain the fewest literals and hence give minimal solutions. Thus, for example, a minimal solution would be

$$\begin{aligned}
T = &MKIJEFL\\
= &D\bar{E} + \bar{A}BDF + BC\bar{E}\bar{F} + \bar{A}CDF + AB\bar{D}EF + ABC\bar{D}E + BCDF
\end{aligned}$$

Table S.13 Continued

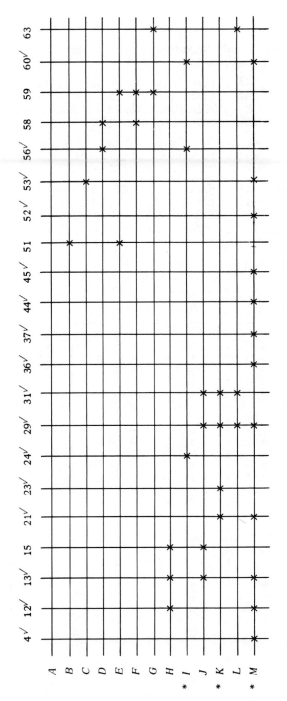

3.5 The truth table for the problem is shown in Table S.14 and the relevant K-maps in Table S.15. In the K-maps, the terms for $A > B$, $A < B$ and $A = B$ are represented by G, L and E respectively. Combining the terms in the usual way, we have the following equations:

$$G = AB\bar{E} + A\bar{D} + C\bar{D}\bar{E}\bar{F} + B\bar{E}\bar{D} + BC\bar{D}\bar{F} + AC\bar{E}\bar{F} + ABCEF$$
$$L = \bar{A}BE + \bar{A}D + \bar{C}DEF + \bar{B}ED + \bar{A}\bar{B}CF + \bar{B}CDF + \bar{A}\bar{C}EF$$
$$E = \bar{A}\bar{B}\bar{C}D\bar{E}\bar{F} + \bar{A}\bar{B}CD\bar{E}F + \bar{A}BCDE\bar{F} + \bar{A}B\bar{C}D\bar{E}\bar{F} + \bar{A}B\bar{C}D\bar{E}\bar{F}$$
$$+ \ ABCD\bar{E}\bar{F} + ABCDEF + AB\bar{C}DE\bar{F}$$

These equations are shown implemented in terms of NAND logic in Fig. S.8. Consider how this would be implemented if only three-input NAND elements were available. Note that the equations could be factored as

$$G = AB\bar{E} + A\bar{D} + \bar{E}\bar{D}(C\bar{F} + B) + C\bar{F}(ABE + B\bar{D} + A\bar{E})$$
$$L = \bar{A}BE + \bar{A}D + ED(\bar{C}F + \bar{B}) + \bar{C}F(\bar{A}\bar{B} + \bar{B}D + \bar{A}E)$$
$$E = \bar{A}\bar{C}D(\bar{B}\bar{E}\bar{F} + BEF) + \bar{A}DC(\bar{B}\bar{E}F + BEF)$$
$$+ \ A\bar{B}\bar{E}(\bar{C}\bar{D}\bar{F} + CDF) + ABD(CEF + \bar{C}\bar{E}\bar{F})$$

Table S.14 Problem 3.5

A	B	C	D	E	F	A>B G	A<B L	A=B E	A	B	C	D	E	F	A>B	A<B	A=B
0	0	0	0	0	0	0	0	1	1	0	0	0	0	1	1	0	0
0	0	0	0	0	1	0	1	0	1	0	0	0	1	0	1	0	0
0	0	0	0	1	0	0	1	0	1	0	0	0	1	1	1	0	0
0	0	0	0	1	1	0	1	0	1	0	0	1	0	0	0	0	1
0	0	0	1	0	0	0	1	0	1	0	0	1	0	1	0	1	0
0	0	0	1	0	1	0	1	0	1	0	0	1	1	0	0	1	0
0	0	0	1	1	0	0	1	0	1	0	0	1	1	1	0	1	0
0	0	0	1	1	1	0	1	0	1	0	1	0	0	0	1	0	0
0	0	1	0	0	0	1	0	0	1	0	1	0	0	1	1	0	0
0	0	1	0	0	1	0	0	1	1	0	1	0	1	0	1	0	0
0	0	1	0	1	0	0	1	0	1	0	1	0	1	1	1	0	0
0	0	1	0	1	1	0	1	0	1	0	1	1	0	0	1	0	0
0	0	1	1	0	0	0	1	0	1	0	1	1	0	1	0	0	1
0	0	1	1	0	1	0	1	0	1	0	1	1	1	0	0	1	0
0	0	1	1	1	0	0	1	0	1	0	1	1	1	1	0	1	0
0	0	1	1	1	1	0	1	0	1	1	0	0	0	0	1	0	0
0	1	0	0	0	0	1	0	0	1	1	0	0	0	1	1	0	0
0	1	0	0	0	1	1	0	0	1	1	0	0	1	0	1	0	0
0	1	0	0	1	0	0	0	1	1	1	0	0	1	1	1	0	0
0	1	0	0	1	1	0	1	0	1	1	0	1	0	0	1	0	0
0	1	0	1	0	0	0	1	0	1	1	0	1	0	1	1	0	0
0	1	0	1	0	1	0	1	0	1	1	0	1	1	0	0	0	1
0	1	0	1	1	0	0	1	0	1	1	0	1	1	1	0	1	0
0	1	0	1	1	1	0	1	0	1	1	1	0	0	0	1	0	0
0	1	1	0	0	0	1	0	0	1	1	1	0	0	1	1	0	0
0	1	1	0	0	1	1	0	0	1	1	1	0	1	0	1	0	0
0	1	1	0	1	0	1	0	0	1	1	1	0	1	1	1	0	0
0	1	1	0	1	1	0	0	1	1	1	1	1	0	0	1	0	0
0	1	1	1	0	0	0	1	0	1	1	1	1	0	1	1	0	0
0	1	1	1	0	1	0	1	0	1	1	1	1	1	0	1	0	0
0	1	1	1	1	0	0	1	0	1	1	1	1	1	1	0	0	1
0	1	1	1	1	1	0	1	0									
1	0	0	0	0	0	1	0	0									

Table S.15 Problem 3.5

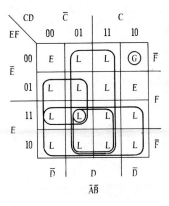

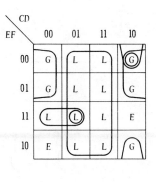

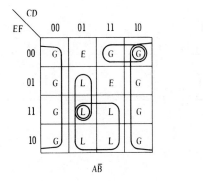

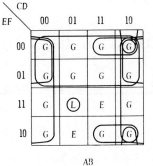

$\overline{A}\overline{B}$ AB

3.6 First we must obtain the OFF array:

$$\text{OFF} = U_n \; \# \; (\text{ON} \cup DC)$$

$$= XXXX \; \# \; \begin{Bmatrix} 0XX1 \\ 10X0 \\ 1100 \\ 111X \\ 1101 \\ 0000 \end{Bmatrix}$$

$$= (((((XXXX \# 0XX1) \# 10X0) \# 1100) \# 111X) \# 1101) \, 0000$$

$$= \left[\begin{Bmatrix} 1XXX \\ XXX0 \end{Bmatrix} \# 10X0 \right] \# 1100 \text{ etc.}$$

$$= \left[(1XXX \# 10X0) \cup (XXX0 \# 10X0) \right] \# 1100 \text{ etc.}$$

$$= \left[\begin{Bmatrix} 0XX0 \\ X1X0 \\ 11XX \\ 1XX1 \end{Bmatrix} \# 1100 \right] \# 111X \text{ etc.}$$

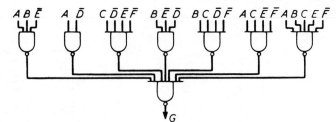

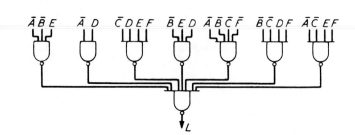

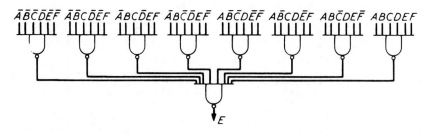

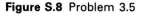

Figure S.8 Problem 3.5

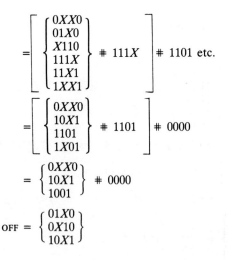

$$= \left[\left\{ \begin{array}{c} 0XX0 \\ 01X0 \\ X110 \\ 111X \\ 11X1 \\ 1XX1 \end{array} \right\} \# 111X \right] \# 1101 \text{ etc.}$$

$$= \left[\left\{ \begin{array}{c} 0XX0 \\ 10X1 \\ 1101 \\ 1X01 \end{array} \right\} \# 1101 \right] \# 0000$$

$$= \left\{ \begin{array}{c} 0XX0 \\ 10X1 \\ 1001 \end{array} \right\} \# 0000$$

$$\text{OFF} = \left\{ \begin{array}{c} 01X0 \\ 0X10 \\ 10X1 \end{array} \right\}$$

Applying the Sharping procedure to determine PIs we have

$$\text{PI}(Z) = U_n \, \# \, \text{OFF}$$

$$= XXXX \, \# \, \left\{ \begin{matrix} 01X0 \\ 0X10 \\ 10X1 \end{matrix} \right\}$$

$$= ((XXXX \, \# \, 01X0) \, \# \, 0X10) \, \# \, 10X1$$

$$= \left[\left\{ \begin{matrix} 1XXX \\ X0XX \\ XXX1 \end{matrix} \right\} \, \# \, 0X10 \right] \# \, 10X1$$

$$= [(1XXX \, \# \, 0X10) \cup (X0XX \, \# \, 0X10)$$
$$\cup \, (XXX1 \, \# \, 0X10)] \, \# \, 10X1$$

$$= \left\{ \begin{matrix} 1XXX \\ X00X \\ XXX1 \end{matrix} \right\} \, \# \, 10X1$$

$$= (1XXX \, \# \, 10X1) \cup (X00X \, \# \, 10X1) \cup (XXX1 \, \# \, 10X1)$$

$$\text{PI}(Z) = \left\{ \begin{matrix} 11XX \\ 1XX0 \\ 000X \\ X000 \\ 0XX1 \\ X1X1 \end{matrix} \right\}$$

The reader is advised to check this result by other methods. To find an irredundant cover we must apply the sharp algorithm as described in the text on page 91.

Pass 1

$$11XX \, \# \, \left\{ \begin{matrix} 1XX0 \\ 000X \\ X000 \\ 0XX1 \\ X1X1 \end{matrix} \right\} = \varphi$$

Pass 2

$$1XX0 \, \# \, \left\{ \begin{matrix} 000X \\ X000 \\ 0XX1 \\ X1X1 \end{matrix} \right\} = \left\{ \begin{matrix} 11X0 \\ 1X10 \\ 101X \end{matrix} \right\}$$

Pass 3

$$000X \, \# \, \left\{ \begin{matrix} 1XX0 \\ X000 \\ 0XX1 \\ X1X1 \end{matrix} \right\} = \varphi$$

Pass 4

$$X000 \, \# \, \left\{ \begin{matrix} 1XX0 \\ 0XX1 \\ X1X1 \end{matrix} \right\} = 0000$$

Pass 5

$$0XX1 \; \# \; \left\{ \begin{array}{l} 1XX0 \\ X000 \\ X1X1 \end{array} \right\} = 00X1$$

Pass 6

$$X1X1 \; \# \; \left\{ \begin{array}{l} 1XX0 \\ X000 \\ 0XX1 \end{array} \right\} = 11X1$$

Thus the cover is given by

$$C = \left\{ \begin{array}{l} 1XX0 \\ X000 \\ 0XX1 \\ X1X1 \end{array} \right\}$$

Note that
(a) the final result will depend on the ordering of the PI cubes in the array;
(b) the algorithm is incomplete in that the don't-care terms are considered as necessary output cubes, i.e. the PIs $X000$ and $X1X1$ are not required.

3.7 The addition table for mod-3 arithmetic is given in problem 1.5. The truth table for the adder is shown in Table S.16. Note that the carry output can be considered as a single output since $C_1 = 0$. The K-maps for the circuit are shown in Table S.17. In this case it is simpler to use NAND gates for the realization and the corresponding circuit is shown in Fig. S.9(a).

The basic half-adder may be used to construct a full-adder circuit as shown in Fig. S.9(b) (note the analogy to the binary half-adder) and extended to form a parallel adder as shown in Fig. S.9(c).

3.8 The truth table is shown in Table S.18; note that there are six don't-care conditions. The output conditions for W, X, Y and Z are plotted on K-maps together with the don't-cares in Table S.19.

Though this is a multi-terminal circuit, little advantage is gained in this case by attempting to choose common prime implicants. The best arrangement is shown on the maps and gives the equations

$$W = A + BD + BC$$
$$X = \bar{B}D + \bar{B}C + B\bar{C}\bar{D}$$
$$Y = \bar{C}\bar{D} + CD$$
$$Z = \bar{D}$$

These may be readily converted into two-level AND/OR gates.

3.9 The full truth table for this problem is shown in Table S.20. There are no don't-care conditions; the output conditions for G_5, G_4, G_3, G_2 and G_1 are shown plotted on K-maps in Table S.21. Though a multi-terminal circuit, once again there are no common prime implicants; the output equations are

$$G_5 = A$$
$$G_4 = \bar{A}B + A\bar{B}$$
$$G_3 = B\bar{C} + \bar{B}C$$
$$G_2 = \bar{C}D + C\bar{D}$$
$$G_1 = D\bar{E} + \bar{D}E$$

Table S.16 Problem 3.7

N_1	N_2	M_1	M_2	S_1	S_2	C_1	C_2
0	0	0	0	0	0	0	0
0	0	0	1	0	1	0	0
0	0	1	0	1	0	0	0
0	0	1	1	X	X	X	X
0	1	0	0	0	1	0	0
0	1	0	1	1	0	0	0
0	1	1	0	0	0	0	1
0	1	1	1	X	X	X	X
1	0	0	0	1	0	0	0
1	0	0	1	0	0	0	1
1	0	1	0	0	1	0	1
1	0	1	1	X	X	X	X
1	1	0	0	X	X	X	X
1	1	0	1	X	X	X	X
1	1	1	0	X	X	X	X
1	1	1	1	X	X	X	X

Table S.17 Problem 3.7

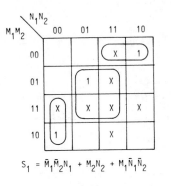

$$S_1 = \bar{M}_1\bar{M}_2N_1 + M_2N_2 + M_1\bar{N}_1\bar{N}_2$$

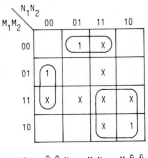

$$S_2 = \bar{M}_1\bar{M}_2N_2 + M_1N_1 + M_2\bar{N}_1\bar{N}_2$$

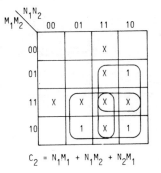

$$C_2 = N_1M_1 + N_1M_2 + N_2M_1$$

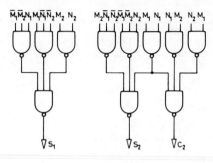

a) Mod 3 half-adder

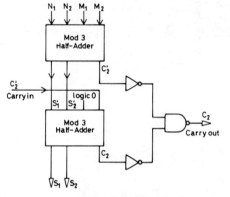

b) Mod 3 full-adder

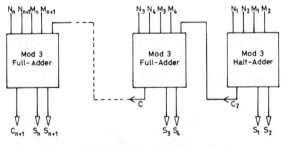

c) Mod 3 parallel adder

Figure S.9 Problem 3.7

Table S.18 Problem 3.8

Decimal	BCD				Excess-three code			
	A	B	C	D	W	X	Y	Z
0	0	0	0	0	0	0	1	1
1	0	0	0	1	0	1	0	0
2	0	0	1	0	0	1	0	1
3	0	0	1	1	0	1	1	0
4	0	1	0	0	0	1	1	1
5	0	1	0	1	1	0	0	0
6	0	1	1	0	1	0	0	1
7	0	1	1	0	1	0	1	0
8	1	0	0	0	1	0	1	1
9	1	0	0	1	1	1	0	0
↓ 15	Don't-care conditions							

Table S.19 Problem 3.8

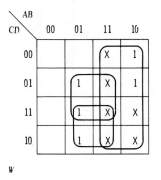

W

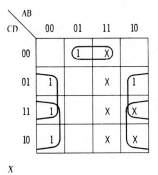

X

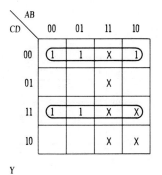

Y

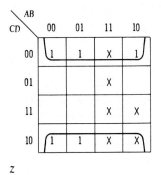

Z

Table S.20 Problem 3.9

Pure binary					Gray code				
A	B	C	D	E	G_5	G_4	G_3	G_2	G_1
0	0	0	0	0	0	0	0	0	0
0	0	0	0	1	0	0	0	0	1
0	0	0	1	0	0	0	0	1	1
0	0	0	1	1	0	0	0	1	0
0	0	1	0	0	0	0	1	1	0
0	0	1	0	1	0	0	1	1	1
0	0	1	1	0	0	0	1	0	1
0	0	1	1	1	0	0	1	0	0
0	1	0	0	0	0	1	1	0	0
0	1	0	0	1	0	1	1	0	1
0	1	0	1	0	0	1	1	1	1
0	1	0	1	1	0	1	1	1	0
0	1	1	0	0	0	1	0	1	0
0	1	1	0	1	0	1	0	1	1
0	1	1	1	0	0	1	0	0	1
0	1	1	1	1	0	1	0	0	0
1	0	0	0	0	1	1	0	0	0
1	0	0	0	1	1	1	0	0	1
1	0	0	1	0	1	1	0	1	1
1	0	0	1	1	1	1	0	1	0
1	0	1	0	0	1	1	1	1	0
1	0	1	0	1	1	1	1	1	1
1	0	1	1	0	1	1	1	0	1
1	0	1	1	1	1	1	1	0	0
1	1	0	0	0	1	0	1	0	0
1	1	0	0	1	1	0	1	0	1
1	1	0	1	0	1	0	1	1	1
1	1	0	1	1	1	0	1	1	0
1	1	1	0	0	1	0	0	1	0
1	1	1	0	1	1	0	0	1	1
1	1	1	1	0	1	0	0	0	1
1	1	1	1	1	1	0	0	0	0

Table S.21 Problem 3.9

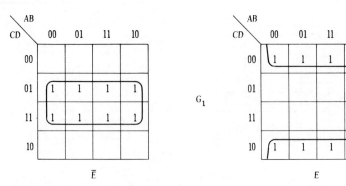

Table S.21 Continued

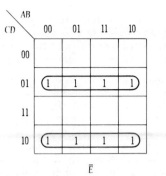

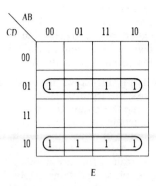

G_2

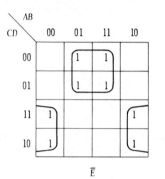

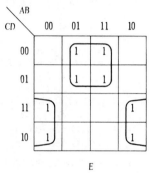

G_3

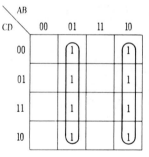

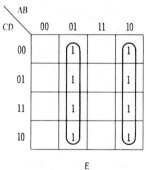

G_4

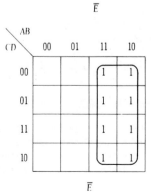

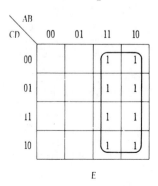

G_5

The form of the equations should be very familiar; they are in fact the exclusive-OR function. This problem illustrates how, quite often, logical design can be short-circuited by close examination of the truth table coupled with experience of logical circuitry. Thus, if we examine Table S.20 it is obvious that digit $G_5 = A$, and digit $G_4 = A \oplus B$, and digit $G_3 = B \oplus C$, etc. Note also that the circuit is iterative and may be extended to any number of digits.

3.10 Let us first solve this problem by simple inspection of the truth table; we may use the truth table for the last problem (Table S.20) remembering that this time the required outputs are $ABCDE$. Note that the most significant digit is unchanged, and hence $A = G_5$; also $B = G_5 \oplus G_4$. The next step is perhaps not so obvious—the output $C = G_3 \oplus B$, also $D = G_2 \oplus C$ and $E = G_1 \oplus D$. Again we have an iterative circuit using the exclusive-OR which may be cascaded for any number of stages. Table S.22 shows the K-maps for the problem, and some algebraic manipulation must be done on the equations to arrive at the same result. On first inspection, the output equations look rather formidable:

$$A = G_5$$
$$B = \bar{G}_5 G_4 + G_5 \bar{G}_4$$
$$C = \bar{G}_5 \bar{G}_4 \bar{G}_3 + G_5 \bar{G}_4 \bar{G}_3 + \bar{G}_5 \bar{G}_4 G_3 + G_5 G_4 G_3$$
$$D = \bar{G}_5 \bar{G}_4 \bar{G}_3 G_2 + \bar{G}_5 \bar{G}_4 G_3 \bar{G}_2 + \bar{G}_5 G_4 \bar{G}_3 \bar{G}_2 + \bar{G}_5 G_4 G_3 G_2 + G_5 G_4 \bar{G}_3 G_2$$
$$\qquad + G_5 \bar{G}_4 G_3 \bar{G}_2 + G_5 \bar{G}_4 \bar{G}_3 \bar{G}_2 + \bar{G}_5 G_4 G_3 G_2$$
$$E = \bar{G}_5 \bar{G}_4 \bar{G}_3 G_2 \bar{G}_1 + \bar{G}_5 \bar{G}_4 G_3 \bar{G}_2 \bar{G}_1 + \bar{G}_5 \bar{G}_4 \bar{G}_3 \bar{G}_2 \bar{G}_1 + \bar{G}_5 G_4 G_3 G_2 \bar{G}_1$$
$$\qquad + \bar{G}_5 G_4 \bar{G}_3 G_2 \bar{G}_1 + G_5 G_4 G_3 \bar{G}_2 \bar{G}_1 + G_5 \bar{G}_4 \bar{G}_3 \bar{G}_2 \bar{G}_1 + G_5 \bar{G}_4 G_3 G_2 \bar{G}_1$$
$$\qquad + \bar{G}_5 \bar{G}_4 \bar{G}_3 \bar{G}_2 G_1 + \bar{G}_5 \bar{G}_4 G_3 G_2 G_1 + \bar{G}_5 G_4 \bar{G}_3 G_2 G_1 + \bar{G}_5 G_4 G_3 \bar{G}_2 G_1$$
$$\qquad + G_5 \bar{G}_4 \bar{G}_3 G_2 G_1 + G_5 \bar{G}_4 G_3 \bar{G}_2 G_1 + G_5 G_4 \bar{G}_3 \bar{G}_2 G_1 + G_5 \bar{G}_4 \bar{G}_3 \bar{G}_2 G_1$$

Now, factoring C, we have

$$C = \bar{G}_3(\bar{G}_5 G_4 + G_5 \bar{G}_4) + G_3(\bar{G}_5 \bar{G}_4 + G_5 G_4)$$

Thus

$$C = \bar{G}_3 B + G_3 \bar{B}$$

Also for D we have

$$D = G_2[\bar{G}_3(\bar{G}_5 \bar{G}_4 + G_5 G_4) + G_3(\bar{G}_5 G_4 + G_5 \bar{G}_4)]$$
$$\qquad + \bar{G}_2[\bar{G}_3(\bar{G}_5 G_4 + G_5 \bar{G}_4) + G_3(\bar{G}_5 \bar{G}_4 + G_5 G_4)]$$

Thus

$$D = G_2 \bar{C} + \bar{G}_2 C$$

Similarly we may factorize for E:

$$E = \bar{G}_1\{\bar{G}_2[\bar{G}_3(\bar{G}_5 G_4 + G_5 \bar{G}_4) + G_3(\bar{G}_5 \bar{G}_4 + G_5 G_4)]$$
$$\qquad + G_2[\bar{G}_3(\bar{G}_5 \bar{G}_4 + G_5 G_4) + G_3(\bar{G}_5 G_4 + G_5 \bar{G}_4)]\}$$
$$\qquad + G_1\{\bar{G}_2[\bar{G}_3(\bar{G}_5 \bar{G}_4 + G_5 G_4) + G_3(\bar{G}_5 G_4 + G_5 \bar{G}_4)]$$
$$\qquad + G_2[\bar{G}_3(\bar{G}_5 G_4 + G_5 \bar{G}_4) + G_3(\bar{G}_5 \bar{G}_4 + G_5 G_4)]\}$$

Hence

$$E = \bar{G}_1 D + G_1 \bar{D}$$

This example illustrates that switching theory must be applied intelligently, and cannot be expected always to yield the final result. Direct implementation of the unfactorized equations would result in a very uneconomical (though correct) circuit. To implement the circuit in NAND logic we can use the

Table S.22 Problem 3.10

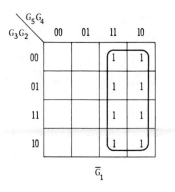

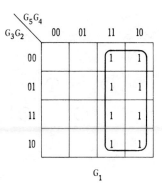

A

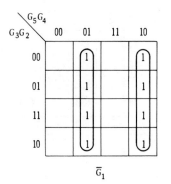

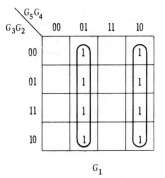

B

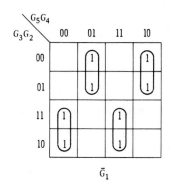

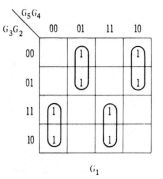

C

Table S.22 Continued

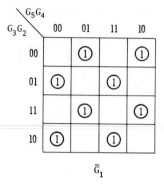

$\bar{G}_1$

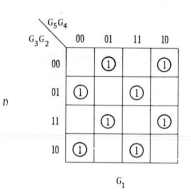

D

G_1

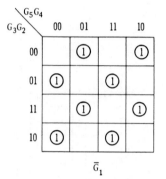

$\bar{G}_1$

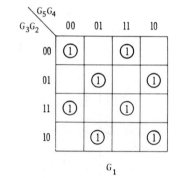

E

G_1

Table S.23 Problem 3.11

W	X	Y	Z		a	b	c	d	e	f	g
0	0	0	0		1	1	1	1	1	1	0
0	0	0	1		0	1	1	0	0	0	0
0	0	1	0		1	1	0	1	1	0	1
0	0	1	1		1	1	1	1	0	0	1
0	1	0	0		0	1	1	0	0	1	1
0	1	0	1		1	0	1	1	0	1	1
0	1	1	0		0	0	1	1	1	1	1
0	1	1	1		1	1	1	0	0	0	0
1	0	0	0		1	1	1	1	1	1	1
1	0	0	1		1	1	1	0	0	1	1
1	0	1	0		X	X	X	X	X	X	X
1	0	1	1		X	X	X	X	X	X	X
1	1	0	0		X	X	X	X	X	X	X
1	1	0	1		X	X	X	X	X	X	X
1	1	1	0		X	X	X	X	X	X	X
1	1	1	1		X	X	X	X	X	X	X

Table S.24 Problem 3.11

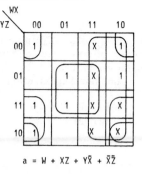

a = W + XZ + YX̄ + X̄Z̄

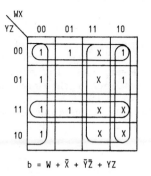

b = W + X̄ + ȲZ̄ + YZ

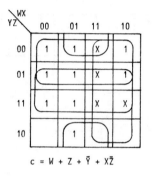

c = W + Z + Ȳ + XZ̄

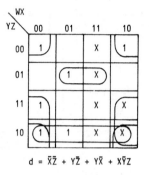

d = X̄Z̄ + YZ̄ + YX̄ + XȲZ

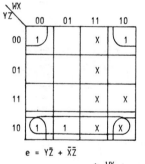

e = YZ̄ + X̄Z̄

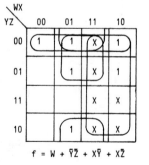

f = W + ȲZ̄ + XȲ + XZ̄

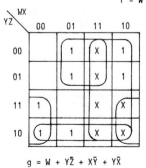

g = W + YZ̄ + XȲ + YX̄

exclusive-OR circuit shown in Fig. S.1, for each stage; this is shown in Fig. S.10. Note that only 16 NAND elements are required compared with 20 using the AND/OR invertor system.

3.11 The truth table for this problem is shown in Table S.23; note that it is a multiple-output circuit with a large number of don't-care conditions.

A K-map method of minimizing is shown in Table S.24 which simply chooses common prime implicant terms. The logic diagram using NAND gates is shown in Fig. S.11.

Note the common, multiple-output, prime implicants—W, $Y\bar{Z}$, $\bar{X}\bar{Z}$, $Y\bar{X}$, $\bar{Y}\bar{Z}$, $X\bar{Z}$ and that the solution is not necessarily minimum but is near optimal.

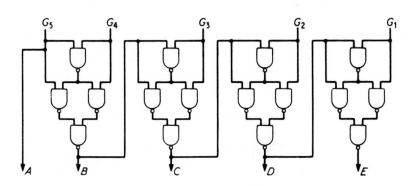

Figure S.10 Problem 3.10

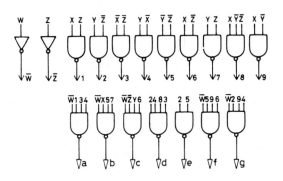

Figure S.11 Problem 3.11

Chapter 4

4.1 The NOR/NAND circuit is shown in Fig. S.12; note the NOT performed on the signal lines for A_L and B_L and that the circuit closely follows the direct logical AND/OR implementation of the function.

4.2 To realize this circuit directly using ROM would require $2^7 \times 4$ bits $= 512$ bits, that is, a ROM store containing 128 words of 4 bits of which only 10 locations would actually be used.

A cascaded configuration may be used by partitioning the truth table as shown in Table S.25 and implementing using two ROMs, as shown in Fig. S.13. Note that the cascaded circuit requires 224 bits as compared with 512 bits for a direct implementation.

However, if we examine the original truth table and note that we are using 7 bits to code only 10 outputs it would seem possible that some of the input variables might be redundant. In fact this is the case and variable bc (or cd) can be ignored. Thus it is possible to implement the circuit using one 32×4-bit ROM with the layout as shown in Table S.25(d).

But it is possible to go further! If we form a new function $H = d.g$ we can reduce the input variables further giving the ROM layout shown in Table S.25(e). In this case the circuit may be realized with one 16×4-bit ROM as shown in Fig. S.13(c), which is the minimum circuit possible.

4.3 (a) This function may be implemented directly as shown in Fig. S.14(a).
(b) With five variables it is obvious that a cascaded circuit is needed. If we choose BC as the first level control variables and AE as the second and then partition as shown in Table S.26(a) and (b), we obtain the network shown in Fig. S.14(b).
(c) Again this function necessitates a cascaded circuit. This time we choose DE as the first level control variables and partition as shown in Table S.26(c). Note that we can simplify by observing that $A' = D'$ and $B' = C'$ which leads to the network shown in Fig. S.15(a). An alternative realization using a 1-out-of-8 MUX is shown in Table S.26(e) and Fig. S.15(b). In this case we have used ABC as the control inputs and combined the inputs D and E using an exclusive-OR function prior to the multiplexer.

As with all multiplexer designs the final circuit will depend on the choice of control variables and the circuits given are not necessarily optimal.

4.4 With 4 inputs, 10 products and 8 outputs the solution is trivial since the function can be completely contained within one PLA chip. Note in this case the similarity to a ROM implementation. Figure S.16 shows the connection diagram using the usual notation of a dot on intersecting lines to represent a diode connection.

4.5 Assuming we partition the input variables X as $X = (X_1 X_2)$ there are three possible assignments: $X_1 = (x_1 x_2)$, $X_2 = (x_3 x_4)$; $X_1 = (x_1 x_3)$, $X_2 = (x_2 x_4)$; and $X_1 = (x_1 x_4)$, $X_2 = (x_2 x_3)$. Choosing $X_1 = (x_1 x_3)$ and $X_2 = (x_2 x_4)$ gives, from Table 4.13, and expressing the ON-terms for output F as a generalized Boolean function,

$$F = X_1^{00} X_2^{00} + X_1^{00} X_2^{01} + X_1^{01} X_2^{00} + X_1^{00} X_2^{10} + X_1^{00} X_2^{11} + X_1^{01} X_2^{11} + X_1^{10} X_2^{00} + X_1^{11} X_2^{01} + X_1^{10} X_2^{11} + X_1^{11} X_2^{10}$$

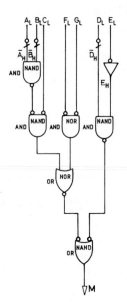

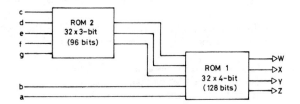

Figure S.12 Problem 4.1

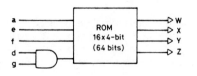

a) Cascaded circuit

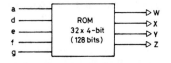

b) ROM after input reduction

c) ROM with gated inputs

Figure S.13 Problem 4.2

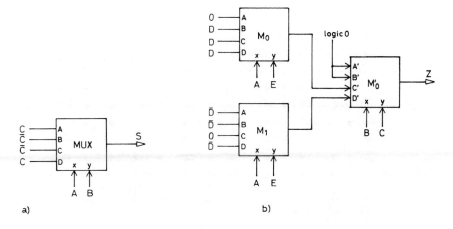

a)

b)

Figure S.14 Problem 4.3

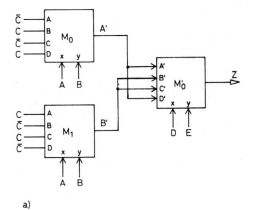

a)

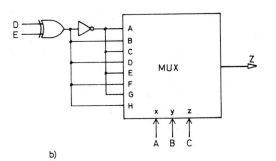

b)

Figure S.15 Problem 4.3(c)

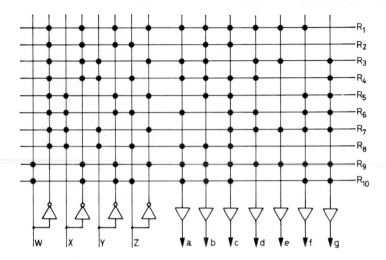

Figure S.16 Problem 4.4

Table S.25 Problem 4.2

(a) Partition of ON terms

		Variables				
a	b	c	d	e	f	g
1	1	1	1	1	1	0
1	1	1	0	0	0	0
1	1	0	1	1	0	1
1	1	1	1	0	0	1
1	1	1	0	0	1	1
1	1	1	1	1	1	1
0	1	1	0	0	0	0
0	1	1	0	0	1	1
1	0	1	1	0	1	1
0	0	1	1	1	1	1

(a) Partition of ON terms

(b) Coded terms

Variables					Coded form		
c	d	e	f	g	A	B	C
0	1	1	0	1	0	0	0
1	0	0	0	0	0	0	1
1	0	0	1	1	0	1	0
1	1	0	0	1	0	1	1
1	1	0	1	1	1	0	0
1	1	1	1	0	1	0	1
1	1	1	1	1	1	1	0

(b) Coded terms

(c) Layout of first level ROM

Input variables					Output			
A	B	C	a	b	W	X	Y	Z
1	0	1	1	1	0	0	0	0
0	0	1	1	1	0	0	0	1
0	0	0	1	1	0	0	1	0
0	1	1	1	1	0	0	1	1
0	1	0	0	1	0	1	0	0
1	0	0	1	0	0	1	0	1
1	1	0	0	0	0	1	1	0
0	0	1	1	1	0	1	1	1
1	1	0	1	1	1	0	0	0
0	1	0	1	1	1	0	0	1

(c) Layout of first level ROM

(d) Layout with reduced inputs

Reduced input variables					Outut			
a	d	e	f	g	W	X	Y	Z
1	1	1	1	0	0	0	0	0
0	0	0	0	0	0	0	0	1
1	1	1	0	1	0	0	1	0
1	1	0	0	1	0	0	1	1
0	0	0	1	1	0	1	0	0
1	1	0	1	1	0	1	0	1
0	1	1	1	1	0	1	1	0
1	0	0	0	0	0	1	1	1
1	1	1	1	1	1	0	0	0
1	0	0	1	1	1	0	0	1

(d) Layout with reduced inputs

Input variables				Output			
a	e	f	$H(d.g)$	W	X	Y	Z
1	1	1	0	0	0	0	0
0	0	0	0	0	0	0	1
1	1	0	1	0	0	1	0
1	0	0	1	0	0	1	1
0	0	1	0	0	1	0	0
1	0	1	1	0	1	0	1
0	1	1	1	0	1	1	0
1	0	0	0	0	1	1	1
1	1	1	1	1	0	0	0
1	0	1	0	1	0	0	1

(e) Layout with gated inputs

Table S.26 Problem 4.3

A	E	D		B	C
1	0	1		1	0
1	1	1	C'	1	0
0	1	1		1	0
0	0	0		1	1
1	0	0	D'	1	1
0	1	0		1	1

(a)

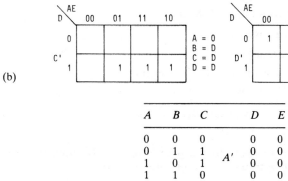

(b)

A	B	C		D	E
0	0	0		0	0
0	1	1	A'	0	0
1	0	1		0	0
1	1	0		0	0
0	0	1		0	1
0	1	0	B'	0	1
1	0	0		0	1
1	1	1		0	1
0	0	1		1	0
0	1	0	C'	1	0
1	0	0		1	0
1	1	1		1	0
0	0	0		1	1
0	1	1	D'	1	1
1	0	1		1	1
1	1	1		1	1

(c)

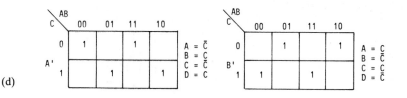

(d)

Table S.26 Continued

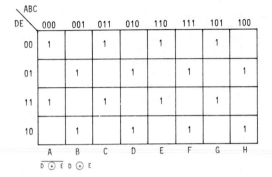

DE \ ABC	000	001	011	010	110	111	101	100
00	1		1		1		1	
01		1		1		1		1
11	1		1		1		1	
10		1		1		1		1

A B C D E F G H

$\overline{D \odot E \ D \odot E}$

Reducing the terms by exhaustively comparing the indices and checking for valid covers we find that the following terms (numbering from the left) will yield effective combinations:

(1, 9) $X_1^{0010} X_2^{0011}$ covering (1, 5, 7, 9)
(4, 8) $X_1^{0011} X_2^{1001}$ covering (2, 4, 8, 10)

which leaves 3, 6 to be covered by

(3, 6) $X_1^{01} X_2^{0011}$ covering (3, 6)

but (3, 6) and (1, 9) may also be combined giving

(3, 6)(1, 9) $X_1^{010010} X_2^{0011}$ covering (1, 3, 5, 6, 7, 9)

Therefore the function can be expressed as

$$F = X_1^{010010} X_2^{0011} + X_1^{0011} X_2^{1001}$$

Note the similarity in principle to the prime implicant covering problem. Translating this equation to the OR-AND-OR form we obtain

$$F = (\bar{x}_1 + \bar{x}_3)(x_2 + \bar{x}_4)(\bar{x}_2 + x_4) + (x_1 + \bar{x}_3)(\bar{x}_1 + x_3)(x_2 + x_4)(\bar{x}_2 + \bar{x}_4)$$

This is in fact the minimal form of the function and only requires two output columns as shown in Fig. S.17.

4.6 Extracting the equation for the circuit we find that

$$F = WX + Z(\bar{X} + WX)$$

when it is obvious that we have a hazard due to $X + \bar{X} \neq 1$; the function is shown plotted on a K-map in Table S.27. To eliminate the fault we need to add the extra loop WZ, which gives

$$F = WX + \bar{X}Z + WZ$$

The function should be implemented directly as a two-level circuit to remain hazard-free.

Table S.27 Problem 4.5

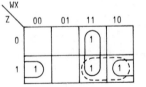

Z \ WX	00	01	11	10
0			1	
1	1		1	1

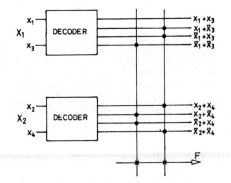

Figure S.17 Problem 4.5

Chapter 5

5.1 This circuit could be designed using normal truth table techniques, and for
and small message words this would yield the most economical result. However, a
5.2 change of message length would mean redesigning the circuit, and for long
messages the problems of factoring the circuits to suit the fan-in factor of the
logic modules could be quite considerable. The circuit is required to detect the
presence of three consecutive 1s in a ten-bit message; for example:

    0011101101   valid
    0011110000   invalid
    0011101110   valid

The state transfer table for this problem is shown in Table S.28; note that six
input states are required necessitating three bits for the state coding. Note also
that only six of the combinations are used, the remaining two being don't-care
conditions. Table S.29 shows the K-maps for the logic gate implementation;
two possible state assignments are given, both giving rather similar results; note
the grouping of common prime implicants for the assignment shown in Table
S.29(b). The next input state equations for W, X and Y are, for the initial
assignment, Table S.29(a)

$$W = wz_n + w\bar{x} + \bar{w}x\bar{y}$$
$$X = xz_n + yz_n$$
$$Y = \bar{w}\bar{x}z_n$$

Table S.28 Problems 5.1 and 5.2

	Input state variables	External input Z_n			Input state variables			External input Z_n	
		0	1		w	x	y	0	1
A	Last digit 0	A_+	B_+	A	0	0	0	000	001
B	Last digit 1	A_+	C_+	B	0	0	1	000	011
C	Last 2 digits 1	A_+	D_+	C	0	1	1	000	010
D	Last 3 digits 1	F_+	E_+	D	0	1	0	100	110
E	Wait for 0	A_+	E_+	E	1	1	0	000	110
F	Output	F_+	F_+	F	1	0	0	100	100

(a) General table   (b) Assigned table for NAND implementation

	Input state variables			External input Z_n	
	w	x	y	0	1
A	0	0	0	000	001
B	0	0	1	000	011
C	0	1	1	000	010
D	0	1	0	111	110
E	1	1	0	000	110
F	1	1	1	111	111

(c) Alternative assignment

Table S.29 Problems 5.1 and 5.2

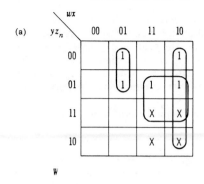

(a) W

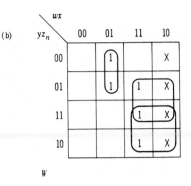

(b) W

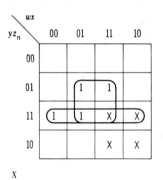

X

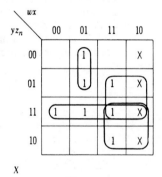

X

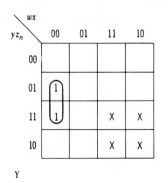

Y

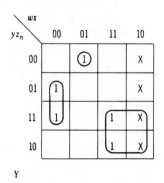

Y

The logic diagram for the cells is shown in Fig. S.18; note that the first cell can be simplified since the input states wxy must indicate that the last digit is 0, i.e. $w\bar{x}\bar{y}$; therefore the output from the first cell depends on the external input z_1. The last cell must be modified because, though the actual output state is given by $w\bar{x}\bar{y}$, we must account for the case when the sequence 0000000111 occurs, and thus we also need to know the input state 'last three digits 1', i.e. $\bar{w}x\bar{y}$; this is accomplished by the additional gating circuitry. Note that the actual cell circuit requires 11 NAND elements, and the complete ten-bit circuit 105 elements.

To design the contact circuit we use Table S.28(a) direct to give the equations

$$A_+ = A\bar{z}_n + B\bar{z}_n + C\bar{z}_n + E\bar{z}_n \qquad D_+ = Cz_n$$
$$B_+ = Az_n \qquad\qquad\qquad\qquad E_+ = Dz_n + Ez_n$$
$$C_+ = Bz_n \qquad\qquad\qquad\qquad F_+ = Fz_n + F\bar{z}_n + D\bar{z}_n = 1 + D\bar{z}_n$$

The circuit may be simplified in the initial and final cells (Fig. S.19). Note that two outputs are required: D and F. The F line may be a straight-through connection.

It is interesting to consider how the circuit may be redesigned to detect the occurrence of *one group only* of three consecutive 1s. Symmetric functions cannot be used to design this circuit. It is normal practice to use one external input per cell, but there is no reason in theory why more should not be used. Table S.30 shows the same circuit allocated with two external inputs z_1z_2. The cell contact equations are

$$A_+ = A\bar{z}_1\bar{z}_2 + Az_1\bar{z}_2 + B\bar{z}_1\bar{z}_2 + Bz_1\bar{z}_2 + C\bar{z}_1\bar{z}_2 + Dz_1\bar{z}_2 + E\bar{z}_1\bar{z}_2 + Ez_1\bar{z}_2$$

Thus

$$A_+ = A\bar{z}_2 + B\bar{z}_2 + C\bar{z}_1\bar{z}_2 + Dz_1\bar{z}_2 + E\bar{z}_2$$
$$B_+ = A\bar{z}_1z_2 + B\bar{z}_1z_2 + C\bar{z}_1z_2 + E\bar{z}_1z_2$$
$$C_+ = Az_1z_2$$
$$D_+ = Bz_1z_2$$
$$E_+ = Cz_1z_2 + Dz_1z_2 + Ez_1z_2$$
$$F_+ = F\bar{z}_1\bar{z}_2 + F\bar{z}_1z_2 + Fz_1\bar{z}_2 + Fz_1z_2 + Cz_1\bar{z}_2 + D\bar{z}_1\bar{z}_2 + D\bar{z}_1z_2$$

Hence

$$F_+ = 1 + Cz_1\bar{z}_2 + D\bar{z}_1$$

Note that the contact circuit is more complicated (27 contacts per cell as against 10) and increases in message length must be in blocks of two digits. A better solution could be obtained by redefining the internal input states. For electronic gate implementation (see Table S.30(b) and S.31) the cell input state equations are

$$W = w\bar{x} + xz_1z_2 + xyz_1 + \bar{w}x\bar{y}\bar{z}_1$$
$$X = xz_1z_2 + \bar{w}z_1z_2$$
$$Y = y\bar{z}_1z_2 + \bar{w}\bar{x}\bar{y}z_2 + wx\bar{z}_1z_2$$

Clearly the implementation of these equations would result in a more economical circuit.

It is important to note the effect of propagation delay through the circuits; the use of more than one input variable will speed up this time. Cascaded electronic circuits of this type would, in general, be too slow for fast systems, and could lead to hazardous operation; the situation could be improved by the use of multiple input cells.

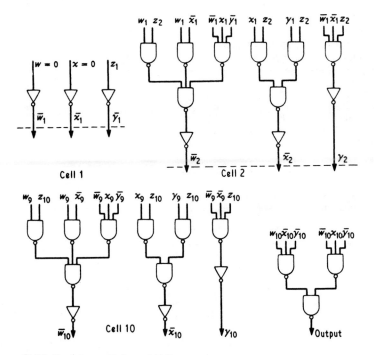

Figure S.18 Problems 5.1 and 5.2

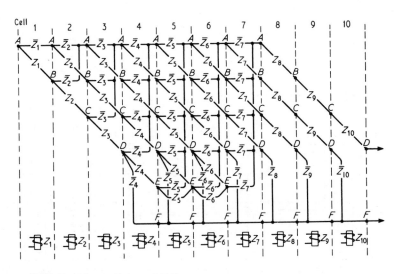

Figure S.19 Problems 5.1 and 5.2

Table S.30 Problems 5.1 and 5.2

Input state variables		External inputs Z_1Z_2			
		00	01	11	10
A	Last digit 0	A_+	B_+	C_+	A_+
B	Last digit 1	A_+	B_+	D_+	A_+
C	Last 2 digits 1	A_+	B_+	E_+	F_+
D	Last 3 digits 1	F_+	F_+	E_+	A_+
E	Wait for 0	A_+	B_+	E_+	A_+
F	Output	F_+	F_+	F_+	F_+

(a)

Input state variables			External inputs Z_1Z_2			
	w	x	y	00	01	11	10
A	0	0	0	000	001	011	000
B	0	0	1	000	001	010	000
C	0	1	1	000	001	110	100
D	0	1	0	100	100	110	000
E	1	1	0	000	001	110	000
F	1	0	0	100	100	100	100

Don't care terms: { 1 0 1 ; 1 1 1 }

(b)

Table S.31 Problems 5.1 and 5.2

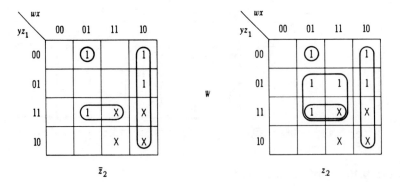

Table S.31 Continued

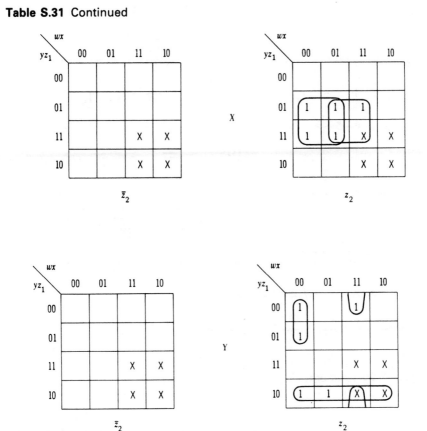

5.3 The truth table for this problem is shown in Table 5.1. From this we see that we want an error output whenever the sum of the 1s is 0-, 2-, or 4-out-of-5; that is, in symmetric notation $S^5_{0,2,4}$ ($ABCDE$). Since the subscripts are in geometrical progression and the next term is larger than the number of variables, we can expect to simplify the circuit by 'folding over'. The full symmetric circuit is given in Fig. S.20(a) with all the required outputs shown. The points marked a, b and c may be 'folded' to the points marked a', b' and c' on the diagrams. The final circuit, Fig. S.20(d), is identical to the iterative cell designed in the main text.

5.4 This circuit can be designed using either symmetric circuits or an iterative approach. For contact networks the symmetric method is easier, and the circuit can be represented as $S^9_{6,8}$ ($ABCDEFGHI$). The full circuit is shown in Fig. S.21; note that this can be simplified since we do not require the outputs for 0–5. Furthermore, the circuit may be 'folded over' to give even greater simplification; the final circuit is shown in Fig. S.21(c).

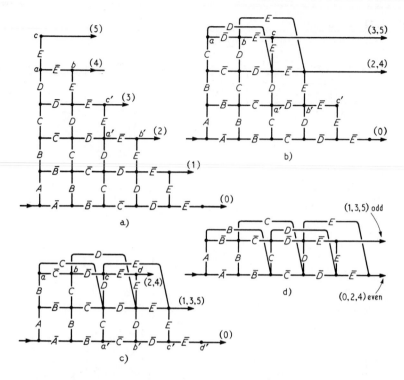

Figure S.20 Problem 5.3

A possible input state transfer table for this problem is shown in Table S.32; for electronic implementation the additional state J is required. The contact circuit equations are

$$A_+ = A\bar{z}_n$$
$$B_+ = Az_n + B\bar{z}_n$$
$$C_+ = Bz_n + C\bar{z}_n$$
$$D_+ = Cz_n + D\bar{z}_n$$
$$E_+ = Dz_n + E\bar{z}_n$$
$$F_+ = Ez_n + F\bar{z}_n$$
$$G_+ = Fz_n + G\bar{z}_n$$
$$H_+ = Gz_n + H\bar{z}_n$$
$$I_+ = Hz_n + I\bar{z}_n$$

Implementation of this circuit is straightforward and yields, after the elimination of redundant contacts, a similar circuit to that obtained with the symmetric design.

5.5 Before performing the matrix expansion, the expression must be factorized, and this is easily done using the map method (see Table S.33);

$$T = A\bar{C}\bar{D}\bar{E} + \bar{A}\bar{B}C\bar{E} + \bar{A}\bar{C}D\bar{E} + ACBE$$
$$= (A + C\bar{B}\bar{E} + \bar{C}D\bar{E})(\bar{A} + \bar{C}D\bar{E} + BCE)$$

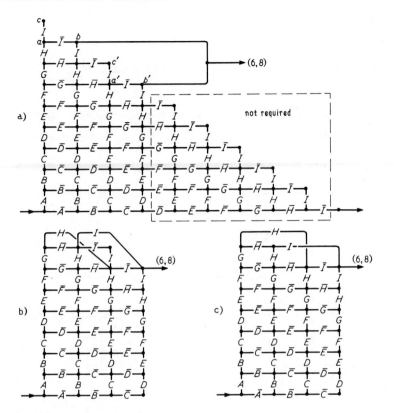

a)

b)

c)

Figure S.21 Problem 5.4

Table S.32 Problem 5.4

Input state variables		External inputs Z	
		0	1
A	Sum zero	A_+	B_+
B	Sum one	B_+	C_+
C	Sum two	C_+	D_+
D	Sum three	D_+	E_+
E	Sum four	E_+	F_+
F	Sum five	F_+	G_+
G	Sume six	G_+	H_+
H	Sum seven	H_+	I_+
I	Sum eight	I_+	J_+
J	Abortive	J_+	J_+

Table S.33 Problem 5.5

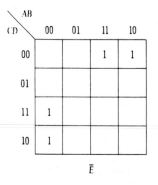

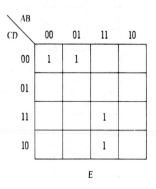

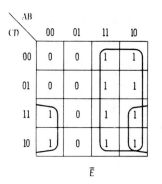

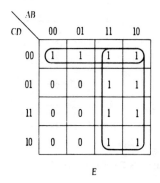

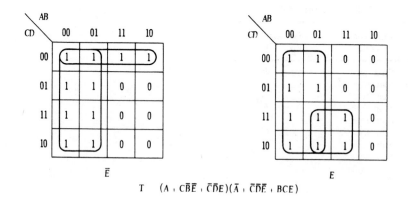

$$T \quad (A + C\bar{B}\bar{E} + \bar{C}\bar{D}E)(\bar{A} + \bar{C}\bar{D}\bar{E} + BCE)$$

Thus

$$
T = \begin{bmatrix}
1 & 0 & A + C\bar{B}\bar{E} + \check{C}\bar{D}E \\
0 & 1 & \bar{A} + \check{C}\bar{D}\bar{E} + BCE \\
A + C\bar{B}\bar{E} + \check{C}\bar{D}E & \bar{A} + \check{C}\bar{D}\bar{E} + BCE & 1
\end{bmatrix}
$$

$$
= \begin{bmatrix}
1 & 0 & A + C\bar{B}\bar{E} & E \\
0 & 1 & \bar{A} + BCE & \bar{E} \\
A + C\bar{B}\bar{E} & \bar{A} + BCE & 1 & \check{C}\bar{D} \\
E & \bar{E} & \check{C}\bar{D} & 1
\end{bmatrix}
$$

Hence

$$
T = \begin{array}{c}
 \\
\begin{array}{ccccc}
1 & 2 & 3 & 4 & 5
\end{array} \\
\begin{array}{c}
1 \\ 2 \\ 3 \\ 4 \\ 5
\end{array}
\begin{bmatrix}
1 & 0 & A & E & \bar{B}\bar{E} \\
0 & 1 & \bar{A} & \bar{E} & BE \\
A & \bar{A} & 1 & \check{C}\bar{D} & C \\
E & \bar{E} & \check{C}\bar{D} & 1 & 0 \\
\bar{B}\bar{E} & BE & C & 0 & 1
\end{bmatrix}
\end{array}
$$

The circuit is shown implemented in Fig. S.22.

5.6 The relay tree will be designed by first assuming that there are no don't-care conditions; Table S.34 shows the relevant stages using the K-map technique. The combinations are first plotted on the maps, and the maps subdivided, indicating the order of subdivision by inserting the appropriate variable in the squares.

The corresponding relay circuit is shown in Fig. S.23(a). If, however, we assume that only the required combinations can occur, the circuit can be simplified (see Table S.35 and Fig. S.23(b)). Note that many alternative solution are possible for this problem.

Note also that the simplified circuit uses 11 change-over contacts, compared with the original circuit which uses 11 change-over and 5 normally open contacts. To design an electronic tree circuit, we split the number of variables thus:

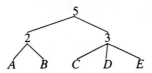

Thus we combine the two variables AB in all possible ways, and the three variables CDE in all possible ways. The outputs are then selected to provide the required combination outputs (see Fig. S.24). Note that the combination $A\bar{B}$ is redundant.

5.7 The first step is to derive the product-of-sums and sum-of-products forms from Table S.36(a); calling the variables a, b, c and d we have

$$Z = \bar{a}\bar{b}c + \bar{a}b\bar{c} + abc + ab\bar{c}$$
$$Z = (a + b + c)(a + \bar{b} + \bar{c})(\bar{a} + \bar{b} + c)(\bar{a} + b + \bar{c})$$

Plotting these terms on an array map (Table S.36(b)) we obtain the input terms for the array, shown in Fig. S.25. Note that there is a choice of terms and we have arbitrarily selected the first one; the array requires 16 cells to realize the function.

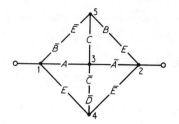

Figure S.22 Problem 5.5

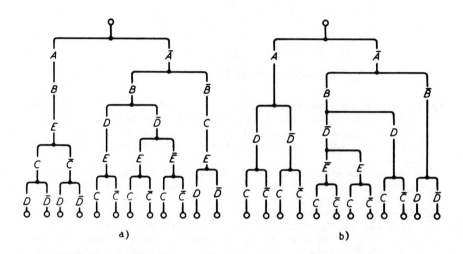

a)   b)

Figure S.23 Problem 5.6

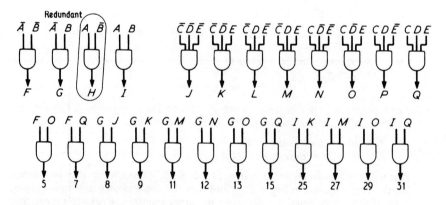

Figure S.24 Problem 5.6

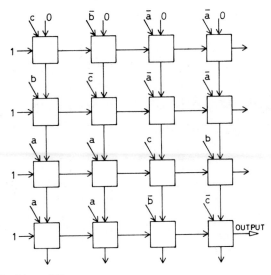

Figure S.25 Problem 5.7

Table S.34 Problem 5.6

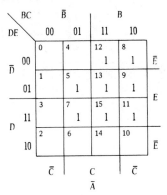

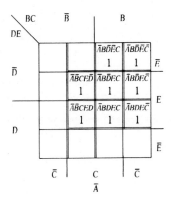

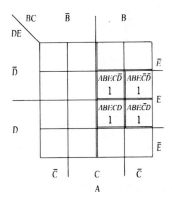

Table S.35 Problem 5.6

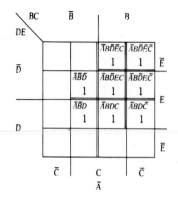

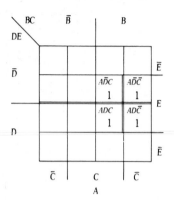

Table S.36 Problem 5.7

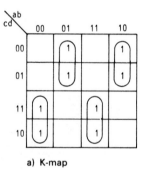

a) K-map

	a+b+c	a+b̄+c̄	ā+b̄+c	ā+b+c̄
āb̄c	c	b̄	ā,b̄,c	ā
āb̄c̄	b	c̄	ā	ā,b,c̄
abc	a,b,c	a	c	b
ab̄c̄	a	a,b̄,c̄	b̄	c̄

b) Array map

Chapter 6

6.1 The state table for this problem is shown in Table S.37; the ring-counter is a five-state device that changes its output, obtained directly from the bistables (rather like a shift-register) on each input pulse. The remaining 27 states can be used as don't-care conditions in the design. This is a rather trivial example and the set and reset conditions can be ascertained by direct inspection of the state table. For example, the condition for setting bistable A is 00001; however, since this is the only state with $E = 1$ (all the others are don't-cares), we can simply use E to set the bistable. All the other conditions follow in the same way to give, for the $SR\text{-}FF$ bistable

$$
\begin{array}{llll}
\text{Set} & S_A = E & \text{Reset} & R_A = A \\
& S_B = A & & R_B = B \\
& S_C = B & & R_C = C \\
& S_D = C & & R_D = D \\
& S_E = D & & R_E = E
\end{array}
$$

The trigger bistable may be treated similarly to give

$$
\begin{aligned}
T_A &= A + E \\
T_B &= A + B \\
T_C &= B + C \\
T_D &= C + D \\
T_E &= D + E
\end{aligned}
$$

Figure S.26 shows the logic diagram for the two cases. Note that in both cases the counters must be initialized by setting a 1 into one of the stages and 0s in the other. Alternatively the don't-care states (the forbidden states) could be used to reset the counter but this would lead to a more complicated circuit; it would, however, be a more reliable one.

6.2 Table S.38 shows the state table for this problem, with the K-maps for the $SR\text{–}FF$ bistable input conditions in Table S.39. Note that there are six don't-care conditions, i.e. $D = (5, 6, 7, 13, 14, 15)$, due to the unused combinations. The input equations are

$$
\begin{array}{llll}
\text{Set} & S_A = \bar{A}B & \text{Reset} & R_A = AB \\
& S_B = CD & & R_B = \bar{C} \\
& S_C = \bar{C}D & & R_C = CD \\
& S_D = \bar{B}\bar{D} & & R_D = D
\end{array}
$$

6.3 The divide-by-5 counter circuit has five states, requiring three bistables, thus leaving three don't-care conditions, i.e. $D = (5, 6, 7)$. Table S.40 shows the

Table S.37 Problem 6.1

Present state					Next state				
A	B	C	D	E	A_+	B_+	C_+	D_+	E_+
1	0	0	0	0	0	1	0	0	0
0	1	0	0	0	0	0	1	0	0
0	0	1	0	0	0	0	0	1	0
0	0	0	1	0	0	0	0	0	1
0	0	0	0	1	1	0	0	0	0

Table S.38 Problem 6.2

Present state				Next state			
5	4	2	1				
A	B	C	D	A_+	B_+	C_+	D_+
0	0	0	0	0	0	0	1
0	0	0	1	0	0	1	0
0	0	1	0	0	0	1	1
0	0	1	1	0	1	0	0
0	1	0	0	1	0	0	0
1	0	0	0	1	0	0	1
1	0	0	1	1	0	1	0
1	0	1	0	1	0	1	1
1	0	1	1	1	1	0	0
1	1	0	0	0	0	0	0

Table S.39 Problem 6.2

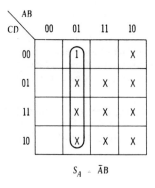

$S_A = \bar{A}B$

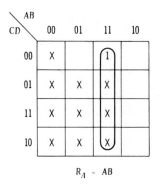

$R_A = AB$

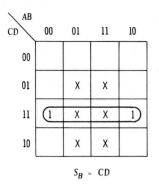

$S_B = CD$

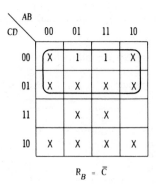

$R_B = \bar{C}$

Table S.39 Continued

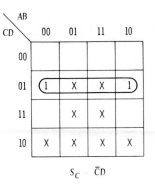

S_C $\bar{C}D$

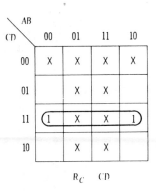

R_C CD

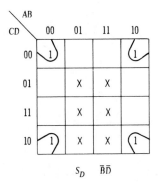

S_D $\bar{B}\bar{D}$

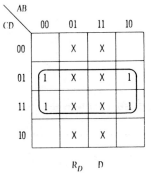

R_D D

Table S.40 Problem 6.3

Present state			Next state		
A	B	C	A_+	B_+	C_+
0	0	0	0	0	1
0	0	1	0	1	0
0	1	0	0	1	1
0	1	1	1	0	0
1	0	0	0	0	0

transition table; note that bistable A is set for every fifth input pulse. If this is used in conjunction with a divide-by-2 circuit, preferably preceding the divide-by-5, a divide-by-10 counter will result.

The K-maps for the JK bistable input conditions are shown in Table S.41 which yield the equations

$$J_A = BC \qquad K_A = 1$$
$$J_B = C \qquad K_B = C$$
$$J_C = \bar{A} \qquad K_C = 1$$

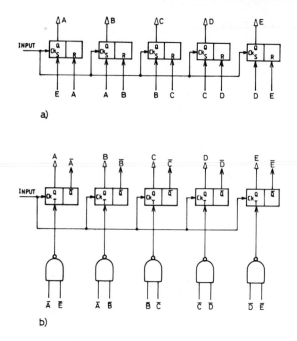

a)

b)

Figure S.26 Problem 6.1

The circuit is shown implemented in Fig. S.27(a), with the divide-by-10 counter in Fig. S.27(b). The waveform diagrams are shown in Fig. S.28 for a negative edge triggered bistable.

6.4 A block diagram of the system is shown in Fig. S.29(a). Because of the need to decode the outputs of the counter a synchronous counter should be used in preference to an asynchronous design. The transition table and the K-maps for a JK bistable implementation are given in Table S.42 which yields the circuit shown in Fig. S.29(b). The truth table for the decoder is given in Table S.43 which after minimization gives the equations

$$W = AB\bar{C} + A\bar{B}C \qquad\qquad X = \bar{A}B + BC + A\bar{B}\bar{C}$$
$$Y = \bar{A}B \qquad\qquad\qquad Z = AB + \bar{A}\bar{B}C$$

Note that in a practical system there are many other factors to be considered —the actual code used for the telephone system—the fact that the counter cycles continuously, but really needs to cycle once and then stop etc.

6.5 The state table for the counter is shown in Table S.44(a) with the K-maps for a JK realization in Table S.44(b); the final circuit is shown in Fig. S.30.

The state machine for the circuit is shown in Fig. S.31 from which the transition table given in Table S.45 can be obtained directly. The final circuit using D-type bistables is shown in Fig. S.32. Note that the equations for A_+ and B_+ can be simplified to

$$D_A = \bar{A} \qquad\text{and}\qquad D_B = \bar{B}\bar{A}X + \bar{B}A\bar{X} + B\bar{A}\bar{X} + BAX$$

Table S.41 Problem 6.3

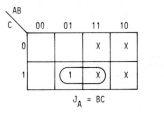

$$J_A = BC$$

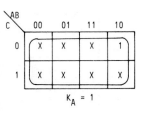

$$K_A = 1$$

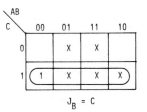

$$J_B = C$$

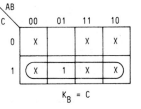

$$K_B = C$$

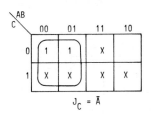

$$J_C = \bar{A}$$

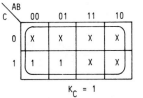

$$K_C = 1$$

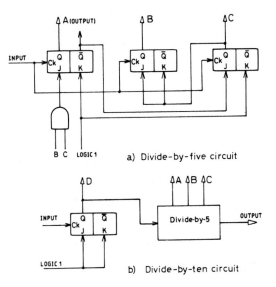

a) Divide–by–five circuit

b) Divide–by–ten circuit

Figure S.27 Problem 6.3

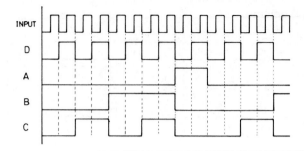

Figure S.28 Problem 6.3

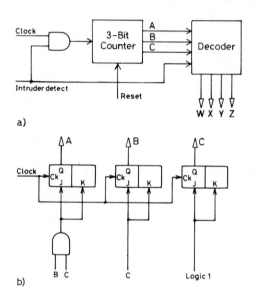

a)

b)

Figure S.29 Problem 6.4

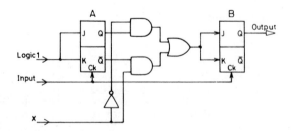

Figure S.30 Problem 6.5

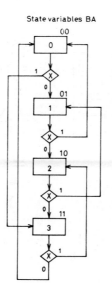

Figure S.31 Problem 6.5

Table S.42 Problem 6.4

A	B	C		A_+	B_+	C_+
0	0	0		0	0	1
0	0	1		0	1	0
0	1	0		0	1	1
0	1	1		1	0	0
1	0	0		1	0	1
1	0	1		1	1	0
1	1	0		1	1	1
1	1	1		0	0	0

(a) Transition table

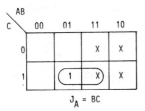

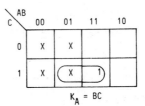

Table S.42 Continued

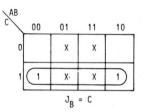

$J_B = C$

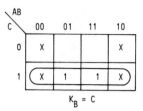

$K_B = C$

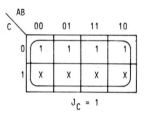

$J_C = 1$

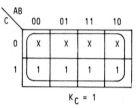

$K_C = 1$

(b) K-maps

Table S.43 Problem 6.4

A	B	C	W	X	Y	Z	Digit
0	0	0	0	0	0	0	0
0	0	1	0	0	0	1	1
0	1	0	0	1	1	0	6
0	1	1	0	1	1	0	6
1	0	0	0	1	0	0	4
1	0	1	1	0	0	0	8
1	1	0	1	0	0	1	9
1	1	1	0	1	0	1	5

Table S.44 Problem 6.5

Present state	Next state Input x		Present state		Next state Input x	
	x = 0	x = 1	B	A	0	1
0	1	3	0	0	01	11
1	2	0	0	1	10	00
2	3	1	1	0	11	01
3	0	2	1	1	00	10

(a)

Table S.44 Continued

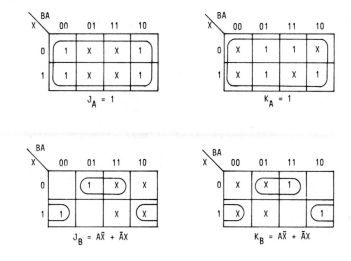

(b)

Table S.45 Problem 6.5

Present state		Input	Next state		Product terms
B	A	x	B_+	A_+	
0	0	0	0	1	$\bar{B}\bar{A}\bar{X}$
0	0	1	1	1	$\bar{B}\bar{A}X$
0	1	0	1	0	$\bar{B}A\bar{X}$
0	1	1	0	0	
1	0	0	1	1	$B\bar{A}\bar{X}$
1	0	1	0	1	$B\bar{A}X$
1	1	0	0	0	
1	1	1	1	0	BAX

In the case of a PLA implementation the product terms would probably be implemented directly since only six are required and these would easily fit into a standard module.

6.6 Figure S.33 shows the complete state diagram for the problem and Table S.46 shows the state table. Note that it is a single serial input, and therefore the only inputs to each state are 0 and 1; also, since it is a five-bit word, we must go back to the initial state after five bits have been examined. It is obvious that there are a large number of redundant states, for example 16, 23, 27, 29 30 and 31 are identical, also 17, 18, 20 and 24, and 19, 21, 22, 25, 26 and 28. If these are called states 16, 17 and 19 respectively, and the other entries replaced in the state table, we find that states (9, 10, 12) and (11, 13, 14) are also equivalent. Replacing these in the table gives (5, 6) equivalent, yielding the reduced state table shown in Table S.47.

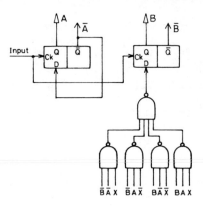

Figure S.32 Problem 6.5

Table S.46 Problem 6.6

| Present state | Input x | | Output | |
| | Next state | | | |
	0	1	0	1
1	2	3	0	0
2	4	5	0	0
3	6^5	7	0	0
4	8	9	0	0
5	10^9	11	0	0
6	12^9	13^{11}	0	0
7	14^{11}	15	0	0
8	16	17	0	0
9	18^{17}	19	0	0
10	20^{17}	21^{19}	0	0
11	22^{19}	23^{16}	0	0
12	24^{17}	25^{19}	0	0
13	26^{19}	27^{16}	0	0
14	28^{19}	29^{16}	0	0
15	30^{16}	31^{16}	0	0
16	1	1	0	0
17	1	1	0	1
18	1	1	0	1
19	1	1	1	0
20	1	1	0	1
21	1	1	1	0
22	1	1	1	0
23	1	1	0	0
24	1	1	0	1
25	1	1	1	0
26	1	1	1	0
27	1	1	0	0
28	1	1	1	0
29	1	1	0	0
30	1	1	0	0
31	1	1	0	0

Table S.47 Problem 6.6

Present state	Input x			
	Next state		Output	
	0	1	0	1
1	2	3	0	0
2	4	5	0	0
3	5	7	0	0
4	8	9	0	0
5	9	11	0	0
7	11	15	0	0
8	16	17	0	0
9	17	19	0	0
11	19	16	0	0
15	16	16	0	0
16	1	1	0	0
17	1	1	0	1
19	1	1	1	0

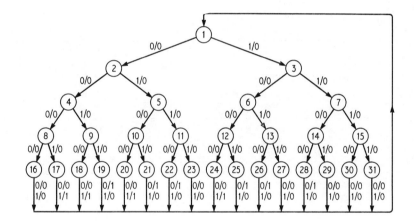

Figure S.33 Problem 6.6

6.7 The Mealy and Moore state diagram is shown in Fig. S.34 with the equivalent ASM chart in Fig. S.35. Note the external input (clock) does not need to be shown on the state machine.

6.8 With a little thought, this circuit could easily be designed intuitively; nevertheless, applying the theory could produce the state diagram shown in Fig. S.36(a), and the corresponding state table in Table S.48(a). This may easily be reduced since it is obvious that states (1, 2, 3, 5) and (4, 6, 7) are identical, giving the reduced state diagram and state table shown in Fig. S.36(b) and Table S.48(b) respectively. The state diagram is obviously that of a gated bistable circuit. Assigning the two states 0 and 1 gives the assigned state

Table S.48 Problem 6.8

	Inputs xy					Output Z			
Present state	Next state								
	00	01	11	10		00	01	11	10
1	2	3	4	5		0	0	1	0
2	2	3	4	5		0	0	1	0
3	2	3	4	5		0	0	1	0
4	6	3	4	7		1	0	1	1
5	2	3	4	5		0	0	1	0
6	6	3	4	7		1	0	1	1
7	6	3	4	7		1	0	1	1
(a)									

	Inputs xy					Output Z			
Present state	Next state								
	00	01	11	10		00	01	11	10
1	1	1	4	1		0	0	1	0
4	4	1	4	4		1	0	1	1
(b)									

	Inputs xy					Output Z			
Present state	Next state								
w	00	01	11	10		00	01	11	10
0	0	0	1	0		0	0	1	0
1	1	0	1	1		1	0	1	1
(c)									

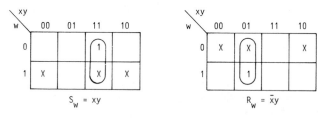

$S_w = xy$ $R_w = \bar{x}y$

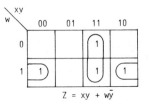

$Z = xy + w\bar{y}$

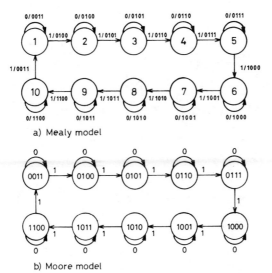

a) Mealy model

b) Moore model

Figure S.34 Problem 6.7

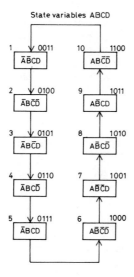

Figure S.35 Problem 6.7

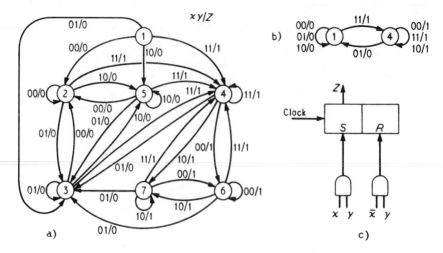

Figure S.36 Problem 6.8

diagram (Table S.48(c)); then, after extracting the input conditions for a set–reset bistable and plotting on a K-map (Table S.48(d)), we have

$$S_w = xy; \qquad R_w = \bar{x}y; \qquad Z = xy + w\bar{y}$$

The circuit is shown in Fig. S.36(c). Note that if a D-type bistable were used the setting condition would be the same as the required output Z and one gate could be saved.

6.9 This problem seems more complicated than it is; it is better to consider the serial full-adder as a sequential circuit with two inputs x and y and a sum output S_0 and allow the carry to be indigenous to the circuit. The first attempt at a state diagram may look something like Fig. S.37(a), but this soon reduces to a two-state diagram; the relevant state tables are shown in Table S.49. The problem has been solved in full, and it is interesting to note the familiar sum equations for the output, i.e.

$$S_0 = \bar{x}\bar{y}z + \bar{x}y\bar{z} + xyz + x\bar{y}\bar{z}$$

The carry logic is combined with the bistable circuit; this may easily be seen from the bistable equation, i.e.

$$z_+ = S + \bar{R}z = xy + (\overline{\bar{x}\bar{y}})z = xy + (x + y)z$$

Thus

$$z_+ = xy + xz + xy$$

Furthermore, note that this is an improvement on the conventional design which uses a combinational full-adder circuit, complete with carry logic, and a one-bit bistable store. In an actual design, the basic clock used for the serial system would also be used to gate the bistables (or input gates) in the synchronous circuit.

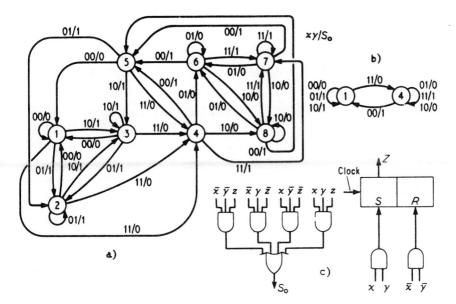

Figure S.37 Problem 6.9

Table S.49 Problem 6.9

Present state	Input xy					Output S			
	Next state								
	00	01	11	10		00	01	11	10
1	1	2	4	3		0	1	0	1
2	1	2	4	3		0	1	0	1
3	1	2	4	3		0	1	0	1
4	5	6	7	8		1	0	1	0
5	1	2	4	3		0	1	0	1
6	5	6	7	8		1	0	1	0
7	5	6	7	8		1	0	1	0
8	5	6	7	8		1	0	1	0
(a)									

Present state	Input xy					Output S			
	Next state								
	00	01	11	10		00	01	11	10
1	1	1	4	1		0	1	0	1
4	1	4	4	4		1	0	1	0
(b)									

	Input xy								
Present state	Next state					Output S			
z	00	01	11	10		00	01	11	10
0	0	0	1	0		0	1	0	1
1	0	1	1	1		1	0	1	0
(c)									

(d)

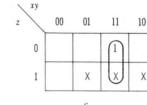

$$S_Z = xy$$

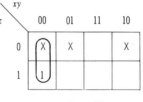

$$R_Z = \bar{x}\bar{y}$$

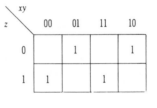

$$S = \bar{x}\bar{y}z + \bar{x}y\bar{z} + xyz + x\bar{y}\bar{z}$$

6.10 Mechanical switches do not make or break cleanly but 'bounce' giving a number of ON/OFF connections before finally settling down. Logic systems require a clean input and this may be obtained by placing the switch in the input circuit of an SR-bistable as shown in Fig. S.38(a). Note the need for pull-up resistors which would be in the order of 1kΩ for a V_{CC} of 5 volts. We now have a clean pulse but it is not synchronized with the clock; to effect this we can simply feed it to a D-type bistable; the waveforms are shown in Fig. S.38(b). Note that the duration of the output (Q) will be some multiple of the clock period and will depend on the input. We are now in a position to design the circuit which is required to produce a *single* clock pulse (SINP) from a clean synchronized pulse (SYNP).

The state machine for the circuit is shown in Fig. S.39. The algorithm looks for the start of a SYNP and when found generates a conditional output pulse (SINP) which lasts for one clock period (remember the ASM proceeds from state to state on each clock pulse). After this nothing will happen until SYNP goes down when the search starts afresh. Only two states are required which can be assigned 0 and 1; the transition table is given in Table S.50, which yields the equations, using D-type bistables,

$$A_D = \text{SYNP}$$
$$\text{SINP} = \bar{A}\ \text{SYNP}$$

The circuit is shown in Fig. S.40.

6.11 The ASM chart for this problem is shown in Fig. S.41(a). Assuming that sensor $S1$ is placed above $S2$ the input sequences are given by

$$\text{UP}\quad \overline{S1}\,\overline{S2} \rightarrow \overline{S1}S2 \rightarrow S1S2 \rightarrow S1\overline{S2} \rightarrow \overline{S1}\,\overline{S2}$$
$$\text{DOWN}\quad \overline{S1}\,\overline{S2} \rightarrow S1\overline{S2} \rightarrow S1S2 \rightarrow \overline{S1}S2 \rightarrow \overline{S1}\,\overline{S2}$$

Note that only balls with the required dimensions will generate $S1S2$; in all other cases the input will fall back to $\overline{S1}\,\overline{S2}$; for example, in the down case

$$\overline{S1}\,\overline{S2} \rightarrow S1\overline{S2} \rightarrow \overline{S1}\,\overline{S2} \rightarrow \overline{S1}S2 \rightarrow \overline{S1}\,\overline{S2}$$

The transition table is given in Table S.51 and the circuit shown implemented using D-type bistables in Fig. S.42. The machine can be considerably simplified by detecting the condition $S1S2$ (for balls with required diameter) and then generating the outputs according to whether $S1\overline{S2}$ (UP) or $\overline{S1}S2$ (DOWN); the state machine is shown in Fig. S.41(b).

6.12 From the transition table shown in Table S.51 we observe that a 16×4-bit word ROM is required and that only 11 out of the available 16 words need to be used. The layout of the ROM is shown in Table S.52; note that the next state and output values are mapped directly into the ROM. The circuit is shown in Fig. S.43 which uses a D-type bistable register to provide the necessary delay.

6.13 To generate a PRBS of length 15 we need a primitive polynomial of order 4, for example, $1 + x^3 + x^4$ with feedback via exclusive-OR gates from the third and fourth stages to the input. The circuit is shown in Fig. S.44 realized with D-type bistables. By analysing the successive states of the register (starting with 0001) the PRBS can be shown to be

$$100110101111000 \mid 100110 \cdots$$
$$\mid \text{repeats}$$

The state diagram is shown in Fig. S.45.

6.14 First we derive the input equations for the JK bistables; these are

$$Jy_1 = xy_2,\ Ky_1 = x$$
$$Jy_2 = x\bar{y}_1,\ Ky_2 = x$$

Then, using the characteristic equation for the JK bistable $Q_+ = J\bar{Q} + \bar{K}Q$, we compute the application and output equations:

$$Y_1 = \bar{y}_1 y_2 x + y_1 \bar{x}$$
$$Y_2 = \bar{y}_1 \bar{y}_2 x + y_2 \bar{x}$$
$$Z_1 = \bar{y}_1 \bar{y}_2$$
$$Z_2 = y_1 \bar{y}_2 x$$
$$Z_3 = y_1 y_2$$

The assigned state table based on these equation is shown in Table S.53(a) with the equivalent ASM chart in Fig. S.46.

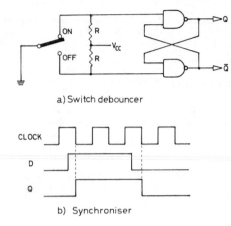

a) Switch debouncer

b) Synchroniser

Figure S.38 Problem 6.10

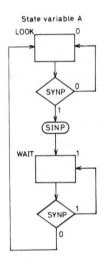

Figure S.39 Problem 6.10

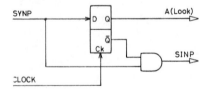

Figure S.40 Problem 6.10

State variables AB

State variable A

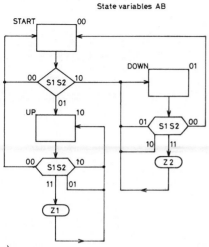

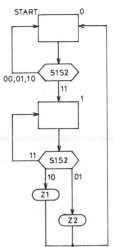

a)

b)

Figure S.41 Problem 6.11

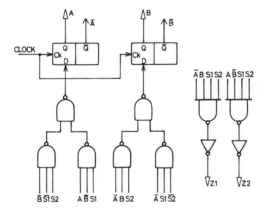

Figure S.42 Problem 6.11

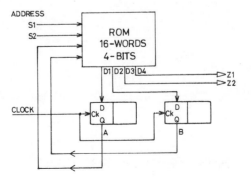

Figure S.43 Problem 6.12

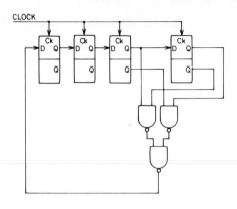

Figure S.44 Problem 6.13

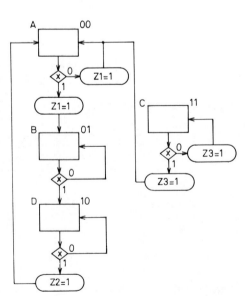

Figure S.45 Problem 6.13

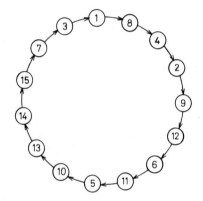

Figure S.46 Problem 6.14

Table S.50 Problem 6.10

Present state A	Input SYNP	Next state A_+	Output SINP
0	0	0	0
0	1	1	1
1	0	0	0
1	1	1	0

Table S.51 Problem 6.11

Present state		Input		Next state		Output state	
A	B	S1	S2	A_+	B_+	Z1	Z2
0	0	0	0	0	0	0	0
0	0	0	1	1	0	0	0
0	0	1	0	0	1	0	0
0	1	0	0	0	0	0	0
0	1	0	1	0	1	0	0
0	1	1	0	0	1	0	0
0	1	1	1	0	1	0	1
1	0	0	0	0	0	0	0
1	0	1	0	1	0	0	0
1	0	0	1	1	0	0	0
1	0	1	1	1	0	1	0

(a)

(b)

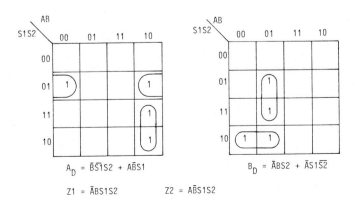

$A_D = \bar{B}\bar{S1}S2 + A\bar{B}S1$

$B_D = \bar{A}BS2 + \bar{A}S1\bar{S2}$

$Z1 = \bar{A}BS1S2$ $Z2 = A\bar{B}S1S2$

Table S.52 Problem 6.12

Address	ROM word			
$ABS1S2$	$D1$	$D2$	$D3$	$D4$
0	0	0	0	0
1	1	0	0	0
2	0	1	0	0
3	X	X	X	X
4	0	0	0	0
5	0	1	0	0
6	0	1	0	0
7	0	1	0	1
8	0	0	0	0
9	1	0	0	0
10	1	0	0	0
11	1	0	1	0
12	X	X	X	X
13	X	X	X	X
14	X	X	X	X
15	X	X	X	X

Table S.53 Problem 6.14

Present state		Input x Next state				Outputs					
		0		1		0			1		
y_1	y_2	y_1'	y_2'	y_1'	y_2'	$Z1$	$Z2$	$Z3$	$Z1$	$Z2$	$Z3$
A	0 0	0	0	0	1	1	0	0	1	0	0
B	0 1	0	1	1	0	0	0	0	0	0	0
C	1 1	1	1	0	0	0	0	1	0	0	1
D	1 0	1	0	0	0	0	0	0	0	1	0

(a)

Present state		Input	Next state		Output		
y_1	y_2	x	y_1'	y_2'	$Z1$	$Z2$	$Z3$
0	0	0	0	0	1	0	0
0	0	1	0	1	1	0	0
0	1	0	0	1	0	0	0
0	1	1	1	0	0	0	0
1	1	0	1	1	0	0	1
1	1	1	0	0	0	0	1
1	0	0	1	0	0	0	0
1	0	1	0	0	0	1	0

(b)

Note that the circuit has only three effective states A, B, D in normal operation; state C is not entered from any other state. When $x = 1$ the circuit operates as a mod-3 counter with $y_1 y_2$ giving the binary count.

To implement using D-type bistables and PLAs it is best to transform the state table into a transition table as shown in Table S.53(b). The product terms comprising the application equations may be read directly and minimized if necessary.

Chapter 7

7.1 From the state table in Table 7.32 we may say that

$$1 \neq 2, \; 1 \neq 5, \; 1 \neq 6, \; 2 \neq 3, \; 3 \neq 5, \; 3 \neq 6$$

These are the initial incompatibles and are easily deduced from the output states. Furthermore states $(2, 4)(4, 5)$ and $(4, 6)$ are identical. The implementation chart is shown in Table S.54(a); the equivalent states are

$$M = (1, 4)(2, 4)(4, 5)(4, 6)(5, 6)$$

Thus

$$M = (3)(1, 4)(2, 4)(4, 5, 6)$$

Note that this is a covering since the machine is incompletely specified; the reduced state table is shown in Table S.54(b). The state assignment is fairly simple for four states since there are only three possible codes. Using the technique of examining the origin of the next states (Table S.54(c)) we find that, if possible, all states should be adjacent! However, applying rule 2, we see that (CD) and (AD) occur as the next states of A, B and D, and thus we would choose these to be adjacent. The partition approach gives one non-trivial partition with the substitution property

$$P_1 = (AD)(C)(B)$$

This means using three bits (three bistables) for the allocation, i.e. two bits to distinguish the blocks and one bit for the elements. However, it is possible that the inclusion of an additional bistable could reduce the amount of combinational logic required and this approach should be investigated. The straightforward allocation of the four states using two bits is shown in Tables S.54(d) and S.55.

7.2 The state table was reduced intuitively in the actual problem 6.6; this should be repeated using the implication chart. The next step is to allocate the internal states and, since there are 13 states, four bits are required. The simple next-state examination is shown in Table S.56(a) with the results of a full adjacency analysis in Table S.56(b).

Using the partition approach the non-trivial partitions are

$$
\begin{aligned}
P_1 &= (1)(2, 3)(4, 5, 7)(8, 9, 11, 15)(16, 17, 19)\\
P_2 &= (1)(2)(3)(4, 5)(7)(8, 9, 11)(15)(16, 17, 19)\\
P_3 &= (1)(2)(3)(4, 7)(5)(8, 11)(9, 15)(16, 17, 19)\\
P_4 &= (1)(2)(3)(4)(5, 7)(8)(9, 11, 15)(16, 17, 19)\\
P_5 &= (1)(2)(3)(4)(5)(7)(8, 9)(16, 17, 19)(11)(15)\\
P_6 &= (1)(2)(3)(4)(5)(7)(9)(8, 11)(16, 17, 19)(15)\\
P_7 &= (1)(2)(3)(4)(5)(7)(8, 15)(9)(11)(16, 17)(19)\\
P_8 &= (1)(2)(3)(4)(5)(7)(8)(9, 11)(15)(16, 17, 19)\\
P_9 &= (1)(2)(3)(4)(5)(7)(8)(9, 15)(11)(16, 17, 19)\\
P_{10} &= (1)(2)(3)(4)(5)(7)(8)(9)(11, 15)(16, 17, 19)\\
P_{11} &= (1)(2)(3)(4)(5)(7)(8)(9)(11)(15)(16, 17)(19)\\
P_{12} &= (1)(2)(3)(4)(5)(7)(8)(9)(11)(15)(16, 19)(17)\\
P_{13} &= (1)(2)(3)(4)(5)(7)(8)(9)(11)(15)(16)(17, 19)
\end{aligned}
$$

Since to implement any of these partitions directly requires at least five bits, we shall once again use the adjacent next-state approach. Table S.57 shows a possible assignment with $(2, 3)(4, 5)(5, 7)(8, 15)(8, 9)(11, 15)(16, 17)$

Table S.54 Problem 7.1

(a)

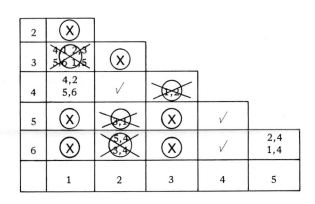

Present state	Inputs x_1x_2				Output T			
	Next state							
	00	01	11	10	00	01	11	10
$A(1, 4)$	C	C	D	A	1	1	0	1
$B(3)$	A	B	D	D	1	1	0	1
$C(2, 4)$	C	D	D	B	0	1	1	1
$D(4, 5, 6)$	C	C	D	A	0	1	1	1

(b)

Next state	Present state
A	A, B, D
B	B, C
C	A, C, D
D	A, B, C, D

(c)

Present state		Inputs x_1x_2				Output T			
y	z	Next state							
		00	01	11	10	00	01	11	10
A 0	0	11	11	10	00	1	1	0	1
B 0	1	00	01	10	10	1	1	0	1
C 1	1	11	10	10	01	0	1	1	1
D 1	0	11	11	10	00	0	1	1	1

(d)

Table S.55 Problem 7.1

(a)

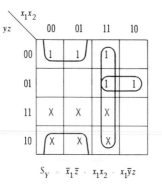

$$S_Y = \bar{x}_1 \bar{z} \cdot x_1 x_2 \cdot x_1 \bar{y} z$$

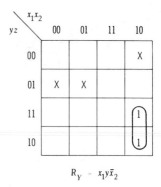

$$R_Y = x_1 y \bar{x}_2$$

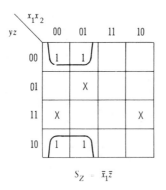

$$S_Z = \bar{x}_1 \bar{z}$$

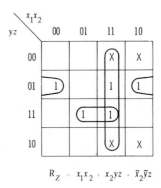

$$R_Z = x_1 x_2 \cdot x_2 y z \cdot \bar{x}_2 \bar{y} z$$

(b)

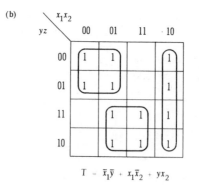

$$T = \bar{x}_1 \bar{y} + x_1 \bar{x}_2 + y x_2$$

Table S.56 Problem 7.2

New state	Present state
1	16, 17, 19
2	1
3	1
4	2
5	2, 3
7	3
8	4
9	4, 5
11	5, 7
15	7
16	8, 11, 15
17	8, 9
19	9, 11

(a) Next-state conditions

State-pair	Weights Type 1	Type 2	Type 3	Type 4	Total
2, 3	0	1	2	0	3
4, 5	0	1	2	0	3
5, 7	0	1	2	0	3
8, 9	0	1	2	0	3
8, 15	4	0	2	0	6
9, 11	0	1	2	0	3
11, 15	4	1	2	0	7
16, 17	8	1	1	0	10
16, 19	8	1	1	0	10
17, 19	8	1	1	0	10

(All other state-pairs score 2 or less)

(b) Adjacency analysis

$(9, 11)(17, 19)$ allocated as adjacent states. The K-maps for the input equations of the JK bistables are shown in Table S.58 and yield

$$J_A = C\bar{D} + \bar{C}Bx + B\bar{C}D + \bar{C}Dx$$
$$K_A = \bar{C}\bar{D} + \bar{C}Dx + \bar{B}x + BCD\bar{x}$$
$$J_B = A\bar{x} + \bar{C}x + \bar{C}D + C\bar{D}$$
$$K_B = \bar{C}Dx + \bar{A}CD\bar{x} + A\bar{C}\bar{x}$$
$$J_C = D$$
$$K_C = \bar{D}$$
$$J_D = \bar{A}\bar{C}$$
$$K_D = C$$
$$T = AC\bar{D}\bar{x} + \bar{A}BC\bar{D}x$$

The logic diagram is shown in Fig. S.47.

Table S.57 Problem 7.2

	Present state A B C D				Input x Next state 0	 1	Output T 0	 1
1	1	1	0	0	0000	0100	0	0
2	0	0	0	0	0001	0101	0	0
3	0	1	0	0	0101	1101	0	0
4	0	0	0	1	0111	1111	0	0
5	0	1	0	1	1111	1011	0	0
7	1	1	0	1	1011	0011	0	0
8	0	1	1	1	0010	0110	0	0
9	1	1	1	1	0110	1110	0	0
11	1	0	1	1	1110	0010	0	0
15	0	0	1	1	0010	0010	0	0
16	0	0	1	0	1100	1100	0	0
17	0	1	1	0	1100	1100	0	1
19	1	1	1	0	1100	1100	1	0
Unused	1	0	0	0				
	1	0	0	1				
	1	0	1	0				

Table S.58 Problem 7.2

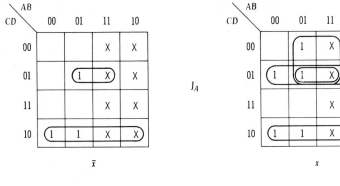

J_A

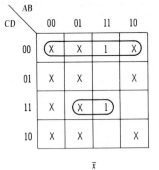

K_A

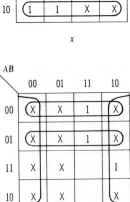

Table S.58 Continued

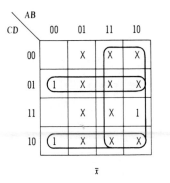

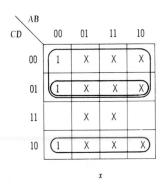

J_B

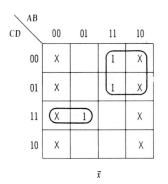

K_B

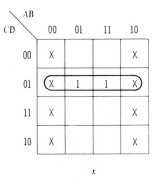

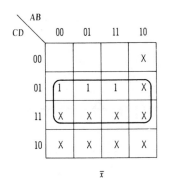

J_C

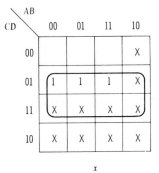

Table S.58 Continued

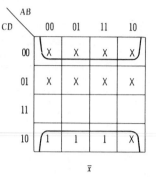

$\bar{x}$

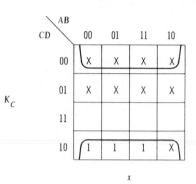

K_C

x

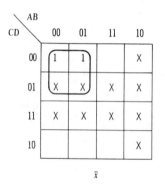

$\bar{x}$

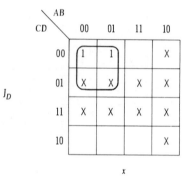

J_D

x

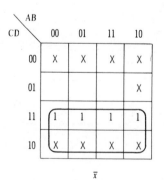

$\bar{x}$

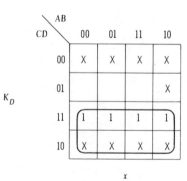

K_D

x

Table S.58 Continued

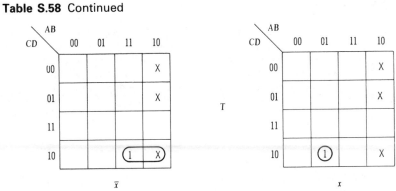

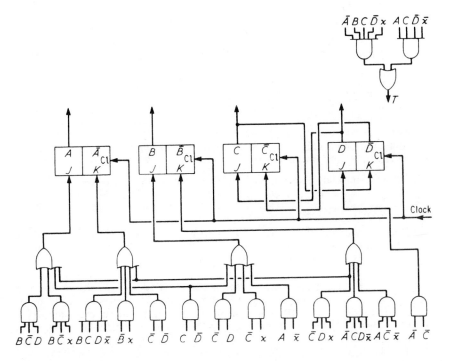

Figure S.47 Problem 7.2

7.3 The reduced state table for this problem is shown in Table S.59(a) and the equivalent state diagram in Fig. S.48. The state assignment is difficult to optimize since there are no non-trivial SP partitions and the adjacency analysis shown in Table S.60 shows no obvious best coding. A good starting point would be to make $(1, 2)(2, 3)(3, 4)(4, 1)(1, 6)(6, 5)(2, 5)$ adjacent giving a score of 81, and this has been done in the assigned state table S.59(b). The K-maps for the JK bistable are shown in Table S.61 and give the following input and output equations:

$$Jy_1 = \bar{x}_1 x_2 y_2 \bar{y}_3 \qquad Ky_1 = x_1 + \bar{y}_2 + x_2 y_1$$
$$Jy_2 = x_1 \bar{y}_3 + x_1 x_2 + y_1 x_2 \qquad Ky_2 = y_3 + \bar{x}_1 \bar{x}_2 + \bar{x}_1 y_1$$
$$Jy_3 = \bar{x}_1 x_2 y_1 \bar{y}_2 \qquad Ky_3 = \bar{x}_1 y_2 + x_2 \bar{y}_2$$
$$T = x_1 y_2 y_3 + \bar{y}_2 \bar{x}_2 y_3$$

7.4 The reduced state table is shown in Table S.62 and the state diagram in Fig. S.49; note that the circuit detects *all* sequences of five digits. There are three non-trivial partitions for this machine, i.e.

$$P_1 = (1)(2)(3)(4, 5)$$
$$P_2 = (1)(2)(3, 4, 5)$$
$$P_3 = (1)(2, 3, 4, 5)$$

Table S.62(c) shows the assigned state table using partition P_3; the K-maps for JK bistables are shown in Table S.63 yielding the input and output equations

$$Jy_1 = \bar{x}_1 x_2 + x_1 \bar{x}_2 \qquad Ky_1 = \bar{x}_1 \bar{x}_2 + x_1 x_2$$
$$Jy_2 = y_3(\bar{x}_1 \bar{x}_2 + x_1 x_2) \qquad Ky_2 = \bar{x}_1 x_2 + x_1 \bar{x}_2$$
$$Jy_3 = \bar{y}_1 \bar{y}_2(\bar{x}_1 \bar{x}_2 + x_1 x_2) \qquad Ky_3 = (\bar{x}_1 x_2 + x_1 \bar{x}_2) + y_2$$
$$T = \bar{y}_3 y_2(\bar{x}_1 \bar{x}_2 + x_1 x_2)$$

Note the common terms in the equations which give an economical circuit configuration (see Fig. S.50). Furthermore, as one would expect, the exclusive-OR circuit is very prominent in these equations.

7.5 The machine, let us call it M, has two non-trivial SP partitions:

$$\pi_1 = (1, 2, 5, 6)(3, 7, 4, 8)$$
$$\pi_2 = (1, 2)(3, 4)(5, 6)(7, 8)$$

Since $\pi_1 . \pi_2 \neq \pi(0)$ and $\pi_1 > \pi_2 > \pi(0)$ a serial decomposition with three submachines M_1, M_2 and M_3 is possible. The state table and assignment for machine M_1 is shown in Table S.64(a); note that it is independent of x.

Machine M_2 is given by $\pi_1 . r_1 = \pi_2$; one possible partition is $r_1 = (1, 2, 3, 4)$ $(5, 6, 7, 8)$. The state table for the machine based on r_1 is shown in Table S.64(b); again it is independent of x. The tail machine M_3 is obtained from the relationship $r_2 . \pi_2 = \pi(0)$ and if possible r_2 should be output consistent. $r_2 = (2, 4, 6, 8)(1, 3, 5, 7)$ satisfies this condition. The state table and coding for machine M_3 together with the system output Z_0 are shown in Table S.64(c). The K-maps for M_3 and the output logic are given in Table S.65(a).

The full equations for the system are

Machine M_1 $Y_1 = \bar{y}_1$, $Z_1 = y_1$
Machine M_2 $Y_2 = Z_1 \bar{y}_2 + \bar{Z}_1 \bar{y}_2$, $Z_2 = y_2$
Machine M_3 $Y_3 = \bar{Z}_1 \bar{Z}_2 \bar{x} \bar{y}_3 + \bar{Z}_1 \bar{Z}_2 x y_3 + \bar{Z}_1 Z_2 \bar{x} y_3 + \bar{Z}_1 Z_2 x \bar{y}_3$
$$+ Z_1 \bar{Z}_2 x \bar{y}_3 + Z_1 \bar{Z}_2 x y_3 + Z_1 \bar{Z}_2 \bar{x} \bar{y}_3$$
$$Z_0 = x \bar{y}_3$$

Note that though the choice of an output-consistent partition r_2 has reduced the

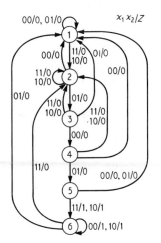

Figure S.48 Problem 7.3

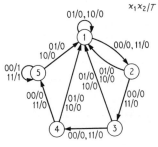

Figure S.49 Problem 7.4

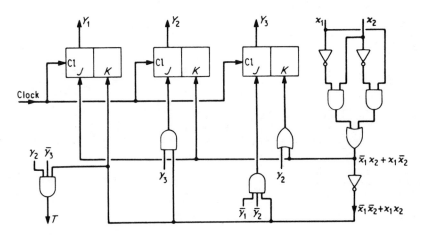

Figure S.50 Problem 7.4

Table S.59 Problem 7.3

	Inputs x_1x_2								
Present state	Next state					Output			
	00	01	11	10		00	01	11	10
1	1	1	2	2		0	0	0	0
2	1	3	2	2		0	0	0	0
3	4	1	2	2		0	0	0	0
4	1	5	2	2		0	0	0	0
5	1	1	6	6		0	0	1	1
6	6	1	2	6		1	0	0	1

(a)

				Inputs x_1x_2				Output T			
	Present state			Next state							
	y_1	y_2	y_3	00	01	11	10	00	01	11	10
1	0	0	0	000	000	010	010	0	0	0	0
2	0	1	0	000	110	010	010	0	0	0	0
3	1	1	0	100	000	010	010	0	0	0	0
4	1	0	0	000	011	010	010	0	0	0	0
5	0	1	1	000	000	001	001	0	0	1	1
6	0	0	1	001	000	010	001	1	0	0	1
Unused	1	0	1								
	1	1	1								

(b)

Table S.60 Problem 7.3

Occurrences/weights

State pair	Type 1	Type 2	Type 3	Type 4	Total weight
1, 2	3/9	6/6	4/4	0/0	19
1, 3	3/9	1/1	4/4	0/0	14
1, 4	3/9	1/1	4/4	0/0	14
1, 5	2/6	1/1	2/2	0/0	9
1, 6	2/6	3/3	2/2	1/2	13
2, 3	2/6	1/1	4/4	0/0	11
2, 4	3/9	1/1	4/4	0/0	14
2, 5	1/3	1/1	2/2	0/0	6
2, 6	1/3	1/1	2/2	1/2	8
3, 4	2/6	0/0	4/4	0/0	10
3, 5	1/3	0/0	2/2	0/0	5
3, 6	2/6	0/0	2/2	0/0	8
4, 5	1/3	0/0	2/2	0/0	5
4, 6	1/3	0/0	2/2	0/0	5
5, 6	2/6	0/0	2/2	0/0	8

Table S.61 Problem 7.3

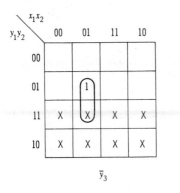

Jy_1

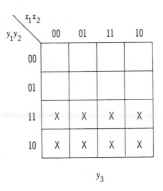

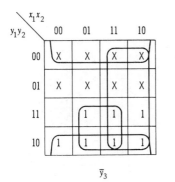

Ky_1

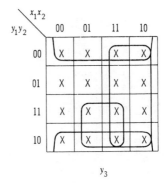

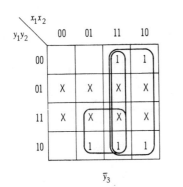

Jy_2

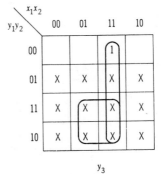

Table S.61 Continued

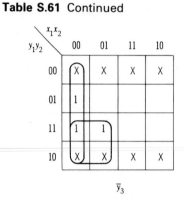

$\bar{y}_3$

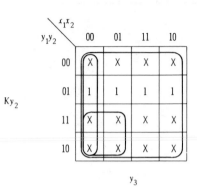

Ky_2

y_3

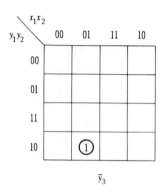

Jy_3

$\bar{y}_3$

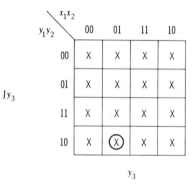

y_3

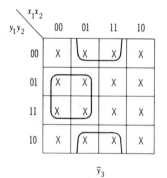

Ky_3

$\bar{y}_3$

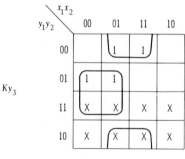

y_3

Table S.61 Continued

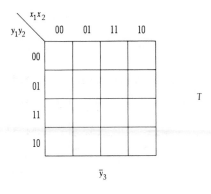

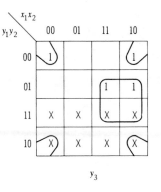

Table S.62 Problem 7.4

Present state	Inputs x_1x_2								
	Next state					Output T			
	00	01	11	10		00	01	11	10
1	2	1	2	1		0	0	0	0
2	3	1	3	1		0	0	0	0
3	4	1	4	1		0	0	0	0
4	5	1	5	1		0	0	0	0
5	5	1	5	1		1	0	1	0

(a)

Next state	Present state
1	1, 2, 3, 4, 5
2	1
3	2
4	3
5	4, 5

(b)

	Present state y_1 y_2 y_3			Inputs x_1x_2					Output T			
				Next state								
	y_1	y_2	y_3	00	01	11	10		00	01	11	10
1	1	0	0	000	100	000	100		0	0	0	0
2	0	0	0	001	100	001	100		0	0	0	0
3	0	0	1	011	100	011	100		0	0	0	0
4	0	1	1	010	100	010	100		0	0	0	0
5	0	1	0	010	100	010	100		1	0	1	0
Unused	1	0	1									
	1	1	0									
	1	1	1									

(c)

Table S.63 Problem 7.4

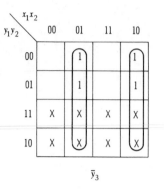

$\bar{y}_3$

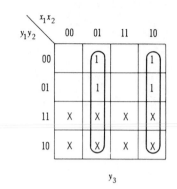

Jy_1

y_3

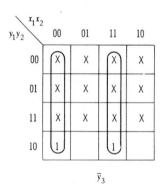

$\bar{y}_3$

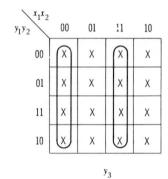

Ky_1

y_3

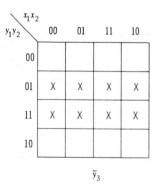

$\bar{y}_3$

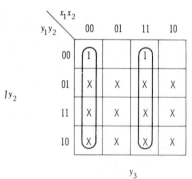

Jy_2

y_3

Table S.63 Continued

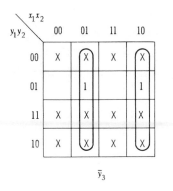

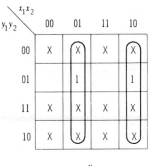

Ky_2

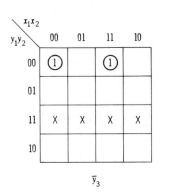

Jy_3

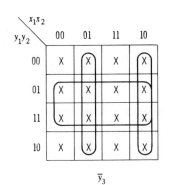

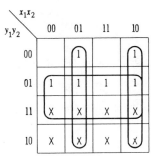

Ky_3

Table S.63 Continued

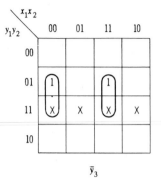
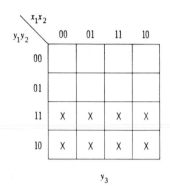

$\bar{y}_3$ y_3

Table S.64 Problem 7.5

Present state	Input x Next state		Output
	0	1	$Z1$
$A(1, 2, 5, 6)$	B	B	0
$B(3, 7, 4, 8)$	A	A	1

(a) Machine $M1$

Present state y_1	Next state y_1'	Output $Z1$
0	1	0
1	0	1

Present state	Input $Z1x$ Next state				Output
	00	01	11	10	$Z2$
$C(1, 2, 3, 4)$	C	C	D	D	0
$D(5, 6, 7, 8)$	D	D	C	C	1

(b) Machine $M2$

Present state y_2	Input $Z1$ Next state		Output $Z2$
	0	1	
0	0	1	0
1	1	0	1

Present state	Input $Z1Z2x$ Next state/$Z0$							
	000	001	010	011	111	110	101	100
$E(2, 4, 6, 8)$	$F/0$	$E/1$	$E/0$	$F/1$	$E/1$	$F/0$	$E/1$	$E/0$
$F(1, 3, 5, 7)$	$E/0$	$F/0$	$F/0$	$E/0$	$F/0$	$E/0$	$E/0$	$F/0$

Present state y_3	Input $Z1Z2x$ Next state							
	000	001	010	011	111	110	101	100
0	1	0	0	1	0	1	0	0
1	0	1	1	0	1	0	0	1

(c) Machine $M3$

Table S.65 Problem 7.5

(a)

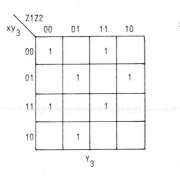 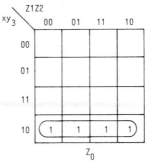

State	Y_1	Y_2	Y_3
1	0	0	1
2	0	0	0
3	1	0	1
4	1	0	0
5	0	1	1
6	0	1	0
7	1	1	1
8	1	1	0

(b)

output logic the M_3 equations are irreducible. The equivalent states are shown in Table S.65(b). Notice how the assignment corresponds to the partition π_2, r_1 and r_2 and that the system must be set to the starting state of 001.

7.6 The circuit equations are

$$Y_1 = \bar{y}_1 \bar{y}_2 + \bar{y}_2 \bar{y}_3 y_4 + \bar{y}_1 \bar{y}_4$$
$$Y_2 = y_1 \bar{y}_2 \bar{y}_4 + \bar{y}_1 y_2 \bar{y}_4 + y_1 y_2 y_4 + y_2 y_3 y_4 + y_1 y_3 y_4$$
$$Y_3 = y_3 y_4 + \bar{y}_2 y_3 + \bar{y}_1 y_3 + y_1 y_2 \bar{y}_3 \bar{y}_4$$
$$Y_4 = y_4$$

The full canonical expansions are

$$Y_1 = \bar{y}_4 \bar{y}_3 \bar{y}_2 \bar{y}_1 + \bar{y}_4 \bar{y}_3 y_2 \bar{y}_1 + \bar{y}_4 y_3 \bar{y}_2 \bar{y}_1 + \bar{y}_4 y_3 y_2 \bar{y}_1 + y_4 \bar{y}_3 \bar{y}_2 \bar{y}_1$$
$$\qquad + y_4 \bar{y}_3 \bar{y}_2 y_1 + y_4 \bar{y}_3 y_2 \bar{y}_1$$
$$Y_2 = \bar{y}_4 \bar{y}_3 \bar{y}_2 y_1 + \bar{y}_4 \bar{y}_3 y_2 \bar{y}_1 + \bar{y}_4 y_3 \bar{y}_2 y_1 + \bar{y}_4 y_3 y_2 \bar{y}_1 + y_4 \bar{y}_3 y_2 y_1$$
$$\qquad + y_4 y_3 \bar{y}_2 y_1 + y_4 y_3 y_2 \bar{y}_1 + y_4 y_3 y_2 y_1$$
$$Y_3 = \bar{y}_4 \bar{y}_3 y_2 y_1 + \bar{y}_4 y_3 \bar{y}_2 y_1 + \bar{y}_4 y_3 \bar{y}_2 \bar{y}_1 + \bar{y}_4 y_3 y_2 \bar{y}_1 + y_4 \bar{y}_3 \bar{y}_2 y_1$$
$$\qquad + y_4 \bar{y}_3 y_2 y_1 + y_4 y_3 \bar{y}_2 y_1 + y_4 y_3 y_2 y_1$$
$$Y_4 = y_4 \bar{y}_3 \bar{y}_2 \bar{y}_1 + y_4 \bar{y}_3 y_2 \bar{y}_1 + y_4 \bar{y}_3 y_2 \bar{y}_1 + y_4 y_3 \bar{y}_2 \bar{y}_1 + y_4 y_3 \bar{y}_2 \bar{y}_1$$
$$\qquad + y_4 y_3 \bar{y}_2 y_1 + y_4 y_3 y_2 \bar{y}_1 + y_4 y_3 y_2 y_1$$

Thus the T matrix is

$$y_4y_3y_2y_1 = \begin{array}{cccccccccccccccc} 0 & 1 & 2 & 3 & 4 & 5 & 6 & 7 & 8 & 9 & 10 & 11 & 12 & 13 & 14 & 15 \end{array}$$

$$T = \begin{bmatrix} 1 & 0 & 1 & 0 & 1 & 0 & 1 & 0 & 1 & 1 & 0 & 0 & 1 & 0 & 0 & 0 \\ 0 & 1 & 1 & 0 & 0 & 1 & 1 & 0 & 0 & 0 & 0 & 1 & 0 & 1 & 1 & 1 \\ 0 & 0 & 0 & 1 & 1 & 1 & 1 & 0 & 0 & 0 & 0 & 0 & 1 & 1 & 1 & 1 \\ 0 & 0 & 0 & 0 & 0 & 0 & 0 & 1 & 1 & 1 & 1 & 1 & 1 & 1 & 1 & 1 \end{bmatrix}$$

and the A matrix is

$$A = \begin{bmatrix} 0 & 1 & 0 & 1 & 0 & 1 & 0 & 1 & 0 & 1 & 0 & 1 & 0 & 1 & 0 & 1 \\ 0 & 0 & 1 & 1 & 0 & 0 & 1 & 1 & 0 & 0 & 1 & 1 & 0 & 0 & 1 & 1 \\ 0 & 0 & 0 & 0 & 1 & 1 & 1 & 1 & 0 & 0 & 0 & 0 & 1 & 1 & 1 & 1 \\ 0 & 0 & 0 & 0 & 0 & 0 & 0 & 0 & 1 & 1 & 1 & 1 & 1 & 1 & 1 & 1 \end{bmatrix}$$

Now we can say that

$$F = \begin{bmatrix} Y_1 \\ Y_2 \\ Y_3 \\ Y_4 \end{bmatrix} = \begin{bmatrix} y_1 \\ y_2 \\ y_3 \\ y_4 \end{bmatrix} \begin{bmatrix} 1 & 0 & 1 & 0 & 1 & 0 & 1 & 0 & 1 & 1 & 0 & 0 & 1 & 0 & 0 & 0 \\ 0 & 1 & 1 & 0 & 0 & 1 & 1 & 0 & 0 & 0 & 0 & 1 & 0 & 1 & 1 & 1 \\ 0 & 0 & 0 & 1 & 1 & 1 & 1 & 0 & 0 & 0 & 0 & 0 & 1 & 1 & 1 & 1 \\ 0 & 0 & 0 & 0 & 0 & 0 & 0 & 1 & 1 & 1 & 1 & 1 & 1 & 1 & 1 & 1 \end{bmatrix}$$

Thus $F = BT$.
 Starting with

we have

$$F = \begin{bmatrix} 0 & 1 & 0 & 1 & 0 & 1 & 0 & 1 & 0 & 1 \\ 0 & 0 & 1 & 1 & 0 & 0 & 1 & 1 & 0 & 0 \\ 0 & 0 & 0 & 0 & 1 & 1 & 1 & 1 & 0 & 0 \\ 0 & 0 & 0 & 0 & 0 & 0 & 0 & 0 & 0 & 0 \end{bmatrix} \cdots$$

Thus the machine will cycle 0–7 and repeat; the same operation will apply to any input from

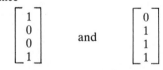

Since

are characteristic vectors of T, whenever this input condition is reached the machine will go into a loop, i.e. starting with

$$\begin{bmatrix} y_1 \\ y_2 \\ y_3 \\ y_4 \end{bmatrix} = \begin{bmatrix} 0 \\ 0 \\ 0 \\ 1 \end{bmatrix}$$

we have

$$F = \begin{bmatrix} 0 & 1 & 1 \\ 0 & 0 & 0 \\ 0 & 0 & 0 \\ 1 & 1 & 1 \end{bmatrix} \cdots$$

This action may be described diagrammatically (see Fig. S.51); decimal notation is used for the vectors of T, i.e.

$$\begin{bmatrix} 0 \\ 0 \\ 0 \\ 1 \end{bmatrix} \equiv 8$$

To find the output after alternate clock pulses we must use the T^2 matrix, i.e.

$$T^2 = \begin{bmatrix} 0 & 1 & 0 & 1 & 0 & 1 & 0 & 1 & 1 & 1 & 1 & 0 & 0 & 0 & 0 & 0 \\ 1 & 1 & 0 & 0 & 1 & 1 & 0 & 0 & 0 & 0 & 0 & 0 & 1 & 1 & 1 & 1 \\ 0 & 0 & 1 & 1 & 1 & 1 & 0 & 0 & 0 & 0 & 0 & 0 & 1 & 1 & 1 & 1 \\ 0 & 0 & 0 & 0 & 0 & 0 & 0 & 0 & 1 & 1 & 1 & 1 & 1 & 1 & 1 & 1 \end{bmatrix}$$

This is obtained by looking up the columns of T in the A matrix and noting the corresponding column numbers; these are then used to select the columns of T which form the T^2 matrix.

We may now use the T^2 matrix to analyse the circuit, i.e.

$$\text{if} \quad \begin{bmatrix} y_1 \\ y_2 \\ y_3 \\ y_4 \end{bmatrix} = \begin{bmatrix} 0 \\ 0 \\ 0 \\ 0 \end{bmatrix} \quad \text{then } F = \begin{bmatrix} 0 & 0 & 0 & 0 & 0 & 0 \\ 0 & 1 & 0 & 1 & 0 & 1 \\ 0 & 0 & 1 & 1 & 0 & 0 \\ 0 & 0 & 0 & 0 & 0 & 0 \end{bmatrix} \cdots \text{etc.}$$

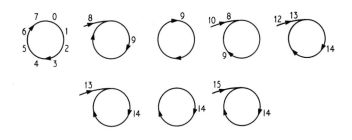

Figure S.51 Problem 7.6

Chapter 8

8.1 The state diagram for this circuit is shown in Fig. S.52, with the primitive flow table in Table S.66(a). The table cannot be reduced but rows (1, 2) and (4, 5, 6) can be merged; the merged table is shown in Table S.66(b). The merged flow table contains only three rows, but since we require all three to be adjacent, it is necessary to use the spare secondary state d' (Table S.66(c) and (d)). The derivation of the excitation and output maps is quite standard (note the elimination of the hazard in the Y_2 map, i.e. the $x_1 y_2$ term) and gives the equations

$$Y_1 = y_1 \bar{y}_2 + x_1 x_2 \bar{y}_2$$
$$Y_2 = x_1 \bar{x}_2 + y_1 \bar{x}_1 + y_2 x_2 + x_1 y_2$$
$$Z = y_1$$

A NAND logic circuit is shown in Fig. S.53.

8.2 The timing diagram for this problem is shown in Fig. S.54, and the primitive flow table in Table S.67(a). As is typically found with counter circuits, the flow table cannot be reduced or merged. Furthermore, in this particular case, the state allocation can be made identical to the required output states, i.e. Gray code (see Table S.67(b) and (c)). From the maps we can extract the following equations:

$$Y_1 = xy_1 + y_1 y_3 + \bar{x} y_2 \bar{y}_3 + y_1 y_2$$
$$Y_2 = y_2 \bar{y}_3 + xy_2 + \bar{x} \bar{y}_1 y_3 + \bar{y}_1 y_2$$
$$Y_3 = \bar{x} y_3 + x \bar{y}_1 \bar{y}_2 + xy_1 y_2 + \bar{y}_1 \bar{y}_2 y_3 + y_1 y_2 y_3$$

Note that the hazard terms are $y_1 y_2$, $\bar{y}_1 y_2$, $y_1 \bar{y}_2 y_3$ and $y_1 y_2 y_3$. The logic diagram for the counter is shown in Fig. S.54(b). Using the alternative method of extracting the set−reset equation for a dc SR bistable, we rearrange the assigned flow table as shown in Table S.68(a); the K-maps for the input conditions are shown in Table S.68(b) and give the following equations:

$$Sy_1 = \bar{x} y_2 \bar{y}_3 \qquad Ry_1 = \bar{x} \bar{y}_2 \bar{y}_3$$
$$Sy_2 = \bar{x} \bar{y}_1 y_3 \qquad Ry_2 = \bar{x} y_1 y_3$$
$$Sy_3 = x \bar{y}_1 \bar{y}_2 + xy_1 y_2 \qquad Ry_3 = xy_1 \bar{y}_2 + x \bar{y}_1 y_2$$

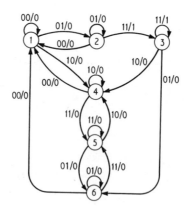

Figure S.52 Problem 8.1

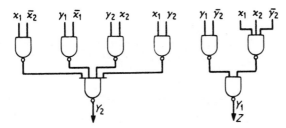

Figure S.53 Problem 8.1

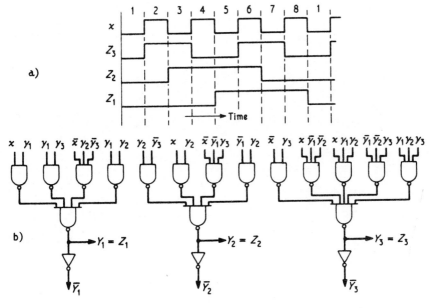

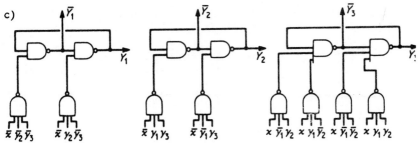

Figure S.54 Problem 8.2

Table S.66 Problem 8.1

Inputs x_1x_2 00	01	11	10	Output Z
①	2	–	4	0
1	②	3	–	0
–	6	③	4	1
1	–	5	④	0
–	6	⑤	4	0
1	⑥	5	–	0

(a)

Inputs x_1x_2 00	01	11	10	Output Z	
①	②	3	4	0	a
–	6	③	4	1	b
1	⑥	⑤	④	0	c

(b)

(c)

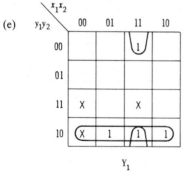

(d)

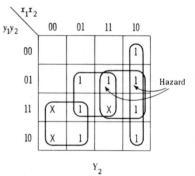

(e)

Y_1 Y_2

(f)

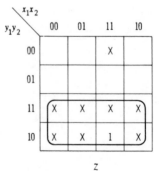

Z

Table S.67 Problem 8.2

Input x 0	1	Output Z_1	Z_2	Z_3	
①	2	0	0	0	a
3	②	0	0	1	b
③	4	0	1	1	c
5	④	0	1	0	d
⑤	6	1	1	0	e
7	⑥	1	1	1	f
⑦	8	1	0	1	g
1	⑧	1	0	0	h

(a)

(b)

y_3 \ y_1y_2	00	01	11	10
0	a	d	e	h
1	b	b	f	g

(c)

y_2y_3 \ xy_1	00	01	11	10
00	① 000	1 000	⑧ 100	2 001
01	3 011	⑦ 101	8 100	② 001
11	③ 011	7 101	⑥ 111	4 010
10	5 110	⑤ 110	6 111	④ 010

(d)

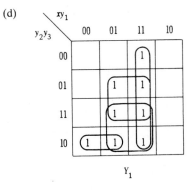

Y_1

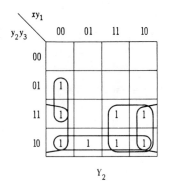

Y_2

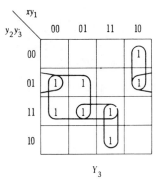

Y_3

Table S.68 Problem 8.2

Present state y_1 y_2 y_3	Input x Next state 0	1
0 0 0	000	001
0 0 1	011	001
0 1 1	011	010
0 1 0	110	010
1 1 0	110	111
1 1 1	101	111
1 0 1	101	100
1 0 0	000	100

(a)

(b)

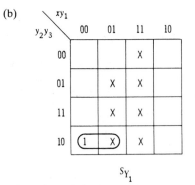

S_{Y_1}

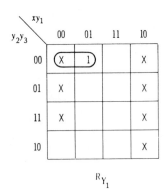

R_{Y_1}

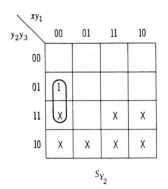

S_{Y_2}

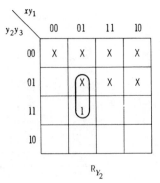

R_{Y_2}

Table S.68 Continued

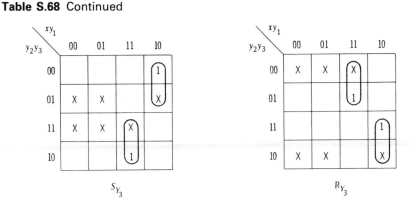

S_{Y_3}

R_{Y_3}

The logic diagram is shown in Fig. S.54(c). Note that 14 NANDS are required for the bistable circuit compared to 19 for the direct implementation of the excitation equations. It is a good exercise to derive the Boolean expressions for Y_1, Y_2 and Y_3 from the bistable circuit and compare with the excitation equations.

8.3 The state diagram for the shift-register stage is shown in Fig. S.55(a), and the primitive flow table in Table S.69(a). The flow table cannot be reduced but rows (1, 2, 3), and (5, 6, 7) may be merged to give Table S.69(b); the state assignment is straightforward (see Table S.69(c) and (d)). The excitation equations, obtained from the K-maps (see Table S.69(e)), are

$$Y_1 = \bar{x}_2 y_2 + x_2 y_1 + y_1 y_2$$
$$Y_2 = \bar{x}_2 y_2 + x_1 x_2 + x_1 y_2$$
$$z = y_1$$

Two hazard terms are required, i.e. $y_1 y_2$ for Y_1 and $x_1 y_2$ for Y_2. Table S.70 shows the flow table and K-maps for the dc bistable version of the circuit; the input equations are

$$Sy_1 = \bar{x}_2 y_2 \qquad Ry_1 = \bar{x}_2 \bar{y}_2$$
$$Sy_2 = x_2 x_1 \qquad Ry_2 = x_2 \bar{x}_1$$

The output equations are obtained in the usual way; a logic diagram for the bistable version of the circuit is shown in Fig. S.55(b).

8.4 The primitive flow table for this circuit is shown in Table S.71(a); note the two sequences for the two directions, i.e. ①→②→③→④ and ⑤→⑥→⑦→⑧. The tables cannot be reduced further, but rows (1, 6), (2, 5), (3, 8) and (4, 7) may be merged to give the flow table shown in Table S.71(b). From the K-maps shown in Table S.71(e), the excitation and output equations are

$$Y_1 = ab + ay_1 + by_1$$
$$Y_2 = \bar{a}b + \bar{a}y_2 + by_2$$
$$Z = \bar{a}\bar{y}_1\bar{y}_2 + b\bar{y}_1 y_2 + ay_1 y_2 + \bar{b}y_1\bar{y}_2$$

There are no hazard terms since all excitation loops overlap.

Table S.69 Problem 8.3

Inputs x_2x_1				Output
00	01	11	10	Z
①	2	–	3	0
1	②	4	–	0
1	–	4	③	0
–	5	④	3	0
6	⑤	7	–	1
⑥	5	–	8	1
–	5	⑦	8	1
1	–	7	⑧	1

(a)

Inputs x_2x_1				Output	
00	01	11	10	Z	
①	②	4	③	0	a
–	5	④	3	0	b
⑥	⑤	⑦	8	1	c
1	–	7	⑧	1	d

(b)

(c)

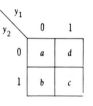

(d)

y_1y_2 \ x_2x_1	00	01	11	10
00	00	00	01	00
01	X	11	01	00
11	11	11	11	10
10	00	X	11	10

(e)

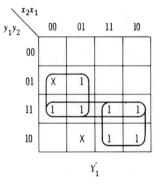

Y'_1

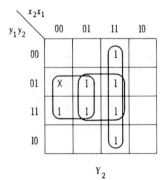

Y_2

Table S.69 Continued

y_1y_2 \ x_2x_1	00	01	11	10
00	0	0	0	0
01	X	X	0	0
11	1	1	1	1
10	X	X	1	1

Z

Table S.70 Problem 8.3

Present state y_1 y_2	Inputs x_2x_1 Next state 00	01	11	10	Output Z
0 0	00	00	01	00	0
0 1	–	11	01	00	0
1 1	11	11	11	10	1
1 0	00	–	11	10	1

(a)

(b)

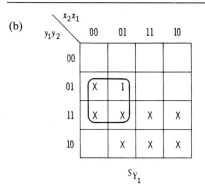

S_{Y_1}

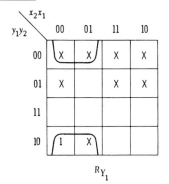

R_{Y_1}

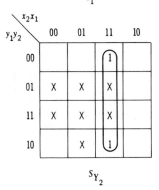

S_{Y_2}

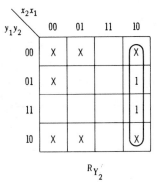

R_{Y_2}'

Table S.71 Problem 8.4

| Inputs ab | | | | Output |
00	01	11	10	Z
①	2	–	6	1
5	②	3	–	1
–	8	③	4	1
1	–	7	④	1
⑤	2	–	6	0
1	–	7	⑥	0
–	8	⑦	4	0
5	⑧	3	–	0
(a)				

| Inputs ab | | | | |
00	01	11	10	
①	2	7	⑥	a
⑤	②	3	6	b
5	⑧	③	4	c
1	8	⑦	④	d
(b)				

(c)

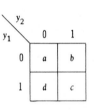

(d)

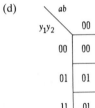

(e)

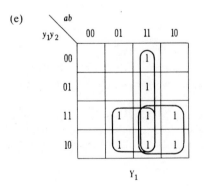

Y_1

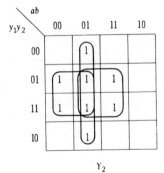

Y_2

Table S.71 Continued

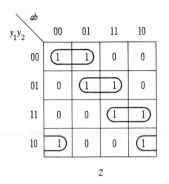

$$z$$

8.5 It is easier to visualize this circuit with a timing diagram (see Fig. S.56) rather than a state diagram. In actual practice the duration of x_2 will be very much longer than x_1, so the difficulties arising from pressing the button twice in succession (Fig. S.56 (10) and (11)) will not occur. Note that the button needs to be pressed at clock frequencies to get this effect! The primitive flow table is shown in Table S.72(a); note that no simplification is possible but rows (1, 6), (2, 9), (4, 5), (7, 10) and (3, 8) may be merged to give Table S.72(b). The state assignment requires the use of spare secondary states (see Table S.72(c) and (d)) and a possible assignment is shown in Table S.72(e). The K-maps for the excitation and output equations are shown in Table S.73; the equations are

$$Y_1 = y_1 y_3 + x_1 y_1 + x_1 x_2 \bar{y}_2 \bar{y}_3 + \bar{x}_1 \bar{x}_2 y_2 y_3$$
$$Y_2 = y_2 \bar{y}_3 + \bar{x}_1 x_2 \bar{y}_3 + x_1 y_2 + y_1 x_2 + \bar{y}_1 y_2 \bar{x}_2$$
$$Y_3 = x_2 y_3 + y_2 y_3 + x_1 \bar{y}_1 y_2$$
$$Z = y_2 y_3$$

There are no hazard terms required and the equations may be implemented in the usual way.

8.6 The state diagram for this problem is shown in Fig. S.57 and should be compared with the synchronous version, Fig. S.36. The primitive flow table (Table S.74(a)) cannot be reduced, but rows (1, 2, 3) and (4, 5, 6) may be merged. The complete design tables are shown in Table S.74(b)–(d). The excitation equation is

$$Z = xy + \bar{y}z + xz$$

and the SR bistable input equations are

$$Sz = xy \qquad Rz = \bar{x}y$$

Note that the bistable implementation results in an identical circuit to that obtained for the synchronous case; this is possible since the bistable output, Z, is not fed back to the input gates.

8.7 The initial state diagram and primitive flow table for the circuit is shown in Fig. S.58(a) and Table S.75(a). The primitive machine reduces to a simple two-row table as shown in Table S.75(b). Assigning y as the state variable (see Table S.75(c)) gives the excitation equations

$$Y = ab + y(a + b)$$

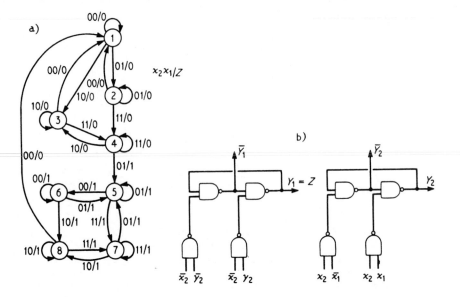

Figure S.55 Problem 8.3

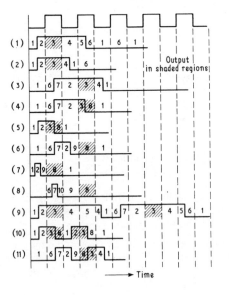

Figure S.56 Problem 8.5

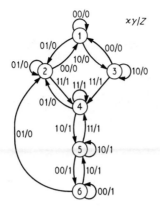

Figure S.57 Problem 8.6

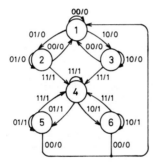

a) Flow-diagram

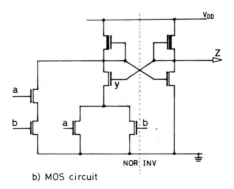

b) MOS circuit

Figure S.58 Problem 8.7

Table S.72 Problem 8.5

Inputs x_1x_2 00	01	11	10	Output Z
①	2	–	6	0
9	②	3	–	0
–	4	③	8	1
1	④	5	–	0
–	4	⑤	6	0
1	–	7	⑥	0
–	2	⑦	10	0
1	–	3	⑧	1
⑨	2	–	8	0
9	–	7	⑩	0

(a)

Inputs x_1x_2 00	01	11	10	Output Z	
①	2	7	⑥	0	a
⑨	②	3	8	0	b
1	4	③	⑧	1	c
1	④	⑤	6	0	d
9	2	⑦	⑩	0	e

(b)

(c)

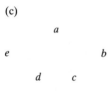

(d)

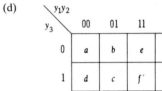

	Inputs x_1x_2				Output
y_1 y_2 y_3	00	01	11	10	Z
a 0 0 0	000	010	100	000	0
d 0 0 1	000	001	001	000	0
c 0 1 1	111	001	011	011	1
b 0 1 0	010	010	011	011	0
e 1 1 0	010	010	110	110	0
f' 1 1 1	101	–	–	–	X
g' 1 0 1	100	–	–	–	X
h' 1 0 0	000	–	110	–	X

(e)

Table S.73 Problem 8.5

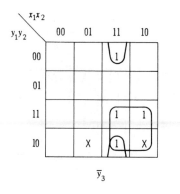

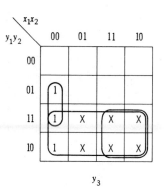

Y_1

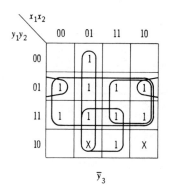

Y_2

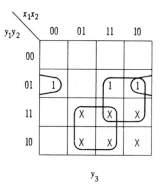

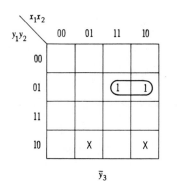

Y_3

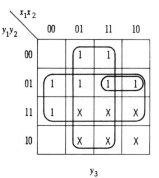

Table S.73 Continued

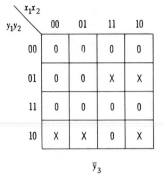

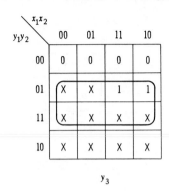

Table S.74 Problem 8.6

Inputs xy				Output
00	01	11	10	Z
①	2	–	3	0
1	②	4	–	0
1	–	4	③	0
–	2	④	5	1
6	–	4	⑤	1
⑥	2	–	5	1
(a)				

Inputs xy				Output
00	01	11	10	Z
①	②	4	③	0
⑥	2	④	⑤	1
(b)				

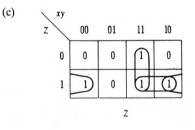

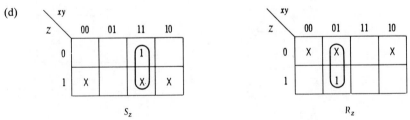

Table S.75 Problem 8.7

| Inputs a, b | | | | Output |
00	01	11	10	Z
①	2	–	3	0
1	②	4	–	0
1	–	4	③	0
–	5	④	6	1
1	⑤	4	–	1
1	–	4	⑥	1
(a) Primitive flow table				

| | Inputs a, b | | | | Output |
y	00	01	11	10	Z
0	①	①	2	①	0
1	1	②	②	②	1
(b) Reduced flow table					

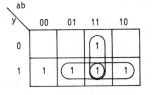

| ab | | | | |
y	00	01	11	10
0			1	
1	1	1	1	1

(c) K-map for excitation equations

If we implement these equations directly using a compound MOS NOR gate and a separate invertor to produce y we arrive at the circuit shown in Fig. S.58(b). Note that the excitation equations are equivalent to a combinational 3-input majority gate circuit, and thus the circuit can easily be extended (within the limits of the technology) by adding extra inputs to the parallel/serial limbs. For example, a 3-input Muller C element would have the excitation equation

$$Y = abc + y(a + b + c)$$

8.8 (a) The uncorrected equations may be read directly from the flow table in Table 8.23. These are

$$Y_1 = \bar{x}_1\bar{x}_2 y_5 + x_1\bar{x}_2 y_3 + y_1\bar{y}_2\bar{y}_3$$
$$Y_2 = y_1\bar{x}_1 x_2 + \bar{y}_3 y_2$$
$$Y_3 = y_1 x_1 x_2 + y_2 x_1 x_2 + y_3\bar{y}_1\bar{y}_4$$
$$Y_4 = \bar{x}_1\bar{x}_2 y_3 + x_1 x_2 y_5 + y_4\bar{y}_5$$
$$Y_5 = x_1\bar{x}_2 y_4 + y_5\bar{y}_1\bar{y}_4$$

Correcting for the loop terms producing hazards we have

$$Y_1 = \bar{x}_1\bar{x}_2 y_5 + x_1\bar{x}_2 y_3 + y_1\bar{y}_2(\bar{y}_3 + x_1\bar{x}_2)$$
$$Y_2 = y_1\bar{x}_1 x_2 + \bar{y}_3 y_2$$
$$Y_3 = x_1 x_2 y_1 + x_1 x_2 y_2 + y_3\bar{y}_4(\bar{y}_1 + x_1 x_2)$$
$$Y_4 = \bar{x}_1\bar{x}_2 y_3 + x_1 x_2 y_5 + y_4(\bar{y}_5 + x_1 x_2)$$
$$Y_5 = x_1\bar{x}_2 y_4 + y_5\bar{y}_1(\bar{y}_4 + x_1\bar{x}_2)$$

These are shown in implemented in Fig. S.59(a); note that the reset terms can be realized directly as they are inverted—as required by the NAND bistables.
(b) The *partition pair* list for the flow table is

$$\pi_1 = (a, e)(c, d) \qquad\qquad \pi_5 = (a, c)(d, e)$$
$$\pi_2 = (c)(a, b) \qquad\qquad \pi_6 = (b, c)(d, e)$$
$$\pi_3 = (e)(a, b)$$
$$\pi_4 = (c)(e)$$

Note that $\pi_5 \geq \pi_4$, giving π_1, π_2, π_3, π_5 and π_6 for the assignment; these can of course be reduced futher to give $r_3 = (a, e)(c, d)$; $r_2 = (a, b)(c, e)$ and $r_1 = (a, b, c)(d, e)$. The *K-sets* are

$$\pi_1 = (a, e)(c, d) \qquad\qquad \pi_4 = (a, b, c)(d, e)$$
$$\pi_2 = (c)(a, b) \qquad\qquad \pi_5 = (a, c)(d, e)$$
$$\pi_3 = (e)(a, b)$$

where $\pi_4 \geq \pi_3$, π_2, π_5, which leaves π_1 and π_4 being required for the assignment.

the *column partitions* are

column 00 $\pi_1 = (a, e)(c, d)$
column 01 $\pi_2 = (c)(e)(a, b)$
column 11 $\pi_3 = (a, b, c)(d, e)$
column 10 $\pi_4 = (a, c)(d, e)$

where $\pi_3 \geq \pi_2$, π_4 which yields π_1 and π_3 being required for the assignment.

Three state variables are needed for the assignment and though many variations are possible we shall choose

$$r_1 = y_1 = (a, b, c)(d, e) = 00011$$
$$r_2 = y_2 = (a, b)(c, d) = 00X01$$
$$r_3 = y_3 = (a, e)(c, d) = 0X011$$

The fully assigned flow table is given in Table S.76(a) which yields the following input equations for SR bistables (shown plotted on K-maps in Table S.76(b)):

$$\text{Set } Y_1 = y_3 \bar{x}_1 \bar{x}_2$$
$$\text{Reset } Y_1 = \bar{y}_3 \bar{x}_1 \bar{x}_2$$
$$\text{Set } Y_2 = y_1 \bar{y}_2 y_3 x_1 \bar{x}_2 + \bar{y}_1 \bar{y}_2 x_1 x_2$$
$$\text{Reset } Y_2 = \bar{x}_1 \bar{x}_2 + \bar{y}_1 y_3 \bar{x}_2 + y_1 y_2 \bar{y}_3 x_1 x_2$$
$$\text{Set } Y_3 = \bar{y}_1 \bar{y}_2 x_2 + y_1 y_2 \bar{y}_3 x_1 x_2$$
$$\text{Reset } Y_3 = \bar{y}_1 y_3 x_1 \bar{x}_2 + y_1 \bar{y}_2 y_3 x_1 \bar{x}_2$$
$$Z = \bar{y}_1 y_2 y_3 + y_1 \bar{y}_2 y_3$$

These equations are shown implemented using NAND bistables in Fig. S.59(b). Note that though one-hot assignment requires more units (27) compared to STT (18) the design and structure is simpler. STT circuits are generally faster than one-hot designs. If the STT assigned circuit is realized in the excitation equation forms hazard detection and correction must be applied.

Table S.76 Problem 8.8

			Input/Output x_1x_2/Z				
y_1	y_2	y_3	00	01	11	10	Z
0	0	0	000/0	001/0	011/0	000/0	0
0	0	1	–	001/0	011/0	–	0
0	1	1	101/1	011/1	011/1	000/1	1
1	0	1	101/1	–	101/1	110/1	1
1	1	0	000/0	110/0	101/0	110/0	0

(a) Assigned flow table

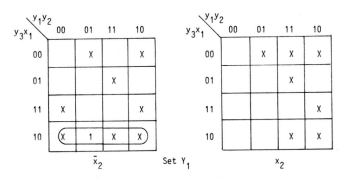

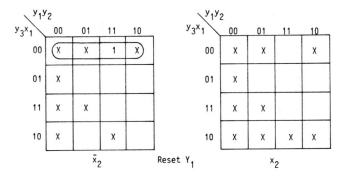

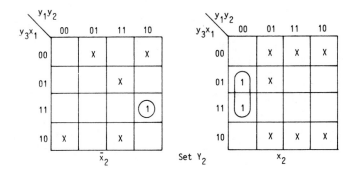

Table S.76 Continued

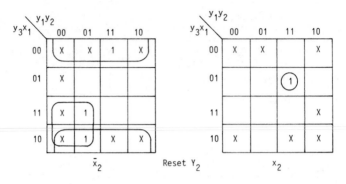

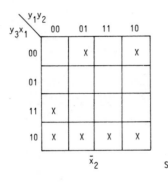

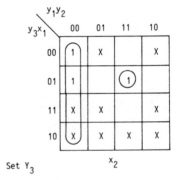

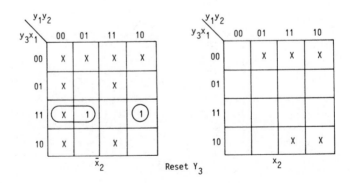

Table S.76 Continued

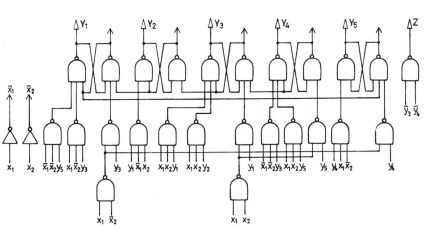

(b) K-maps

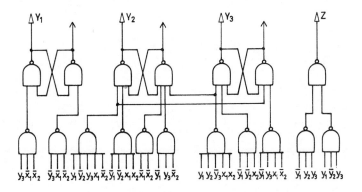

a) One-hot implementation

b) STT implementation

Figure S.59 Problem 8.8

Chapter 9

9.1 The F-matrix and the derived G_D-matrix are shown in Table S.77. From the G_D-matrix, tests t_5 and t_7 are seen to be essential and cover between them faults $f_1, f_3, f_5, f_8, f_9, f_{11}, f_{14}, f_4, f_7$ and f_{13}; the remaining faults may be covered using tests t_3 and t_6. Therefore the full test set is 101/1, 111/0, 010/0 and 110/1. Checking using the Boolean Difference we have:

$$Z = \bar{A}C + A\bar{C} + A\bar{B}$$

error in A $Z_A = AC + \bar{A}\bar{C} + \bar{A}\bar{B}$ shown in Table S.78(a)

error in B $Z_B = \bar{A}C + A\bar{C} + AB$ shown in Table S.78(b)

error in C $Z_C = \bar{A}\bar{C} + AC + A\bar{B}$ shown in Table S.78(c)

For an error in A the Boolean Difference gives the characteristics of the circuit when A s-at-0 or s-at-1. Thus, gathering those terms containing A ($C1$ s-at-0) we have ABC, $AB\bar{C}$, $A\bar{B}\bar{C}$ equivalent to tests t_7, t_6 and t_4 and $\bar{A}$ ($C1$ s-at-1) gives $\bar{A}BC$, $\bar{A}B\bar{C}$, $\bar{A}\bar{B}\bar{C}$ equivalent to tests t_3, t_2 and t_0. Similarly for an error in B we have for B ($C2$ s-at-0) ABC and for $\bar{B}$ ($C2$ s-at-1) $A\bar{B}C$ which confirms the essential tests t_7 and t_5 respectively.

Similarly an error in C yields the tests t_1, t_3 and t_7 for $C3$ s-at-0 and t_0, t_2 and t_6 for $C3$ s-at-1.

Note that though the circuit exhibits fan-out from $C4$ and the paths reconverge, fault masking does not occur because there are an equal number of signal conversions.

9.2 The good circuit is given by

$$Z = \bar{A}B + B\bar{C} + \bar{B}C \quad \text{and} \quad \bar{Z} = ABC + \bar{B}\bar{C}$$

and the faulty circuit with an error in B by

$$Z_B = \bar{A}\bar{B} + \bar{B}\bar{C} + BC \quad \text{and} \quad \bar{Z}_B = B\bar{C} + A\bar{B}C$$

These functions are shown plotted on K-maps in Table S.79(a) with the primitive cubes in Table S.79(b).

Now intersecting the cubes of PF with $\bar{P}$ to obtain $\bar{D}$ and $\overline{PF}$ with P to obtain D we obtain

$PF1$	$00X1$	$PF1$	$00X1$	$PF2$	$X001$	$PF2$	$X001$
$\overline{P1}$	$X000$	$\overline{P2}$	1110	$\overline{P1}$	$X000$	$\overline{P2}$	1110
	$000\bar{D}$		$\varphi\varphi1\bar{D}$		$X00\bar{D}$		$1\varphi\varphi\bar{D}$

$PF3$	$X111$	$PF3$	$X111$
$\overline{P1}$	$X000$	$\overline{P2}$	1110
	$X\varphi\varphi\bar{D}$		$111\bar{D}$

$\overline{PF1}$	$X100$	$\overline{PF1}$	$X100$	$\overline{PF1}$	$X100$
$P1$	$01X1$	$P2$	$X101$	$P3$	$X011$
	$010D$		$X10D$		$X\varphi\varphi D$

$\overline{PF2}$	1010	$\overline{PF2}$	1010	$\overline{PF2}$	1010
$P1$	$01X1$	$P2$	$X101$	$P3$	$X011$
	$\varphi\varphi1D$		$1\varphi\varphi D$		$101D$

Table S.77 Problem 9.1

(a) F-matrix

A B C				f_0	$\vee f_1$ C1/0	f_2 C1/1	$\vee f_3$ C2/0	$\vee f_4$ C2/1	$\vee f_5$ C3/0	f_6 C3/1	$\vee f_7$ C4/0	$\vee f_8$ C4/1	$\vee f_9$ C5/0	f_{10} C5/1	$\vee f_{11}$ C6/0	f_{12} C6/1	$\vee f_{13}$ C7/0	$\vee f_{14}$ C7/1
0 0 0			t_0	0	0	1	0	0	0	1	0	0	1	0	1	0	0	1
0 0 1			t_1	1	1	1	1	1	0	1	0	1	1	1	1	0	0	1
0 1 0			t_2	0	0	1	0	0	0	1	0	0	1	0	1	0	0	1
0 1 1			t_3	1	1	0	1	1	0	1	0	1	1	1	1	0	0	1
* 1 0 0			t_4	1	0	1	1	1	1	1	0	1	1	0	1	1	0	1
1 0 1			t_5	1	1	1	1	0	1	1	0	1	1	1	1	1	0	1
1 1 0			t_6	1	0	1	1	1	1	0	0	1	1	0	1	1	0	1
* 1 1 1			t_7	0	1	0	1	0	1	0	0	1	1	0	1	0	0	1

(b) G_D-matrix

	f_0f_1	f_0f_2	f_0f_3	f_0f_4	f_0f_5	f_0f_6	f_0f_7	f_0f_8	f_0f_9	f_0f_{10}	f_0f_{11}	f_0f_{12}	f_0f_{13}	f_0f_{14}
t_0		1				1		1			1			1
t_1							1		1	1	1	1	1	
t_2		1				1					1	1		
t_3		1			1							1		
t_4	1			1					1					
t_5	1				1				1					
t_6	1				1	1								
t_7			1				1	1		1	1	1	1	1

Table S.78 Problem 9.1

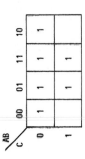

=

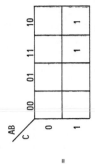

=

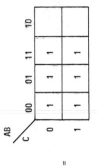

=

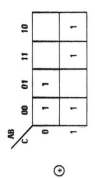

(+)

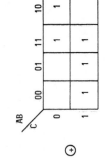

(+)

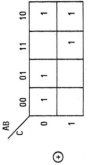

(+)

(a) Error in A

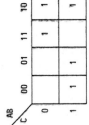

(b) Error in B

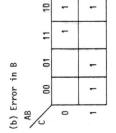

(c) Error in C

Table S.79 Problem 9.2

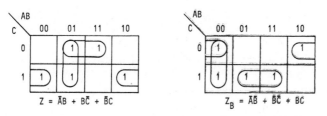

Z = ĀB + BC̄ + B̄C   Z_B = ĀB + B̄C̄ + BC

(a) K-maps

	A	B	C	Z		A	B	C	Z
p_1	0	1	X	1	PF_1	0	0	X	1
p_2	X	1	0	1	PF_2	X	0	0	1
p_3	X	0	1	1	PF_3	X	1	1	1
$\check{P}_1$	X	0	0	0	$\overline{PF}_1$	X	1	0	0
$\check{P}_2$	1	1	1	0	$\overline{PF}_2$	1	0	1	0

(b) Primitive cubes

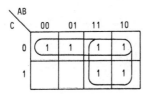

(c) Boolean difference

which gives (ignoring those terms with φ elements) the following D-cubes of failure:

000Ď, X00Ď, 111Ď
010D, X10D, 101D

Checking using the Boolean difference we have from Table S.79(c) $dZ/dZ_B =$ C̄ + A which gives the following tests:

B s-at-1   AB̄C̄   and   B s-at-0   ĀBC̄
           AB̄C   ABC̄
           ĀB̄C̄   ABC

9.3 The four-variable Reed–Muller expansion is given by

$$Z = f(x_1, x_2, x_3, x_4) = C_0 \oplus C_1 x_4 \oplus C_2 x_3 \oplus C_3 x_3 x_4 \oplus C_4 x_2$$
$$\oplus C_5 x_2 x_4 \oplus C_6 x_2 x_3 \oplus C_7 x_2 x_3 x_4 \oplus C_8 x_1 \oplus C_9 x_4 x_1 \oplus C_{10} x_1 x_3$$
$$\oplus C_{11} x_1 x_3 x_4 \oplus C_{12} x_1 x_2 \oplus C_{13} x_1 x_2 x_4 \oplus C_{14} x_1 x_2 x_3$$
$$\oplus C_{15} x_1 x_2 x_3 x_4$$

The derivation of the binary coefficient C is shown in Table S.80 and yields the following expression for the function:

$$Z = x_3 \oplus x_3 x_4 \oplus x_2 x_4 \oplus x_2 x_3 \oplus x_2 x_3 x_4 \oplus x_1$$
$$\oplus x_1 x_3 \oplus x_1 x_2 \oplus x_1 x_2 x_3$$

The function is shown realized in AND/EXOR gates in Fig. S.60.

9.4 The error states for the assignment are

$$\text{Stuck- at } 0, \quad 000 \rightarrow (011, 101, 110)$$
$$\text{Stuck- at } 1, \quad 111 \rightarrow (001, 010, 100)$$

The required transitions are shown in the state diagram in Fig. S.61(a) with the full transition table in Table S.81(a). Plotting the input terms for JK bistables on K-maps (Table S.81(b)) yields the equations

$$J_A = \bar{B} + \bar{C}; \qquad K_A = B + C$$
$$J_B = \bar{A} + \bar{C}; \qquad K_B = A + C$$
$$J_C = \bar{A} + \bar{B}; \qquad K_C = A + B$$
$$Z = AB + BC + AC$$

Note that in the equation for the output it is necessary to generate an output for the error terms 011, 101 and 110. The circuit realized in NAND gates is shown in Fig. S.61(b).

9.5 From Table 9.23 we can obtain the p-sets as follows:

column $\bar{x}$		column x	
$p_1 = (AB)$		$p_4 = (AC)$	
$p_2 = (C)$		$p_5 = (B)$	
$p_3 = (D)$			

This gives the r-partitions:

$r_1 = (AB)(CD)$		$r_4 = (AC)(BD)$	
$r_2 = (C)(ABD)$		$r_5 = (B)(ACD)$	
$r_3 = (D)(ABC)$			

which yields the assigned table shown in Table S.82. Deriving next-state partitions we have

$\bar{x}$		$\bar{x}$	
$n_1^1 = (D)(ABC)$		$n_1^2 = (ABC)(D)$	
$n_2^1 = (AB)(CD)$		$n_2^2 = (\emptyset)(ABCD)$	
$n_3^1 = (C)(ABD)$		$n_3^2 = (D)(ABC)$	
$n_4^1 = (AB)(CD)$		$n_4^2 = (B)(ACD)$	
$n_5^1 = (D)(ABC)$		$n_5^2 = (AC)(BD)$	

and

$n_1^1 = r_3 \therefore f_1^1 = y_3$		$n_1^2 = (p_4 p_5)(p_3) \therefore f_1^2 = y_4 + y_5$	
$n_2^1 = r_1 \therefore f_2^1 = y_1$		$n_2^2 = (\emptyset)(ABCD) \therefore f_2^2 = 0$	
$n_3^1 = r_2 \therefore f_3^1 = y_2$		$n_3^2 = r_3 \therefore f_3^2 = y_3$	
$n_4^1 = r_1 \therefore f_4^1 = y_1$		$n_4^2 = r_5 \therefore f_4^2 = y_5$	
$n_5^1 = r_3 \therefore f_5^1 = y_3$		$n_5^2 = r_4 \therefore f_5^2 = y_4$	

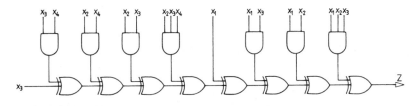

Figure S.60 Problem 9.3

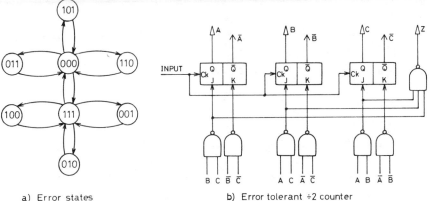

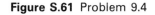
a) Error states

b) Error tolerant ÷2 counter

Figure S.61 Problem 9.4

From this we may deduce that

$$Y_1 = y_3\bar{x} + y_4x + y_5x$$
$$Y_2 = y_1\bar{x}$$
$$Y_3 = y_2\bar{x} + y_3x$$
$$Y_4 = y_1\bar{x} + y_5x$$
$$Y_5 = y_3\bar{x} + y_4x$$

where the partitioning variables are for $\bar{x}$ $(y_1y_2y_3)$ and for x $(y_3y_4y_5)$.

9.6 Repeating the design for the machine in problem 9.5 but using a 2-out-of-4 code for state assignment we arrive at the assigned state table shown in Table S.83; note that the code 1100, the $K1$ assignment, is not used. Letting $B1 = y_1y_2$ and $B2 = y_3y_4$ and extracting the OFF-sets for $B1$ and the ON-sets for $B2$ we have

$$Y_1 = [(y_2 + y_3)(y_2 + y_4)(y_1 + y_4) + \bar{x}][(y_1 + y_3) + x]$$
$$Y_2 = [(y_1 + y_3) + \bar{x}][(y_2 + y_3)(y_2 + y_4)(y_1 + y_4) + x]$$
$$Y_3 = y_1y_4\bar{x} + y_1y_3\bar{x} + y_2y_4\bar{x} + y_1y_4x + y_2y_3x$$
$$Y_4 = y_2y_3\bar{x} + y_1y_3x + y_2y_4x$$

It will be obvious that though fewer state variables are used considerably more logic is required to realize the circuit.

Table S.80 Problem 9.3

	x_1 x_2 x_3 x_4	Z	
f_0	0 0 0 0	0	$C_0 = f_0 = 0$
f_1	0 0 0 1	0	$C_1 = f_0 \oplus f_1 = 0$
f_2	0 0 1 0	1	$C_2 = f_0 \oplus f_2 = 1$
f_3	0 0 1 1	0	$C_3 = f_0 \oplus f_1 \oplus f_2 \oplus f_3 = 1$
f_4	0 1 0 0	0	$C_4 = f_0 \oplus f_4 = 0$
f_5	0 1 0 1	1	$C_5 = f_0 \oplus f_1 \oplus f_4 \oplus f_5 = 1$
f_6	0 1 1 0	0	$C_6 = f_0 \oplus f_2 \oplus f_4 \oplus f_6 = 1$
f_7	0 1 1 1	1	$C_7 = f_0 \oplus f_1 \oplus f_2 \oplus f_4 \oplus f_3 \oplus f_6 \oplus f_5 \oplus f_7 = 1$
f_8	1 0 0 0	1	$C_8 = f_0 \oplus f_8 = 1$
f_9	1 0 0 1	1	$C_9 = f_0 \oplus f_8 \oplus f_1 \oplus f_9 = 0$
f_{10}	1 0 1 0	1	$C_{10} = f_0 \oplus f_2 \oplus f_8 \oplus f_{10} = 1$
f_{11}	1 0 1 1	0	$C_{11} = f_0 \oplus f_1 \oplus f_2 \oplus f_3 \oplus f_8 \oplus f_9 \oplus f_{10} \oplus f_{11} = 0$
f_{12}	1 1 0 0	0	$C_{12} = f_0 \oplus f_8 \oplus f_4 \oplus f_{12} = 1$
f_{13}	1 1 0 1	1	$C_{13} = f_0 \oplus f_1 \oplus f_4 \oplus f_8 \oplus f_9 \oplus f_5 \oplus f_{12} \oplus f_{13} = 0$
f_{14}	1 1 1 0	0	$C_{14} = f_0 \oplus f_2 \oplus f_4 \oplus f_8 \oplus f_6 \oplus f_{10} \oplus f_{12} \oplus f_{14} = 1$
f_{15}	1 1 1 1	1	$C_{15} = f_0 \oplus f_1 \oplus f_2 \oplus f_3 \oplus f_4 \oplus f_5 \oplus f_6 \oplus f_7 \oplus f_8 \oplus f_9$ $\oplus f_{10} \oplus f_{11} \oplus f_{12} \oplus f_{13} \oplus f_{14} \oplus f_{15} = 0$

Table S.81 Problem 9.4

Present state A B C	Next state A_+ B_+ C_+	Output Z
0 0 0	1 1 1	0
0 0 1	0 1 1	0
0 1 0	1 0 1	0
1 0 0	1 1 0	0
1 1 1	0 0 0	1
0 1 1	0 0 1	1
1 0 1	0 1 0	1
1 1 0	1 0 0	1

(a) Transition table

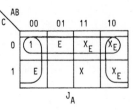

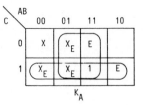

Table S.81 Continued

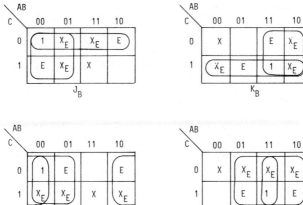

(b) K-maps

Table S.82 Problem 9.5

	Present state					Next state									
						$x = 0$						$x = 1$				
	y_1	y_2	y_3	y_4	y_5	Y_1	Y_2	Y_3	Y_4	Y_5	Y_1	Y_2	Y_3	Y_4	Y_5
A	1	0	0	1	0	0	1	0	1	0	1	0	0	0	1
B	1	0	0	0	1	0	1	0	1	0	1	0	0	1	0
C	0	1	0	1	0	0	0	1	0	0	1	0	0	0	1
D	0	0	1	0	0	1	0	0	0	1	0	0	1	0	0

Table S.83 Problem 9.6

	Present state				Next state							
					$x = 0$				$x = 1$			
	y_1	y_2	y_3	y_4	Y_1	Y_2	Y_3	Y_4	Y_1	Y_2	Y_3	Y_4
A	1	0	0	1	0	1	1	0	1	0	1	0
B	1	0	1	0	0	1	1	0	1	0	0	1
C	0	1	1	0	0	1	0	1	1	0	1	0
D	0	1	0	1	1	0	1	0	0	1	0	1

9.7 The destination sets for the flow table are

$$d_1 = (AB); \quad d_2 = (C); \quad d_3 = (A); \quad d_4 = (BC)$$

giving the r-partitions

$$r_1 = (AB)(C); \quad r_2 = (C)(AB); \quad r_3 = (A)(BC); \quad r_4 = (BC)(A)$$

The assigned flow table is shown in Table S.84. Extracting the next-state partitions gives

$$\bar{x}_1 \bar{x}_2 \quad n_1^1 = (AB)(C) \qquad \bar{x}_1 x_2 \quad n_1^2 = (AB)(C)$$
$$n_2^1 = (C)(AB) \qquad\qquad n_2^2 = (C)(AB)$$
$$n_3^1 = (AB)(C) \qquad\qquad n_3^2 = (\emptyset)(ABC)$$
$$n_4^1 = (C)(AB) \qquad\qquad n_4^2 = (ABC)(\emptyset)$$

$$x_1 x_2 \quad n_1^3 = (A)(BC) \qquad x_1 \bar{x}_2 \quad n_1^4 = (ABC)(\emptyset)$$
$$n_2^3 = (BC)(A) \qquad\qquad n_2^4 = (\emptyset)(ABC)$$
$$n_3^3 = (A)(BC) \qquad\qquad n_3^4 = (A)(BC)$$
$$n_4^3 = (BC)(A) \qquad\qquad n_4^4 = (BC)(A)$$

Thus

$$f_1^1 = y_1 \qquad f_1^2 = y_1 \qquad\qquad f_1^3 = y_3 \qquad f_1^4 = 1(y_3 + y_4)$$
$$f_2^1 = y_2 \qquad f_2^2 = y_2 \qquad\qquad f_2^3 = y_4 \qquad f_2^4 = 0$$
$$f_3^1 = y_1 \qquad f_3^2 = 0 \qquad\qquad f_3^3 = y_3 \qquad f_3^4 = y_3$$
$$f_4^1 = y_2 \qquad f_4^2 = 1(y_1 + y_2) \qquad f_4^3 = y_4 \qquad f_4^4 = y_4$$

Note that the partitioning variables are $y_1 y_2$ for $\bar{x}_1 \bar{x}_2$ and $\bar{x}_1 x_2$ and $y_3 y_4$ for $x_1 x_2$ and $x_1 \bar{x}_2$; note also that in the case of $f_i^j = 1$ all the partitioning variables for that input must be included. The full set of excitation equations are

$$Y_1 = \bar{x}_1 \bar{x}_2 y_1 + \bar{x}_1 x_2 y_1 + x_1 x_2 y_3 + x_1 \bar{x}_2 y_4 + x_1 \bar{x}_2 y_3$$
$$Y_2 = \bar{x}_1 \bar{x}_2 y_2 + \bar{x}_1 x_2 y_2 + x_1 x_2 y_4$$
$$Y_3 = \bar{x}_1 \bar{x}_2 y_1 + x_1 x_2 y_3 + x_1 \bar{x}_2 y_3$$
$$Y_4 = \bar{x}_1 \bar{x}_2 y_2 + \bar{x}_1 x_2 y_1 + \bar{x}_1 x_2 y_2 + x_1 x_2 y_4 + x_1 \bar{x}_2 y_4$$

The circuit is shown realized in NAND gates in Fig. S.62.

Table S.84 Problem 9.7

Present state				Next state Inputs $x_1 x_2$																
y_1	y_2	y_3	y_4	0 0				0 1				1 1				1 0			
A	1	0	1	0	1	0	1	0	1	0	0	1	1	0	1	0	1	0	1	0
B	1	0	0	1	1	0	1	0	1	0	0	1	0	1	0	1	1	0	0	1
C	0	1	0	1	0	1	0	1	0	1	0	1	0	1	0	1	1	0	0	1

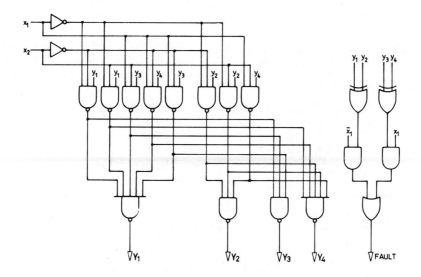

Figure S.62 Problem 9.7

Chapter 10

10.1 If the states are assigned in an arbitrary manner, say $A = 000$, $B = 001$, $C = 010$, $D = 011$ and $E = 100$, we arrive at the transition table shown in Table S.85. Note that the ASM has been converted directly without state-reduction. The product terms are given by

$$R_1 = \overline{\text{CLR}}\ \overline{\text{ADD}}\ \bar{Y}_1\bar{Y}_2\bar{Y}_3 \qquad R_7 = \text{CLR}\ \bar{Y}_1 Y_2 \bar{Y}_3$$
$$R_2 = \overline{\text{CLR}}\ \text{ADD}\ \bar{Y}_1\bar{Y}_2\bar{Y}_3 \qquad R_8 = \text{ADD}\ \bar{Y}_1 Y_2 \bar{Y}_3$$
$$R_3 = \text{CLR}\ \overline{\text{ADD}}\ \bar{Y}_1\bar{Y}_2\bar{Y}_3 \qquad R_9 = \text{CAR}\ \bar{Y}_1 Y_2 Y_3$$
$$R_4 = \text{CLR}\ \text{ADD}\ \bar{Y}_1\bar{Y}_2\bar{Y}_3 \qquad R_{10} = \overline{\text{CAR}}\ \bar{Y}_1 Y_2 Y_3$$
$$R_5 = \bar{Y}_1\bar{Y}_2 Y_3 \qquad R_{11} = \text{CLR}\ Y_1 \bar{Y}_2 \bar{Y}_3$$
$$R_6 = \overline{\text{CLR}}\ \overline{\text{ADD}}\ \bar{Y}_1 Y_2 \bar{Y}_3 \qquad R_{12} = \overline{\text{CLR}}\ Y_1 \bar{Y}_2 \bar{Y}_3$$

Thus

$$
\begin{aligned}
Y_1 &= R_9 + R_{12} \\
&= \text{CAR}\ \bar{Y}_1 Y_2 Y_3 + \overline{\text{CLR}}\ Y_1 \bar{Y}_2 \bar{Y}_3 \\
Y_2 &= R_2 + R_4 + R_5 + R_7 + R_8 + R_{10} \\
&= \overline{\text{CLR}}\ \text{ADD}\ \bar{Y}_1\bar{Y}_2\bar{Y}_3 + \text{CLR}\ \text{ADD}\ \bar{Y}_1\bar{Y}_2\bar{Y}_3 + \bar{Y}_1\bar{Y}_2 Y_3 \\
&\quad + \text{CLR}\ \bar{Y}_1 Y_2 \bar{Y}_3 + \text{ADD}\ \bar{Y}_1 Y_2 \bar{Y}_3 + \overline{\text{CAR}}\ \bar{Y}_1 Y_2 Y_3 \\
Y_3 &= R_2 + R_3 + R_{11} \\
&= \overline{\text{CLR}}\ \text{ADD}\ \bar{Y}_1\bar{Y}_2\bar{Y}_3 + \text{CLR}\ \overline{\text{ADD}}\ \bar{Y}_1\bar{Y}_2\bar{Y}_3 + \text{CLR}\ Y_1 \bar{Y}_2 \bar{Y}_3 \\
\text{RSAC} &= \text{RSOF} = R_3 + R_{11} \\
&= \text{CLR}\ \overline{\text{ADD}}\ \bar{Y}_1\bar{Y}_2\bar{Y}_3 + \text{CLR}\ Y_1 \bar{Y}_2 \bar{Y}_3 \\
\text{SUM} &= \text{COF} = R_2 \\
&= \overline{\text{CLR}}\ \text{ADD}\ \bar{Y}_1\bar{Y}_2\bar{Y}_3
\end{aligned}
$$

The stick diagram for the PLA implementation is shown in Fig. S.63. Note that the PLA is effectively a NOR/NOR circuit and consequently the inverse of the inputs is used to generate the product terms; these are then inverted in the buffer amplifiers.

If we now assign the states using a 2-out-of-4 code which ensures that the inverses of the zero terms fall in the don't-care set, thus: $A = 1001$, $B = 0101$, $C = 0110$, $D = 0011$ and $E = 1010$, we can generate a monotonic next-state assignment (see Table S.86).

Looking at the ON terms only we have the products

$$R_1 = \overline{\text{CLR}}\ \overline{\text{ADD}}\ Y_1 Y_4 \qquad R_7 = \text{CLR}\ Y_2 Y_3$$
$$R_2 = \overline{\text{CLR}}\ \text{ADD}\ Y_1 Y_4 \qquad R_8 = \text{ADD}\ Y_2 Y_3$$
$$R_3 = \text{CLR}\ \overline{\text{ADD}}\ Y_1 Y_4 \qquad R_9 = \text{CAR}\ Y_3 Y_4$$
$$R_4 = \text{CLR}\ \text{ADD}\ Y_1 Y_4 \qquad R_{10} = \overline{\text{CAR}}\ Y_3 Y_4$$
$$R_5 = Y_2 Y_4 \qquad R_{11} = \text{CLR}\ Y_1 Y_3$$
$$R_6 = \overline{\text{CLR}}\ \overline{\text{ADD}}\ Y_2 Y_3 \qquad R_{12} = \overline{\text{CLR}}\ Y_1 Y_3$$

Thus

$$
\begin{aligned}
Y_1 &= R_1 + R_6 + R_9 + R_{12} \\
Y_2 &= R_3 + R_4 + R_5 + R_7 + R_8 + R_{10} + R_{11} \\
Y_3 &= R_2 + R_4 + R_5 + R_7 + R_8 + R_9 + R_{10} + R_{12} \\
Y_4 &= R_1 + R_2 + R_3 + R_6 + R_{11} \\
\text{SUM} &= \text{COF} = R_2 \\
\text{RASC} &= \text{RSOF} = R_3 + R_{11}
\end{aligned}
$$

The realization requires one less input to the AND matrix thereby reducing the length of the array. The width remains the same as the number of product

Table S.85 Problem 10.1

Input state CLR ADD CAR			Present state Y_1 Y_2 Y_3	Next state Y_1 Y_2 Y_3	Output state RSAC RSOF SUM COF				Product term
0	0	X	0 0 0 (A)	0 0 0 (A)	0	0	0	0	R_1
0	1	X	0 0 0 (A)	0 1 1 (D)	0	0	1	1	R_2
1	0	X	0 0 0 (A)	0 0 1 (B)	1	1	0	0	R_3
1	1	X	0 0 0 (A)	0 1 0 (C)	0	0	0	0	R_4
X	X	X	0 0 1 (B)	0 1 0 (C)	0	0	0	0	R_5
0	0	X	0 1 0 (C)	0 0 0 (A)	0	0	0	0	R_6
0	1	X	0 1 0 (C)	0 1 0 (C)	0	0	0	0	R_7
1	0	X	0 1 0 (C)	0 1 0 (C)	0	0	0	0	R_8
1	1	X	0 1 0 (C)	0 1 0 (C)	0	0	0	0	
X	X	1	0 1 1 (D)	1 0 0 (E)	0	0	0	0	R_9
X	X	0	0 1 1 (D)	0 1 0 (C)	0	0	0	0	R_{10}
1	X	X	1 0 0 (E)	0 0 1 (B)	1	1	0	0	R_{11}
0	X	X	1 0 0 (E)	1 0 0 (E)	0	0	0	0	R_{12}

Table S.86 Problem 10.1

Input state CLR ADD CAR			Present state Y_1 Y_2 Y_3 Y_4	Next state Y_1 Y_2 Y_3 Y_4	Output state RSAC RSOF SUM COF				Product terms
0	0	X	1 0 0 1	1 0 0 1	0	0	0	0	R_1
0	1	X	1 0 0 1	0 0 1 1	0	0	1	1	R_2
1	0	X	1 0 0 1	0 1 0 1	1	1	0	0	R_3
1	1	X	1 0 0 1	0 1 1 0	0	0	0	0	R_4
X	X	X	0 1 0 1	0 1 1 0	0	0	0	0	R_5
0	0	X	0 1 1 0	1 0 0 1	0	0	0	0	R_6
0	1	X	0 1 1 0	0 1 1 0	0	0	0	0	R_7
1	0	X	0 1 1 0	0 1 1 0	0	0	0	0	R_8
1	1	X	0 1 1 0	0 1 1 0	0	0	0	0	
X	X	1	0 0 1 1	1 0 1 0	0	0	0	0	R_9
X	X	0	0 0 1 1	0 1 1 0	0	0	0	0	R_{10}
1	X	X	1 0 1 0	0 1 0 1	1	1	0	0	R_{11}
0	X	X	1 0 1 0	1 0 1 0	0	0	0	0	R_{12}

terms are equal. It may be possible, however, to reduce the number of product terms using a multiple output minimization.

10.2 Figure S.64(a) shows the parity cell circuit and Fig. S.64(b) the symmetric network. Note in both cases the need to insert extra MOS devices to ensure that the output is not grounded.

10.3 The logic diagram for the JK bistable shown simulated in Table 10.7 is given in Fig. S.65. The simulation simply tests the master–slave action of the bistable by putting $J = K = 1$ and checking that the circuit toggles, as shown in the timing diagram (note the one-unit delay, relative to clock, in both $N3$ and Q outputs).

A full logical test of the circuit would require the truth table for the JK-bistable to be checked out using different forms of the CLK command.

The reason for the one-unit delay in NAND gates $N3$ and Q is to create an unequal delay round the feedback loops. If both gates had the same delay the NAND latches would give rise to oscillation (critical race situation); for example, if at $t = 0$ the initial conditions were $N1 = N2 = 1$ with $Q = 0$ and $NQ = 1$ then $Q_+ = 1010101 \cdots$. Some simulators correct for this condition by detecting an oscillatory state and then holding one of the nodes constant until the circuit settles out.

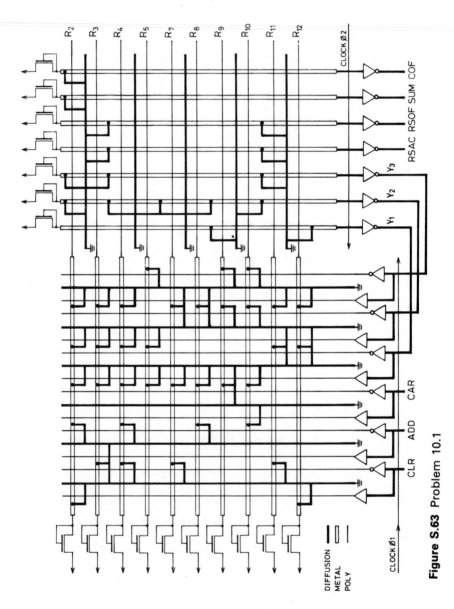

Figure S.63 Problem 10.1

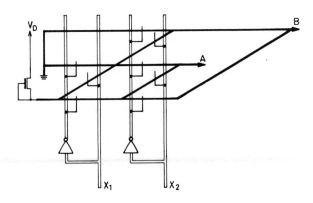

a) Parity cell

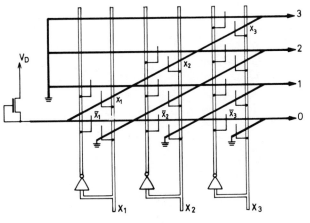

b) Symmetric circuit

Figure S.64 Problem 10.2

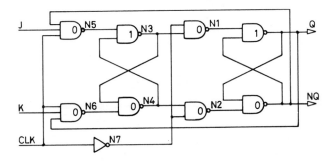

Figure S.65 Problem 10.3

Index